ANNUAL REVIEW OF
FLUID MECHANICS

ANNUAL REVIEW OF FLUID MECHANICS

MILTON VAN DYKE, *Co-Editor*
Stanford University

J. V. WEHAUSEN, *Co-Editor*
University of California, Berkeley

JOHN L. LUMLEY, *Associate Editor*
Pennsylvania State University

VOLUME 9

1977

ANNUAL REVIEWS INC.   4139 EL CAMINO WAY   PALO ALTO, CALIFORNIA 94306

ANNUAL REVIEWS INC.
Palo Alto, California, USA

International Standard Book Number: 0-8243-0709-7
Library of Congress Catalog Card Number: 74-80866

REPRINTS

The conspicuous number aligned in the margin with the title of each article in
this volume is a key for use in ordering reprints. Available reprints are priced at
the uniform rate of $1 each postpaid. The minimum acceptable reprint order
is 10 reprints and/or $10.00, prepaid. A quantity discount is available.

FILMSET BY TYPESETTING SERVICES LTD, GLASGOW, SCOTLAND
PRINTED AND BOUND IN THE UNITED STATES OF AMERICA

PREFACE

Just occasionally we ask ourselves (as seems quite proper) whether the *Annual Review of Fluid Mechanics* is worth the effort. Most of our colleagues declare themselves in favor of the principle underlying it as forthrightly as they would defend Fatherhood; and on cross-examination we often find that they not only have the volumes on their bookshelves but have read and used one or more articles from each, and recommended them to their students and co-workers. This should perhaps serve to justify all our planning, correspondence, editing, and proofreading, and the dedicated labor of our contributors. But we would welcome more objective reassurance.

A quantitative test of the utility of a technical article—imperfect though it certainly is—is the extent to which it has been cited in the literature. Garfield (1976)* observes that "most articles are cited only a few times. Most of this occurs within two or three years after publication." From *Science Citations* we find that the 17 articles that appeared in the first volume of this *Review* in 1969 have, on the average, been cited as follows:

Year:	1970	1971	1972	1973	1974	1975
Average citations:	1.9	2.9	3.4	3.5	3.1	2.1

Thus each article has been referred to an average of 17 times in the course of six years, and the rate is declining only slowly. This is nearly an order of magnitude greater than "a few times." Recalling how many articles we read with profit but fail to cite, we rest provisionally reassured by these statistics as we struggle with next year's Volume 10 and its successor.

THE EDITORS AND THE EDITORIAL COMMITTEE

---

* Garfield, E. 1976. Current comments: Highly cited articles. 26. Some classic papers of the late 19th and early 20th centuries. *Current Contents*, May 24, 1976, p. 5.

CONTENTS

ANNUAL REVIEWS INC. is a nonprofit corporation established to promote the advancement of the sciences. Beginning in 1932 with the *Annual Review of Biochemistry,* the Company has pursued as its principal function the publication of high quality, reasonably priced Annual Review volumes. The volumes are organized by Editors and Editorial Committees who invite qualified authors to contribute critical articles reviewing significant developments within each major discipline.

Annual Reviews Inc. is administered by a Board of Directors whose members serve without compensation.

Annual Reviews are published in the following sciences: Anthropology, Astronomy and Astrophysics, Biochemistry, Biophysics and Bioengineering, Earth and Planetary Sciences, Ecology and Systematics, Energy, Entomology, Fluid Mechanics, Genetics, Materials Science, Medicine, Microbiology, Nuclear Science, Pharmacology and Toxicology, Physical Chemistry, Physiology, Phytopathology, Plant Physiology, Psychology, and Sociology. The *Annual Review of Neuroscience* will begin publication in 1978. In addition, two special volumes have been published by Annual Reviews Inc.: *History of Entomology* (1973) and *The Excitement and Fascination of Science* (1965).

1975

Robert T. Jones

Ann. Rev. Fluid Mech. 1977. 9 : 1–11

RECOLLECTIONS FROM AN   ×8094
EARLIER PERIOD IN
AMERICAN AERONAUTICS

R. T. Jones

Ames Research Center, Moffett Field, California 94040

Nineteen twenty-nine was an exciting year for American aeronautics. The flood of World War I surplus airplanes was receding, and manufacturers all over the US were bringing out new models to capture an assured market. Aviation was going to have a bright future; its commercial success was expected to follow that of the automobile. The only question was who would be the Henry Ford of the new era.

All during the late twenties the weekly magazine *Aviation* appeared on the local newsstand in my hometown, Macon, Missouri. *Aviation* carried technical articles by eminent aeronautical engineers such as B. V. Korvin-Krovkovsky, Alexander Klemin, and others. Included in both *Aero Digest* and *Aviation* were notices of forthcoming *NACA Technical Reports* and *Notes*. These could be procured from the Government Printing Office usually for ten cents and sometimes even free simply by writing NACA Headquarters in Washington. The contents of these reports seemed much more interesting to me than the regular high school and college curricula, and I suspect that my English teachers may have been quite perplexed by the essays I wrote for them on aeronautical subjects.

Leaving the University of Missouri after one year, I took a job with Charles Fower, who operated a flying circus based at Macon. Fower and his wife, Marie Meyer, had for several years flown Standard J-1 airplanes at county fairs and exhibitions throughout the Middle West. Now, early 1929, only one Standard was left and it was in rather poor shape with tattered fabric and a leaky radiator. I patched the wings using cotton from the local dry goods store and on a calm evening we took off for Marshall, Missouri, to get a new set of wings from the Nicholas-Beazley Airplane Company.

Serving as crew for the Marie Meyer Flying Circus, I was supposed to see that the new wings were properly fitted. After the job was done and all the multitude of wires were tight we took off to test the rigging. On the last bounce before becoming airborne the outboard, forward interplane strut snapped outward in an extreme curve and the leading edge of the wing drooped precariously. Fortunately, the Standard has large powerful ailerons, and these together with the rudder enabled us to circle slowly and make a wheel landing. Though the information may no longer be of service, it seems that in rigging a two-bay biplane one should start at the center and work outward. Tightening the inboard wires last may place an undue strain on the outer bays.

When we arrived at Marshall, Nicholas-Beazley was just ready to start production on an advanced three-passenger airplane designed by Walter H. Barling, a well-known World War I British aeronautical engineer and designer of the famous "Barling Bomber." In spite of my somewhat questionable performance as crew for the Standard, Fower recommended me to Mr. Nicholas, and I started to work on the new production line at Nicholas-Beazley.

The introduction of the Barling NB-3 was accompanied by considerable fanfare. Somehow, Nicholas-Beazley had persuaded all their suppliers to place ads simultaneously announcing the NB-3. Barling's design was a low-wing cantilever monoplane of all-metal construction (except for fabric covering). By what must have been extraordinary skill in structural design Barling had kept the empty weight of the airplane below 700 pounds, with the result that it performed well carrying three passengers with a 60-hp engine.

During the boom that preceded the financial crash, *Aviation* and *Aero Digest* started coming out in color and at one point listed nearly one hundred aircraft manufacturers. Airplanes were being made in Little Rock, Arkansas (*Command-Aire*); Lexington, Kentucky (the *Kentucky Cardinal*); and Colorado Springs

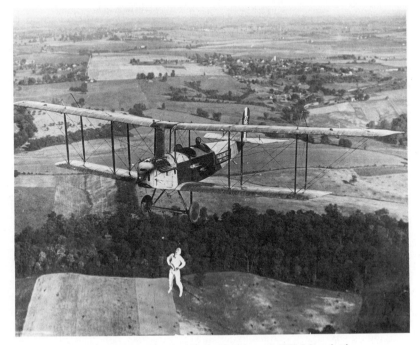

Figure 1   Bertie Brooks hanging by his teeth. OX-5 Standard.

(*Alexander Eaglerock*). Many of the designs were the work of enthusiasts rather than professionally trained engineers, and often a company had to rely on Washington-based consultants to overcome the hazards of the US Department of Commerce "Approved Type Certificate."

The wings of World War I trainers such as the Standard and the JN-4 were rather thin (6–7%) and required lots of bracing. It seems that the early wind-tunnel tests on which these designs were based were made at rather low values of the Reynolds number, a regime in which thin highly cambered sections show favorable properties. Later tests made at higher speeds with larger models revealed, however, that much thicker profiles such as the Clark Y, USA 35, and the G-387 could be used. Many of the aircraft designs of the 1920s simply took advantage of this knowledge and substituted a single-bay biplane with thicker profiles for the older two-bay designs. The cantilever Barling used a slightly modified Goettingen section approximately 18% thick.

An outstanding design of that period was the *Alexander Eaglerock Bullet*. The *Bullet* was a very clean, low-wing cantilever monoplane with a retractable landing gear and the advertised ability to carry "four people and a dog." Unfortunately,

Figure 2   Charles Fower and Marie Meyer flying down Broadway in St. Louis, ca. 1926.

this remarkable airplane never reached quantity production. Flight tests disclosed a much dreaded phenomenon, the "flat spin" from which recovery was evidently impossible.

Beginning in 1920, NACA began collecting and disseminating in a uniform notation aerodynamic characteristics of airfoils from laboratories around the world. By 1929 NACA had published data on nearly one thousand different airfoil shapes. Each report ended with a series of summary plots intended to show which airfoils were "optimum" with respect to certain performance criteria. Unfortunately, the points on these plots scatter rather widely, no doubt because of varying conditions of the tests. Most of the tests were made at low Reynolds numbers and in tunnels with turbulent streams. In spite of the deficiency in the aeronautical laboratories of its day, the NACA collection was of great service to aircraft designers in providing at least an approximate quantitative idea of the behavior of different shapes.

The invention of the variable-density wind tunnel by Max M. Munk at the

ca. 1929

NACA Langley Laboratory essentially overcame the difficulty inherent in earlier wind-tunnel tests and permitted small models to be tested at full-scale values of the Reynolds number. At about the same time, Munk introduced a significant advance in airfoil theory in the form of a linearization which permitted the calculation of airfoil characteristics directly in terms of easily identified parameters of the shape. Previous airfoil theories of Joukowski, Chaplygin, and von Mises were more accurate (none included viscosity) but were considerably more complex, being based on the artful application of conformal mapping to derive special shapes. Von Mises' theory did indeed encompass all shapes expressible by a series of complex coefficients, but the shape is not explicit. At about the same time Munk introduced his theory for the air forces on an airship. Both the thin-airfoil theory and the airship theory may be thought of as theories based on extreme proportions, the airfoil thin and

Max M. Munk (1947)

slightly cambered and the airship long and slender so that the cross-flow in planes perpendicular to the long axis is approximately two-dimensional. These linearizations proved extremely valuable in later years when airfoil theory was extended to the near-sonic and supersonic speed ranges.

In modern times the importance of simplified theories seems to have been diminished by the prodigious capacity of the electronic computer. Approximate theories appear merely as poor substitutes for accurate calculations which may incorporate a multitude of complex interactions. I hope that the drive toward simplification in theoretical work will not be completely forgotten, however, since only in this way can we arrive at deductions that embrace a wide class of phenomena.

Although many airfoil shapes had been tested, there seems to have been little systematic variation of parameters. One series of Joukowski sections had been tested by Ackeret and Schrenk, but again the Reynolds number had been rather low. The completion of the NACA variable-density wind tunnel together with Munk's newly formulated thin-airfoil theory made possible the testing of a systematic series of wing sections at full-scale Reynolds numbers. Munk's analysis permitted him to derive shapes that in theory would have a stable center of pressure travel. Such airfoils have a slight upward camber near the trailing edge. (The effect of the reflexed trailing edge on airfoil stability had been discovered experimentally by W. R. Turnbull in Canada some twenty years earlier.) The new airfoils were named appropriately "M" sections. M-6 and M-12 had quite good characteristics, and I selected the M-12 for use in a small racer design while at Nicholas-Beazley.

Munk's work during this period received special recognition from Dr. Joseph S. Ames, who was chairman of the NACA, in a report entitled *Resume of the Advances in Theoretical Aerodynamics Made by Max. M. Munk (NACA TR No. 213)*. Basic work in aerodynamic theory at NACA declined rather abruptly following Munk's departure from the Laboratory in 1926.

One project deserving to be remembered from this period is the Guggenheim Safe Airplane Competition. At that time it was believed that safe flight would depend on the ability to land slowly and take off from restricted areas. The Guggenheim competition required an airplane that could take off and clear a 35-ft obstacle in less than 500 ft. The maximum permissible landing speed was 35 mph, and the landing run in still air could not exceed 100 ft. Moreover, the airplane was required to demonstrate stable flight in gusty air for five minutes with hands off the controls at all throttle settings from 45 to 100 mph. At least two entries, one by Handley Page of England, were able to meet these requirements. The contest was won by the Curtiss Tanager, a cabin biplane having full-span slots and flaps, with lateral control provided by free-floating tip ailerons extending out from the lower wing.

The years just preceding the depression of the thirties have been characterized as an era of wild speculation. To me it seemed more an era of high activity and enthusiasm. At Nicholas-Beazley we were for a short time building and selling a plane every day. The NB-3 set several altitude and distance records. We stayed up late nights in the small cafes in Marshall designing "flivver" airplanes to be stamped out of metal. For a period of several months, I worked from early morning until midnight in the N.-B. engineering office designing a small racing plane for the 1930 air races.

The loss of confidence following the 1929 crash put a stop to all this activity. Only a few of the aircraft companies (e.g. Cessna, Beech) survive from this period.

Following the election of Roosevelt in 1932, concerted efforts were made to revive the failing aircraft industry. Early in 1933 James A. Farley issued the following announcement: "By authority of President Roosevelt's Executive Council, a bulletin is being sent to the head of every executive agency of the Government directing the use of air mail for all but the most urgent Government messages."

Among the imaginative "New Dealers" appointed by Roosevelt was Eugene Vidal, Director of Aeronautics, Department of Commerce. Vidal believed, as many of us did, that the progress of aeronautics would depend on the development of an inexpensive small airplane for individual use. Vidal instituted a design competition for a "$700 light airplane." Vidal's specifications were remarkably close to our earlier Barling monoplane which cost $3500. Presumably, real mass production could reduce this figure. The idea created considerable interest in aviation circles—not all favorable—and brought out some ingenious designs. One of these, by the well-known engineer Waldo Waterman, called the *Arrowmobile,* was an all-wing arrangement having considerable sweepback with weathercock stability provided by vertical fins at the wing tips. The winner of the competition was a more conservative Hammond design. It seems that the Government had agreed to buy a certain number of prototypes of the winning design at approximately $3000.

To many people deeply involved in the problems of the day, the idea of spending money for aeronautical research seemed wasteful. One of the ideas used to sell the virtues of aeronautical development in the period was the concept of "spin-off." It was claimed that the stimulation of aeronautical activity would in the long run lead to more visible practical benefits to the taxpayer. I do not recall a specific claim that a better frying pan would result, but a potential 25–30% reduction in the weight of the automobile was mentioned.

NACA and its Langley Laboratory suffered badly in the depression but managed to survive. I have been told that at one point a bill was introduced in the Congress to abolish the NACA. Mr. John F. Victory, NACA's Executive Secretary, fortunately kept close watch on such legislative happenings and within hours had summoned enough support in the form of telegrams from aircraft manufacturers all over the US to defeat the measure.

In 1932, the NACA budget reached a peak of about one million dollars. By 1934, this had declined to $690,000, and NACA employees had voluntarily accepted a 15% reduction in salary. By any analysis, the level of activity that was sustained by these limited budgets seems remarkable. There was at the Langley Laboratory a "full scale" 30' × 60' wind tunnel, a 20-ft-diameter propeller tunnel, the 20-atmosphere variable-density wind tunnel, a mile-long seaplane towing basin, an active flight research section, and several smaller wind tunnels.

To cope with widespread unemployment, the Public Works Administration under Harold Ickes opened up a number of temporary scientific positions in Government. In late 1934, I was thus enabled to secure a temporary (nine months) appointment as scientific aide at the Langley Laboratory of NACA. My first assignment was in the 7' × 10' Atmospheric Wind Tunnel with Carl J. Wenzinger,

Thomas A. Harris, Robert Platt, and others under Fred E. Weick, who was Assistant Chief of Aerodynamics.

Our main task in the $7' \times 10'$ wind tunnel was the development of high-lift devices and lateral controls to improve the safety of flight at low speeds. Most accidents had resulted from stalls and spins, especially during forced landings. Many of the engineers at Langley were pilots and hence were acquainted with the practical as well as the theoretical aspects of such problems.

Fred Weick had built (with assistance from the wind-tunnel group) a stall-proof airplane called the "W-I." The W-I was a high-wing monoplane with an ungainly looking fixed slat supported ahead of the leading edge and extending from tip to tip along the wing. On one occasion, the engine failed just after takeoff, and Fred had to bring it down in the NACA tennis court. The W-I never flew again but fortunately Fred was unhurt.

One of Weick's ideas for improving the safety of landing was the tricycle landing gear. The conventional "tail-dragger" with fixed wheels ahead of the center of gravity was, of course, inherently unstable. By placing the main wheels behind the center of gravity and allowing the front wheel to pivot, stability could be achieved. In spite of its simplicity the idea encountered some resistance, and one of my first tasks was to try to show from dynamical calculations that passengers in the rear of an airplane would not be thrown out of their seats when the airplane pitched down onto the nose wheel.

A persistent fault of the tricycle gear was shimmy of the nose wheel. Weick gave this problem to Arthur Kantrowitz, who had recently joined the staff. By experimenting with models, Arthur found that a swiveling nose wheel could actually shimmy at essentially zero forward speed, i.e. in a purely kinematic way due to a

Fred Weick   Henry J. E. Reid

characteristic mode of distortion of the tire. Dynamic shimmy could be prevented by allowing the wheel to slide laterally on a slightly curved axle—a very neat solution, I thought. The designers, however, seemed to prefer a more direct approach, using hydraulic dampers.

In the early days of flying, inherent stability was of the utmost importance, and *NACA Technical Report No. 1* by Jerome C. Hunsaker and E. B. Wilson (1915) was devoted to this subject. Under Fred Weick's guidance, these studies were continued in the thirties at Langley. Following a tendency that persists even to this day, we did not read the earlier work carefully enough, and I am afraid that we repeated some of the mistakes that had been made and corrected years before. Thus, in calculating the stability coefficient N_p, the yawing moment due to rolling, I made what I thought was a very clever use of Munk's theory of the twisted elliptic wing. According to my theory, the downgoing wing developed more lift and, hence, more drag. Charles Zimmerman eventually noticed that my value of N_p had the wrong sign. The downgoing wing pulls forward not back; and when I took account of the correct resolution of velocity vectors, the forward thrust appeared. It seems that E. B. Wilson had, in 1918, found this error in earlier work of J. C. Hunsaker and L. Bairstow and wrote *NACA TR No. 26* to give the correct version of the theory.

In the early thirties, the Langley Laboratory was the acknowledged US center of aeronautical research. Much of what I had learned about aeronautics had been gleaned from *NACA Reports* and *Technical Notes*. I could hardly have wished for a better fortune than to find myself among these engineers who were so involved in the advancement of the art. Because of this, it is perhaps difficult for me to make a purely objective assessment, but others have confirmed my impression of the extraordinary group at Langley. At that time, the inflation of the language had not yet reached the point where we were called "scientists." Even the director of the Laboratory, H. J. E. Reid, was termed "Engineer in Charge." I, of course, did not even qualify as an engineer; and sometime later when it appeared that the lowest professional or engineering grade called for a certain academic preparation, it was necessary for me to take the next higher grade where the academic requirement, though presumed, was not mentioned.

Eastman N. Jacobs, one of the most skillful and innovative American aerodynamicists, had come to Langley in 1925 and his activities were invariably a center of interest there. Jacobs had a wide appreciation of science but did not devote much time to theoretical studies. Rather, he used his theoretical understanding to devise intelligent experiments. Thus, in 1932, he and James M. Shoemaker tested thrust augmentors for jet propulsion. Many years later, after Campini's jet airplane had flown, Jacobs was instrumental in encouraging work on jet propulsion at Langley. Jacobs is, of course, best known for his development of the low-drag laminar-flow airfoil.

As noted earlier, several of the engineers at Langley devoted their spare time to building and flying airplanes. Soon after coming to the lab, Jacobs built a small monoplane powered by an Ace motorcycle engine. Having a one-wheel landing gear, the design was not highly regarded by the professional pilots at the lab. However, Jacobs took off in this machine for his first solo flight. It seems that

Jacobs' flight plan was disclosed prematurely to the laboratory personnel and they gathered around the small grass field to see what they were sure was an impending crash. Evidently, the presence of this audience only brought out Jacobs' best skill, and it is said that he made a perfect landing.

Hurricanes were rather frequent along the Virginia coast, and the one that struck in 1933 nearly destroyed the laboratory. By that time, Jacobs had acquired a Pitcairn biplane (*Mailwing*) and happened to be in Norfolk when the hurricane struck. He tied the airplane down hard with its tail into the wind and waited for the center of the hurricane to arrive. As soon as the wind abated slightly, he took off, following the "eye" of the hurricane until he reached higher ground where he landed and in his words "saved the airplane."

Flying at Langley was often a mixture of fun and aerodynamic experiment. Robert Platt, with whom I worked, maintained a World War I Fokker D-VII which had beautiful flying qualities and two seats. He and I used it frequently to test ideas of stability and control. In one of our experiments we determined that the airplane remained stable and controllable at extreme angles of sideslip.

Following Munk's departure in the late twenties, theoretical work at the laboratory had declined, but was revived somewhat later by Dr. Theodore Theodorsen, Carl Kaplan, and Edward Garrick. In his *Theory of Airfoils of Arbitrary Shape* Theodorsen found a way to determine the von Mises coefficients by successive approximation starting with any two-dimensional or cylindrical shape.

When Eastman Jacobs discovered that the maintenance of laminar flow depended on a prescription of the pressure distribution rather than the shape, he disappeared from the laboratory for several days, and one day he called me over to his house to help him unravel Theodorsen's theory. We decided it could not be used that way

Eastman Jacobs   Theodore Theodorsen

and I devised a simple extension of Munk's theory to serve this purpose. Thin-airfoil theory proved too inaccurate, however, and H. J. Allen developed a more satisfactory theory based on a linearization that started from a Joukowski airfoil having some thickness.

One of the most important contributions of the Theodorsen group was the theory of oscillating airfoils with hinged flaps—related to the problem of flutter. Garrick subsequently extended this theory to cover propulsion of a flapping airfoil. Some time later, I became interested in the extension of this theory to the three-dimensional wing.

Eastman Jacobs represented the laboratory at the 1935 Volta Congress on high-speed aeronautics. Following his return, during lunch-time conversations he and Arthur Kantrowitz tried somewhat unsuccessfully to explain the principles of super-sonic flow to me. Being familiar with Laplace's equation and its smooth stream-lines, I found it difficult to believe that the streamlines could make sharp bends at Mach waves. Quite a few years later, I found a way to make the supersonic streamlines smooth by sweeping the leading edge of the wing behind the Mach cone.

It was at the 1935 Volta Congress that Busemann had introduced the idea of sweeping the wings to diminish the wave drag at supersonic speeds. Busemann utilized the "independence principle" but kept the wing ahead of the Mach cone so that the cross-flow was still supersonic. Evidently, Busemann had put too many ideas in this one paper, for neither Jacobs nor von Kármán remembered his suggestion when I proposed sweeping the wings some ten years later. In my version, the wing was swept behind the Mach cone to get a purely subsonic type of flow and thus to eliminate the wave drag entirely for infinite aspect ratio. For the independence principle, I had relied on an earlier paper of Munk entitled *The Relative Effects of the Dihedral and the Sweepback of Airplane Wings (NACA TN 177, 1924)*. Fortunately, before my paper was published, Robert Hess at Langley found Busemann's earlier paper and I was able to refer to it.

The first tests of swept wings at Langley were made by Robert Gilruth, M. C. Ellis, and Clinton Brown. Ellis and Brown tested the independence principle by placing a length of streamlined wire in their supersonic tunnel. Gilruth obtained the first accurate results by attaching wings to a body dropped from a high altitude.

It is not widely known that the first experiments in the US designed to produce power from thermonuclear fusion were initiated at Langley some time before the Manhattan atomic bomb project. The Langley experiments were the idea of Arthur Kantrowitz and Eastman Jacobs, who made use of earlier theoretical work on fusion by Hans Bethe. It is interesting that Kantrowitz and Jacobs attempted to initiate fusion by magnetic confinement of a plasma in a toroidal field—a technique often seen in more recent attempts. By arguing that fusion power could become impor-tant for aircraft propulsion they were able to secure an appropriation of $5000 to carry on the work. Both Kantrowitz and Jacobs spent many hours glass blowing and constructing coils, and it is said that during the final test one of them held in the circuit breakers of the variable-density-tunnel power supply to get more current. Unfortunately, the experiment was defeated by the stubborn (and still persistent) tendency of the plasma to become unstable.

Ann. Rev. Fluid Mech. 1977. 9:13–32

STEADY NON-VISCOMETRIC   ✳8095
FLOWS OF VISCOELASTIC
LIQUIDS

A. C. Pipkin

Division of Applied Mathematics, Brown University, Providence, Rhode Island 02912

R. I. Tanner

Department of Mechanical Engineering, University of Sydney, Sydney, Australia

1   NON-VISCOMETRIC FLOWS

Figure 1 shows a phenomenon that cannot be explained on the basis of material properties measured in standard viscometers. A liquid is being drawn upward into a tube whose orifice is not submerged in the liquid. The liquid is mainly a mixture of glycerin and water, but it contains a small amount of a substance composed of very long-chain molecules, a polymeric material.

The mechanical properties of polymer melts and solutions are highly complex. Experimentalists and theoreticians who attempt to characterize these properties often restrict attention to very simple flows, in which limited aspects of the mechanical response of the material can be isolated for study. Steady shearing flows, such as those in the capillary, Couette, cone-and-plate, and other standard viscometers, are particularly simple. The theory of these so-called *viscometric* flows is the subject of a book by Coleman, Markovitz & Noll (1966), and more recent theoretical and experimental work has been reviewed by Pipkin & Tanner (1972).

In the present review we discuss some flows in a category that at first appears to be only slightly broader than viscometric flows. We restrict attention to flows in which the velocity gradient is constant in time, after the motion begins, and uniform in space, throughout the flow region. It might appear that there could hardly be a much simpler class of flows, but the work of Giesekus (1962a,b) has shown that there is a fascinating variety in such motions. We discuss the experimental evidence that by now exists, which shows that the material properties exhibited in some of these motions can differ drastically from anything one might guess from viscometric data.

The evident stability of the flow in Figure 1 illustrates this difference. If the shearing viscosity $\eta_s(\gamma)$ as a function of the shear rate γ is known, one might suppose by analogy with the Newtonian case that the extensional viscosity would be $3\eta_s(\varepsilon)$ at the rate of extension ε. In a crude stability analysis, one might argue that if a neck begins to form in the strand of liquid, the extension rate will increase there. If the extensional viscosity increases as the rate of extension does, the neck will be strengthened and the flow will be stabilized. However, the shearing viscosities of

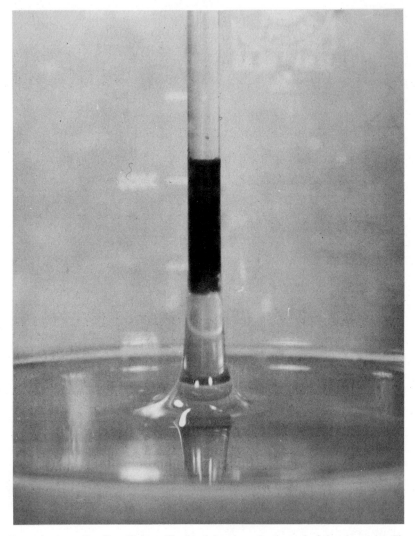

Figure 1 Ascending Free Siphon. The black band marks the end of the glass tube. The fluid is a solution of 1% Separan AP-30 in a mixture of 75% glycerin and 25% water. (Photo courtesy of D. V. Boger.)

polymer melts and solutions are usually very rapidly *decreasing* functions of the shear rate. If the extensional viscosity were also a decreasing function of the rate of extension, a viscoelastic liquid under tension would be even less stable than a Newtonian liquid. We are led by this contradiction to conclude that the behavior of a liquid in an extensional motion is not very closely connected to its behavior in shearing.

For materials containing long-chain molecules, there is a large qualitative difference between flows in which particles can separate at an exponential rate, as in extensional flows, and those in which this does not occur, as in viscometric flows. In Section 2 we discuss a classification of flows as *strong* or *weak* on this basis. Although we do not review the literature on molecular theories of polymer melts and solutions, we indicate the significance of this classification in molecular terms.

The viscometric flows lie on the borderline between strong and weak flows. In Sections 3 to 5 we discuss some flows in which the stress response of the material is not like that in viscometric flow. The flow in the eccentric disc rheometer (Section 3) is a weak flow in which there is no steady growth in particle separations. Although the velocity gradient in this flow is constant in time and space, the material properties that the fluid exhibits are those usually associated with small-amplitude sinusoidal oscillations. In Sections 4 and 5 we discuss extensional motions, which are strong flows. The extensional viscosities measured in such flows can be orders of magnitude larger than the shearing viscosities observed in viscometric motions.

In addition to discussing what has been learned about material properties, we mention various problems that have arisen in the design and analysis of experiments. We do not discuss constitutive equations to any great extent; interested readers may wish to refer to the books by Lodge (1964), Pipkin (1972), Huilgol (1975), and Walters (1975), or to a review article by Rivlin & Sawyers (1971).

2   A KINEMATIC CLASSIFICATION OF FLOWS

We wish to make a distinction among various flows on the basis of the rate of separation of neighboring particles. We say that the flow is *strong* at a given material element if there are particles in it whose separation grows at an exponential rate, and *weak* otherwise. The weak flows include both oscillatory flows and steady shearing motions. In shearing flows, the distance between particles on the same slip surface does not change, and although the distance between particles on different slip surfaces may grow arbitrarily large, the growth is roughly linear in time rather than exponential.

The distinction between strong and weak flows is unimportant when considering the properties of fluids composed of small molecules, but it appears to be significant in molecular theories of polymer networks (Lodge 1956, 1964; Yamamoto 1956, 1957, 1958) and solutions containing long-chain molecules (Takserman-Krozer 1963, Peterlin 1966, Bird et al 1971). The classification used here is a simplification of that proposed by Giesekus (1962a) and is closely related to that of Tanner (1975) and Hinch (1975).

The length of an extended polymer chain is so large that in a solution it is necessary to take into account the difference in velocity of the solvent flow at the two ends of the chain. To lowest order, the velocity gradient $v_{i,j} = L_{ij}$ can be considered constant over the length of the chain, and the velocity field can be taken to be of the form $v_i = L_{ij}x_j$. In this review we restrict attention to flows in which L_{ij} is constant in time as well as space.

In the crudest approximation, the ends of a polymer chain can be regarded as convected at the local velocity v_i of the solvent. If the chain has one end at the origin and the other at the point x_i, the extension of the chain caused by the motion is determined by solving the system of equations $\dot{x}_i = L_{ij}x_j$. The type of, motion that ensues depends on the Jordan canonical form of the matrix L_{ij} (Tanner & Huilgol 1975) and in particular on the eigenvalues of the matrix. Giesekus (1962a) has examined all of the possibilities in great detail.

The motion is *strong* if some eigenvalue has a positive real part, and *weak* otherwise. With restriction to isochoric motions ($L_{ii} = 0$), the sum of the three eigenvalues is zero. Consequently, in weak flows all eigenvalues have zero real part. If they are distinct, one is zero and the other two are conjugate imaginaries, and the motion is oscillatory, with no steady growth in particle separations. In weak flows with all three eigenvalues equal to zero, the degeneracy allows linear or quadratic growth in particle separation. The flows in this category are the simple shearing (viscometric) flows and the double shearing flows mentioned by Noll (1962).

In steady simple shearing the velocity field has the form

$$u = \gamma y, \qquad v = w = 0. \tag{1}$$

[Here $(v_i) = (u, v, w)$ and $(x_i) = (x, y, z)$.] The particle trajectories are

$$x(t) = x(0) + \gamma y(0)t, \qquad y(t) = y(0), \qquad z(t) = z(0). \tag{2}$$

If this is the trajectory of one end of a molecule, the other end being fixed at the origin, then the molecule would become infinitely stretched [if $y(0) \neq 0$] and oriented parallel to the x, z-plane (the slip surfaces) in the limit of large times. Retractive forces in the polymer chain prevent the infinite stretching, and Brownian motion sometimes moves the end of the polymer chain into the region of negative velocity, so the actual motion of a chain in such a flow is a tumbling motion with alternate stretching and contraction.

Steady extensional flow is a strong flow, which we discuss further in Section 4. If ε is the constant rate of extension, the velocity field is

$$u = \varepsilon x, \qquad v = -\varepsilon y/2, \qquad w = -\varepsilon z/2, \tag{3}$$

and the particle trajectories are

$$x(t) = x(0)\exp(\varepsilon t), \qquad y(t) = y(0)\exp(-\varepsilon t/2), \qquad z(t) = z(0)\exp(-\varepsilon t/2). \tag{4}$$

Molecules are powerfully stretched and oriented by such a flow. Brownian motion prevents perfect orientation, but it cannot lead to tumbling motion as in shearing flow.

In Section 3 we discuss the flow in the eccentric-disc rheometer. The velocity field in this instrument is given approximately by

$$u = -\Omega(y - az/h), \qquad v = \Omega x, \qquad w = 0. \tag{5}$$

The particle trajectories are circles in the planes $z =$ constant:

$$x(t) = x(0) \cos \Omega t - y^*(0) \sin \Omega t, \qquad y^*(t) = x(0) \sin \Omega t + y^*(0) \cos \Omega t, \tag{6}$$

where

$$y^* = y - az/h. \tag{7}$$

This is a weak flow with no unbounded growth in particle separations. The distance between particles on different planes $z =$ constant oscillates sinusoidally, although the velocity field is time-independent.

The effect of retractive forces on extended molecules can be modeled in the simplest way by supposing that in a fluid at rest the end-to-end vector y_i of a polymer chain would return to zero exponentially: $\dot{y}_i = -y_i/T$. For a fluid in motion, taking convection into account, one may assume that

$$\dot{y}_i = L_{ij} y_j - y_i/T. \tag{8}$$

The nature of the time-dependence of the vector y_i depends on the eigenvalues of the matrix $L_{ij} - T^{-1}\delta_{ij}$. If λ is an eigenvalue of L_{ij}, then $\lambda - T^{-1}$ is an eigenvalue of the former. In this model, the molecule can become more extended as time progresses only if L_{ij} has an eigenvalue with a real part exceeding $1/T$. Despite the obvious deficiencies of this model, it suggests that molecules do not become fully extended in viscometric flows, and do not grow large even in extensional flows unless the rate of extension ε exceeds a critical value $1/T$. The classification scheme proposed by Tanner (1975) is based on this idea.

3   A WEAK FLOW: ECCENTRIC-DISC FLOW

A common method of testing viscoelastic materials is to measure the stress needed to produce a small, sinusoidally oscillating deformation. The stress can be decomposed into a part in phase with the strain and a part $90°$ out of phase, in phase with the strain rate. By dividing these parts by the strain amplitude, one obtains values of the elastic-storage modulus $G'(\omega)$ and the viscous loss modulus $G''(\omega)$. The dynamic viscosity is related to the loss modulus by $\eta(\omega) = G''(\omega)/\omega$. These moduli are strongly dependent on the frequency of oscillation ω. Ferry's book (1970) gives such data for a wide variety of polymers.

The methods usually employed are somewhat inconvenient, especially at low frequencies, because of the need to measure phase angles accurately. Gent (1960) described an ingenious method of testing solid polymers that avoids this problem. Maxwell & Chartoff (1965) used the same principle in a device adapted for the testing of polymer melts (Figure 2). In this instrument, a *steady* rotational flow is used to measure the frequency-dependent properties of the material, and the measurement of phase angle is replaced by the simpler task of measuring steady forces.

The device consists of parallel discs that rotate at the same angular velocity Ω about axes perpendicular to the discs but not coincident. When the distance h

between the plates is sufficiently small, the fluid can be held in the gap by surface tension. In some cases the lower disc is replaced by a rotating cup (Payvar & Tanner 1973).

A fluid element starting at the right-hand side in Figure 2 is sheared from a top-outward position to a top-inward position as the discs rotate through 180°. The elastic force required for the shearing is in the positive y-direction on both sides. The viscous force, related to the rate of shear, is parallel to the x-direction.

Gent (1960) recognized that the stress components in the test specimen would be

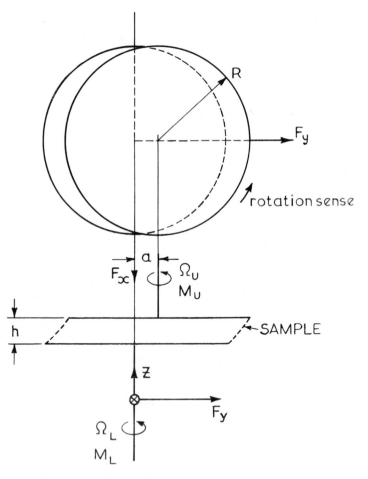

Figure 2   Eccentric Disc Rheometer. The illustration shows the "free" edge configuration. (Another possibility is the drowned edge formed by replacing the lower disc by a shallow cup whose diameter is greater than that for the upper disc.) In normal operation $\Omega_U \simeq \Omega_L = \Omega$.

uniform and constant in time, and he showed that the forces needed to hold the top disc in place would be related to the viscoelastic moduli by

$$F_x = A(a/h)G''(\Omega)$$

and   (9)

$$F_y = A(a/h)G'(\Omega).$$

Here A is the contact area, a is the distance between the axes of rotation, h is the gap width, and Ω is the angular velocity. This result requires that the shear amplitude a/h be small enough that nonlinear effects can be neglected.

After Maxwell and Chartoff's paper brought the instrument to the attention of fluid dynamicists, Blyler & Kurtz (1967) analyzed the flow by assuming a velocity field of the form (5), using a special viscoelastic constitutive equation. Bird & Harris (1968, 1970) used the same velocity field with a much more complex nonlinear integral constitutive equation.

Although theoretical justification of the relations (9) would require the strain amplitude a/h to be infinitesimal, Gross & Maxwell (1972) have shown that the measured stress remains proportional to a/h up to values of this strain of the order of 0.5. They emphasized that the limit of linear behavior depends on the strain amplitude, not the strain rate. This is of interest because the model used by Bird and Harris requires a small value of the strain-rate amplitude $a\Omega/h$ for linearity.

The unimportance of the strain rate can be seen in a different way. With knowledge of the steady-shearing viscosity $\eta_s(\gamma)$ from viscometric experiments, one might be tempted to estimate the apparent viscosity for the eccentric-disc flow by evaluating η_s at $\gamma = a\Omega/h$. This contradicts (9), according to which the viscous stress divided by the shear-rate amplitude is $\eta(\Omega) = G''(\Omega)/\Omega$, a function of frequency alone. The true relation between the steady-shearing and dynamic viscosities is more subtle (Merz & Cox 1958). As innumerable experiments have shown, the steady-shearing viscosity $\eta_s(\gamma)$ is about equal to the dynamic viscosity $\eta(\omega)$ at a frequency $\omega = \gamma$. The inference is that even in steady shearing the stress is determined by some oscillatory motion. Presumably this is the alternate stretching and contraction of the long-chain molecules as they are tumbled at an angular velocity determined by the shear rate.

Huilgol (1969) recognized that the velocity field (5) would satisfy the equations of motion, with neglect of inertia, whatever the constitutive equation of the fluid might be. For the uniformity of the velocity gradient implies that in the steady state the stress must also be uniform (apart from the reaction pressure for an incompressible liquid, which is not determined by the deformation history). The material response is described by the dependence of the five independent stress components on the shear amplitude and the frequency. Huilgol showed the extent to which these functions are related to the three functions of the shear rate that can be measured in viscometric flows, and later (1970) he discussed their relation to the response functions for nearly viscometric flows (Pipkin & Owen 1967).

Because of inertial effects, differences in the angular velocities of the discs, and edge effects, the velocity field (5) cannot be exactly correct. If edge effects and velocity lag are neglected but inertia is taken into account, the velocity field has the form

$$u = -\Omega[y - y_0(z)], \qquad v = \Omega[x - x_0(z)], \qquad w = 0, \tag{10}$$

with a line of centers that is not the straight line $x_0 = 0$, $y_0 = az/h$ assumed in (5). Abbott & Walters (1970) determined the line of centers exactly for the case of a Newtonian liquid (thus discovering a previously unknown exact solution of the Navier-Stokes equation) and for a linearly viscoelastic liquid with arbitrary moduli. They found that the line of centers departs from straightness by an amount proportional to the offset a and proportional in lowest order to the Reynolds number $\rho\Omega h^2/\eta$. Under ordinary operating conditions this Reynolds number is so small that inertial effects are entirely negligible.

The viscoelastic constitutive equation used by Abbott and Walters has the form

$$\sigma_{ij}(t) = -p(t)\sigma_{ij} + \int_{-\infty}^{t} G(t-s)C'_{ij}(s,t)ds, \tag{11}$$

where $p(t)$ is the reaction pressure. $G(t)$ is the shearing-stress relaxation modulus; the complex viscosity $(G' + iG'')/i\omega$ is the Fourier transform of $G(t)$. Here C_{ij} is the strain at time s relative to the state of time t, and $C'_{ij} = DC_{ij}/Ds$. Although this is essentially a small-strain relation, a finite strain measure must be used since displacements and rotations are large in the present application (see Pipkin 1972, for example).

There is an interesting paradox concerning the flow between discs of finite radii. Except near the edges, the rate of energy dissipation is $\sigma_{xz}a\Omega/h$, with σ_{xz} uniform. The total dissipation is then $V\sigma_{xz}a\Omega/h$, with V the effective volume of the flow region. If we define the effective contact area between the fluid and either disc as $A = V/h$, then the total rate of dissipation is seen to be equal to $aF_x\Omega$, with F_x given by (9). However, if the effective contact area is a circle centered on the axis of rotation and the stress is uniform in this region, then no work is done by the disc on the fluid. This way of stating the paradox contains its solution, of course; the region in which the stress is tolerably uniform is not centered on the axis of rotation.

The rate of energy dissipation can be used to estimate the importance of edge effects in flows with a drowned edge, in which there is fluid outside the gap. Payvar & Tanner (1973) have reported that in such flows, with fluids of known viscosity, measured values of F_x are larger than (9) would indicate, by a factor of the order of $2h/R$, where R is the disc radius; A is taken to be πR^2. Macosko & Davis (1974) have also mentioned observations of F_x larger than expected. This can be explained by supposing that the flow region V has a radius of the order of $R+h$, as Payvar and Tanner have pointed out. The effective contact area $A = V/h$ for use in (9) is then larger than the disc area by a term of the order of $2h/R$.

In normal operation of the rheometer, one of the discs is driven at a prescribed angular velocity Ω and the other is allowed to rotate freely. Except when $a = 0$, this leads to a small difference $\Delta\Omega$ between the angular velocities of the discs. If the top disc, say, is turned with a moment M and the lower disc rotates freely about its axis, then the power supplied is $M\Omega$. This must be equal to the rate of dissipation $aF_x\Omega$, so that the moment is $M = aF_x$. Then either the effective contact area is not a circle centered on the axis of rotation, or the velocity field does not have the ideal form (5). In the case of a very large offset with a drowned edge, the latter alternative seems preferable. Payvar & Tanner (1973) assumed that the velocity field is then a torsional flow $v_\theta = z\Delta\Omega/h$ superimposed on the basic flow (5). Within the linearly viscoelastic approximation (11), the stress fields of the two flows are additive. The moment required to produce the torsional flow is

$$M = (\pi R^4/2)\eta(0)\Delta\Omega/h. \tag{12}$$

The steady-shearing viscosity $\eta(0)$ appears because the torsional flow is viscometric. By setting this moment equal to aF_x, with F_x given by (9), the angular velocity lag is found:

$$\Delta\Omega/\Omega = 2(a/R)^2\eta(\Omega)/\eta(0) = \beta \quad \text{(say)}. \tag{13}$$

This agrees with Payvar and Tanner's observations, which covered a range of β values up to about 15%. Davis & Macosko (1974) performed a more detailed calculation and obtained $\beta/(1+\beta)$ as the relative lag; this is more realistic at high offsets. They also gave the result for the case in which a known torque is supplied by friction in the bearing of the passive disc.

Within the linearly viscoelastic approximation (11), the exact solution accounting for velocity lag and inertia simultaneously has the form (10) with a linearly varying angular velocity $\Omega(z)$. This might be of interest as another exact solution of the Navier-Stokes equation, but the analysis has not been published.

Other interesting work not discussed here can be found in the book by Walters (1975).

4   A STRONG FLOW: STEADY EXTENSION

In the steady extensional flow (3) of an isotropic liquid, all shearing-stress components are zero and $\sigma_{yy} = \sigma_{zz}$, by symmetry. For an incompressible liquid the stress response is then completely defined by the dependence of $\sigma_{xx} - \sigma_{yy}$ on the rate of extension ε and the time t elapsed since the stretching began:

$$\sigma_{xx} - \sigma_{yy} = \varepsilon\eta(\varepsilon, t). \tag{14}$$

The initial response of a polymer melt is largely elastic, and might more appropriately be described in terms of the dependence of the stress on the strain εt. For this reason, the stress generally increases as time progresses, during the initial stages of the motion (Figure 3). If the rate of extension is large, this initial stage may end in fracture of the specimen. For example, it is a simple matter to fracture some silicone liquids ("silly putty") in one's own hands.

When rupture does not occur, the stretching viscosity $\eta(\varepsilon, t)$ must eventually approach a limiting value,

$$\eta(\varepsilon, \infty) = \eta_T(\varepsilon). \tag{15}$$

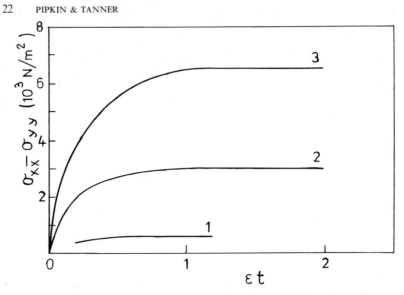

Figure 3 Elongational stress versus strain for Butyl 035 at 100°C (low elongation rates). *Curve 1*, $\varepsilon = 2.85 \times 10^{-4}$ sec^{-1}; *Curve 2*, $\varepsilon = 1.48 \times 10^{-3}$ sec^{-1}; *Curve 3*, $\varepsilon = 2.82 \times 10^{-3}$ sec^{-1}. Note that the elongational stress attains steady state.

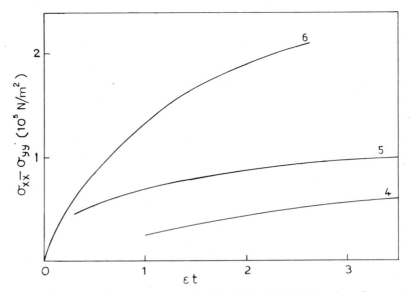

Figure 4 Elongational stress versus strain for Butyl 035 at 100°C (higher elongation rates). Elongation rates are as follows: *Curve 4*, $\varepsilon = 2.48 \times 10^{-2}$ sec^{-1}; *Curve 5*, $\varepsilon = 4.51 \times 10^{-2}$ sec^{-1}; *Curve 6*, $\varepsilon = 1.35 \times 10^{-1}$ sec^{-1}. Note that the elongational stress may not be approaching a steady state at the higher rates.

The limiting value is called the steady extensional viscosity, or Trouton viscosity. Trouton (1906) found that the extensional viscosity of mixtures of pitch and tar is independent of ε and about equal to $3\eta_0$, the value for an incompressible Newtonian fluid with shear viscosity η_0.

Concern with the properties of the Trouton viscosity began in the 1930s in connection with the important problem of spinning synthetic fibers from molten liquid. Nitschmann (1949) described a thread-spinning experiment for determining elongational viscosity, which probably measures some average value of the stretching viscosity $\eta(\varepsilon, t)$. Ziabicki & Kedzierska (1960) and Ziabicki (1961) kept the subject of spinning in view, and by now there is a large literature on the subject.

Spinning experiments can rarely be used for the unambiguous determination of the steady-state viscosity $\eta_T(\varepsilon)$, because each fluid element experiences a highly unsteady stretching and there is usually not enough time for transient elastic effects to reach steady state. Ballman's (1965) work on polystyrene was the first in which the strain rate was kept constant during the motion. A tensile test was carried out on a bar of this extremely viscous liquid, with the ends of the specimen

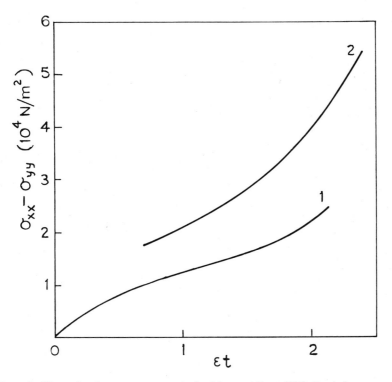

Figure 5 Elongational stress versus strain for Natsyn 410 at 80°C. Symbols represent elongation rates as indicated: *Curve 1*, $\varepsilon = 6.37 \times 10^{-4}$ sec^{-1}; *Curve 2*, $\varepsilon = 1.57 \times 10^{-3}$ sec^{-1}. Note that the elongational stress-strain curves are S-shaped.

moving apart exponentially so as to keep the velocity gradient $\partial u/\partial x$ constant. Essentially the same technique has been used by Cogswell (1969, 1972), Goldberg et al (1969), Vinogradov et al (1970a,b; 1972; 1975), and Stevenson (1972), who obtained the results shown in Figures 3 to 6. Usually the weight of the specimen is supported by floating it in a liquid of the same or slightly higher density, as Trouton did.

With this technique only limited extensions can be achieved, since the length of the specimen increases in proportion to $\exp(\varepsilon t)$, and the rate of extension cannot be much larger than about 0.1 sec^{-1}. Meissner (1969, 1971, 1972) introduced a substantially improved method of testing. Each end of a strand of liquid is drawn between a pair of gears that rotate at constant angular velocity. The two ends are pulled in opposite directions, so that the axial velocities are U and $-U$, say, at the locations of the gears, $x = L$ and $x = -L$. The axial velocity in the strand is then $u = Ux/L$, giving a constant extension rate $\varepsilon = U/L$. With this method, Meissner (1971) has drawn samples out to 90 times their original length and has achieved extension rates of 1 sec^{-1}. The data in Figure 7 were obtained by Meissner (1971).

In order to measure the steady-state viscosity $\eta_T(\varepsilon)$, the fluid must be subjected to a constant rate of extension for so long that the stress reaches a constant value, or a constant stress for so long that the rate of extension becomes constant. All reported cases in which it is certain that a steady state was reached involve fluids of very high viscosity and rates of extension not greater than about 0.1 sec^{-1}. In this limited range, the Trouton viscosity has invariably been found to be either nearly constant or, at the highest rates of extension, an increasing function of ε. Although these rates of extension are very low, the steady-shearing viscosities $\eta_s(\gamma)$ of the same materials are usually significantly lower at a shear rate $\gamma = 0.1$ sec^{-1} than they are in the limit of zero shear rate. Consequently, the ratio of the extensional

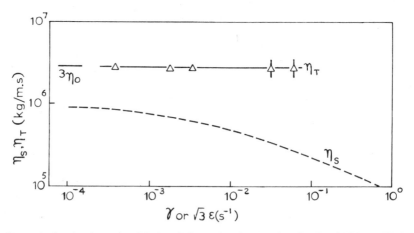

Figure 6   Steady elongational (η_T) and shear viscosity (η_S) data for Butyl 035 at 100°C. The abscissa is the square root of the second invariant of the rate-of-strain tensor.

viscosity to the shearing viscosity is an increasing function of the strain rate even when the Trouton viscosity is still constant, as in the data of Stevenson (1972) shown in Figure 6.

It is easy to obtain much higher rates of extension, but the experiments in which this is done always involve flows with a nonconstant rate of extension. Spinning experiments typically involve extension rates of the order of 0.1 to 10 sec^{-1}. Recent measurements of this kind include those reported by Chen et al (1972), Spearot & Metzner (1972), Acierno et al (1974), Mewis & Metzner (1974), Weinberger & Goddard (1974), and Moore & Pearson (1975). The convergent die-entry flows studied by Cogswell (1969) involve extension rates of the order of 1 to 100 sec^{-1} Jet-thrust experiments involve extension rates of the order of 10 to 10^3 sec^{-1} (Metzner & Metzner 1970, Kizior & Seyer 1974, Oliver & Bragg 1974). Rates in the range 100 to 800 sec^{-1} are obtained in the triple jet apparatus proposed by Oliver & Bragg (1974). In two jets colliding head-on, or the reverse, the rate of extension has been estimated as 50 sec^{-1} for a melt (Mackley & Keller 1973) and $8 \cdot 10^3$ for a polymer solution (Frank et al 1971); mechanical measurements on these interesting flows have not been reported.

Except in the case of spinning, analyses of these flows must make use of approximations and assumptions that are subject to doubt. Furthermore, it is usually clear that the apparent extensional viscosity η_e, say, in these flows is not the steady-state Trouton viscosity. However, the general trend is toward higher values of η_e as the rate of extension increases, and the rise is often so drastic that no error of

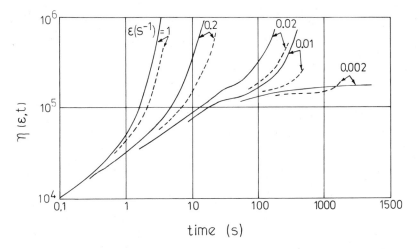

time (s)

Figure 7   Growth of stretching viscosity with time t during elongation at constant rate ε suddenly applied at $t = 0$. *Dashed lines:* data of Meissner (1971) for a low-density branched polyethylene ("Melt I" at 150°C; density at 20°C = 918 kg m^{-3}; melt index = 1.33; $M_w =$ 4.82 × 10^5 = 28.1 M_n). *Full curves:* predictions of Lodge (1964) rubberlike-liquid equations, with memory-function constants chosen to fit the data at $\varepsilon = 0.001$ sec^{-1}. The units of $\eta(\varepsilon, t)$ are kg ms^{-1}

approximation could explain it. The results are often stated in terms of the ratio of the apparent extensional viscosity to the steady-shearing viscosity η_s at a shear rate equal to the rate of extension. Oliver & Bragg (1974) have obtained a ratio of 1400 at $\varepsilon = 200$ sec^{-1} in jet-thrust experiments on a polyethylene oxide solution, and Metzner & Metzner (1970) report a ratio of 29,000 at $\varepsilon = 380$ sec^{-1}.

It is generally agreed that the observed increases in extensional viscosity are caused by alignment of the long-chain polymer molecules along the direction of stretching. This may explain the data of Cogswell (1969) on polypropylene, from which he deduced that the extensional viscosity *decreases* as the rate of extension increases, because this polymer has large side groups that may inhibit orientation.

Since polymer molecules are long, flexible chains, the degree of extension and orientation in a solution of such molecules depends on the rate of extension of the solution. Both extension and orientation grow larger as the rate of extension increases, and both effects increase the extensional viscosity of the solution. To understand the effect of orientation alone, as modified by Brownian motion, one may consider the results for rigid dumbbells obtained by Bird et al (1970). They found that the excess of the Trouton viscosity over that of the solvent increases by a factor of two as the extension rate increases from zero to infinity.

There is a much greater increase in the extensional viscosity because the distance from one end of the molecule to the other grows larger as the extension rate increases. This effect by itself can be understood by considering the viscosity of a suspension of rigid, needle-shaped particles, fully aligned along the direction of stretching. Batchelor (1970, 1971) has obtained theoretical expressions for the extensional viscosities of dilute and concentrated suspensions of such particles in Newtonian liquids. When the distance between particles is small in comparison to their length, the excess of the Trouton viscosity over that of the solvent, $3\eta_0$, is given approximately by

$$(\eta_T - 3\eta_0)/3\eta_0 = (4/9)(L/D)^2 C/ln(\pi/C), \qquad (16)$$

where L/D is the length-to-diameter ratio for the particles and C is their volume concentration. The length-to-diameter ratio for a randomly coiled polymer chain is effectively unity; when fully extended by an extensional flow field, it can be enormous.

Batchelor's results have been tested in spinning experiments. Spinning can be analyzed without significant error because the velocity field is essentially one-dimensional. From analysis of the data, using mass and momentum balances, one can deduce the extension rate and the tensile stress at each point along the thread-line, without any prior knowledge of the connection between them. The ratio of stress to extension rate at each point defines an apparent extensional viscosity η_e there. If the fluid is Newtonian, or in any case in which the extensional viscosity is independent of the rate of extension, the apparent viscosity should be the same at every point along the threadline and equal to the Trouton viscosity η_T. Weinberger & Goddàrd (1974) and Mewis & Metzner (1974) have used suspensions of chopped glass fibers in a Newtonian liquid, and have found that the extensional viscosity is indeed independent of the rate of extension. For the two suspensions tested by Weinberger and Goddard, the extensional viscosities were in reasonable agreement with Batchelor's theory. Mewis and Metzner tested several suspensions with different volume concentrations and particle aspect ratios, and found good agreement with the theory. The highest value of $\eta_T/3\eta_0$ observed was about 260 (with η_0 the shear viscosity of the suspending agent) for particles with length-to-diameter ratio 586 at a volume concentration of about 1%.

In summary, there is ample reason to believe that the Trouton viscosity of a polymer solution must be an increasing function of the extension rate, with a high but finite value in the limit of large extension rates. However, no experiment on a polymer solution has achieved a steady state at high rates of extension.

The molecular picture for polymer melts is much more complicated. In this case the polymer chains form a network. When deformed rapidly, the network structure may remain intact and the deformation is elastic (recoverable). When deformed gently, with the aid of the random agitation from Brownian motion molecules can become disentangled from one another and a state of steady flow can ensue. This is a simplified summary of the conclusions reached by Vinogradov et al (1970a,b; 1972; 1975) from their experimental results. The data of Stevenson (1972) in Figures 3 to 6 are for a polymer melt. It is seen that at a low rate of extension, the stress increases to a steady-state value (Figure 3). At a higher rate, it may be necessary to terminate the experiment before it is clear whether or not a steady state would be reached (Figure 4). At the highest rates (Figure 5), it appears that elastic deformation is taking place, and it is not clear that the stress would ever approach a limiting value. In this case the highly aligned molecules may form crystallites; this is known to occur in highly stretched natural rubber (Wood 1946). Crystallization has also been observed in melts (Mackley & Keller 1973) and even in solutions (Frank et al 1971) at extremely high rates of extension.

Many of the constitutive equations that attempt to describe the large-deformation behavior of solutions and melts are elaborations on the Lodge (1956, 1964) model. The predictions of the Trouton viscosity given by a large number of constitutive equations are summarized in the review article by Dealy (1971). The Lodge model in its simplest form is sufficient to illustrate the kinds of results that are predicted. This model has a form similar to (11), but with a different measure of finite strain ($-C^{-1}$ in place of C). For an extensional flow (3) starting from rest at time zero, this model gives a stretching viscosity

$$\eta(\varepsilon, t) = \int_0^t G(s)[2\exp(2\varepsilon s) + \exp(-\varepsilon s)]\,ds. \tag{17}$$

The shearing-stress relaxation modulus $G(t)$ can be represented as a linear combination of exponentials. For illustration we take it to be a single exponential, $G(t) = G_0 \exp(-t/T)$. With this form, one obtains

$$\frac{\eta(\varepsilon, t)}{G_0} = 2\,\frac{\exp(2\varepsilon - T^{-1})t - 1}{2\varepsilon - T^{-1}} + \frac{1 - \exp(-\varepsilon - T^{-1})t}{\varepsilon + T^{-1}}. \tag{18}$$

In the limit of zero rate of extension, this yields the result that would be given by linear viscoelasticity theory:

$$\eta(0,t) = 3\eta_0[1 - \exp(-t/T)], \tag{19}$$

where $\eta_0 = G_0 T$ is the steady-shearing viscosity. In this case, and whenever the rate of extension is less than $1/2T$, the stretching viscosity approaches a limiting Trouton viscosity as time progresses:

$$\eta_T(\varepsilon)/\eta_0 = 2(1 - 2\varepsilon T)^{-1} + (1 + \varepsilon T)^{-1}. \tag{20}$$

The Trouton viscosity approaches infinity as the rate of extension approaches the critical value $\varepsilon_c = 1/2T$. For this reason, some investigators have suggested that this is the maximum rate of extension for the material. Denn & Marrucci (1971) have shown experimentally that this is not so, and have used a model related to that used here to show what does happen when the rate of extension exceeds the critical value. We see that, at the critical rate of extension, the stretching viscosity given by (18) is asymptotic to $2\eta_0 t/T$ at large times, and no steady state is reached. At higher rates of extension, the stretching viscosity increases exponentially as time progresses.

At strain rates close to the critical rate but below it, the time t_c required to reach the steady state is of the order of that given by setting $\eta(\varepsilon_c, t_c)$ equal to $\eta_T(\varepsilon)$. With the simple model used here, this time is

$$t_c = T/|1 - 2\varepsilon T|. \tag{21}$$

When ε is slightly larger than the critical value, this is the time at which the exponential rise in stretching viscosity begins. For the data of Meissner (1971) shown in Figure 7, the stretching viscosities at various strain rates follow a common curve, close to $\eta(0, t)$, until some more or less well-defined time that depends on the extension rate, after which the stretching viscosity rises very rapidly. The data for the two lowest rates of extension suggest that the longest relaxation time of the material is of the order of $T = 300$ sec. If this value is used in (21), the times at which the rapid rises begin are predicted tolerably well for all of the extension rates shown in Figure 7.

The solid lines in Figure 7 were computed by Chang & Lodge (1972) by using an expression for $G(t)$ with five exponentials, with the longest relaxation time arbitrarily taken to be 100 sec. The reasonable agreement between the computed curves and the data presumably might be improved by taking one of the relaxation times to be 300 sec.

The stress in the Lodge model can diverge to infinity because in this model, molecules are infinitely extensible and the tension in a molecule is proportional to its extension. For real, finitely extensible molecules, the tensile stress probably levels off at a high value when the molecules approach their maximum extension. Convincing experimental evidence in this range is not available.

5   BIAXIAL AND OTHER EXTENSIONAL FLOWS

Stretching motions of the form (3) with ε negative can be produced by stretching a sheet of material, so that $-\varepsilon/2$ represents the stretching rate in the plane of the sheet.

The Lodge model suggests that the Trouton viscosity (20) should at first decrease as $|\varepsilon/2|$ increases, and then become larger again, approaching infinity at $|\varepsilon/2| = 1/2T$, which is the same as the critical extension rate for uniaxial elongation.

In inflation of a circular sheet with clamped edges, the flow at the center of the sheet is an equal biaxial extension. Denson & Gallo (1971) have proposed a method of achieving a nearing constant rate of extension at the center. This method has been used by Joye et al (1972, 1973) and by Maerker & Schowalter (1974). Both of these groups found that the Trouton viscosity at first decreases as the rate of extension increases. Maerker and Schowalter found that, at higher rates of extension, the viscosity passes through a minimum and then increases sharply, as (20) would suggest. White (1975) has done a theoretical study of the flow problem in equal biaxial extension starting from rest.

In any steady velocity field of the form $v_i = \varepsilon_i x_i$ (no sum on i), with the sum of the extension rates equal to zero for isochoric flow, the only relevant material properties are the dependence of the two independent normal stress differences on ε_i and t. In the flows that we have discussed, $\varepsilon_2 = \varepsilon_3$ and thus $\sigma_{yy} = \sigma_{zz}$ by symmetry. In strip biaxial tests, a sheet is stretched in the x-direction while its width in the z-direction is held constant, so that $\varepsilon_3 = 0$ and $\varepsilon_2 = -\varepsilon_1 = -\varepsilon$, say. The normal stress difference $\sigma_{xx} - \sigma_{yy}$, divided by ε, defines the apparent viscosity for this test. The difference $\sigma_{xx} - \sigma_{zz}$ might also be measured, with difficulty, but we are unaware of any data on this.

Strip biaxial tests have been reported by Peng & Landel (1974), who also report tests on equal biaxial extension. They find that the apparent viscosity decreases as the rate of extension increases, with very much the same form of dependence in both tests. Denson & Crady (1974) have used inflation of rectangular strips, in which the strip becomes more or less cylindrical, and they report that the strain rate along the axial direction is less than 1% of that along the circumferential direction. At the lowest rates of extension they find that the extensional viscosity is of the order of $4\eta_0$, the value for an incompressible Newtonian liquid, and they find that the viscosity decreases as the rate of extension increases. Chung & Stevenson (1975) have proposed simultaneous extension and inflation of tubes as a method of producing arbitrary combinations of extension rates.

These motions do not have uniform velocity gradient, of course, but they can be made materially steady in the sense that the velocity gradient at a particle is constant if viewed from a system of axes that rotate with the local rotation of the sheet. The velocity gradient is approximately uniform through the thickness of the sheet, but not exactly. There is only one class of inhomogeneous motions in which every fluid element undergoes steady stretching and the equations of motion are satisfied exactly. The particle trajectories are described in polar coordinates by

$$r = r_0 e^{\varepsilon t}, \qquad \theta = \theta_0 e^{-\varepsilon t}, \qquad z = z_0 e^{-\varepsilon t}. \tag{22}$$

Radial filaments remain radial and stretch at the rate ε, and axial filaments contract at that rate; the relevant material functions would be those appropriate to strip biaxial tests. However, this strange flow does not appear to be suitable for experimental use. Thus, it appears that the experimentalist is practically limited to homogeneous motions, which do satisfy the equations of motion exactly (Coleman & Noll 1962), or to motions that are only approximately steady stretching motions.

6 CONCLUDING REMARKS

We have discussed two types of fluid motion, both with uniform, time-independent velocity gradient. These two types illustrate the distinction between strong flows, in which neighboring particles can separate at an exponential rate, and weak flows, in which no exponential separation occurs. The experimental evidence indicates that the relation between stress and strain rate for strong flows is qualitatively different from that for weak flows. The so-called viscometric flows, which are used in standard shearing viscometers, are weak flows; the material properties measured in viscometric flows are much more closely related to properties relevant to small oscillations than to the properties exhibited in steady extension.

ACKNOWLEDGMENTS

We are grateful to Dr. David Boger of the Chemical Engineering Department, Monash University, Clayton, Victoria, for permission to use Figure 1. Figures 3 to 6 are adapted from the article by J. M. Stevenson (1972) and Figure 7 is adapted from that by J. Meissner (1971). We are grateful for permission to use these figures.

This work was supported by a grant from the Advanced Research Projects Agency to Brown University. We gratefully acknowledge this support.

Literature Cited

Abbott, T. N. G., Walters, K. 1970. Rheometrical flow systems. II. Theory for the orthogonal rheometer, including an exact solution of the Navier-Stokes equations. *J. Fluid Mech.* 40:205–13

Acierno, D., Titomanlio, G., Nicodemo, L. 1974. Elongational flow of dilute polymer solutions. *Rheol. Acta* 13:532–37

Ballman, R. L. 1965. Extensional flow of polystyrene melt. *Rheol. Acta* 4:137–40

Batchelor, G. K. 1970. Slender-body theory for particles of arbitrary cross-section in Stokes flow. *J. Fluid Mech.* 44:419–40

Batchelor, G. K. 1971. The stress generated in a non-dilute suspension of elongated particles by pure straining motion. *J. Fluid Mech.* 46:813–29

Bird, R. B., Harris, E. K. 1968. Analysis of steady state shearing and stress relaxation in the Maxwell orthogonal rheometer. *AIChE J.* 14:758–61

Bird, R. B., Harris, E. K. 1970. Analysis of steady state shearing and stress relaxation in the Maxwell orthogonal rheometer: corrigenda and addenda. *AIChE J.* 16:149

Bird, R. B., Johnson, M. W., Stevenson, J. F. 1970. Molecular theories of elongational viscosity. *Proc. Int. Congr. Rheol., 5th*, 4:159–68

Bird, R. B., Warner, H. R., Evans, D. C. 1971. Kinetic theory and rheology of dumbbell suspensions with Brownian motion. *Adv. Polym. Sci.* 8:1–90

Blyler, L. L., Kurtz, S. J. 1967. Analysis of the Maxwell orthogonal rheometer. *J. Appl. Polym. Sci.* 11:127–31

Chang, H., Lodge, A. S. 1972. Comparison of rubberlike-liquid theory with stress-growth data for elongation of a low-density branched polyethylene melt. *Rheol. Acta* 11:127–29

Chen, I.-J., Hagler, G. E., Abbott, L. E., Bogue, D. C., White, J. L. 1972. Interpretation of tensile and melt spinning experiments on low density and high density polyethylene. *Trans. Soc. Rheol.* 16:473–94

Chung, S. C.-K., Stevenson, J. F. 1975. A general elongational flow experiment: inflation and extension of a viscoelastic tube. *Rheol. Acta* 14:832–41

Cogswell, F. N. 1969. Tensile deformations

in molten polymers. *Rheol. Acta* 8:187–94

Cogswell, F. N. 1972. Measuring the extensional rheology of polymer melts. *Trans. Soc. Rheol.* 16:383–403

Coleman, B. D., Markovitz, H., Noll, W. 1966. *Viscometric flows of non-Newtonian fluids*. New York: Springer

Coleman, B. D., Noll, W. 1962. Steady extension of incompressible simple fluids. *Phys. Fluids* 5:840–43

Davis, W. M., Macosko, C. W. 1974. Mechanical equilibrium for eccentric rotating disks. *AIChE J.* 20:600–2

Dealy, J. M. 1971. Extensional flow of non-Newtonian fluids—a review. *Polym. Eng. Sci.* 11:433–45

Denn, M. M., Marrucci, G. 1971. Stretching of viscoelastic liquids. *AIChE J.* 17:101–3

Denson, C. D., Crady, D. L. 1974. Measurements on the planar extensional viscosity of bulk polymers: The inflation of a thin, rectangular polymer sheet. *J. Appl. Polym. Sci.* 18:1611–17

Denson, C. D., Gallo, R. J. 1971. Measurements on the biaxial extension viscosity of bulk polymers: the inflation of a thin polymer sheet. *Polym. Eng. Sci.* 11:174–76

Ferry, J. D. 1970. *Viscoelastic properties of polymers*. 2nd ed. New York: Wiley

Frank, F. C., Keller, A., Mackley, M. R. 1971. Polymer chain extension produced by impinging jets and its effect on polyethylene solutions. *Polymer* 12:467–73

Gent, A. N. 1960. Simple rotary dynamic testing machine. *Brit. J. Appl. Phys.* 11:165–67

Giesekus, H. 1962a. Strömungen mit konstantem Geschwindigkeitsgradienten und die Bewegung von darin suspendierten Teilchen. *Rheol. Acta* 2:101–12, 112–22

Giesekus, H. 1962b. Die rheologische Zustandsgleichung elastoviskosen Flüssigkeiten—insbesondere von Weissenberg-Flüssigkeiten—für allgemeine und stationäre Fliessvorgänge. *Z. Angew. Math. Mech.* 42:32–61

Goldberg, W., Bernstein, B., Lianis, G. 1969. The exponential extension rate history, comparison of theory with experiment. *Int. J. Non-Linear Mech.* 4:277–300

Gross, L. H., Maxwell, B. 1972. The limit of linear viscoelastic response in polymer melts as measured in the Maxwell orthogonal rheometer. *Trans. Soc. Rheol.* 16:577–601

Hinch, E. J. 1975. Polymères et lubrification. *Centre Natl. Rech. Sci. Colloques Int. No. 233*, p. 241

Huilgol, R. R. 1969. On the properties of the motion with constant stretch history occurring in the Maxwell rheometer. *Trans. Soc. Rheol.* 13:513–26

Huilgol, R. R. 1970. Relations between certain non-viscometric and viscometric material functions. *Trans. Soc. Rheol.* 14:425–37

Huilgol, R. R. 1975. *Continuum Mechanics of Viscoelastic Liquids*. New York: Wiley

Joye, D. D., Poehlein, G. W., Denson, C. D. 1972, 1973. A bubble inflation technique for the measurement of viscoelastic properties in equal biaxial extensional flow. *Trans. Soc. Rheol.* 16:421–45; 17:287–302

Kizior, T. E., Seyer, F. A. 1974. Axial stress in elongational flow of fiber suspension. *Trans. Soc. Rheol.* 18:271–85

Lodge, A. S. 1956. A network theory of flow birefringence and stress in concentrated polymer solutions. *Trans. Faraday Soc.* 52:120–30

Lodge, A. S. 1964. *Elastic Liquids*. New York: Academic

Mackley, M. R., Keller, A. 1973. Flow induced crystallization of polyethylene melts. *Polymer* 14:16–20

Macosko, C. W., Davis, W. M. 1974. Dynamical mechanical measurements with the eccentric rotating disks flow. *Rheol. Acta* 13:814–29

Maerker, J. M., Schowalter, W. R. 1974. Biaxial extension of an elastic liquid. *Rheol. Acta* 13:627–38

Maxwell, B., Chartoff, R. P. 1965. Studies of a polymer melt in an orthogonal rheometer. *Trans. Soc. Rheol.* 9:41–52

Meissner, J. 1969. Rheometer zur Untersuchung der deformationsmechanischen Eigenschaften von Kunststoff-Schmelzen unter definierter Zugbeanspruchung. *Rheol. Acta* 8:78–88

Meissner, J. 1971. Dehnungsverhalten von Polyäthylen-Schmelzen. *Rheol. Acta* 10:230–42

Meissner, J. 1972. Development of a universal extensional rheometer for the uniaxial extension of polymer melts. *Trans. Soc. Rheol.* 16:405–20

Merz, E. H., Cox, W. P. 1958. Rheology of polymer melts—a correlation of dynamic and steady flow measurements. *Am. Soc. Testing Mat. Spec. Tech. Publ. No. 247*, p. 178

Metzner, A. B., Metzner, A. P. 1970. Stress levels in rapid extensional flows of polymeric fluids. *Rheol. Acta* 9:174–81

Mewis, J., Metzner, A. B. 1974. The rheological properties of suspensions of fibres in Newtonian fluids subjected to extensional deformations. *J. Fluid Mech.* 62: 593–600

Moore, C. A., Pearson, J. R. A. 1975. Experimental investigation into an isothermal spinning threadline: extensional rheology of a Separan AP 30 solution in glycerol and water. *Rheol. Acta* 14:436–46

Nitschmann, H. 1949. The viscosity anomaly causing the spinning of liquids. *Proc. Int. Congr. Rheol., 1st,* II:32

Noll, W. 1962. Motions with constant stretch history. *Arch. Ration. Mech. Anal.* 11: 97–105

Oliver, D. R., Bragg, R. 1974. The triple jet: a new method for measurement of extensional viscosity. *Rheol. Acta* 13:830–35

Payvar, P., Tanner, R. I. 1973. Velocity lag, axial thrust, and edge effects in the eccentric disc rheometer. *Trans. Soc. Rheol.* 17:449–63

Peng, St. T. J., Landel, R. F. 1974. Response of bulk polymer under motion with constant stretch histories. *Rheol. Acta* 13:548

Peterlin, A. 1966. Hydrodynamics of macromolecules in a velocity field with longitudinal gradient. *J. Polym. Sci. Part B. Polym. Lett.* 4:287–91

Pipkin, A. C. 1972. *Lectures on Viscoelasticity Theory.* New York: Springer

Pipkin, A. C., Owen, D. R. 1967. Nearly viscometric flows. *Phys. Fluids* 10:836–43

Pipkin, A. C., Tanner, R. I. 1972. A survey of theory and experiment in viscometric flows of viscoelastic liquids. *Mech. Today* 1:262–321

Rivlin, R. S., Sawyers, K. N. 1971. Nonlinear continuum mechanics of viscoelastic fluids. *Ann. Rev. Fluid Mech.* 3:117–46

Spearot, J. A., Metzner, A. B. 1972. Isothermal spinning of molten polyethylenes. *Trans. Soc. Rheol.* 16:495–518

Stevenson, J. F. 1972. Elongational flow of polymer melts. *AIChE J.* 18:540–47

Takserman-Krozer, R. 1963. Behavior of polymer solutions in a velocity field with parallel gradient, III, IV. *J. Polym. Sci. Part A* 1:2477–86, 2487–94

Tanner, R. I. 1975. Stresses in dilute solutions of bead-nonlinear-spring macromolecules. II. Unsteady flows and approximate constitutive relations. *Trans. Soc.*

Rheol. 19:37–65

Tanner, R. I., Huilgol, R. R. 1975. On a classification scheme for flow fields. *Rheol. Acta* 14:959–62

Trouton, F. T. 1906. On the coefficient of viscous traction and its relation to that of viscosity. *Proc. R. Soc. London Ser. A* 77:426–40

Vinogradov, G. V. 1975. Viscoelastic and fracture phenomena in uniaxial extension of high-molecular linear polymers. *Rheol. Acta* 14:942–54

Vinogradov, G. V., Fikhman, V. D., Radushkevich, B. V. 1972. Uniaxial extension of polystyrene at true constant stress. *Rheol. Acta* 11:286–91

Vinogradov, G. V., Radushkevich, B. V., Fikhman, V. D. 1970a. Extension of elastic liquids: polyisobutylene. *J. Polym. Sci. Part A-2* 8:1–17

Vinogradov, G. V., Radushkevich, B. V., Fikhman, V. D., Malkin, A. Ya. 1970b. Viscoelastic and relaxation properties of a polystyrene melt in axial extension. *J. Polym. Sci. Part A-2* 8:657–78

Walters, K. 1975. *Rheometry.* London: Chapman & Hall

Weinberger, C. B., Goddard, J. D. 1974. Extensional flow behavior of polymer solutions and particle suspensions in a spinning motion. *Int. J. Multiphase Flow* 1:465–86

White, J. L. 1975. Theoretical consideration of biaxial stretching of viscoelastic fluid sheets with application to plastic sheet forming. *Rheol. Acta* 14:600–11

Wood, L. A. 1946. Crystallization phenomena in natural and synthetic rubbers. *Adv. Colloid Sci.* 2:57–93

Yamamoto, M. 1956, 1957, 1958. The viscoelastic properties of network structure. I. General formalism. *J. Phys. Soc. Jpn.* 11: 413–21; II. Structural viscosity. *J. Phys. Soc. Jpn.* 12:1148–58; III. Normal stress effect. *J. Phys. Soc. Jpn.* 13:1200–11

Ziabicki, A. 1961. Mechanical aspects of fibre spinning process in molten polymers, III. *Kolloid-Z.* 175:14–27

Ziabicki, A., Kedzierska, K. 1960. Mechanical aspects of fibre spinning process in molten polymers, I, II. *Kolloid-Z.* 171: 51–61, 111–19

Ann. Rev. Fluid Mech. 1977. 9:33–54

COMPRESSIBLE TURBULENT   ✵8096
SHEAR LAYERS

P. Bradshaw[1]

Department of Aeronautics, Imperial College, London, SW7 2BY, England

1 INTRODUCTION

It is generally accepted that the direct effects of density fluctuations on turbulence are small if the root-mean-square density fluctuation is small compared with the absolute density: this is Morkovin's hypothesis (Favre 1964, p. 367). This means that the turbulence structure of boundary layers and wakes at free-stream Mach numbers M_e less than about 5, and of jets at Mach numbers less than about 1.5, is closely the same as in the corresponding constant-density flow. By "turbulence structure" we mean dimensionless properties like correlation coefficients and spectrum shapes: the skin-friction coefficient $c_f \equiv \tau_w / \frac{1}{2}\rho_e U_e^2$ and other ratios of turbulence quantities to mean flow quantities are greatly affected by the influence of mean density changes on the mean motion. The effect of mean density variations in x or y on the turbulence structure is *not* covered by Morkovin's hypothesis, but is often negligible at the lower Mach numbers if streamwise pressure gradients are small. Therefore assumptions about turbulence structure that give good results in calculation methods for constant-density flow will, if properly scaled, give good results in compressible boundary layers or wakes for $M_e < 5$, compressible jets for $M < 1.5$, or low-speed flows with density variations comparable to these. A main object of this review is to substantiate, and qualify, these statements and to discuss the changes in turbulence structure that occur in hypersonic boundary layers, $M_e > 5$, say.

Basic equations for compressible shear layers are given by Howarth (1953) and Lin (1959). More recent treatments (Favre 1971, Cebeci & Smith 1974, Rubesin & Rose 1973, Bilger 1975) use "mass-weighted" variables, which remove density fluctuations from the time-averaged equations of motion but not from the turbulence or from the response of measuring instruments (although Laufer, in Birch et al 1972, p. 462, suggests that pitot tubes probably yield mass-averaged velocities). It seems probable that the difference between conventional and mass-weighted averages rises more slowly with Mach number than current errors in measuring either. Of problems

---

[1] The author is grateful for a number of helpful comments or contributions, especially from Professor H. Fernholz, Dr. P. J. Finley, Dr. L. C. Squire, and Professor J. L. Stollery.

related to the present topic, transition is reviewed by Reshotko (1975), combustion by Libby & Williams (1976), and aeroacoustics by Ffowcs Williams in this volume.

2   MORKOVIN'S HYPOTHESIS OF THE NEGLIGIBLE EFFECT OF DENSITY FLUCTUATIONS

In compressible turbulent flows the velocity, temperature, and density (and in high-speed flows, the pressure) all fluctuate; in analytical work it is more convenient (Chu & Kovasznay 1958) to treat the equivalent fluctuations of vorticity, entropy (or total temperature), and acoustic pressure. Morkovin's important contribution (Favre 1964) was to use the limited data available in 1961 to show that in non-hypersonic boundary layers the acoustic mode is negligible (see also Laufer in Favre 1964) and the entropy (total-temperature) mode is very small for conventional rates of heat transfer. For a recent discussion of mode interactions, with reference to hot-wire calibration problems, see Laderman & Demetriades (1974).

It follows from Morkovin's findings that

$$p'/\bar{p} \ll 1, \qquad T_0'/\bar{T}_0 \ll 1 \tag{1}$$

so that

$$\rho'/\bar{\rho} \approx -T'/\bar{T} \approx (\gamma-1)M^2 u/U, \tag{2}$$

where dashes and overbars denote fluctuations and means, and the total temperature is $T_0 \equiv T + U^2/2c_p$. Since the fractional velocity fluctuation u/U is small, $\rho'/\bar{\rho}$ is small as long as $(\gamma-1)M^2$ is not large compared with unity, which is a convenient definition of "non-hypersonic." Because velocity fluctuations are small in the outer layer of a boundary layer where the Mach number is large, $(\overline{\rho'^2})^{1/2}/\bar{\rho}$ is nowhere greater than 0.1 even at $M_e = 5$. At higher Mach numbers the wall is usually strongly cooled, and Morkovin's "strong Reynolds analogy" (the assumption of negligible total-temperature fluctuations) breaks down, but because of wall cooling the general level of static temperature or density fluctuations increases only slowly with Mach number: see for instance figures 14 and 17 of Laderman and Demetriades. However, $(\overline{p'^2})^{1/2}/\bar{p}$ increases, and the turbulence structure may be altered by vorticity-pressure interaction.

At present there is adequate evidence that the turbulence structure in boundary layers at $M_e < 5$ is virtually the same as at low speeds, that (1) and (2) are obeyed, and that the coefficient of correlation between u and T is close to -1 as a consequence of (2): see for example Morkovin's original paper, Kistler (1959), Demetriades (1968a), Demetriades & Laderman (1973), Johnson (1974), and Rose (1974). For an order-of-magnitude analysis of the terms in the time-averaged equations of motion see Bradshaw & Ferriss (1971). In free mixing layers, where $\sqrt{\overline{u^2}}/U$ reaches 0.3, density fluctuations are larger and the limit of validity of Morkovin's hypothesis (expressed as $\sqrt{\overline{\rho'^2}}/\bar{\rho} = 0.1$) is reached at $M_e \approx 1.5$ in a mixing layer with constant total temperature (Section 8).

The influences of compressibility that Morkovin's hypothesis does not treat are the effect of viscosity fluctuations, which can usually be ignored (Bradshaw & Ferriss

1971) except in regions where mean viscous stresses are important, and the effects of spatial gradients of mean density. A turbulent eddy is likely to be affected by transverse variations in density if the fractional change in density over its width is significant. Therefore the larger eddies, which effect the entrainment of free-stream fluid, may be significantly affected by $\partial \bar{\rho}/\partial y$ even at $M_e < 5$ in a boundary layer. The strong dependence of intermittency factor on Mach number is discussed in Section 6. The effect on the Reynolds-stress structural parameters such as $\overline{uv}/\overline{u^2}$ seems to be small, and the rise in dimensionless entrainment velocity with M_e (Green 1968) seems to be a result of inappropriateness of the usual non-dimensionalizing factor rather than a true effect of compressibility on eddy structure. It is known that low-speed mixing layers between two streams of different density have almost the same spreading rate as in a homogeneous fluid, so there is some independent evidence that the effect of $\partial \bar{\rho}/\partial y$ need not be large. The largest eddies (with wavelengths of order δ) are likely to be identifiable for a streamwise distance of at least 20δ in a boundary layer, or rather less in a free shear layer. In the experiments of Lewis et al (1972) a boundary layer at $M_e = 4$ decelerated to $M_e \approx 2.7$ (a density change of a factor of 4 in the free stream and more in the boundary layer) in a distance of about 20δ, and much larger streamwise density gradients can occur in rapid compressions. The influence of $\partial \bar{\rho}/\partial x$ on the turbulence inferred from this and other experiments is large (Bradshaw 1974) although it is often difficult to distinguish from effects of streamline curvature (Bushnell & Alston 1972, Bradshaw 1973).

In constant-pressure boundary layers at any Mach number, the shear stress reaches a maximum at the wall and asymptotes to zero at the outer edge. For this and no deeper reason, the change of shear-stress profile shape with Mach number is small, too small to be detectable with current measurements. Strictly the Reynolds shear stress in conventionally averaged variables is $-\bar{\rho}\overline{uv} - \overline{\rho' uv}$, while in mass-averaged variables the second term disappears. Here we neglect the second term for simplicity of notation: it is always fairly small. If $-\bar{\rho}\overline{uv}/\tau_w$ is a unique function of y/δ, it follows that changes in turbulence structure are best shown by plotting $\bar{\rho}\overline{u^2}/\tau_w$, etc., rather than $\overline{u^2}/U_e^2$ or $\overline{u^2}/u_\tau^2$ as used by some authors. There is little point in taking turbulence measurements in constant-pressure boundary layers at $M_e < 5$ to check Morkovin's hypothesis, except possibly for intermittency measurements. A good review of existing turbulence measurements is by Sandborn (1974), who also concludes that the structural measurements for $M_e < 5$ agree with low-speed data to within the likely experimental error. However, turbulence measurements in strong pressure gradients, preferably on flat surfaces, are badly needed to elucidate the apparent effects of $\partial \bar{\rho}/\partial x$ mentioned above.

Measurements in shock/boundary-layer interactions by Rose & Murphy (1973), Rose & Childs (1974), and Rose & Johnson (1975) show strong increases in turbulence intensity accompanying compression. Shock/boundary-layer interactions have been reviewed by Green (1970), Korkegi (1971), and Hankey & Holden (1975): for a recent experiment at $M_e \approx 7$ see Horstman et al (1975), and for work on three-dimensional interactions see Korkegi (1975) and Oskam et al (1975).

3   INNER-LAYER SCALING AND THE VAN DRIEST "TRANSFORMATION"

In compressible flow the usual "law of the wall" becomes

$$U/u_\tau = f[u_\tau y/v_w, \; Q_w/(\rho_w c_p u_\tau T_w) \equiv B_q, \; u_\tau/a_w \equiv M_\tau] \tag{3}$$

where $u_\tau = (\tau_w/\rho_w)^{1/2}$ and a_w and Q_w are the speed of sound and (outward) rate of heat transfer at the wall. Arguments that turbulent transfer of heat and momentum become independent of viscosity and conductivity if $u_\tau y/v_w$ is large enough lead to "mixing length" formulas for temperature and velocity [see Rotta (1960) for a careful discussion]. Integration of the mixing-length formulas, $\partial U/\partial y = (-\overline{uv})^{1/2}/(Ky)$, $\partial T/\partial y = -Q/[\bar{\rho} c_p(-\overline{uv})^{1/2} K_\theta y]$, gives

$$T = C_1 T_w - K Q_w/(K_\theta c_p \tau_w) - K U^2/(2 K_\theta c_p) \tag{4}$$

$$U/u_\tau = (C_1^{1/2}/R)\sin(RU^*/u_\tau) - H[1 - \cos(RU^*/u_\tau)] \tag{5}$$

where

$$U^*/u_\tau = (1/K)\ln(u_\tau y/v_w) + C \tag{6a}$$

$$R = M_\tau[(\gamma-1)K/(2K_\theta)]^{1/2} \tag{6b}$$

$$H = Q_w/(\tau_w u_\tau) = B_q/[(\gamma-1)M_\tau^2] \tag{6c}$$

and where K and K_θ are about 0.41 and 0.45, respectively. Equations (4) and (5) are valid for $40v/(\tau_w/\bar{\rho})^{1/2} < y < 0.1\delta$, say. The constants of integration C_1 and C depend on conditions in the viscous sublayer and are expected to be functions of B_q and M_τ. The Van Driest "transformation" [2] is the inverse of (5),

$$\frac{U^*}{u_\tau} = \frac{1}{R}\left\{\sin^{-1}\left[\frac{R(U/u_\tau + H)}{(C_1 + R^2 H^2)^{1/2}}\right] - \sin^{-1}\left[\frac{RH}{(C_1 + R^2 H^2)^{1/2}}\right]\right\}. \tag{7}$$

However, it is found in practice that if (7) is applied to the whole region from the edge of the sublayer to $y = \delta$, then U^* can be fitted by

$$U^*/u_\tau = (1/K)\ln(u_\tau y/v_w) + C + \Pi(x)w(y/\delta)/K \tag{8}$$

as at low speeds, where $w(1) = 2$ and $w \approx (1 - \cos \pi y/\delta)$. Compressibility effects can then be discussed in terms of C, C_1, and the "wake parameter" Π as functions of M_τ and B_q.

It is frequently assumed that $C_1 = 1$, $C = $ constant (say 5.2) in all compressible flows, but this cannot be exactly true. A 10% change in skin-friction coefficient corresponds to a change of about 1.3 in C, which is therefore about the smallest change worth considering in practice and also the uncertainty in current values. Data for C_1 are scarce and unreliable. Bradshaw (1976b) secured good predictions

---

[2] Quotation marks are used to distinguish between true transformations for the whole shear layer and the results that Van Driest (1956a) derived for the inner layer only.

of c_f by writing (7) with (8) at $y = \delta$ and taking

$$C_1 = 1, \qquad C = 5.2 + 95M_\tau^2 + 30.7B_q + 226B_q^2 \qquad (9)$$

(see Section 4). The locus on which $C = 5.2$, according to (9), coincides in part with popular experimental ranges, small B_q for $M_\tau < 0.1$ and a range $0.03 < -B_q < 0.1$ for $M_\tau \approx 0.1$. The correlations of Rotta (1960) for C and C_1 were based on old data.

At low speeds one finds $\Pi \approx 0.58$ for constant-pressure boundary layers with $U_e\theta/v_e \equiv R_\theta > 5000$, but Π falls at lower R_θ (Coles & Hirst 1969). In compressible flow Π is in principle a function of M_τ and B_q as well, but data correlations by Fernholz (1971a) and Squire (1971) suggest that in constant-pressure flows Π is nearly the same function of the empirically chosen Reynolds number $\rho_e U_e\theta/\mu_w$ as at low speeds, at least for $M_e < 5$. Maise & McDonald (1968) have said that Π becomes larger on cold walls; but Gran, Lewis & Kubota (1974) found almost identical $\Pi(x)$ in two boundary layers in a strong pressure gradient at $M_e \approx 4$, with the wall respectively at 0.9 and 0.5 times the external-stream total temperature. The behavior of Π at $M_e > 5$, especially with heat transfer, is less certain. A simpler parameter than Π is the index $1/n$ of an optimum power-law fit to the velocity profile, but, unlike Π, n varies slowly with Re even at high Re. Johnson & Bushnell (1970) have given an extensive data correlation for n. Both they and Bushnell et al (1975) found large n (implying low Π) at low Reynolds number on flat plates, as at low speeds, but the latter authors found unusually *small* values of n at low Reynolds numbers on tunnel walls. Possible explanations are the persistence of transitional disturbances on the plates (see Morrisette et al in Bertram 1968) or the effect of the history of pressure gradient on tunnel walls. Tunnel-wall results show some consistency because tunnel designs do not vary very widely.

Comparisons of outer-layer data are made difficult by uncertainties in the definition or measurement of δ. In particular $\partial U/\partial y$ is very small in the outer part of hypersonic boundary layers, most of the total-pressure gradient being attributable to $\partial \bar{p}/\partial y$. Bushnell & Morris (1971) have shown that δ_{995} is only about half the thickness of the pitot-pressure profile at $M_e = 20$. A suitable generalization is the distance from the surface at which the total pressure P, or more accurately the pressure difference $P - p$, reaches 0.99 of its maximum value: at low speeds, this equals δ_{995}.

The mixing length $l \equiv (-\overline{uv})^{1/2}/(\partial U/\partial y)$ is not a true eddy length scale, but in slowly changing boundary layers it is close to the dissipation length parameter, $L \equiv (-\overline{uv})^{3/2}/(\text{turbulent energy dissipation rate})$, which is a representative length scale of the energy-containing eddies. $L/\delta = f(y/\delta)$ at low speeds, and Morkovin's hypothesis suggests that the same will be true at high speeds if mean-density gradients do not affect the turbulence. Therefore l/δ in turn should be the same function of y/δ as at low speeds, if the high-speed flow is also slowly changing (e.g. zero pressure gradient). Maise & McDonald (1968) and Sivasegaram & Whitelaw (1971) found that this is so, for $M_e < 5$ at least. Horstman & Owen (1972), Fischer et al (1971), Bushnell & Morris (1971), and other investigators have found at worst a moderate decrease in l/δ at hypersonic speeds with increasing Mach number.

However, l/δ increases at low Reynolds number as Π decreases (at least on flat plates, if not tunnel walls), which may partly disguise a decrease with increasing Mach number. If we discount these doubts, the Mach-number invariance of l/δ (and thus L/δ) is our most conclusive check on Morkovin's hypothesis at present. Maise and McDonald also found that in the outer layer the eddy viscosity, $v_\tau \equiv (-\overline{uv})/(\partial U/\partial y)$ correlated well as $v_\tau/(U_e \delta_i^*) = f(y/\delta)$ where δ_i^* is the "incompressible" or "kinematic" displacement thickness, $\int_0^\infty (1 - U/U_e)\,dy$. This is an empirical result that cannot be easily related to Morkovin's hypothesis. The decrease of v_τ with increasing Mach number is rather stronger than that of l (e.g. Horstman & Owen 1972).

Assumptions of Mach-number independence of l/δ, L/δ, or $v_\tau/(U_e \delta_i^*)$ are made in nearly all compressible-flow calculation methods based on partial differential equations, sometimes quite uncritically. Shear-layer calculation methods in general have been reviewed by Reynolds (1976). Good presentations of eddy-viscosity methods for compressible flow have been given by Herring and Mellor (see Bertram 1968) and Cebeci (1971). Shang et al (1973) and Shang (1974) compared different eddy-viscosity assumptions. Models that are based on the Reynolds-stress transport equations and that rely on Mach-number independence of L have been described by Bradshaw & Ferriss (1971; see also Bradshaw et al 1976) and Varma et al (1974), and eddy-viscosity transport equations have been used by Gibson & Spalding (1972), Launder and Spalding (see Birch et al 1972), Libby (see Birch et al 1972), and Wilcox and Alber (see Landis & Hordemann 1972). The many integral methods (Beckwith 1970) normally assume Mach-number independence, or simple correlation, of the entrainment rate (e.g. Green 1972). Green et al (1972) have presented an integral version of a transport-equation method.

In the viscous sublayer, l is frequently correlated by empirical formulas of which the most popular is Van Driest's (1956b):

$$l/(Ky) = 1 - \exp\left[-(\tau_{\text{total}}/\bar{\rho})^{1/2} y/(A^+ v)\right]. \tag{10}$$

If $K = 0.41$, then $A^+ = 25.6$ gives $C = 5.2$ in incompressible flow; and in compressible flow, A^+ will be a function of B_q and M_τ. To calculate the temperature profile we need a similar formula for the temperature mixing length l_θ which replaces $K_\theta y$. Meier et al (1974) applied (10) to l and l_q $[\equiv (l l_\theta)^{1/2}]$ with A^+ assumed independent of B_q and M_τ and the corresponding constant in the l_q formula, A_q^+, taken as $1.3A^+$ to give agreement with the limited experimental data for temperature profiles. Bradshaw (1976b) tabulated the value of A^+ and A_q^+ required to reproduce the values of C and C_1 given by (9) for a range of M_τ and B_q, and Rotta (1974) correlated A^+ as a function of B_q from measurements at low speeds. Various workers (e.g. Reda et al 1975) have shown that the effect of small uniformly distributed roughness on the logarithmic-law constant C can be correlated as the same function of $u_\tau k/v_w$ as at low speeds: larger excrescences are treated by Gaudet & Winter (1973). Work on "cross-hatching," grooves at roughly the free-stream Mach angle which appear spontaneously on certain types of ablating surfaces, has been reviewed by Swigart (1974). Squire (1969) has given a correlation of the additive constant C as a function of V_w/u_τ and M_e in transpired boundary layers, and later work was reported by Squire (1971), Dunbar & Squire (1971), Squire & Verma (1973), and Thomas (1974). It appears that, as at low speeds, the turbulence structure in the outer layer is unaltered by moderate transpiration, so that the only modifications needed to a calculation method concern the wall boundary condition.

4 SKIN-FRICTION FORMULAS FOR COMPRESSIBLE BOUNDARY LAYERS

Hopkins & Inouye (1971) showed that all the formulas for c_f as a function of M_e, T_w/T_e and $U_e\theta/v_e$ that were available to them could be reduced to the form introduced by Spalding & Chi (1964), $F_c c_f = f(F_\theta R_\theta)$, where F_c and F_θ are functions of M_e and T_w/T_e, and the function f is the incompressible-flow skin-friction formula (with $F_c = F_\theta = 1$). This is a convenient form for discussion.

4.1 Formulas Based on the Van Driest "Transformation"

Writing (8) at $y = \delta$ and using (7) yield a relation between U_e/u_τ and $u_\tau\delta/v_w$, with M_τ and B_q as parameters, from which it is easy to deduce c_f as a function of $U_e\delta/v_e$, M_τ, and B_q. To convert to $U_e\theta/v_e$, θ can be obtained by integrating over the whole profile, but nearly all the published formulas of this type use empirical or semiempirical relations for θ/δ or its equivalent. Van Driest I (1951), Van Driest II (1956a; see also Görtler & Tollmien 1955), and Spalding & Chi agree on F_c but have $F_\theta = (\mu_e/\mu_w)(\rho_w/\rho_e)^{1/2}$, μ_e/μ_w, and $(T_e/T_w)^{0.702}$ $(T_{aw}/T_w)^{0.772}$, respectively. Van Driest II and Spalding & Chi agree with each other, and with data for $M_e < 5$ at least, on adiabatic walls (because $\mu \propto T^{0.702}$ approximately): however, Spalding & Chi underpredicts the rise in c_f on cold walls and is low by about 20% at $T_w/T_{aw} = 0.4$. Fernholz (1971a) used a form in which variations of C and Π are allowed for explicitly instead of being hidden in F_θ; his analysis is closely related to the Van Driest transformation but cannot be easily cast into the same form.

4.2 Intermediate-Temperature Correlations

If an incompressible-flow c_f formula is used with fluid properties evaluated at a temperature T' intermediate between T_w and T_e, then $F_\theta = F_c^{-\omega}$, where $\mu \propto T^\omega$, and $F_c = T'/T_e$. Intermediate-temperature concepts have been well reviewed by Coles (1964) but have no physical basis, and the connection between F_c and F_θ sacrifices a degree of freedom compared with formulas of the Van Driest type.

4.3 Formulas Based on Transformations for the Whole Layer

The most general and flexible of the proposed transformations is that of Coles (1964); any objections to Coles' work would apply with greater force to the earlier transformations (discussed by Coles), and we therefore discuss only the main points of Coles' transformation here. He seeks a transformation of independent and dependent variables, x, y, U, V, p, ρ, τ, ψ (stream function), etc., to a new system $\bar{x}$, $\bar{y}$, $\bar{U}$, $\bar{V}$,... in which $\bar{\rho}$ is constant. The simplest conditions for such a transformation to exist for the continuity and momentum equations are that the quantities $\partial \bar{\psi}/\partial \psi \equiv \sigma$, $\partial \bar{x}/\partial x \equiv \xi$, and $\bar{\rho}\partial\bar{y}/(\rho\partial y) \equiv \eta$ shall be functions of x only (earlier workers generally assumed $\sigma = 1$). The crucial step is to assume that these simplest conditions are unique and that they also apply to all equations necessary to describe the turbulence, specifically the shear stress. Morkovin's hypothesis provides no justification for this. If the transformation is valid in the viscous sublayer (and we recall that the Van Driest "transformation" is not, in general), the viscous-stress formula and the fact that velocity profiles are identical at corresponding stations in the two flows (the mapping defined above being independent of y) yield

$$\bar{c}_f \bar{R}_\theta = (\mu_e \rho_e / \mu_w \rho_w) c_f R_\theta. \tag{11}$$

This result is independent of σ, ξ, and η as long as they are functions of x alone. To derive a skin-friction formula a relation for σ is required and this must be empirical. Coles assumes, with some justifying arguments and data analysis, that $\sigma = \bar{\mu}/\mu_s$ where μ_s is the viscosity at the arbitrarily defined edge of the sublayer in the compressible flow. This is in effect an intermediate-temperature assumption. By using the law of the wall in the transformed flow he deduces $\bar{T}/T_s$ and thence

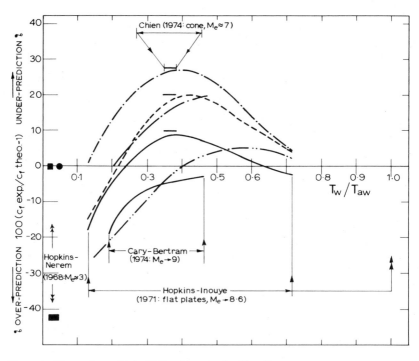

Figure 1   Comparison of skin-friction formulas for $M_e < 9$.
Van Driest II: ————; Sommer-Short: – – – –; Spalding-Chi: – · — · — · —;
Coles: — · · — · · — · ·.

$\bar{\mu}/\mu_s$. The resulting skin-friction formula is an implicit one: the steps in its evaluation have been well summarized by Hopkins and Inouye, and charts have been given by Miles & Kim (1968).

This author's view is that because of their arbitrary nature transformations should be held in no more respect than data-fit correlations like the skin-friction formulas discussed above, particularly since transformations for more general flows (e.g. Baronti & Libby 1966, Economos 1970, Lewis et al 1970) are even more arbitrary. The critique of Crocco (1963) is more friendly than the present one but his preferred transformation is into a low-speed variable-property flow (consistent with the criticism of Lewis et al that Coles' transformation does not correctly represent viscosity gradients): it can be argued that Morkovin's hypothesis does as much. Küster (1972) discussed extensions of the Coles transformation, and Jeromin (1968) presented a transformation for boundary layers with injection. There are a number of correlations for skin-friction or shape parameter that do not fit exactly into the above categories: examples are the adiabatic-wall c_f correlation of Winter & Gaudet (1970) and the shape-parameter correlations of Stollery (1976).

4.4 Accuracy of Skin-Friction Formulas

Of the two most painstaking reviews of c_f formulas carried out in recent years, Hopkins & Inouye (1971) prefer Van Driest II, while Cary & Bertram (1974) prefer Spalding & Chi (but remark that only Coles' formula is acceptable for $M_e > 10$). Figure 1 summarizes the results of these and other reviewers for the error in predictions made by the most common formulas, plotted against wall-temperature ratio. The error does not seem to depend strongly on M_e for $M_e < 9$. Typical scatter in results is at least $\pm 10\%$, so that the shapes of the curves (drawn by the present author) are not to be taken too seriously. Cary and Bertram and Hopkins and Inouye have reviewed data for $3 < M_e < 9$; Chien (1974) and Hopkins & Nerem (1968) plotted their own results at $M_e \approx 7$ and $M_e \approx 3$, respectively. We have represented Chien's scattered data by a single symbol group at midrange.

Figure 1 seems to show that the preference of Hopkins and Inouye for Van Driest II (reinforced by the original data of Hopkins et al 1972 and Keener & Hopkins 1972, 1973) is justified, although like all the other theories it fails to predict the fall in c_f that apparently occurs on very cold walls. The preference for Spalding & Chi among workers concerned only with very cold walls is explained by Figure 1, but it is clear that Spalding & Chi's performance on moderately cooled walls is as poor as, or poorer than, Sommer and Short's "Intermediate-Temperature" formula: the cold-wall correlations of both these formulas are based on free-flight data of Sommer & Short (1955). Coles' predictions for very cold walls are even less accurate than Van Driest's, and the Coles predictions reported by Hopkins et al (1972) show an unusually rapid decrease of c_f with increasing R_θ, even at high R_θ, suggesting that the Reynolds-number trend provided by the sublayer hypothesis is not correct.

Cary and Bertram have shown that Coles' predictions for data at $M = 10$ to 13, $T_w/T_{aw} = 0.15$ to 0.2, are much better and less scattered than Van Driest or Spalding & Chi, both of which are now starting to underpredict even on these very cold walls.

The data at very high Mach numbers are too scattered for useful conclusions to be drawn: some data have been obtained in helium, and comparison on $(\gamma - 1)M_e^2$ rather than M_e is necessary, though not perhaps sufficient, to relate air and helium data. A comparison of the common formulas with tunnel-wall measurements (in nitrogen) at $M_e \approx 20$, $T_w/T_{aw} \approx 0.2$ and $R_\theta < 6000$ was made by Harvey & Clark (1972), who prefer Coles and Van Driest II to Spalding & Chi. Van Driest II underpredicts the measurements of Harvey and Clark by an average of about 25%, while Coles again shows a strong trend with Reynolds number, overpredicting by 40% at $R_\theta \approx 2500$ but by only 5% at $R_\theta \approx 5300$. In principle one would expect formulas that neglect low-Reynolds-number effects to underpredict at low R_θ.

There are discrepancies between different reviewers' results because of differences in the method of finding R_θ. Holden (in Hankey & Holden 1975) and Cary and Bertram have carefully discussed the alternative methods of deriving R_θ but have come to no firm conclusion. Low-Reynolds-number effects and memories of transition are likely to have affected many of the hypersonic data sets. Coleman et al (1973), who deduced c_f and R_θ from St, have shown that only Fernholz's method predicts the sharp rise in c_f with decreasing R_θ in the range 4000 to 6000. However, the Reynolds-analogy factor St/(0.5 c_f) rises rapidly with decreasing Reynolds number in low-speed flow, and the derivation of c_f from St in low-Reynolds-number hypersonic flow is suspect for this reason alone. Cary and Bertram's deductions of c_f from St gave significantly different results from direct c_f measurements and partly accounted for their preference for Spalding & Chi. Cary himself (1970) commented that heat-transfer data cannot in general be used to validate c_f formulas because of uncertainty in the Reynolds-analogy factor, and it seems wisest to accept this conclusion even today. This author's impression is that inference of c_f from St, and correlations on R_x instead of R_θ, have caused considerable confusion. R_x correlations are particularly uncertain at supersonic speeds because the transition Reynolds number is always high and usually ill documented.

Hopkins and Inouye distinguished flat-plate and tunnel-wall data, which show significant differences at the higher Mach numbers and also at low Reynolds numbers (Bushnell et al 1975). Clearly, flat plates are in principle closer to conditions on real vehicles.

It is not worth speculating on the physical reasons for the discrepancies between all these rather unphysical formulas. Fernholz's formula, although it uses various wholly empirical correlations, is probably the closest at present to the spirit of the Van Driest transformation (Van Driest's own choice of F_θ is crude). Unfortunately Fernholz's formula has not been compared with data on the basis outlined above, and its predicted increase of c_f with decrease in wall temperature is smaller than that of other formulas: further development is needed before it can be used over the full range of T_w/T_{aw}. In the long run the "exact" Van Driest transformation (7), with correlations for C and C_1 in terms of the wall-layer parameters M_τ and B_q, and for Π in terms of outer-layer parameters and Reynolds number (Bradshaw 1976b) offers the best vehicle for experimental data, whether profiles or just surface measurements. The name "Van Driest III" is suggested. It offers an escape from arbitrary assumptions about F_θ, which simply augment the existing confusion over low-Reynolds-number effects.

5   HEAT-TRANSFER AND TEMPERATURE-PROFILE FORMULAS

5.1   *Reynolds Analogy Factor*

Heat-transfer formulas for constant-pressure constant-wall temperature flow are usually linked to a skin-friction formula by means of a Reynolds analogy factor $F \equiv St/(0.5 c_f) \equiv Q_w U_e/[c_p(T_w - T_{aw})\tau_w]$. In compressible flow the dependence of F on R_θ (i.e. on c_f) has been generally ignored, except by Cary & Bertram (1974). Chi & Spalding (1966) suggested $F = 1.16$ as a mean of the available data, mainly at nonhypersonic Mach numbers, and Cary and Bertram endorsed this suggestion "for Mach numbers less than approximately 5 and near-adiabatic wall conditions." For $M_e > 5$ and moderate cooling Cary (1970) recommended a version of Von Karman's formula yielding $F \approx 1.1$, while for very cold walls $(T_w/T_{aw} < 0.3)$ Cary and Bertram and Keener & Polek (1972) suggest that F scatters about 1.0. Data plotted by Chien (1974) for $0.1 < T_w/T_{aw} < 0.55$ are scattered between $F = 1.0$ and 1.2 with no clear trend with T_w. $F = 1.1$ is an adequate average of the data plotted by Chien and of the recommendations quoted above.

5.2   *Turbulent Prandtl Number*

Recent low-speed data (see Launder in Bradshaw 1976a), and the measurements of Meier et al (1974) at $M_e \approx 4.9$ and Horstman & Owen (1972) at $M_e \approx 7$, agree that the turbulent Prandtl number exceeds unity in the viscous sublayer where it is a function of $u_\tau y/v_w$, reaches a constant value, K/K_θ, of about 0.9 in the inner layer and out to $y/\delta \approx 0.4$, and falls well below unity in the outermost part of the boundary layer where it is a function of y/δ. In general Pr_t in the outer layer must depend on boundary-layer history, and in compressible flow values in the sublayer and outer layer may depend on M_τ and B_q even if Morkovin's hypothesis is valid. Use of a turbulent Prandtl number based on total temperature implies that turbulent processes depend on the velocity of convection past the observer, and this is untrue.

5.3   *Temperature-Profile Formulas*

The temperature profile (4) is valid only in the inner layer, but quadratic temperature profiles are widely used for the whole of constant-pressure boundary layers on constant-temperature walls, the unspoken argument being that a formula that is correct in the inner layer and forced to be correct in the free stream can hardly go disastrously wrong in between. The most general form of the Crocco temperature profile is

$$(T_0 - T_w)/(T_{aw} - T_w) = U/U_e + (U/U_e)^2(T_{0e} - T_{aw})/(T_{aw} - T_w). \tag{12}$$

For the case $T_{aw} = T_{0e}$ [recovery factor $r \equiv (T_{aw} - T_e)/(T_{0e} - T_e) = 1$] the second term on the right disappears and we obtain a straight line on a plot of $(T_0 - T_w)/(T_{aw} - T_w)$ against U/U_e. For a typical value of $r = 0.9$, the plot falls below the straight line by a maximum of 0.04, whereas the real temperature profile on a cold wall [e.g. Figure 8 of Laderman & Demetriades (1974)] can lie markedly below the straight line, and it is becoming generally accepted that (12) is not very accurate after all, especially in boundary layers with nonuniform wall-temperature history such as those on tunnel walls or real vehicles (Fernholz 1971b, Feller 1973, Gates 1973). On an adiabatic wall T_0 must exceed T_{0e} somewhere to compensate for low T_0 near the surface, and (12) does not permit this. The use of (4) throughout the layer to define the complete Van Driest transformation can be justified pragmatically, at least if $T_w \approx$ constant: even if the correct temperature profile were used, the transformed outer-layer profile would still be liable to depend on Mach number or heat transfer.

Values of C_1 in (4) can be deduced from the inner part of the profile of Laderman and Demetriades and from Figure 5 of Owen et al (1975): the latter profile follows the linear version of (12) much more closely than do the Owen-Horstman (1972) data for the same model. C_1 seems to lie between 0.95 and 1.05, but the differences between the two experiments by Owen's group and the rapid departure of Laderman's and Demetriades' temperature profile from (12) deter one from being too specific. It is suggested that any further measurement or analysis of temperature profiles at high or low speeds should be directed to improving the correlations for Pr_t and C_1 (which depends on the distribution of Pr_t in the sublayer): the main use of correlations for T itself is in reducing pitot measurements. Measurements of the effect of upstream history on Pr_t—not just on T—would be valuable.

6   TURBULENCE STRUCTURE OF HYPERSONIC BOUNDARY LAYERS

In boundary layers, the growth of the viscous sublayer as a fraction of boundary-layer thickness is the biggest single influence on turbulent structure at given y/δ: at $M_e = 5$, the start of the hypersonic range, the sublayer is ten times as thick as in low-speed flow at the same R_θ. The stages in breakdown of the inner-layer analysis as M_e increases are probably (a) dependence of C and C_1 on B_q and M_τ; (b) significant variation of total shear stress across the thickening viscous sublayer, causing C and C_1 to depend additionally on $v_w/(\rho_w u_\tau^3)\partial\tau/\partial y$; (c) dependence of K and K_θ on M_τ and B_q, because of the effect of large mean density gradients or of density fluctuations. No reliable evidence for changes in K or K_θ has been found. Possibly this stage is preceded by (d) occupation of the whole inner-layer region by the sublayer so that the region of validity of (5) shrinks to zero.

In the outer layer, changes in structure may be caused by (a) mean density gradients: $\bar{\rho} \propto y$ is a fair approximation for the middle region of typical boundary layers around Mach 10; (b) density fluctuations: $(\overline{\rho'^2})^{1/2}/\bar{\rho}$ typically reaches 0.1 by $M_e = 10$; and (c) low-Reynolds-number effects (Section 7).

Some experiments have found large pressure gradients normal to a flat surface, and controversy has resulted: for a discussion, see Shang (1974). Either the pressure difference across the layer is balanced by flow turning (since the Reynolds stress $-\overline{\rho v^2}$ is zero at $y = 0$ and $y = \delta$) or the measurements are wrong!

The three-fold or four-fold increase in density across the outer part of the shear layer, normally occupied by the large eddies, is probably responsible for the two-fold or three-fold decrease in standard deviation of the intermittent interface position from its low-speed value [Klebanoff (1955), $M_e = 0.05$, $\sigma/\delta = 0.14$; Owen & Horstman (1972), $M_e = 6.7$, $\sigma/\delta = 0.08$; Laderman & Demetriades (1974), $M_e = 9.4$, $\sigma/\delta = 0.07$]. The flatness-factor $[\overline{u^4}/(\overline{u^2})^2]$ measurements of Yanta & Lee (1974) at $M_e = 3$ imply an intermittency factor, $3(\overline{u^2})^2/\overline{u^4}$ approximately, which falls off near the edge of the boundary layer less rapidly than at low speeds. Kistler's (1959) observations of oscilloscope traces at $M_e \approx 4.5$ also suggest considerable changes in intermittency statistics, but only the hot-wire signal (ρu approximately) rather than the pure velocity signal was observed. It is not safe to equate interface deviation with large-eddy size or strength, because low-speed smoke pictures show that many of the interface excursions are thin tongues of fluid which clearly do not contain much turbulent kinetic energy: these tongues may well fail to penetrate into the high-density fluid of a hypersonic external stream, while the more coherent energy-containing eddies are likely to be less affected. None of the other turbulence properties changes as spectacularly as intermittency. It is not likely that the change in intermittency is a low-Reynolds-number effect; in constant-density flow σ/δ is, if anything, larger at low Reynolds numbers. However, the changes in intermittency almost certainly control the low-Reynolds-number effects.

The only measurements of $\overline{u^2}$ in hypersonic flows appear to be those of Laderman & Demetriades (1974) and Owen et al (1975). Both, notably the former, yield values of $\rho \overline{u^2}/\tau_w$, which are below low-speed data in the inner part of the boundary layer by a factor of three or more. This implies local values of the "shear parameter" $\tau/(\rho \overline{u^2})$ of 1 to 1.5, which is implausibly high compared with the low-speed value of 0.4. Strictly Morkovin's hypothesis requires constancy of $-\overline{uv}/\overline{u^2}$, whereas $\tau/\bar{\rho}$ is $-(\overline{uv} + \overline{\rho'uv}/\rho)$: however the most generous estimate of $\overline{\rho'uv}$ cannot account for so large a difference. Laderman and Demetriades have suggested that sound radiation may perhaps decrease $\overline{u^2}$; but one cannot see how this or any other likely compressibility effect can greatly *increase* the shear parameter, which is a measure of the efficiency of generation of shear stress by the turbulence. Most of the sources of error in hot-wire measurements lead to low readings, and one is inclined to doubt the measurements of $\overline{u^2}$.

Laderman (1974), using the techniques of Laderman and Demetriades, has measured u, T', and p' in transitional-turbulent adiabatic flow on a cone at $M_e = 7$. Values of $\overline{u^2}$ are higher than those of the latter authors. The "Strong Reynolds Analogy" (2) fits the T' data to within the experimental scatter of about $\pm 20\%$: $(\overline{T'^2})^{1/2}/\overline{T}$ reaches 0.3 at $y/\delta \approx 0.7$.

Figure 31 of Raman (1974) shows the unsatisfactory state of our data for surface pressure fluctuations. In low-speed constant-pressure flow $(\overline{p'2})_w^{1/2} \equiv \tilde{p}_w$ is about $2.6\tau_w$, virtually independent of Reynolds number. The high-speed data mainly suggest a slow increase with Mach number, but are very scattered and do not permit any effect of Reynolds number, or indeed of heat transfer, to be detected. Laderman & Demetriades (1974) have deduced pressure fluctuations within a boundary layer at $M_e = 9.4$ from hot-wire measurements. The pressure fluctuation profile has a weak maximum at $y/\delta \approx 0.6$; the free-stream value agrees well with an extrapolation of the measurements of Laufer (1964), while the indicated wall value is about $4\tau_w$, lower than the most probable curve through the data plotted by Raman but not disastrously so. Raman's own data appear to suffer from poor frequency response.

Smith & Driscoll (1975) have made preliminary electron-beam measurements of p' and ρ' in an adiabatic wall flow at $M_e = 16$; $(\overline{p'^2})^{1/2}/\bar{p}$ reaches 1.0 and p' has a highly non-Gaussian probability distribution.

Morkovin's hypothesis, and the hypothesis that $\partial \bar{p}/\partial y$ does not affect eddy behavior, together imply that $\tilde{p}_w/\tau_w$ is independent of Mach number because the contribution to $\tilde{p}_w$ from given y is related to $\bar{\rho} u^2$ (or $\bar{\rho} u \partial U/\partial y$) and the profiles of $\bar{\rho} u^2$ and l/δ are roughly independent of Mach number. The moderate increase of $\tilde{p}_w/\tau_w$ with M_e, even below $M_e = 5$, may merely reflect the increasing importance of the acoustic mode or, at least, of compressibility effects in the *propagation* —rather than generation—of the pressure fluctuations.

7   LOW-REYNOLDS-NUMBER EFFECTS

The location of viscous influence on the outer layer (Falco 1974, Murlis 1975) is likely to be the "viscous superlayer" or irrotational-turbulent interface. The appropriate Reynolds number for scaling at given M_e is $(\tau_w/\rho_e)^{1/2} \delta/v_e \equiv R_v$: the velocity scale follows from the arguments about intensity scaling in Section 2, but the intermittency measurements discussed in Section 6 suggest that the length scale will be affected by compressibility. This is confirmed by the empirical finding of Fernholz (1971a) that in adiabatic boundary layers the size of the outer-layer "wake parameter" Π correlates on $\rho_e U_e \theta/\mu_w$ for M_e up to at least five: R_v increases rapidly with M_e at given $\rho_e U_e \theta/\mu_w$, so the length scale decreases rapidly. Morkovin's hypothesis excludes density fluctuations but viscosity fluctuations or mean density gradients may be responsible. Either can be correlated adequately by the mean density or temperature ratio across the outer layer, say $T_{0.2}/T_e$ where suffix 0.2 denotes conditions at $y/\delta = 0.2$, the nominal inner edge of the outer layer: mixing-length arguments suffice to relate viscosity (temperature) fluctuations to mean temperature gradients. It is therefore plausible that Π is a universal function of $R_v f(T_{0.2}/T_e)$, i.e. that the length scale is $f\delta$. The function f can be obtained by forcing agreement with Fernholz's correlation for adiabatic walls. Details are given by Bradshaw (1976b) and suggest that most of the existing hypersonic data, except the results of Laderman & Demetriades (1974), suffer from significant low-Reynolds-number effects. Values of $f(T_{0.2}/T_e)$ decrease with increasing Mach number, more rapidly than the standard deviation of the interface.

It is important to note that the lowest Reynolds number at which fully turbulent flow can occur will be determined by the viscosity at or near the wall and will not correlate in the same way as viscous effects in the outer layer. In particular it will not coincide with the Reynolds number at which Π becomes zero, which happens, approximately and fortuitously, at low speeds.

At low speeds, reverse transition in strong favorable pressure gradients begins when τ_w rises so much that the excess of energy dissipation over production in the viscous sublayer cannot be replaced by energy transport from the outer layer, whose *absolute* intensity is almost unchanged. For a recent review see Fernholz in Bradshaw (1976a). In high-speed flow the effect of negative $\partial \bar{p}/\partial x$ in reducing the absolute intensity throughout the shear layer may dominate. Narasimha & Viswanath (1975) found that complete reverse transition occurs when the pressure-drop across a sudden expansion exceeds 75 times the upstream wall shear stress. The absence of viscosity from this correlation suggests that $\partial \bar{p}/\partial x$ is more important at high Mach numbers than "sublayer bankruptcy." There is a continuing controversy about Reynolds-number dependence of pressure rise to separation, specifically for the case of flow up a wedge or plane ramp. The data, reviewed and plotted by Appels (1974), Law (1974), Roshko & Thomke (1975), and Settles et al (1975), show that the wedge angle to induce significant separation decreases with increasing Reynolds number in the range $U_e \delta/v_e < 5 \times 10^5$ [say $U_e(\theta/v_e) < 2 \times 10^4$ at $M_e \approx 5$] but thereafter increases according to most workers; some, e.g. Settles et al, find no significant Reynolds number effect beyond about 10^5. Some small differences between the experiments may be caused by the effects of lateral divergence (i.e. increasing circumference) in axisymmetric flows (Coleman & Stollery 1974). The main differences arise from difficulty in defining separation, and the controversy would probably be settled if the contending parties analyzed one another's results. In low-speed flow it is generally accepted that pressure rise to separation increases with Reynolds number, except possibly in transitional or low-Reynolds-number flows.

8   FREE SHEAR LAYERS AND DUCT FLOWS

The most important feature of free shear layers, with the exception of the mixing layer or "half-jet," is that typical velocity and temperature differences decrease rapidly with streamwise distance, so that, for instance, an axisymmetric jet or wake becomes effectively incompressible a few tens of nozzle or body diameters downstream, unless the initial Mach number is extremely large. Experiments on axisymmetric wakes have been reported by Demetriades (1968a,b), who found $U/U_e \approx$ 0.1, $\Delta T/T_e \approx 0.3$ only 20 diameters behind a blunt-based rod at $M_e = 3$; by the free-flight range group at DREV, Quebec (e.g. Heckman 1973); and by Avidor & Schneiderman (1975). Measurements in a two-dimensional, bluff-body wake are presented by Demetriades (1970). Work on supersonic jets has concentrated mainly on the behavior of the shock-expansion system.

The spreading rate of a plane mixing layer decreases with increasing stream Mach number (see Birch and Eggers in Birch et al 1972, p. 11; and Harvey & Hunter 1975). However, the spreading rate of low-speed variable-density jets is almost independent of density ratio at least up to $\bar{\rho}_1/\bar{\rho}_2 = 7$ (Brown & Roshko 1974), which corresponds to the density ratio across an isoenergetic compressible flow at $M_e \approx 5$. Therefore the effect of Mach number is a real compressibility effect, resulting from a breakdown of Morkovin's hypothesis (see Bradshaw in Birch et al

1972, p. 39). In the maximum-intensity region of a low-speed mixing layer $(\overline{u^2})^{1/2}/U \approx$ 0.3, and (2) shows that the density fluctuations approach Morkovin's limit of, say, 0.1 at $M_e \approx 1.5$. The start of significant changes in σ above $M_e \approx 1.5$ is plausible in view of this: however, equally large density fluctuations occur in a low-speed variable-density mixing layer with $\rho_1/\rho_2 = 1.3$, so that the breakdown of Morkovin's hypothesis is caused by pressure fluctuations, not density fluctuations. Oh (see Murthy 1975) has developed a semiempirical model of the fluctuating pressure-dilatation term in the turbulent energy equation, $\overline{p'\,\partial u_i/\partial x_i}$, which rests on the behavior of eddy Mach waves and is as yet unsupported by direct experiment. The turbulence measurements of Ikawa & Kubota (1974) at $M_e = 2.5$, at distances less than 50 nozzle boundary-layer thicknesses from the nozzle lip, show no spectacular changes from low-speed flow except a general decrease in turbulent intensity. Limited results at $M_e = 5$ are given by Wagner (1973) in a mixing layer described as "nearly fully developed." Harvey & Hunter (1975) show that the spreading rate at $M_e = 19$, very close to the nozzle exit, is roughly $\frac{1}{4}$ of the low-speed value, with $(\overline{\rho'^2})^{1/2}/\bar{\rho} \approx 0.15$.

Fletcher et al (1974) have presented an extensive review of supersonic base flows, nominally devoted to heat-transfer problems but in fact forming a useful entry to the literature as a whole: see also Dussauge & Gaviglio (1975).

The main interest in compressible internal flows is in the effects of the acceleration, sometimes leading to reverse transition, produced by strong heating of low-speed pipe flows (e.g. Perkins & McEligot 1975).

Appendix   *A selection from the "Eurovisc" Compressible Boundary Layer Catalog.* This catalog of tabulated data, edited by H. Fernholz (TU, Berlin) and P. J. Finley (Imperial College, London), will be published by AGARD in 1977. The following is a minimum list, less than one half of the total, of experiments that cannot be ignored, although none is perfect. The editors have given preference throughout to studies in which both total-temperature profiles and c_f were measured directly, while recognizing the unreliability of most c_f measurements by whatever method. As seen below, care has been taken to distinguish the different ways in which a given pressure gradient (including zero) can be obtained, and the full catalog presents a still more detailed classification.

Authors	M_e	$R_\theta \times 10^{-3}$	T_w/T_{aw}	T_0 profile[a]	$c_f^{a,b}$
Group IA : Zero pressure gradient with zero-pressure-gradient history					
Type IA1, flat plates					
Coles (1953)	2–4.5	2–10	1.0	X	F
Hastings & Sawyer (1970)	4	2–25	1.0	$\sqrt{}$	F
Keener & Hopkins (1972)	6.3	2.5–9	0.3–0.5	$\sqrt{}$	F
Watson et al (1973)	10	1.4–13	1.0	($\sqrt{}$)	F
Mabey et al (1974)	2.5–4.5	5–30	1.0	$\sqrt{}$	F
Type IA2, cylindrical bodies					
Peake et al (1971)	4	8–50	1.0	X	P
Horstman & Owen (1972)	7.2	5–10	0.5	$\sqrt{}$	F
(See also Owen et al 1975)					

Group IB: Zero pressure gradient with varied histories (e.g. nozzles)

Moore & Harkness (1965)	2.7	180–700	1.0	X	F
Thomke (1969)	2–5	100–350	1.0	X	P
Voisinet & Lee (1972)	4.9	7–58	0.3, 0.7, 1.0	√	F
Hopkins & Keener (1972)	7.2	16–62	0.3, 0.5	√	F
Gates (1973)	4, 4.9	8–31	~1.0	√	—
Winter & Gaudet (1970)	0.2–2.8	14–80	1.0	√	F

Cases for which we have been unable to obtain numerical data

Gran et al (1974)	4	~10	0.5	√	(S)
Laderman & Demetriades (1974)	9.4	37	0.4	√	(V)
Martelucci & Langanelli (1974)	7.9	0.8–5	0.5?	√	—

Group IIA: Pressure gradients on a flat surface (reflected wave)

Type IIA, 1/3/5, favorable pressure gradients

Voisinet & Lee (1973)	3.8–4.6	6–60	0.3, 0.8, 1.0	√	F

Types IIA, 3.4, adverse/favorable pressure gradient

Lewis et al (1972)	2.4–3.7	~10	1.0	√	(S)

Types IIA, 2/4/6, adverse pressure gradients

Zwarts (1970)	4–3	35–70	1.0	X	P
Waltrup & Schetz (1973)	2.4–1.9	18	1.0	√	F
Peake et al (1971)	4–2	8–50	1.0	X	P
Voisinet & Lee (1975, unpublished)	4.9–4.1	6–70	0.3, 0.8, 1.0	√	F

Group IIB: Pressure gradients on a longitudinally curved surface (simple wave)

Types IIB, 1/3/5, favorable pressure gradients (all nozzles)

Fischer et al (1971)	20	5–10	1.0	(√)	—
Beckwith et al (1971)	19.5	3.5–5	0.2	√	(F)
Kemp & Owen (1972)	19–45	0.9–5.5	0.3–1	√	F

Types IIB, 2/4/6, adverse pressure gradients

Sturek & Danberg (1971), Sturek (1974)	3.5–2.8	20–40	1.0	√	P

Mixed B3/B4. Adverse then favorable pressure gradient

Winter et al (1968)	0.6–2.8	8–40	1.0	X	S

Various: Group II A/B/C

Thomann (1968)	Wall heat transfer measurements only in various pressure gradients starting from $M = 2.5$. Includes IIC— pure normal pressure gradient.

[a] Parentheses around an entry imply that the information is of small quantity or of questionable quality.

[b] F = floating element; P = preston tube; S = stanton tube; V = from velocity profile.

Literature Cited

Appels, C. 1974. Incipient separation of a compressible turbulent boundary layer. *Von Kármán Inst., Brussels, Tech. Note 99; AGARD Conf. Proc. CP-168, 1975*

Avidor, J. M., Schneiderman, A. M. 1975. Experimental investigation of high Reynolds number compressible axisymmetric turbulent wakes. *AIAA J.* 13:485–89

Baronti, P. O., Libby, P. A. 1966. Velocity profiles in turbulent compressible boundary layers. *AIAA J.* 4:193–202

Beckwith, I. E. 1970. Recent advances in research on compressible turbulent boundary layers. *NASA Spec. Publ. SP-228*, pp. 355–416

Beckwith, I. E., Harvey, W. D., Clark, F. L.

1971. Comparisons of turbulent boundary layer measurements at Mach number 19.5 with theory and an assessment of probe errors. *NASA Tech. Note D-6192*
Bertram, M. H., ed. 1968. Compressible turbulent boundary layers. *NASA Spec. Publ. SP-216*
Bilger, R. W. 1975. A note on Favre averaging in variable-density flows. *Project SQUID TR-UCSD-6-PU(AD-A010226)*
Birch, S. F., Rudy, D. H., Bushnell, D. M., eds. 1972. Free turbulent shear flow. *NASA Spec. Publ. SP-321*
Bradshaw, P. 1973. The effect of streamline curvature on turbulent flow. *AGARDograph 169*
Bradshaw, P. 1974. The effect of mean compression or dilatation on the turbulence structure of supersonic boundary layers. *J. Fluid Mech.* 63:449–64
Bradshaw, P., ed. 1976a. *Topics in Applied Physics,* Vol. 12: *Turbulence.* Heidelberg: Springer
Bradshaw, P. 1976b. A skin-friction law for compressible turbulent boundary layers based on the full Van Driest transformation. *Imperial Coll. Aeronaut. Rep. 76-02*
Bradshaw, P., Ferriss, D. H. 1971. Calculation of boundary-layer development using the turbulent energy equation: compressible flow on adiabatic walls. *J. Fluid Mech.* 46:83–110
Bradshaw, P., Mizner, G. A., Unsworth, K. 1976. Calculation of compressible turbulent boundary layers on straight-tapered swept wings. *AIAA J.* 14:399
Brown, G. L., Roshko, A. 1974. On density effects and large structure in turbulent mixing layers. *J. Fluid Mech.* 64:775–816
Bushnell, D. M., Alston, D. W. 1972. Calculation of compressible adverse pressure gradient turbulent boundary layers. *AIAA J.* 10:229–30
Bushnell, D. M., Cary, A. M., Holley, B. B. 1975. Mixing length in low Reynolds number compressible turbulent boundary layers. *AIAA J.* 13:1119–21
Bushnell, D. M., Morris, D. J. 1971. Eddy viscosity distributions in a Mach 20 turbulent boundary layer. *AIAA J.* 9:764–66
Cary, A. M. 1970. Summary of available information on Reynolds analogy for zero-pressure-gradient, compressible, turbulent-boundary-layer flow. *NASA Tech. Note D-5560*
Cary, A. M., Bertram, M. H. 1974. Engineering prediction of turbulent skin friction and heat transfer in high-speed flow. *NASA Tech. Note D-7507*
Cebeci, T. 1971. Calculation of compressible

turbulent boundary layers with heat and mass transfer. *AIAA J.* 9:1091–97
Cebeci, T., Smith, A. M. O. 1974. *Analysis of Turbulent Boundary Layers.* New York: Academic
Chi, S. W., Spalding, D. B. 1966. Influence of temperature ratio on heat transfer to a flat plate through a turbulent boundary layer in air. *Int. Heat Transfer Conf., 4th,* 2:41–49
Chien, K.-Y. 1974. Hypersonic, turbulent skin-friction and heat-transfer measurements on a sharp cone. *AIAA J.* 12:1522–26
Chu, B. T., Kovasznay, L. S. G. 1958. Nonlinear interactions in a viscous heat-conducting compressible gas. *J. Fluid Mech.* 3:494–514
Coleman, G. T., Osborne, C., Stollery, J. L. 1973. Heat transfer from a hypersonic turbulent boundary layer on a flat plate. *J. Fluid Mech.* 60:257–71
Coleman, G. T., Stollery, J. L. 1974. Incipient separation of axially symmetric hypersonic turbulent boundary layers. *AIAA J.* 12:119–20
Coles, D. 1953. Measurements in the boundary layer on a flat plate in supersonic flow. *Jet Propulsion Lab. Tech. Reps. 20-69, 20-70, 20-71,* Pasadena, Calif.
Coles, D. 1964. The turbulent boundary layer in a compressible fluid. *Phys. Fluids* 7:1403–23
Coles, D., Hirst, E. A., eds. 1969. *Proc. AFOSR-IFP-Stanford Conf. Computation of Turbulent Boundary Layers, 1968.* Thermosci. Div., Stanford Univ., Calif.
Crocco, L. 1963. Transformation of the compressible turbulent boundary layer with heat exchange. *AIAA J.* 1:2723–31
Demetriades, A. 1968a. Turbulence measurements in an axisymmetric compressible wake. *Phys. Fluids* 11:1841–52
Demetriades, A. 1968b. Turbulent front structure of an axisymmetric compressible wake. *J. Fluid Mech.* 34:465–79
Demetriades, A. 1970. Turbulence measurements in a supersonic two-dimensional wake. *Phys. Fluids* 13:1672–78
Demetriades, A., Laderman, A. J. 1973. Reynolds stress measurements in a hypersonic boundary layer. *AIAA J.* 11:1594–96
Dunbar, D. I. A., Squire, L. C. 1971. Correlations of concentration, temperature and velocity profiles in compressible turbulent boundary layers with foreign gas injection. *Int. J. Heat Mass Transfer* 14:27–40
Dussauge, J. P., Gaviglio, J. 1975. Comportement d'un ecoulement turbulent de

proche sillage, a vitesse supersonique. *Rech. Aerosp.* No. 1975-3: 145–54

Economos, C. 1970. A modified form of the Coles compressibility transformation. *AIAA J.* 8: 2284–86 (see also *NASA Contract. Rep. CR-1680*)

Falco, R. E. 1974. Some comments on turbulent boundary layer structure inferred from the movements of a passive contaminant. *AIAA Paper 74-99*

Favre, A., ed. 1964. *The Mechanics of Turbulence.* New York: Gordon & Breach

Favre, A. 1971. Equations statistiques aux fluctuations turbulents dans les ecoulements compressibles; cas des vitesses et des temperatures. *C. R. Acad. Sci. Paris Ser. A* 273: 1087

Feller, W. V. 1973. Effect of upstream wall temperatures on hypersonic tunnel wall boundary-layer profile measurements. *AIAA J.* 11: 556–58

Fernholz, H. 1971a. Ein halbempirisches Gesetz für die Wandreibung in Kompressiblen turbulenten Grenzschichten bei isothermer und adiabater Wand. *Z. Angew. Math. Mech.* 51: 146–47

Fernholz, H. 1971b. Departures from a fully developed turbulent velocity profile on a flat plate in compressible boundary layers. *Fluid Dyn. Trans.* 6: Part II, pp. 161–77

Fischer, M. C., Maddalon, D. V., Weinstein, L. M., Wagner, R. D. 1971. Boundary-layer pitot and hot-wire surveys at $M_\infty = 20$. *AIAA J.* 9: 826–34

Fletcher, L. S., Briggs, D. G., Page, R. H. 1974. Heat transfer in separated and reattached flows—a review. *Isr. J. Technol.* 12: 236 (see also *AIAA Paper 70-767*)

Gates, D. F. 1973. Measurements of upstream history effects in compressible turbulent boundary layers. *Nav. Ordnance Lab. Tech. Rep. 73-152*

Gaudet, L., Winter, K. G. 1973. Measurements of the drag of some characteristic aircraft excrescences immersed in turbulent boundary layers. *AGARD Conf. Proc. 124* (see also *R. Aircr. Establishment Tech. Note Aero 1538*)

Gibson, M. M., Spalding, D. B. 1972. A two-equation model of turbulence applied to the prediction of heat and mass transfer in wall boundary layers. *ASME Paper 72-HT-15*

Görtler, H., Tollmien, W., eds. 1955. *50 Jahre Grenzschichtforschung.* Braunschweig: Vieweg

Gran, R. L., Lewis, J. E., Kubota, T. 1974. The effect of wall cooling on a compressible turbulent boundary layer. *J. Fluid Mech.* 66: 507–28

Green, J. E. 1968. The prediction of turbulent boundary layer development in compressible flow. *J. Fluid Mech.* 31: 753–78

Green, J. E. 1970. Interaction between shock waves and turbulent boundary layers. *Prog. Aerosp. Sci.* 11: 235–340

Green, J. E. 1972. Application of Head's entrainment method to the prediction of turbulent boundary layers and wakes in compressible flow. *R. Aircr. Establishment Tech. Rep. 72079*

Green, J. E., Weeks, D. J., Brooman, J. W. F. 1972. Prediction of turbulent boundary layers and wakes in compressible flow by a lag-entrainment method. *R. Aircr. Establishment Tech. Rep. 72231*

Hankey, W. L., Holden, M. S. 1975. Two-dimensional shock wave–boundary layer interactions in high speed flows. *AGARD-ograph 203*

Harvey, W. D., Clark, F. L. 1972. Measurements of skin friction on the wall of a hypersonic nozzle. *AIAA J.* 10: 1256–58

Harvey, W. D., Hunter, W. W. 1975. Experimental study of a free turbulent shear flow at Mach 19 with electron-beam and conventional probes. *NASA Tech. Note D-7981*

Hastings, R. C., Sawyer, W. G. 1970. Turbulent boundary layers on a large flat plate at $M = 4$. *Aeronaut. Res. Counc. R & M 3678*

Heckman, D. 1973. Summary of re-entry physics research program on turbulent wakes. *Def. Res. Estab. Valcartier, Quebec, R-697/73*

Hopkins, E. J., Inouye, M. 1971. An evaluation of theories for predicting turbulent skin friction and heat transfer on flat plates at supersonic and hypersonic Mach numbers. *AIAA J.* 9: 993–1003

Hopkins, E. J., Keener, E. R. 1972. Pressure gradient effects on hypersonic turbulent skin friction and boundary layer profiles. *AIAA J.* 10: 1141–42

Hopkins, E. J., Keener, E. R., Polek, T. E., Dwyer, H. A. 1972. Hypersonic turbulent skin friction and boundary layer profiles on non-adiabatic flat plates. *AIAA J.* 10: 40–48 (see also *NASA Tech. Note D-6907, 1972*)

Hopkins, R. A., Nerem, R. M. 1968. An experimental investigation of heat transfer from a highly-cooled turbulent boundary layer. *AIAA J.* 6: 1912–18

Horstman, C. C., Owen, F. K. 1972. Turbulence properties of a compressible boundary layer. *AIAA J.* 10: 1418–24

Horstman, C. C. et al 1975. Shock-wave-induced turbulent boundary layer separation at hypersonic speeds. *AIAA Paper 75-4*

Howarth, L. ed. 1953. *Modern Developments in Fluid Dynamics, High Speed Flow.* Oxford: Clarendon

Ikawa, H., Kubota, T. 1974. An experimental investigation of a two-dimensional, self-similar, supersonic turbulent mixing layer with zero pressure gradient. *AIAA Paper 74-20*

Jeromin, L. O. F. 1968. A transformation for compressible turbulent boundary layers with air injection. *J. Fluid Mech.* 31:65–94

Johnson, C. B., Bushnell, D. M. 1970. Power law velocity-profile-exponent variations with Reynolds number, wall cooling, and Mach number in a turbulent boundary layer. *NASA Tech. Note D-5753*

Johnson, D. A. 1974. Turbulence measurements in a Mach 2.9 boundary layer using laser velocimetry. *AIAA J.* 12:711–14

Keener, E. R., Hopkins, E. J., 1972. Turbulent boundary layer velocity profiles on a non-adiabatic flat plate at Mach number 6.5. *NASA Tech. Note D-6907*

Keener, E. R., Hopkins, E. J. 1973. Van Driest generalization applied to turbulent skin friction and velocity profiles measured on the walls of a Mach 7.4 wind tunnel. *AIAA J.* 12:1784–85

Keener, E. R., Polek, T. E. 1972. Measurements of Reynolds analogy for a hypersonic turbulent boundary layer on a non-adiabatic flat plate. *AIAA J.* 10:845–46

Kemp, J. H., Owen, F. K. 1972. Experimental study of nozzle wall boundary layers at Mach numbers 20 to 47. *NASA Tech. Note D-6965; AIAA J.* 10:813–19

Kistler, A. L. 1959. Fluctuation measurements in a supersonic turbulent boundary layer. *Phys. Fluids* 2:290–96

Klebanoff, P. S. 1955. Characteristics of turbulence in a boundary layer with zero pressure gradient. *NACA Rep. 1247*

Korkegi, R. 1971. Survey of viscous interactions associated with high Mach number flight. *AIAA J.* 9:771–84

Korkegi, R. 1975. Comparison of shock-induced two- and three-dimensional incipient turbulent separation. *AIAA J.* 13:534–35

Küster, H. J. 1972. Ein Integral verfahren zur Berechnung zweidimensionaler Incompressibiler turbulenter Grentschichten mit Druckgradient und Wärmeübergang auf der Basis einer Kompressibilitätstransformation. Dr.-Ing. thesis. Tech. Univ., Berlin

Laderman, A. J. 1974. Hypersonic viscous flow over a slender cone: Part II: turbulence structure of the boundary layer. *AIAA Paper 74-534*

Laderman, A. J., Demetriades, A. 1974. Mean and fluctuating flow measurements in the hypersonic boundary layer over a cooled wall. *J. Fluid Mech.* 63:121–44

Landis, R. B., Hordemann, G. J., eds. 1972. *Proc. Heat Transfer and Fluid Mech. Inst., 1972*

Laufer, J. 1964. Some statistical properties of the pressure field radiated by a turbulent boundary layer. *Phys. Fluids* 7:1191–97

Law, C. H. 1974. Supersonic, turbulent boundary layer separation. *AIAA J.* 12:794–97

Lewis, J. E., Gran, R. L., Kubota, T. 1972. An experiment on the adiabatic, compressible turbulent boundary layer in adverse and favourable pressure gradients. *J. Fluid Mech.* 51:657–72

Lewis, J. E., Kubota, T., Webb, W. J. 1970. Transformation theory for the adiabatic compressible turbulent boundary layer with pressure gradient. *AIAA J.* 8:1644–50

Libby, P. A., Williams, F. A. 1976. *Ann. Rev. Fluid Mech.* 8:351–76

Lin, C. C., ed. 1959. *High Speed Aerodynamics and Jet Propulsion,* Vol. 5. Princeton, NJ: Princeton Univ. Press

Mabey, D. G., Meier, H. U., Sawyer, W. G. 1974. Experimental and theoretical studies of the boundary layer on a flat plate at Mach numbers from 2.5 to 4.5. *R. Aircr. Establishment Rep. 74127*

Maise, G., McDonald, H. 1968. Mixing length and kinematic eddy viscosity in a compressible boundary layer. *AIAA J.* 6:73–80

Martellucci, A., Langanelli, A. L. 1974. Hypersonic viscous flow over a cone. I. *AIAA Paper 74-533*

Meier, H. U., Voisinet, R. L. P., Gates, D. F. 1974. Temperature distributions using the law of the wall for compressible flow with variable turbulent Prandtl numbers. *AIAA Paper 74-596*

Miles, J. B., Kim, J. H. 1968. Evaluation of Coles' turbulent compressible boundary layer theory. *AIAA J.* 6:1187–89

Moore, D. R., Harkness, J. 1965. Experimental investigations of the compressible turbulent boundary layer at very high Reynolds number. *AIAA J.* 3:631–38

Murlis, J. 1975. *The structure of a turbulent boundary layer at low Reynolds number.* PhD thesis. Imperial Coll., London

Murthy, S. N. B., ed. 1975. *Turbulent Mixing*

in Nonreactive and Reactive Flows. New York: Plenum

Narasimha, R., Viswanath, P. R. 1975. Reverse transition at an expansion corner in supersonic flow. *AIAA J.* 13:693–95

Oskam, B., Vas, I. E., Bogdonoff, S. M. 1975. An exploratory study of a three-dimensional shock-wave boundary layer interaction at Mach 3. *Princeton Univ. Dep. Aerosp. Mech. Sci. Rep. 1227; AGARD CP-168*

Owen, F. K., Horstman, C. C. 1972. On the structure of hypersonic turbulent boundary layers. *J. Fluid Mech.* 53:611–36

Owen, F. K., Horstman, C. C., Kussoy, M. I. 1975. Mean and fluctuating flow measurements of a fully-developed non-adiabatic hypersonic boundary layer. *J. Fluid Mech.* 70:393–413

Peake, D. J., Brakmann, G., Romeskie, J. M. 1971. Comparisons between some high Reynolds number turbulent boundary layer experiments at Mach 4 and various recent calculation procedures. *AGARD Conf. Proc. 93*

Perkins, H. C., McEligot, D. M. 1975. Mean temperature profiles in heated laminarizing air flows. *J. Heat Transfer* 97:589–93

Raman, K. R. 1974. A study of surface fluctuations in hypersonic turbulent boundary layers. *NASA Contract. Rep. CR-2386*

Reda, D. C., Ketter, F. C., Fan, C. 1975. Compressible turbulent skin friction on rough and rough/wavy walls. *AIAA J.* 13:553–54

Reshotko, E. 1975. A program for transition research. *AIAA J.* 13:261–65 (see also following papers)

Reynolds, W. C. 1976. *Ann. Rev. Fluid Mech.* 8:183–208

Rose, W. C. 1974. Turbulence measurements in a compressible boundary layer. *AIAA J.* 12:1060–64 (see also *NASA Tech. Note D-7092,* 1973)

Rose, W. C., Childs, M. E. 1974. Reynolds-stress measurements in a compressible boundary layer within a shock-wave-induced adverse pressure gradient. *J. Fluid Mech.* 65:177–88

Rose, W. C., Johnson, D. A. 1975. Turbulence in a shock-wave boundary layer interaction. *AIAA J.* 13:884–89

Rose, W. C., Murphy, J. D. 1973. Ratio of Reynolds shear stress to turbulence kinetic energy in a boundary layer. *Phys. Fluids.* 16:935–37

Roshko, A., Thomke, G. J. 1975. Flare-induced separation lengths in supersonic, turbulent boundary layers. *AIAA Paper 75-6*

Rotta, J. C. 1960. Turbulent boundary layers with heat transfer in compressible flow. *AGARD Rep. 281.* See also *Z. Flugwiss.* 7:264–74 (1959)

Rotta, J. C. 1974. Die turbulente Grenzschicht an einer stark geheizten ebenen Platte bei Unterschallströmung. *Wärme Stoffübertrag.* 7:133–44

Rubesin, M. W., Rose, W. C. 1973. The turbulent mean-flow, Reynolds-stress and heat-flux equations in mass-averaged dependent variables. *NASA Tech. Memo X-62248*

Sandborn, V. A. 1974. A review of turbulence measurements in compressible flow. *NASA Tech. Memo X-62337*

Settles, G. S., Bogdonoff, S. M., Vas, I. E. 1975. Incipient separation of a supersonic turbulent boundary layer, at moderate to high Reynolds numbers. *AIAA Paper 75-7*

Shang, J. S. 1974. Computation of hypersonic turbulent boundary layer with heat transfer. *AIAA J.* 12:883–84

Shang, J. S., Hankey, W. L., Dwoyer, D. L. 1973. Numerical analysis of eddy viscosity models in supersonic turbulent boundary layers. *AIAA J.* 11:1677–83

Sivasegaram, S., Whitelaw, J. H. 1971. The prediction of turbulent, supersonic, two-dimensional boundary layer flows. *Aeronaut. Q.* 22:274–94

Smith, J. A., Driscoll, J. F. 1975. The electron-beam fluorescence technique for measurements in hypersonic turbulent flows. *J. Fluid Mech.* 72:695–719

Sommer, S. C., Short, B. J. 1955. Free-flight measurements of turbulent boundary layer skin friction in the presence of severe aerodynamic heating at Mach numbers from 2.8 to 7.0. *NACA Tech. Note 3391*

Spalding, D. B., Chi, S. W. 1964. The drag of a compressible turbulent boundary layer on a smooth flat plate with and without heat transfer. *J. Fluid Mech.* 18:117–43

Squire, L. C. 1969. A law of the wall for compressible turbulent boundary layers with air injection. *J. Fluid Mech.* 37:449–56

Squire, L. C. 1971. Eddy viscosity distributions in compressible turbulent boundary layers with injection. *Aeronaut. Q.* 23:169–82

Squire, L. C., Verma, V. K. 1973. The calculation of compressible turbulent boundary layers with fluid injection. *ARC CP 1265*

Stollery, J. L. 1976. Supersonic turbulent

boundary layers: some comparisons between experiment and a simple theory. *Aeronaut. Q.* 27:77–98

Sturek, W. B. 1974. Turbulent boundary layer shear stress distributions for compressible adverse pressure gradient flow. *AIAA J.* 12:375–76

Sturek, W. B., Danberg, J. E. 1971. The supersonic turbulent boundary layer in an adverse pressure gradient—experiment and data tabulation. *AGARD Conf. Proc. 93*

Swigart, R. J. 1974. Cross-hatching studies —a critical review. *AIAA J.* 12:1301–18

Thomann, H. 1968. Effect of streamwise wall curvature on heat transfer in a turbulent boundary layer. *J. Fluid Mech.* 33:283–92

Thomas, G. D. 1974. Compressible turbulent boundary layers with combined air injection and pressure gradient. *ARC Paper No. 35,483*

Thomke, G. J. 1969. Boundary layer skin friction characteristics in the supersonic test section of the Douglas Aerophysics Laboratory 4 ft. trisonic wind tunnel. McDonnell Douglas Astronautics Co., Santa Monica, Calif. Unpublished

Van Driest, E. R. 1951. Turbulent boundary layer in compressible fluids. *J. Aeronaut. Sci.* 18:145–60

Van Driest, E. R. 1956a. The problem of aerodynamic heating. *Aeronaut. Eng. Rev.* 15: No. 10, p. 26

Van Driest, E. R. 1956b. On turbulent flow near a wall. *J. Aeronaut. Sci.* 23:1007–11, 1036

Varma, A. J., Beddini, R. A., Sullivan, R. D., Donaldson, C. duP. 1974. Application of an invariant second order closure model to compressible turbulent shear layers.

AIAA Paper 74-592

Voisinet, R. L. P., Lee, R. E. 1972. Measurements of a Mach 4.9 zero pressure gradient boundary layer with heat transfer. *Nav. Ordnance Lab. Tech. Rep. 72-232*

Voisinet, R. L. P., Lee, R. E. 1973. Measurements of a supersonic favorable-pressure gradient turbulent boundary layer with heat transfer, Part I. *Nav. Ordnance Lab. Tech. Rep. 73-224*

Wagner, R. D. 1973. Mean flow and turbulence measurements in a Mach 5 free shear layer. *NASA Tech. Note D-6366*

Waltrup, P. J., Schetz, J. A. 1973. Supersonic turbulent boundary layer subjected to adverse pressure gradients. *AIAA J.* 11:50–58

Watson, R. D., Harris, J. E., Anders, J. B. 1973. Measurements in a transitional/turbulent Mach 10 boundary layer at high Reynolds number. *AIAA Paper 73-165*

Winter, K. G., Gaudet, L. 1970. Turbulent boundary layer studies at high Reynolds numbers at Mach numbers between 0.2 and 2.8. *Aeronaut. Res. Counc. R & M 3712*

Winter, K. G., Rotta, J. C., Smith, K. G. 1968. Studies of the turbulent boundary layer on a waisted body of revolution in subsonic and supersonic flow. *Aeronaut. Res. Counc. R & M 3633*

Yanta, W. J., Lee, R. E. 1974. Determination of turbulence transport properties with the laser doppler velocimeter and conventional time-averaged mean flow measurements at Mach 3. *AIAA Paper 74-575*

Zwarts, F. 1970. *The compressible turbulent boundary layer in a pressure gradient.* PhD thesis. Mech. Eng. Dept., McGill Univ., Montreal

Ann. Rev. Fluid Mech. 1977. 9:55–86
Copyright © 1977 by Annual Reviews Inc. All rights reserved

ON THE LIQUIDLIKE   ×8097
BEHAVIOR OF FLUIDIZED BEDS

J. F. Davidson, D. Harrison, and J. R. F. Guedes de Carvalho
Department of Chemical Engineering, University of Cambridge,
Cambridge CB2 3RA, England

A bed of particles can be partly or fully supported by an upward flow of fluid through the interstices between the particles; when the bed is fully supported it is said to be fluidized. When the fluid flow is the minimum to give full support, i.e. the buoyant weight is only just overcome by the fluid pressure drop, the bed is said

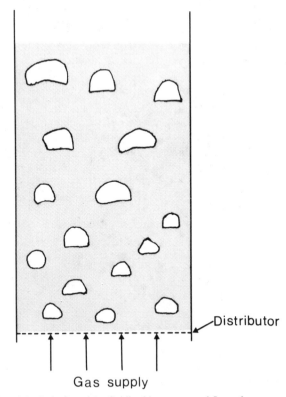

Distributor

Gas supply

Figure 1   Bed of particles fluidized by an upward flow of gas.

55

to be incipiently fluidized and the superficial velocity is U_{mf}. Further increases in fluid flow cause the particle bed to expand; if the fluid is gas, in most cases the gas passes through the particle bed in the form of gas bubbles (illustrated in Figure 1), and the top surface of the bed (shown in Figure 2) resembles that of a boiling liquid. This is aggregative fluidization. A gas-solids system has other properties of a liquid. A bubbling fluidized bed may be stirred easily; light objects

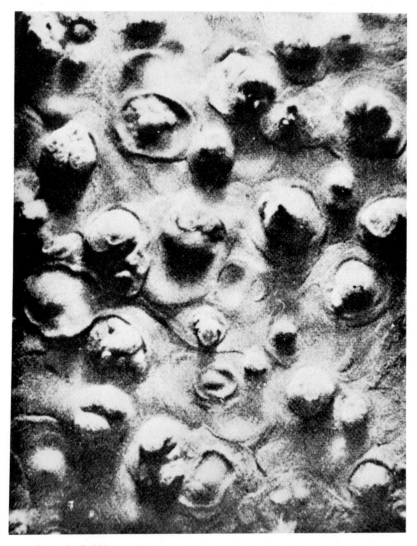

Figure 2   Bubbles breaking surface at the top of a bed of sand fluidized by air.

float on the bed surface and dense objects sink. If there is a hole in the side of the containing vessel, the particles flow out like a liquid jet. For most particle beds fluidized by liquids, the bed expands smoothly when $U > U_{mf}$ and liquid bubbling is not observed; this is particulate fluidization.

In this account of the liquidlike behavior of gas-fluidized beds we turn first to experiment, which may be considered in two ways. First, the appearance and properties of gas bubbles in fluidized beds have qualitative similarities to those of gas bubbles in liquids; and in this connection we give a brief account of bubble size and shape, formation, rising velocity, and coalescence. Second, we discuss how certain bubble properties, particularly bubble shape, help to explain the viscosity of the particulate phase of a fluidized bed.

We then review theory for the motion of a bubble in a viscous liquid and discuss the extent to which this provides an adequate description of a gas bubble in a fluidized bed. Next, we consider the extent to which the motion of particles and the interstitial fluid can be described by treating the particulate phase as an inviscid liquid, and we give the general equations for particle and fluid motion. We conclude, from a comparison of theory and experiment, that particle collisions provide the major contribution to fluidized-bed viscosity.

1   SOME PROPERTIES OF BUBBLING FLUIDIZED BEDS

Bubble Size and Shape

Figure 3 (Jones 1965) shows photographs of air bubbles of different sizes rising in paraffin oil, and Figure 4 (Rowe 1971) shows X-ray photographs of air bubbles in fluidized beds of different materials. The similarity between the spherical-cap bubbles in the two systems is striking, but they are not identical; and in particular the spherical cap is not the same for a bubble in liquid of low viscosity, such as water. In a fluidized bed the bubble shape depends on the nature of the particles, and Rowe (1971) found a fairly simple relationship between particle size and the wake fraction of the bubble (which is a measure of bubble shape). It is clear that large air bubbles in water generally have a much smaller ratio of height to horizontal width. We return to the relation between bubble shape and bed viscosity in Section 2.

Bubble Formation

The ways in which bubbles form in fluidized beds depends upon the methods of gas injection to the bed. Gas may be introduced, for example, through a porous plate or a multi-orifice distributor. When air is blown steadily through an orifice into a liquid, a stream of bubbles is formed. The same is found when air is blown at low velocity through a tube immersed in an incipiently fluidized bed, as indicated in Figure 5. At low orifice velocities single (independent) bubbles form and as we show in Section 4, the quantitative similarities of behavior between fluidized beds and gas-liquid systems provide powerful evidence for the application of inviscid-flow theory to gas fluidization.

Bubble Velocity

The rising velocity of bubbles in fluidized beds was the first bubble property to receive extensive experimental investigation, and the results give good support to the theory that bubbles in fluidized beds rise as if they were in a liquid of low viscosity. The rising velocity of a single bubble injected into an incipiently fluidized bed (Davidson & Harrison 1963) may be compared with data for large gas bubbles rising in liquids; a review of the latter has been given by Wegener & Parlange

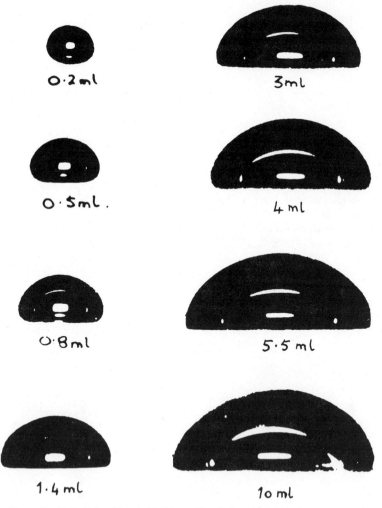

Figure 3   Air bubbles rising in liquid paraffin of viscosity 2.93 poise (Jones 1965).

(1973). For large bubbles in liquids of low viscosity, the effects of viscosity and surface tension are small, and the rising velocity is given by two semiempirical equations,

$$u_b = 0.71(gD_e)^{1/2} \tag{1}$$

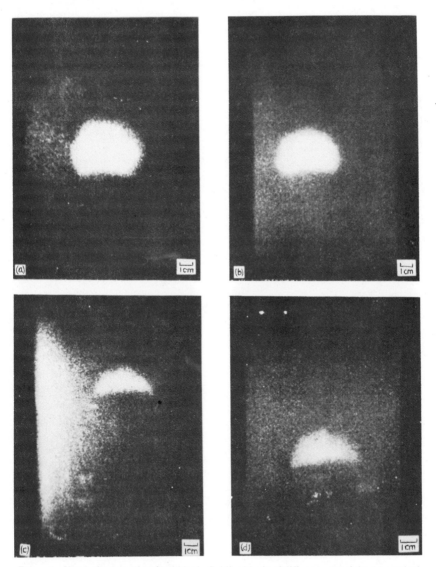

Figure 4   X-ray photographs of bubbles in fluidized beds of different materials: (*a*) crushed coal, (*b*) magnesite, (*c*) synclyst, (*d*) acrylic granules (Rowe 1971).

and

$$u_b = 0.35(gD)^{1/2}, \qquad (2)$$

where D_e is the diameter of the sphere having the same volume as the bubble, D is the diameter of the containing vessel, and g is the gravitational acceleration. Equation (1) applies when $D \gg D_e$, that is, a bubble rising in a large volume of liquid; and Equation (2), to a bubble or slug for which D_e is comparable with D or greater. Equations of the form of (1) and (2) may be derived by considering potential flow around the nose of a bubble, but the numerical coefficients come from experiments with ordinary liquids like water.

Figure 6 shows a comparison between Equations (1) and (2) and experimental results for single bubbles injected into various fluidized beds. These results give evidence from a wide range of particle size and density. The general validity of Equation (1) is also strongly supported by the very extensive data of Toei et al (1965), who measured bubble velocities and bubble heights in a range of particles using X-ray photography and capacitance techniques. Figure 7 shows their results, which have been correlated by $u_b = 0.83(gh)^{1/2}$, where h is the bubble height, and therefore if the bubbles are geometrically similar, u_b is proportional to $D_e^{1/2}$. Toei et al found the following expression for the bubble volume: $V_b = 2.0h^3$, which by substitution in their expression for the bubble velocity gives $u_b = 0.66(gD_e)^{1/2}$, in good agreement with Equation (1).

The results in Figures 6 and 7 show considerable scatter, but nevertheless they suggest that the particulate phase behaves as an inviscid liquid. Each particle in a fluidized bed moves primarily under the influence of the fluid; interparticle forces would appear to be of secondary importance. We note that if bubble rise were governed by viscous forces, as has been suggested as a possibility by Nguyen et al (1973), u_b would be proportional to h^2, rather than to $h^{1/2}$, as shown in Figure 7.

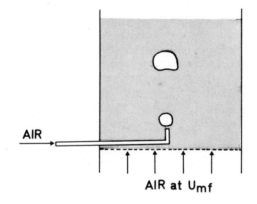

AIR

AIR at U_{mf}

Figure 5   Apparatus for experiments on bubble formation at an orifice in a fluidized bed.

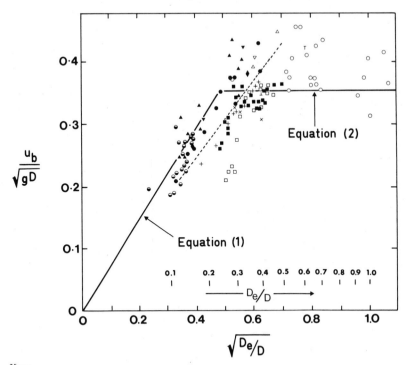

Key:

	Particles	Diameter (μm)	References
+	Sand	175	
×	Glass beads	150	Davidson & Harrison (1963, p. 34)
⊥	Sand	400	
T	Seeds	1700	
●	Ballotini	60–550	
▲	Silver sand	72–500	
▼	Acrylic granules	121	Rowe & Partridge (1965)
◆	Synclyst	52	
▽	Magnesite	240	
△	Crushed coal	410	
□	Ballotini (A)	230	Angelino et al (1964)
■	Ballotini (B)	230	
○	Coke	344	Park et al (1969)
◑	Alkalized alumina	1760	Cranfield & Geldart (1974)
---	Ballotini	300–400	Rowe & Matsuno (1971) (experimental correlation)

Figure 6   Velocity of a single bubble injected into a fluidized bed.

Bubble Coalescence

Figure 8 shows an X-ray ciné sequence of the coalescence of two bubbles in a fluidized bed of silver sand (Rowe 1971). The mechanism of coalescence always appears to be in-line: When a following bubble approaches a leading bubble closely enough, it is accelerated, undergoes some elongation in shape, and is gathered into the rear of the leading bubble. Coalescence then occurs. Similar observations (see Calderbank 1967) have been reported on the coalescence of large gas bubbles in liquids where surface tension forces may be expected to be small.

2   THE VISCOSITY OF FLUIDIZED BEDS: EXPERIMENTAL EVIDENCE

The measurement of fluidized-bed viscosity has attracted a number of experimentalists, whose work has been reviewed by Grace (1970). "Viscosities" must be suspect that are derived from techniques used to measure the viscosity of Newtonian liquids (e.g. the resistance of the "fluid" to rotating paddles); such techniques may disturb the fluidized bed itself. Although such results give qualitative information,

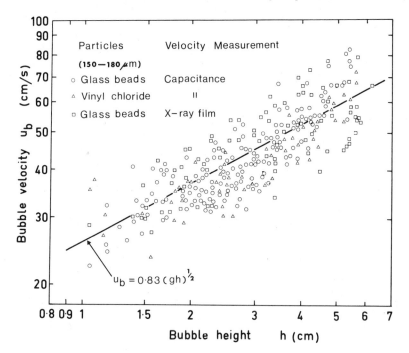

Figure 7   Rising velocity of single bubbles as a function of bubble height in a 0.1 × 0.1 m square column (Toei et al 1965).

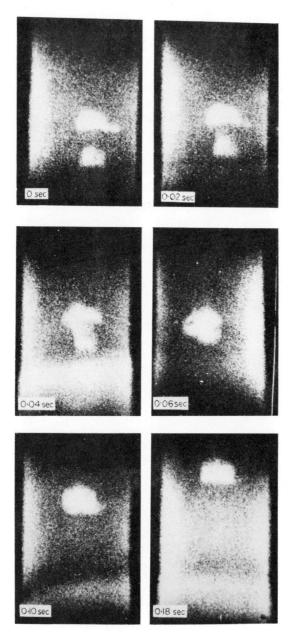

Figure 8   X-ray photographs showing in-line coalescence of two bubbles (Rowe 1971).

they are an unsatisfactory basis for quantitative measurements of fluidized-bed viscosity. However, two studies [by Schügerl et al (1961), using a Couette viscometer, and by Hagyard & Sacerdote (1966), using a torsion pendulum viscometer], which employed very low shear rates in order to minimize bed disturbance, provide experimental viscosities in the range 1–10 poise for a variety of bed materials from smooth glass beads to silicon carbide particles. A reasonable estimate for the viscosity of an ordinary fully fluidized bed, therefore, may be put at about 10 poise. A gas-fluidized bed consists of both bubble and particulate phases, but as the bubbles are essentially particle-free, the interparticle forces that give rise to shear stresses are effectively confined to the particulate phase; we are thus concerned with the viscosity of the particulate phase, which is assumed to be a continuum on a macroscopic scale.

Grace (1970) has pointed out that since a fluidized bed, in any case, is subject to shear stresses caused by the bubble motion, the effective viscosity of the particulate phase might best be inferred from bubble properties. Both Stewart (1968) and Grace have adopted this approach, which has a considerable advantage over others, that is, the bed is not subject to external shear stresses.

Table 1   Apparent viscosities (Grace 1970) for systems studied by Rowe & Partridge (1965)

Particles		Apparent viscosities (poise)		
Material	Size (μm)	Calculated from bubble shapes (Grace 1970)	Based on results of Schügerl et al (1961) [from Stewart (1968)]	Calculated by Stewart (1968)
Ballotini	550	9.5	—	26
	460	—	12	18
	220	8.5	9	8
	170	7.5	8	8.5
	140	8	8	2.5
	120	8.5	8	7
	82	9	6	1.5
	60	7	4	7.5
Silver sand	500	12	—	24
	460	10	14	18
	330	13	12	14
	230	7	10	6.5
	140	9	8	4.5
	82	8.5	7	3.5
	72	8	5	Negative result
Synclyst catalyst	52	4	—	4
Magnesite	240	9	10	15

Stewart compared the theories of Davidson (1961), Jackson (1963), and Murray (1965) with measurements by Reuter (1963) on the pressure field around a three-dimensional half-bubble rising up a plane, almost vertical wall. If Reuter's experimental bubble velocities are used in Jackson's analysis, the latter predicts about 65% of the experimental pressure rise close to the nose of the bubble; and Stewart suggested that a possible explanation for this discrepancy might lie in the presence of viscous stresses in a fluidized bed above a rising bubble. Accordingly, he equated the additional pressure $(6\mu_p u_b/a)$ needed inside the bubble (to counteract possible viscous stresses) to the difference at the bubble nose between (a) Reuter's total pressure increase to the bubble and (b) Jackson's term for the pressure required to balance the momentum of the particles relative to the bubble. Hence

$$\frac{6\mu_p u_b}{a} = \frac{\rho_p g h}{2} - \frac{\rho_p u_b^2}{2} \tag{3}$$

where μ_p and ρ_p are the effective viscosity and the bulk density of the particulate phase, respectively, and u_b is the rising velocity of a single bubble of radius a and height h. Stewart calculated μ_p for a variety of systems from Equation (3), using values of u_b, ρ_p, and h obtained from the X-ray work of Rowe & Partridge (1965). Table 1 shows that these calculated bed viscosities are in fair agreement with the experimental measurements of Schügerl et al.

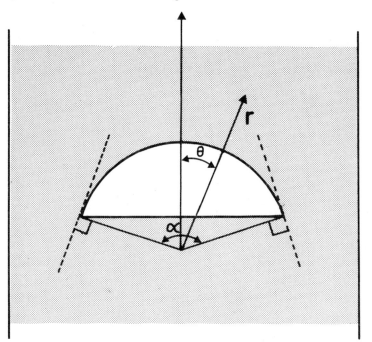

Figure 9   Spherical-cap bubble, definition of angle α.

Thus, Stewart's work provides some evidence that viscous forces influence the shape and rising velocity of bubbles in fluidized beds, but it is not conclusive. First, on the experimental side, Reuter's measurements obtained for bubbles rising along a flat inclined wall do not justify a strict comparison with a three-dimensional bubble rising in isolation. Second, Stewart's Figure 8 shows that Davidson's inviscid theory comes closest to predicting the pressure-probe results in front of the bubble. This is further confirmed by later work on the pressure field around a two-dimensional bubble by Littman & Homolka (1973).

Grace (1970) has estimated bed viscosities from the shape of bubbles. A convenient method of describing the shape of a spherical-cap bubble, shown in Figure 9, is by means of the angle α defined by drawing normals to the cap of the sphere. Grace has collected observations of α by many authors for bubbles in liquids, and the results, shown in Figure 10, indicate that α is uniquely determined by the bubble

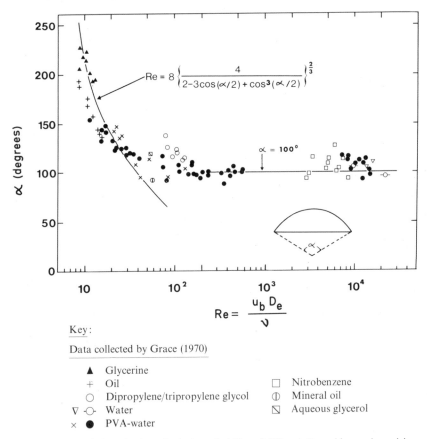

Key:

Data collected by Grace (1970)

   ▲   Glycerine
   +   Oil   ◻ Nitrobenzene
   ○   Dipropylene/tripropylene glycol   ⊕ Mineral oil
▽ -○- Water   ◩ Aqueous glycerol
   × ● PVA-water

Figure 10   Included angles for spherical-cap bubbles of different Reynolds numbers rising in liquids.

Reynolds number $Re = u_b D_e / v$, where v is the kinematic viscosity of the liquid. The limit of these results as $Re \to 0$ is consistent with the theoretical limit that when $Re \to 0$ the bubble must be spherical, even though surface tension is zero (Batchelor 1967). The data in Figure 10 are selected for conditions in which the effect of surface tension is negligible, which can be arranged by experimenting with bubble diameters greater than about 20 mm; the data are from apparatuses large enough so that the wall effect is small. Figure 3 shows that as bubble size is increased, the importance of surface tension decreases and bubble shape approaches the spherical-cap form. Rippin & Davidson (1967) have shown that inviscid flow theory gives α as about $100°$ at high Re, and Figure 10 indicates that this is confirmed by experiment. Theory is available [see Davidson et al (1977) and Section 3] to predict the relation between α and Re for the transition region $5 < Re < 100$, in which both inertial and viscous forces are significant; this theory gives the curve shown in Figure 10, in reasonable agreement with experiment.

To calculate effective bed viscosities Grace measured α from X-ray photographs like those in Figure 4 and then used Figure 10 to infer a Reynolds number and hence the effective viscosity. Table 1 gives bed viscosities deduced by Grace. They are in good agreement with values obtained by other methods; the agreement with Schügerl's results is particularly good and significantly better than those derived by Stewart. These viscosities—of order 4–13 poise—are in the range now generally accepted. Murray (1967), using rather different considerations, concluded that the shear viscosity of a fluidized bed was normally about 10 poise or a little less. Viscosities of this order are very high, about 10^4–10^5 times higher than that of typical fluidizing gases (e.g. air) at room temperature. Frankel & Acrivos (1967) have collected data which show that if shearing of the gas in the bed interstices were the only factor, the bed viscosity might be at most 100 times that of the fluidizing gas. The fact that observed bed viscosities are so much higher must be caused by interparticle collisions.

Fluidized beds therefore present a fascinating conflict of experimental evidence. The well-established formulas for bubble velocity [Equations (1) and (2)] support a picture of the particulate phase as an inviscid liquid, whereas the careful study of bubble shape points to a substantial bed viscosity. We first examine this conflict of evidence by considering bubble motion in a viscous liquid.

3 BUBBLE MOTION IN A VISCOUS LIQUID

In gas-fluidized beds in which bubbles are present the gas momentum resulting from bubble motion is negligible in comparison with the particulate motion; the relevant equations of motion are discussed in Section 5. A useful approximate solution is that in which the voidage is constant, and the motion of the particulate phase around the bubble is approximated by an irrotational flow. Thus the particulate motion around a sphere of radius a is given by the velocity potential

$$\phi = -u_b \left(r + \frac{a^3}{2r^2} \right) \cos \theta, \tag{4}$$

where u_b is the streaming velocity, and r and θ are the polar coordinates shown in Figure 9. The drag on a bubble in a liquid is given by Levich (1962), and his result has been confirmed by a rigorous study of bubble motion in a viscous liquid by Moore (1963). The rate of dissipation of energy of the system due to motion outside the sphere (see Levich) is given by

$$E = 12\pi\mu a u_b^2, \tag{5}$$

where μ is the shear viscosity of the particulate phase. Hence the drag force is

$$F = 12\pi\mu a u_b. \tag{6}$$

Now the buoyancy force $\rho g \pi D_e^3/6$ must balance the drag force, and hence from Equation (6)

$$\frac{2a}{D_e} = \frac{gD_e^2}{36 v u_b}. \tag{7}$$

By geometry, from Figure 9, the volume of a spherical-cap bubble is given in terms of α and a by

$$\frac{\pi D_e^3}{6} = \pi a^3 \left(\frac{1}{3}\cos^3\frac{\alpha}{2} - \cos\frac{\alpha}{2} + \frac{2}{3} \right). \tag{8}$$

The drag coefficient $C_D = (\rho g \pi D_e^3/6)(4/\pi D_e^2)(2/\rho u_b^2)$, and $\mathrm{Re} = u_b D_e/v$; and together with equations (7) and (8) we have

$$\left[\frac{4}{2 - 3\cos\frac{\alpha}{2} + \cos^3\frac{\alpha}{2}} \right]^{1/3} = \frac{2a}{D_e} = \frac{\mathrm{Re}\, C_D}{48}. \tag{9}$$

Another relation between rising velocity and the sphere radius a may be obtained from the condition of constant pressure within the bubble. For irrotational flow defined by Equation (4), Davies & Taylor (1950) showed by considering flow near the nose of the bubble that for constant pressure within the bubble

$$u_b = \tfrac{2}{3}(ga)^{1/2}. \tag{10}$$

Eliminating a from Equations (9) and (10), and using the definition of C_D, gives

$$C_D = \left(\frac{288}{\mathrm{Re}} \right)^{1/2}. \tag{11}$$

Combining Equations (9) and (11) we have

$$\mathrm{Re} = 8 \left[\frac{4}{2 - 3\cos\frac{\alpha}{2} + \cos^3\frac{\alpha}{2}} \right]^{2/3} \tag{12}$$

which defines the shape of the bubble. This equation is shown in Figure 10.

Equation (1) corresponds to $C_D = 2.6$; Equation (11) gives $C_D = 2.6$ at $\mathrm{Re} = 42.6$, and the corresponding angle α from Equation (12) is $100°$, which agrees well with observed values of α for large bubbles in inviscid liquids. Therefore, Re = 42.6 is an upper limit for the validity of Equations (11) and (12).

At very low Reynolds numbers the Hadamard result $C_D = 16/\text{Re}$ is an exact solution to the problem of flow around a bubble in liquid with zero surface tension; this gives the same value of C_D as Equation (11) when Re = 0.89. However, from Equation (12), $\alpha = 360°$ (i.e. the bubble becomes spherical) when Re = 8, so this must be a lower limit to the validity of this analysis. For Re < 8 a transition must occur to the Hadamard result as the boundary layer becomes thicker.

We may note, therefore, in the range 8 < Re < 42, that the above analysis gives approximate answers to a very complex problem. It gives remarkably good predictions of bubble shape (Figure 10) and of drag coefficient (Figure 11). The latter plot is for gas-liquid bubbles large enough for surface tension effects to be negligible.

This approach may be compared with that of Parlange (1969), who used the flow pattern of Figure 12, with irrotational flow outside the sphere of radius a. He assumed the dissipation of energy to be the same as though the sphere were filled with liquid having the motion of Hill's spherical vortex (Lamb 1932);

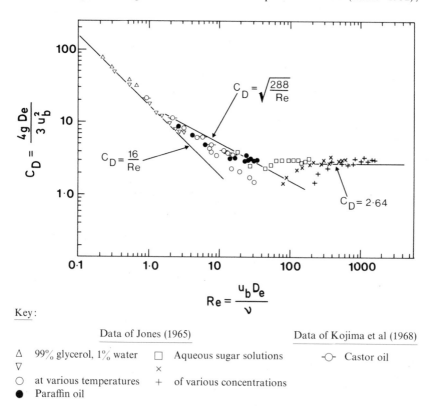

Key:

Data of Jones (1965)		Data of Kojima et al (1968)

$\triangle$ 99% glycerol, 1% water $\square$ Aqueous sugar solutions $-\!\bigcirc\!-$ Castor oil
$\triangledown$ $\times$
$\bigcirc$ at various temperatures $+$ of various concentrations
$\bullet$ Paraffin oil

Figure 11 Drag coefficient as a function of Reynolds number for bubbles rising in liquids.

consequently the drag force—calculated from Harper and Moore's (1968) result—was 2.5 times that given by Equation (6). Grace (1970) pointed out that Parlange's result does not give the correct bubble shape, whereas Figure 10 shows that Equations (9) and (11) do give a reasonably good prediction. The differences between these two results lie only in our neglect of energy dissipation in the wake, and this can be justified in the following way. Circulation within the vortex below the bubble must be less vigorous than assumed by Parlange because (a) there is no shear stress on the liquid at the free surface forming the base of the bubble and (b) the boundary layer in the liquid just outside and just inside the radius a must weaken the circulation within the wake. As energy dissipation is proportional to the square of velocity, it is probably a better approximation to neglect wake dissipation, as in the above analysis, than to overestimate it, as does Parlange. The correct description must lie between these two approaches, and it awaits the solution of the Navier-Stokes equation with boundary conditions at the bubble surface of constant normal stress and zero shear. It is no small difficulty that the shape of the bubble itself is part of the answer to the problem.

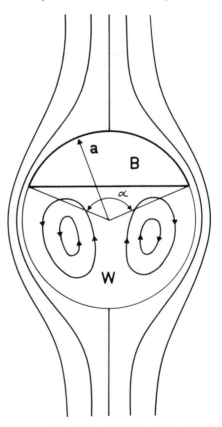

Figure 12   Spherical-cap bubble with closed wake.

Figure 11 shows that Equation (11) is in good agreement with experiment in the transition region $5 < \text{Re} < 100$. In a fluidized bed, a 5-cm-diameter bubble, say, which may be observed to rise steadily at about 30 cm sec^{-1} in a particulate phase of viscosity in the range of 4–13 poise, has a bubble Reynolds number of about 20. The range $5 < \text{Re} < 100$ in fact covers many bubble experiments in fluidized beds (see Table 1), as well as the change of shape from spherical cap to spherical shape (Haberman & Morton 1953). Theory here shows that bubble velocity $u_b \propto D_e$ in this region $5 < \text{Re} < 100$; the bubbles are in transition from viscous-dominated ($u_b \propto D_e^2$) to inertia-dominated $u_b \propto D_e^{1/2}$ [Equation (1)]. But Figure 11 shows that the transition region covers a relatively small change in C_D, from about 4 to 2.6. Now $u_b \propto C_D^{-1/2}$, and therefore the variation in u_b within this region is 25% at most, less than the experimental scatter in Figure 6. Thus very high bed viscosities are not inconsistent with the observed agreement between inviscid formulas for bubble rise [equations (1) and (2)] and experiment. From a practical point of view these equations are very useful; we see that the application of inviscid theory can be justified, if only barely.

4 THEORY FOR MOTION OF PARTICLES AND INTERSTITIAL FLUID

Further insight into why bubbles, with respect to velocity and formation, behave as if they were in liquids of low viscosity can be obtained by considering the motion of a bubble suddenly released in an incipiently fluidized bed.

Rowe (1964) considered the behavior of a cylindrical or spherical cavity that was initially at rest in an incipiently fluidized bed and then suddenly released. Figure 13a shows the streamlines of the fluidizing fluid prior to release of the

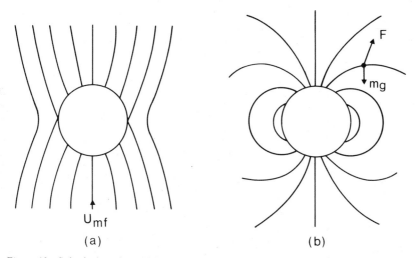

Figure 13 Spherical cavity suddenly released in an incipiently fluidized bed: (*a*) fluid streamlines at release and (*b*) streamlines of induced particle motion.

spherical bubble; the release of the cavity leaves the particles free to move under the influence of gravity and of the forces due to the fluidizing fluid. Rowe considered the forces on individual particles around the bubble immediately following the removal of restraints on its motion; these are shown in Figure 13b. F is a drag force due to the flow of fluid through the interstices of the bed, and mg is the particle weight. Rowe showed that these forces generate particle motion given by the streamlines in Figure 13b, and that these are precisely the streamlines for potential flow of a sphere. Thus the initial motion of every particle is like an element of ideal fluid with no interparticle forces.

We now present an alternative approach to this problem, which also serves to introduce the equations of bubble motion.

Figure 14 shows the coordinate system for a bubble suddenly released in an incipiently fluidized bed, r being measured from origin 0 which has the velocity of the bubble u. We assume the particulate phase does behave as an incompressible inviscid liquid whose pressure P is therefore given by Bernoulli's equation for unsteady motion (Lamb 1932)

$$\frac{P}{\rho_p} + \frac{\partial \phi}{\partial t} + gr \cos \theta = 0, \tag{13}$$

where ϕ is the velocity potential for the particle motion and ρ_p is the bulk density of the particulate phase. We have neglected the term $\frac{1}{2}q^2$ in Equation (13) because

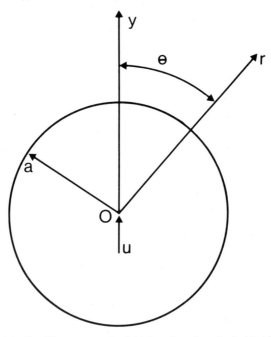

Figure 14   Coordinate system for initial motion of a spherical bubble.

the particle velocity q is small for the initial motion. At a great distance above or below the bubble, $\phi = 0$ and $P = -\rho_p gr \cos \theta$, the required condition for incipient fluidization. For potential flow around a sphere of radius a moving with velocity u, from Equation (4),

$$\phi = -\left(\frac{a^3 u}{2r^2}\right)\cos\theta.$$ (14)

In Equation (13) $\partial\phi/\partial t$ is the derivative at a fixed point in space, calculable from Equation (14) in terms of $\dot{u}$, $\dot{r} = -u\cos\theta$ and $\dot{\theta} = u\sin\theta/r$. At the start, $u = 0$, so we need consider only the term in $\dot{u}$ and $\partial\phi/\partial t = -(a^3\dot{u}/2r^2)\cos\theta$, which gives, with Equation (13),

$$\frac{P}{\rho_p} = \left(\frac{a^3\dot{u}}{2r^2} - gr\right)\cos\theta.$$ (15)

Now the pressure P must be constant at the bubble surface $r = a$, and using this condition in Equation (15) gives $du/dt = 2g$. In other words, the bubble moves off with an acceleration $2g$. This is expected because the effective inertia due to the particle motion around the bubble is half the displaced mass.

Substituting $du/dt = 2g$ into Equation (15) gives the pressure distribution required by the particle motion

$$P = \rho_p g\left(\frac{a^3}{r^2} - r\right)\cos\theta.$$ (16)

This analysis has treated the particulate phase as a continuum, and clearly this is only valid provided the "elements" containing particles are much larger than the particles themselves. It follows therefore that the bubbles must also be much larger than the particles in the bed.

Now let us turn to the pressure distribution generated ˙by the percolation of fluidizing fluid through the stationary particles, leading to the streamlines in Figure 13a. Since the voidage is constant, the fluid pressure p_f must satisfy Laplace's equation div grad $p_f = 0$. A solution, which moreover satisfies the requirement of constant pressure at $r = a$, is $p_f = C(a^3/r^2 - r)\cos\theta$, where C is a constant. For incipient fluidization, far from the bubble, we have $-dp_f/dy = \rho_p g$, with y as the vertical distance from a fixed origin. This gives C and hence

$$p_f = \rho_p g\left(\frac{a^3}{r^2} - r\right)\cos\theta.$$ (17)

The identity of Equations (16) and (17) shows that the fluid pressure is precisely that required to produce the irrotational motion of the particles. Therefore we can expect the initial motion of the particles to be as if they were elements of an ideal liquid, without interparticle forces.

A very similar result follows for two-dimensional motion, for which the pressure distribution is

$$P = p_f = \rho_p g\left(\frac{a^2}{r} - r\right)\cos\theta$$ (18)

and $\dot{u} = \dot{g}$. The initial motion is the same as for a bubble suddenly released in a liquid of low viscosity, and direct experimental verification of this similarity was provided by Partridge & Lyall (1967), who photographed the initial motion of a bubble released in a two-dimensional fluidized bed.

The theory of bubble formation at an orifice in a fluidized bed follows a similar analysis. Figure 15 shows successive stages during the growth of a bubble being formed at an orifice in stationary liquid. We assume the bubble is always spherical and its growth is determined by the gas-volume flow rate G that is constant. Consequently, the bubble volume V_b at time t after the start is given by

$$V_b = 4\pi a^3/3 = Gt, \tag{19}$$

where a is the bubble radius. The upward bubble motion is determined by a balance between the buoyancy force $\rho g V_b$, ρ being the liquid density, and the inertia of the liquid surrounding the bubble. This inertia effect is calculated from potential-flow theory [see Lamb (1932)]: when a sphere displacing a mass M of liquid is given an acceleration α, the force needed is $\frac{1}{2}M\alpha$. Neglecting viscous forces and the inertia of the bubble gas, the equation of motion for the bubble is

$$\rho V_b g = \frac{d}{dt}\left[\frac{1}{2}\rho V_b \frac{ds}{dt}\right], \tag{20}$$

which is obtained by balancing the buoyancy force against the rate of change of upward momentum; s is the vertical distance of the bubble center above the orifice. Eliminating V_b between (19) and (20) and integrating with respect to time give $ds/dt = gt$, with an initial condition $ds/dt = 0$ at $t = 0$. A second integration gives

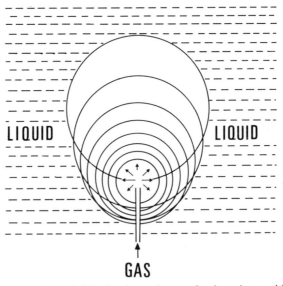

LIQUID   LIQUID

GAS

Figure 15   Sequence of a gas bubble forming at the top of a pipe submerged in a liquid.

$$s = \tfrac{1}{2}gt^2, \tag{21}$$

with $s = 0$ at $t = 0$.

Equations (19) and (21) give the growth rates of a and s; Figure 15 indicates the way in which the bubble grows and rises during formation. Initially, the base of the bubble moves down and envelopes the orifice; for larger values of t the upward motion predominates, and bubble detachment occurs when $a = s$. This gives the growth time of the bubble $(6G/\pi)^{1/5}/g^{3/5}$, and hence the bubble volume at detachment

$$V_b = \left(\frac{6}{\pi}\right)^{1/5} \frac{G^{6/5}}{g^{3/5}} = 1.138 \, \frac{G^{6/5}}{g^{3/5}}. \tag{22}$$

Figure 16 summarizes the experimental data of Harrison & Leung (1961) on bubble formation at a single orifice in an incipiently fluidized bed. These data were obtained using progressively larger orifices with increasing orifice flow rates, to avoid multiple bubble formation which occurs when the orifice velocity is so high that, after detachment, the next bubble expands rapidly and coalesces with the bubble above. The agreement between Equation (22) and the data in Figure 16 is powerful evidence that the fluidized bed behaves like a gas-liquid system. There is, however, a need for similar data from beds of larger particles in which the effect of leakage between bubble and the particulate phase is likely to be more significant than with the particles used by Harrison and Leung.

5 EQUATIONS OF BUBBLE MOTION

The phenomenon of bubble motion in fluidized beds is complex, and it is well to recall that even the motion of a gas bubble in an incompressible Newtonian fluid continues to present unsolved problems. Jackson (1963) and Anderson & Jackson (1968) linked the formation of bubbles to an instability of the state of uniform fluidization. They showed that this state was always unstable to small perturbations and that the growth rate of such a disturbance in gas-solids systems was an order of magnitude greater than in liquid-solids systems. This pointed to the basic difference between aggregative and particulate fluidization. Recently El-Kaissy & Homsy (1976), from a quantitative study of the instability wave motion that occurs above incipient fluidization, suggested that the advanced stages of wave break-up are characterized by clusters of low voidage not unlike the beginnings of bubbling.

Studies of bubble motion normally start from the assumption that a fluidized-bed motion exists in which a void is completely free of particles and separated from the rest of the bed—the particulate phase—by a sharp interface. It is further assumed that the void remains constant in size and shape and rises with constant velocity through a bed that is uniformly fluidized at a large distance above the void. A steady-flow problem follows if reference is made to the rest frame of the void. Buyevich (1975) has recently published an interesting model that allows bubble growth or shrinkage during the rise of the bubble; spherical and spherical-cap bubbles are considered as limiting cases.

Bubbles are a feature of gas-fluidized systems, and studies of bubble motion normally neglect terms in the equations of motion for (*a*) the buoyancy forces on each particle due to the displaced fluid, (*b*) the added or virtual mass for each particle due to the motion of the fluid around the particle, and (*c*) the fluid momentum. In addition, viscous effects in the gas are ignored and it is assumed that wake effects are confined to a finite region near the bubble. Finally, in spite of direct experimental evidence (Schügerl et al 1961), interparticle forces are assumed to be negligible.

Table 2 summarizes the three approaches that have been made to the statement of the equations of bubble motion: Davidson (1961), Jackson (1963), and Murray (1965). A detailed comparison of these theories has been given by Jackson (1971).

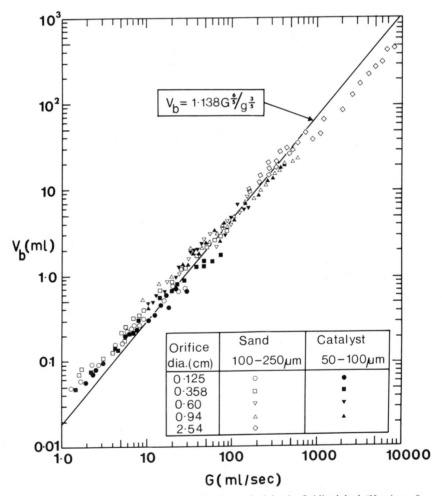

$$V_b = 1 \cdot 138 G^{\frac{6}{5}} / g^{\frac{3}{5}}$$

Orifice dia.(cm)	Sand 100–250μm	Catalyst 50–100μm
0·125	○	●
0·358	□	■
0·60	▽	▼
0·94	△	▲
2·54	◇	

Figure 16 Bubble formation at an orifice in an incipiently fluidized bed (Harrison & Leung 1961).

Table 2 Equations of motion for a bubble[a]

	Davidson (1961)	Jackson (1963)	Murray (1965)	
Voidage	Constant $= \varepsilon_{mf}$	Dependent variable	$\varepsilon \to 0$	
Particle motion	As for inviscid liquid; pressure P from Bernoulli's theorem	$\operatorname{div} \mathbf{v} = 0$ (i) $\operatorname{grad} p = -\rho_p(\mathbf{v} \cdot \operatorname{grad})\mathbf{v} - \rho_p g\mathbf{i}$ (ii) As for inviscid liquid	$\operatorname{div} \mathbf{v} = 0$ (i) $\operatorname{grad} p = cU\dfrac{\partial \mathbf{v}}{\partial x} - \dfrac{\mathbf{i}}{F}$ (ii) Linearized, with constants c and F	
Fluid motion	$\operatorname{div} \mathbf{u} = 0$ $\mathbf{u} = \mathbf{v} - \dfrac{1}{\beta_0} \operatorname{grad} p_f$	$\operatorname{div} \mathbf{u} = 0$ (iii) $\mathbf{u} = \mathbf{v} - \dfrac{1}{\beta} \operatorname{grad} p$ (iv)	$\operatorname{div} \mathbf{u} = 0$ (iii) $\mathbf{u} = \mathbf{v} - F \operatorname{grad} p$ (iv)	

[a] $\mathbf{v}$ = particle velocity; $\mathbf{u}$ = interstitial fluid velocity; ε = free volume/total volume; ρ_p = bulk density of particulate phase at incipient fluidization when $\varepsilon = \varepsilon_{mf}$. $\operatorname{grad} p_f = -\rho_p g$.

Davidson (1961)

This analysis assumes that the particulate phase behaves as an incompressible fluid, so that $\varepsilon = \varepsilon_{mf}$ at all points outside the bubble. The particle flow field resembles that observed experimentally, and the approach yields useful results that are simple in form. Two important consequences of this theory follow:

1. Since div $\mathbf{v} = 0$, div grad $p_f = 0$, and hence the fluid pressure is unaffected by the particle motion; if the boundary conditions are fixed, the pressure distribution is the same for stationary particles as for moving particles.
2. The particulate-phase pressure P, obtained by applying Bernoulli's theorem, must equal p_f. If not, interparticle forces are indicated, or $\varepsilon \neq \varepsilon_{mf}$.

Jackson (1963)

This treatment has a mass acceleration term in the equation for the particle motion (Table 2) and it further allows for the percolation of fluidizing fluid and variation of voidage around the bubble. The form of the voidage function β is $K(1-\varepsilon)^2/\varepsilon^2$ where K is the Carman-Kozeny constant. On the other hand, this analysis makes no allowance for interparticle pressures that must occur if $\varepsilon < \varepsilon_{mf}$.

Jackson's analysis predicts the existence of a region of increased bed voidage above the bubble, and this has been confirmed experimentally by Lockett & Harrison (1967) using a capacitance technique. The agreement, though, is not good quantitatively; experiment would suggest that there is a variation of voidage from ε_{mf} over a larger region near the bubble than theory predicts. Nevertheless, Jackson's equations represent the most complete description of bubble motion yet available. He suggests himself that the Davidson and Jackson solutions are best thought of as complementary, and he emphasizes their essential similarity. In Davidson's treatment the constant-pressure condition is satisfied exactly everywhere on the bubble surface, whereas the particle momentum balance is only satisfied approximately near the bubble nose. Conversely, Jackson's equations satisfy the particle momentum balance over the whole flow field, but the constant-pressure condition at the bubble surface is satisfied only approximately near the bubble nose.

Murray (1965)

Murray showed that by relaxing the condition of strict constancy of pressure on the bubble surface, a solution of Davidson's equations could be found which also satisfied an Oseen-type approximation to more complete equations of motion. As a consequence of linearization, only three of Murray's equations (Table 2) are independent. Murray's solution satisfies the Oseen equations everywhere outside the bubble, but they are in error near the bubble itself, which is a region of special interest. Furthermore, experiment (Lockett and Harrison) does not support the assumption in this approach that for theoretical purposes it is reasonable to confine all voidage changes to a thin "boundary layer" region close to the bubble; changes in voidage have been detected a bubble diameter away from the edge of a bubble.

6 THE EQUATIONS OF MOTION AND EXPERIMENT

Theory and experiment in fluidized beds may be tested, as Jackson (1971) points out, in several ways: measurements of bubble rising velocity, or measurements near bubbles of "cloud" formation, gas pressure distribution, and voidage fraction.

As Figures 6 and 7 show, measurements of bubble rise velocity are certainly consistent with the Davies & Taylor (1950) result [Equation (1)], but these results do not provide a sensitive test of theory because the bubble volume is uncertain, and $u_b \propto D_e^{1/2}$. Reuter's (1963) measurements on the pressure field near a bubble have been mentioned (Section 2), and these results lie between the curves obtained from Davidson's and Jackson's analyses. Qualitative agreement between experimental voidage fractions and Jackson's theory has been described in Section 5. However, although studies on bubble velocity, pressure distribution, and voidage fraction provide much insight, the most striking comparisons between theory and experiment have come from work on "cloud" formation.

Figure 17 shows a bubble formed by injecting a pulse of nitrogen dioxide into a two-dimensional bed of glass ballotini. It is clear that the gas penetrates a finite distance, forming a "cloud"; this penetration was first predicted by theory (Davidson 1961), and it is a crucial feature of all the theories outlined for systems for which $u_b/u_{mf} > 1$, where u_{mf} is the interstitial-gas velocity at incipient fluidization. Slugs also form "clouds" when they rise faster than the interstitial fluid; a typical "cloud" is shown in Figure 18.

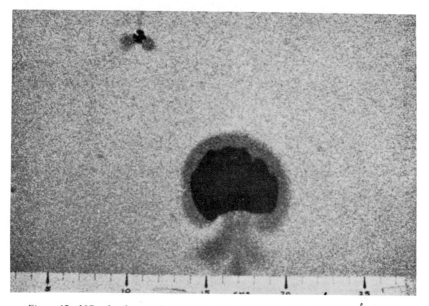

Figure 17   NO$_2$ cloud around a two-dimensional bubble $u_b/u_{mf} = 2.4$ (Rowe 1971).

Figure 18   NO$_2$ cloud around a two-dimensional slug $u_s/u_{mf} = 1.7$ (Stewart & Davidson 1967).

Davidson & Harrison (1963) show that, for a two-dimensional bubble, fluid on the axis penetrates to a radius r_c that defines the top of the "cloud" beyond which fluid from the bubble does not penetrate; r_c is given by

$$\frac{r_c^2}{a^2} = \frac{u_b + u_{mf}}{u_b - u_{mf}}. \tag{23}$$

The corresponding result for two-dimensional slug flow, given by Hovmand & Davidson (1971), is

$$\exp\left(\pi Y/b\right) = \frac{u_s + u_{mf}/2}{u_s - u_{mf}}, \tag{24}$$

where Y is the penetration distance of the "cloud" above the nose of the slug, b is half the column width, and u_s is the slug velocity.

Figures 19 and 20 show how the measured "cloud" penetration compares with theory for two-dimensional bubbles and slugs. These results are a striking confirmation of theory, and the quantitative agreement supports the hypothesis that as a first approximation the particulate phase has a constant voidage and that it moves like an inviscid liquid.

The evidence from slug flow systems is important because (a) the slug velocity and particle flow pattern near the nose can be predicted entirely from theory because the relevant diameter, that of the container, is known; and (b) the wake of a slug is comparatively small and far below the nose; the theory for flow near the nose is therefore more likely to hold in practice for a slug than for a bubble.

Following Stewart & Davidson (1967), the analysis procedure for slug flow

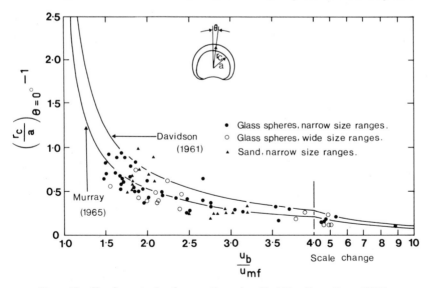

Figure 19   Cloud penetration for two-dimensional bubbles. From Rowe (1971).

is to solve Laplace's equation $\nabla^2 p_f = 0$ with suitable boundary conditions for the region above the slug envelope, to calculate the pressure P using Bernoulli's theorem, and then to evaluate the interparticle pressure $p_p = P - p_f$. For the two-dimensional case, Stewart & Davidson (1967) obtained the computed values for the interparticle pressure shown in Figure 21; x, y are Cartesian coordinates with origin at the slug nose. Values of $p_p/\rho_p gb$ are shown where ρ_p is the bulk density of the particulate phase and b is half the column width. Above the slug nose small negative values of p_p are found, of the order of a tenth of the pressure head due to a height b of particulate phase. In this region the particles would in practice separate and the voidage fraction would increase locally. However, there is a region for which the predicted p_p is positive. This can be regarded as a de-fluidized zone in which there will be a tendency for shear stresses to develop. The finite values of p_p at the wall may account for the development of particle

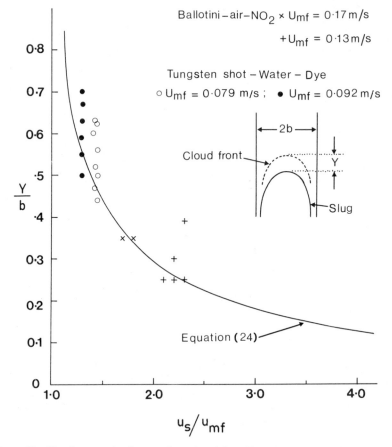

Figure 20   Cloud penetration for two-dimensional slugs. From Stewart & Davidson (1967).

bridges and possibly for the formation of wall slugs which rise alongside the walls (Hovmand & Davidson 1971).

Hoath & Collins (1973) solved a similar problem using Jackson's method and obtained the same prediction of "cloud" penetration ahead of a slug as Stewart and Davidson. They also found, in contrast to Stewart and Davidson, an increase in voidage everywhere in the field near the slug, although the increase of voidage near the wall was less than on the slug axis.

Littman & Homolka (1973) described extensive pressure measurements around a two-dimensional bubble rising in an incipiently fluidized bed. Figure 22 shows

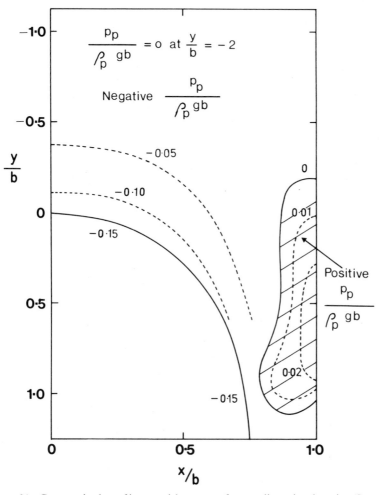

Figure 21   Computed values of interparticle pressure for two-dimensional motion (Stewart & Davidson 1967).

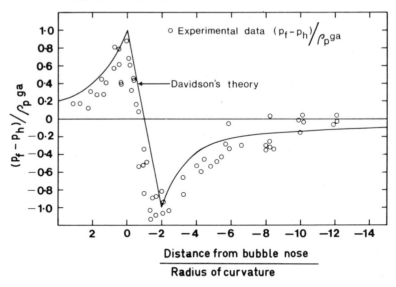

Figure 22   Pressure along the vertical axis of a two-dimensional bubble (Littman & Homolka 1973).

their normalized plot of experimental values of $p_f - p_h$ against distance from the bubble nose measured along the vertical axis of symmetry, where p_f is the actual fluid pressure and p_h is the hydrostatic pressure. The agreement with Davidson's theory is good, but Littman and Homolka found that pressures given by Jackson's analysis did not fit the data. The discrepancy is puzzling, and is without satisfactory explanation at the present time. Littman and Homolka point out that bubble velocity is a term in Jackson's pressure-field equation and the value chosen markedly affects predictions of p_f. They also observe that the particle motion above the bubble roof is predominantly vertical, with little relative motion horizontally between the particles and the bubble, and they suggest that this may have the effect of making Darcy's law a good approximation to the local pressure gradient.

7   THE NATURE OF FLUIDIZED BED VISCOSITY

We note in Section 2 that the high viscosity of a gas-fluidized bed is unlikely to be due to the "concentration" effect analyzed by Frankel & Acrivos (1967); even with a "concentration" effect the predictions of apparent viscosity are some two or three orders of magnitude too small. Two other experimental results are also relevant. First, Rigby et al (1970) have measured the apparent viscosity of a water-fluidized bed by two techniques and found values in the range 0.5–3 poise at water flow rates 2–3 times that at incipient fluidization. Above that, to flow rates of $10U_{mf}$, the measured viscosities were fairly constant at between 0.2 and

1.5 poise. These viscosities are much greater than that of the fluidizing water (0.01 poise). Second, Goldschmidt & Le Goff (1963) and Graham & Harvey (1966) have measured the electrical conductivity of gas-fluidized beds of coke and charcoal and have found an appreciable conductivity at flow rates above incipient fluidization. This must indicate that there are continuous paths of particle contact through the bed.

Therefore the evidence points strongly to fluidized bed viscosity being caused by interparticle collisions. It would thus seem appropriate that Jackson's equation for particle motion (Table 2) should be modified to include apparent viscosity as follows:

$$\rho_p(\mathbf{v} \cdot grad)\mathbf{v} + \mathbf{i}\rho_p g = -\operatorname{grad} p + \mu_{\mathrm{app}} \nabla^2 \mathbf{v}. \tag{25}$$

The effect of particle collisions on fluidized behavior is not easy to analyze, but here we explore an interesting approach put forward by Bagnold (1954).

In Bagnold's experiments dispersions of spherical grains were sheared in Newtonian fluids of varying viscosity in the annular space between two concentric drums, with the inner drum stationary. Bagnold postulated a dispersive grain pressure—or interparticle pressure—to explain his observations, and measured values were 50 to 700 times the differential pressure due to rotation, depending on the grain concentration. We now consider his experiments on systems in which grain inertia dominates and which have voidage fractions comparable with incipiently fluidized beds.

An apparent viscosity μ_{app} can be defined by $\tau/(dU/dy)$ where τ, the grain shear stress in dynes cm^{-2}, and dU/dy, the shear strain rate in sec^{-1}, are available from Bagnold's Figure 3 for a grain concentration ($\lambda = 17$) which corresponds to a voidage $\varepsilon = 0.38$, i.e. approximately incipient fluidization. The experimental data cover the ranges $\tau = 40$–600 dynes cm^{-2} and $dU/dy = 7.6$–41.5 sec^{-1}, and hence the apparent viscosities at these shear rates are 5–15 poise. As shown in Section 2, this is the range of viscosities found for gas-fluidized beds (4–13 poise), and it provides support for the view that Bagnold's mechanics are applicable to such systems. Also, these shear rates seem to be relevant for typical rising bubbles, e.g. for a 2-cm-diameter bubble rising at 30 cm sec^{-1} the shear rate would be of the order of 15 sec^{-1}. However, his theory—that the shear stress is due to intergranular collisions dissipating energy—leads to the result $\tau = $ constant $(dU/dy)^2$ for a given voidage, whereas most investigators have supposed that a fluidized bed behaves like a Newtonian liquid and $\tau = $ constant (dU/dy). None of the published measurements of fluidized bed viscosity is in our view adequate to discriminate between the square law proposed by Bagnold and the linear relation that would apply if the fluidized bed behaved like a Newtonian liquid. This suggests a good topic for research, but the experiments will need to be more subtle than those that have yet been devised: in particular, a more accurate measurement of rate of shear strain is required; previous workers have assumed a no-slip condition at a solid surface, but from simple observations, e.g. for bubbles moving up plane surfaces, it is clear that fluidized particles do indeed slip over a solid surface.

Literature Cited

Anderson, T. B., Jackson, R. 1968. *Ind. Eng. Chem. (Fundam.)* 7:12–21

Angelino, H., Charzat, C., Williams, R. 1964. *Chem. Eng. Sci.* 19:289–304

Bagnold, R. A. 1954. *Proc. R. Soc. London Ser. A* 225:49–63

Batchelor, G. K. 1967. *An Introduction to Fluid Dynamics.* Cambridge: The University Press

Buyevich, Y. A. 1975. *Int. J. Multiphase Flow* 2:337–51

Calderbank, P. H. 1967. *Chem. Eng. No.* 212:209–33

Cranfield, R. R., Geldart, D. 1974. *Chem. Eng. Sci.* 29:935–47

Davidson, J. F. 1961. *Trans. Inst. Chem. Eng.* 39:230–32

Davidson, J. F., Harrison, D. 1963. *Fluidised Particles.* Cambridge: The University Press

Davidson, J. F., Harrison, D., Darton, R. C., LaNauze, R. D. 1977. In *Chemical Reactor Theory, A Review,* ed. L. Lapidus, N. R. Amundson, Chap. 10. Englewood Cliffs, NJ: Prentice Hall

Davies, R. M., Taylor, G. I. 1950. *Proc. R. Soc. London Ser. A* 200:375–90

El-Kaissy, M. M., Homsy, G. M. 1976. *Int. J. Multiphase Flow.* 2:379–95

Frankel, N. A., Acrivos, A. 1967. *Chem. Eng. Sci.* 22:847–53

Goldschmidt, D., Le Goff, P. 1963. *Chem. Eng. Sci.* 18:805–6

Grace, J. R. 1970. *Can. J. Chem. Eng.* 48:30–33

Graham, W., Harvey, F. A. 1966. *Can. J. Chem. Eng.* 44:17–19

Haberman, W. L., Morton, R. K. 1953. *David Taylor Model Basin Rep. No. 802*

Hagyard, T., Sacerdote, A. M. 1966. *Ind. Eng. Chem. (Fundam.)* 5:500–8

Harper, J. F., Moore, D. W. 1968. *J. Fluid Mech.* 32:367–91

Harrison, D., Leung, L. S. 1961. *Trans. Inst. Chem. Eng.* 39:409–14

Hoath, M. T., Collins, R. 1973. *Proc. Int. Symp. on Fluidization and Its Applications, Toulouse, France,* pp. 254–62

Hovmand, S., Davidson, J. F. 1971. In *Fluidization,* ed. J. F. Davidson, D. Harrison, Chap. 5. New York and London: Academic

Jackson, R. 1963. *Trans. Inst. Chem. Eng.* 41:13–28

Jackson, R. 1971. See Hovmand & Davidson 1971, Chap. 3

Jones, D. R. M. 1965. PhD dissertation. Univ. Cambridge, Cambridge, England (Fig. 9)

Kojima, E., Akehata, T., Shirai, T. 1968. *J. Chem. Eng. Jpn.* 1:45

Lamb, H. 1932. *Hydrodynamics.* Cambridge: The University Press. 6th ed.

Levich, V. G. 1962. *Physicochemical Hydrodynamics.* Englewood Cliffs, NJ: Prentice-Hall

Littman, H., Homolka, G. A. J. 1973. *Chem. Eng. Sci.* 28:2231–43

Lockett, M. J., Harrison, D. 1967. In *Proceedings of the Symposium on Fluidization,* ed. A. A. H. Drinkenburg, pp. 257–67. Amsterdam: Netherlands Univ. Press

Moore, D. W. 1963. *J. Fluid Mech.* 16:161–76

Murray, J. D. 1965. *J. Fluid Mech.* 22:57–80

Murray, J. D. 1967. *Rheol. Acta* 6:27–30

Nguyen, X. T., Leung, L. S., Weiland, R. H. 1973. See Hoath & Collins 1973, pp. 230–39

Park, W. H., Kang, W. K., Capes, C. E., Osberg, G. L. 1969. *Chem. Eng. Sci.* 24:851–65

Parlange, J.-Y. 1969. *J. Fluid Mech.* 37:257–63

Partridge, B. A., Lyall, E. 1967. *J. Fluid Mech.* 28:429–31

Reuter, H. 1963. *Chem. Ing. Tech.* 35:98–103, 219–28

Rigby, G. R., Van Blockland, G. P., Parks, W. H., Capes, C. E. 1970. *Chem. Eng. Sci.* 25:1729–41

Rippin, D. W. T., Davidson, J. F. 1967. *Chem. Eng. Sci.* 22:217–28

Rowe, P. N. 1964. *Chem. Eng. Sci.* 19:75–77

Rowe, P. N. 1971. See Hovmand & Davidson 1971, Chap. 4, pp. 138, 155, 182

Rowe, P. N., Matsuno, R. 1971. *Chem. Eng. Sci.* 26:923–35

Rowe, P. N., Partridge, B. A. 1965. *Trans. Inst. Chem. Eng.* 43:157–75

Schügerl, K., Merz, M., Fetting, F. 1961. *Chem. Eng. Sci.* 15:1–99

Stewart, P. S. B. 1968. *Trans. Inst. Chem. Eng.* 46:60–66

Stewart, P. S. B., Davidson, J. F. 1967. *Powder Tech.* 1:61–80

Toei, R., Matsuno, R., Kojima, H., Nagai, Y., Nakagawa, K., Yu, S. 1965. *Mem. Fac. Eng. Kyoto* 27:475–89

Wegener, P. P., Parlange, J.-Y. 1973. *Ann. Rev. Fluid Mech.* 5:79–100

Ann. Rev. Fluid Mech. 1977. 9:87–111
Copyright © 1977 by Annual Reviews Inc. All rights reserved

HISTORY OF BOUNDARY-   ×8098
LAYER THEORY

Itiro Tani

National Aerospace Laboratory, 1880 Jindaiji, Chofu, Tokyo, Japan

GENESIS AND EARLIER DEVELOPMENTS

Introduction

The boundary-layer theory began with Ludwig Prandtl's paper *On the motion of a fluid with very small viscosity*, which was presented at the Third International Congress of Mathematicians in August, 1904, at Heidelberg and published in the Proceedings of the Congress in the following year. This paper marked an epoch in the history of fluid mechanics, opening the way for understanding the motion of real fluids. Nevertheless, the genesis of the boundary-layer theory stood in sublime isolation: nothing similar had ever been suggested before, and no publications on the subject followed except a small number of papers due to Prandtl's students for almost two decades.

The equations of motion of a viscous fluid were established in the first half of the last century by Navier (1823), Poisson (1831), Saint-Venant (1843), and Stokes (1845), having attained the form that is now called the Navier-Stokes equations. Stokes used the equations to consider the small oscillations of a sphere in a viscous fluid by assuming that there is no slip, that is, no relative tangential velocity, at the surface of the sphere. Confusion had prevailed before as to the conditions to be satisfied by the fluid at the wall of the solid boundary: Stokes (1845) seems to have been initially inclined to the hypothesis of no slip, but when calculations on flow through a pipe gave results at variance with experiments known to him at that time, he hesitated between the no-slip and slip hypotheses. In his 1851 paper, however, he decided to adopt the former on the grounds that this would mean regarding the friction between solid and fluid as of the same nature as the friction between fluids, and also that this would lead to satisfactory agreement with experiments. Later, it was found that the calculations based on the same hypothesis for flow through a pipe, begun by Stokes (1845) and repeated by various authors, also gave good agreement with subsequent experimental results.

The solutions obtained by Stokes, however, were confined to rather special cases, where it was possible to solve the Navier-Stokes equations exactly, because the nonlinear terms were either negligibly small or identically vanishing. This having not been the case in the majority of the problems met in practice, it was necessary

87

to introduce some approximations for solution. The simplest was, of course, to neglect the viscosity of the fluid, but this brought about nothing but the d'Alembert paradox, according to which a solid body of any shape placed in a uniform stream experiences no resistance. This failure was particularly disturbing since the viscosity was considered to produce only small effects in the motion of such fluids as air or water. According to the 1888 edition of the *Encyclopaedia Britannica,* "hydro-dynamics" was the branch of science that dealt with the mathematical theory of the motion of fluids, neglecting viscosity, while it was in the branch of "hydraulics" that hydrodynamical questions of practical application were investigated.

The mathematical difficulties of integrating the equations of a viscous fluid made it compelling to neglect the nonlinear terms. This approximation, justified only for slow motions, was made unavoidably also for faster motions, but with the optimistic hope that these solutions might give a better representation of the flow than those obtained by neglecting the viscosity (Bassett 1888). It was a relief to find that the solutions predicted at least nonzero resistance, although far too small in magnitude. It was almost universally agreed that there is no slip at the solid wall in the case of slow motions. The views divided, however, on fast motions. Some authors adopted the no-slip condition also for fast motions, but do not seem to have thought about the necessarily continuous variation of velocity starting from zero at the wall (Lighthill 1963). Other authors suggested that there is a slip at the wall and that the slip is resisted by a frictional force depending on the relative velocity. A number of attempts were made to express the law of friction in the form of an empirical formula applicable to fast as well as slow motions (Unwin 1888).

Prandtl's Paper

In the paper of 1905, Prandtl started from the clear recognition that the most important question concerning the flow of a fluid of small viscosity is the behavior of the fluid at the wall of the solid boundary. It appears that the flow is almost irrotational until comparatively close to the wall, so that the variation of velocity from the value corresponding to irrotational motion to the zero velocity demanded by the condition of no slip at the wall takes place within a thin layer adjacent to the wall. The smaller the viscosity, the thinner is the transition layer. But the steep velocity gradient, in spite of the small viscosity, produces marked effects, which are comparable in magnitude with those due to the inertia force, if the thickness of the transition layer is proportional to the square root of the kinematic viscosity. Thus, the effects of viscosity are significant only within a thin transition layer, which is called the *boundary layer.*[1] Outside this layer, the flow is essentially free of viscosity and is described by an irrotational motion to a high degree of accuracy.

[1] It may be noted that Prandtl used the term *Grenzschicht* (boundary layer) only once and the term *Übergangsschicht* (transition layer) several times in the paper. The term Grenzschicht has come into more definite use since the paper of Blasius (1908). Later, Prandtl (1925a) wrote that the term Grenzschicht, usually employed by specialists, appears by no means to be happy, but should be continued since it has already been introduced.

The small thickness of the boundary layer permits certain approximations for the governing equations within the boundary layer: the variation of pressure normal to the wall is negligibly small, and the variation of velocity along the wall is much smaller than its variation normal to it. In the case of flow in two dimensions the effect of moderate curvature of the wall is negligibly small, so that x and y may be taken as the distances along and normal to the wall, respectively, and u and v as the corresponding velocity components. The x-component of the Navier-Stokes equations is then simplified to the form

$$\frac{\partial u}{\partial t} + u\frac{\partial u}{\partial x} + v\frac{\partial u}{\partial y} + \frac{1}{\rho}\frac{\partial p}{\partial x} = v\frac{\partial^2 u}{\partial y^2},$$

where t is the time, p the pressure, ρ the density, and v the kinematic viscosity. The pressure p is regarded as a function of x and t and prescribed by the irrotational motion outside the boundary layer. The equation is parabolic, although the original Navier-Stokes equations are elliptic. Thus it can be integrated step by step in the direction of x when u is known at a fixed value of x for all values of y and t, the upstream influence being suppressed to the order of approximation. Prandtl considered the solution of the equation for the simple case $p = $ constant, that is, the case of a semi-infinite thin flat plate placed parallel to a stream of uniform velocity U, and obtained a rough estimate $1.1\rho v^{1/2}l^{1/2}U^{3/2}$ for the frictional resistance exerted on the two sides of unit width of a plate of length l. This was the first theoretical analysis of the frictional resistance, although the numerical coefficient 1.1 was later corrected by Blasius (1908) to 1.33.

A remarkable consequence of the investigation from the standpoint of application was, according to Prandtl, that "in certain cases, *the flow separates from the surface* at a point entirely determined by external conditions. A fluid layer, which is set in rotation by the friction on the wall, is thus forced into the free fluid and, in accomplishing a complete transformation of the flow, plays the same role as the Helmholtz separation layers. A change in the viscosity constant μ simply changes the thickness of the transition[2] layer (proportional to the quantity $\sqrt{\mu l/\rho U}$), everything else remaining unchanged. It is therefore possible to pass to the limit $\mu = 0$ and still retain the same flow figure" (1928 translation of Prandtl 1905). Without going into a mathematical analysis, Prandtl explained the plausible reason for flow separation with the increase of pressure in the streamwise direction. He also deduced that "the treatment of a given flow process is resolved into two components mutually related to one another. On the one hand, we have the free fluid, which can be treated as nonviscous according to the Helmholtz vortex laws, while, on the other hand, we have the transition layers on the solid boundaries, whose motion is determined by the free fluid, but which, in their turn, impart their characteristic impress to the free flow by the emission of vortex[2] layers" (1928 translation of Prandtl 1905). Prandtl closed the paper with confirmation of the theory by photographs of flows obtained in a small hand-operated water tank.

---

[2] Here the translation of *Wirbelschicht* has been corrected.

Prototype of Concept

Prandtl's paper is an extraordinary paper in at least three aspects. First, it is extraordinary in the sense that an unprecedentedly novel, but fully matured, idea emerged in a single paper. Of course, brief mention of the existence of a boundary layer and its connection with frictional resistance had already been scattered in the literature up to that time, but it had amounted to very little compared with Prandtl's achievement. There had been no boundary-layer equations and no explanation of flow separation, as is seen below.

In a paper on the prediction of the required engine power of proposed ships, Rankine (1864) considered the frictional resistance to be due to "the direct and indirect effects of the adhesion between the skin of the ship and the particles of water which glide over it; which adhesion, together with the stiffness of the water, occasions the production of a vast number of small whirls or eddies in the layer of water immediately adjoining the ship's surface. The velocity with which the particles of water whirl in those eddies bears some fixed proportion to that with which those particles glide over the ship's surface: hence the actual energy of the whirling motion impressed on a given mass of water at the expense of the propelling power of the ship, being proportional to the square of the velocity of whirling motion, is proportional to the square of the velocity of gliding." Thus, Rankine visualized the formation of a boundary layer adjacent to the ship's surface. However, the argument leading to the quadratic law of resistance is relevant only to a surface of appreciable roughness. According to Loitsianskii (1970), the Russian chemist Dmitrii Mendeleyev clearly distinguished between smooth and rough surfaces in his monograph entitled *On the Resistance of Fluids and the Problem of Flight* (St. Petersburg, 1880). He recognized the important role played by "a thin layer of fluid adjacent to the solid surface and carrying along the neighboring layers" (translated) in generating frictional resistance of a smooth surface. He also considered the resistance of a rough surface to be of the same nature as the resistance experienced by a plate at right angles to the stream.

Experiments of Froude (1872) on a thin flat plate towed through still water made it clear that the frictional resistance does not vary as the length but at a smaller rate. This result was considered to be due to the fact that the rear portions of the surface are in contact with water that has been set in motion by the front portions, and therefore cannot experience as much frictional force as the front portions. Froude thus anticipated the existence of a boundary layer growing in thickness with the distance downstream. In a subsequent paper, Froude (1874) pointed out that the frictional force must have its counterpart in the loss of momentum of the fluid that has passed along the surface of the plate. Prandtl (1927b) quoted Froude as "the first English author to refer the frictional resistance of a flat plate to the layers of fluid in intense shear near the surface." Judging from the summary of his lecture delivered at the British Association for the Advancement of Science in 1869, Froude seems to have had arrived at some concept of the boundary layer before carrying out the systematic towing experiments.

In considering the free convection from a heated vertical plate in still air, Lorenz

(1881) assumed that the flow is parallel to the plate ($v = 0$) and that the velocity u and temperature T depend only on y, where x and y are distances along and normal to the plate, respectively, with the origin at the lower edge of the plate. The momentum equation was simplified to $0 = g(T - T_0)/T_0 + v d^2 u/dy^2$, and the energy equation to $u(T - T_0)/x = \kappa d^2 T/dy^2$, where T_0 is the ambient temperature, g the acceleration due to gravity, and κ the thermometric conductivity. The transformation of variables $y = \alpha y'$, $u = \beta u'$, $T - T_0 = (T_w - T_0)\theta'$ was made so as to reduce the equations to a form expressed purely in terms of the nondimensional variables y', u', and θ'. This led to the result $\alpha^4 = v\kappa x T_0/g(T_w - T_0)$, $\beta^2 = \kappa g x(T_w - T_0)/v T_0$, indicating that the thickness of the boundary layer increases as $x^{1/4}$, while the maximum velocity increases as $x^{1/2}$. In spite of the inconsistent approximation for the governing equations (in which the convection terms were wholly neglected in the momentum equation, but approximately taken into account in the energy equation), the results are in agreement with those obtained by the subsequent, more consistent treatment (Schmidt & Beckmann 1930). Lorenz expressed the solution of the non-dimensional equations as power series in the variable $z = 1 - \exp(-y')$ and obtained the rate of heat transfer from the plate, which turned out to be 36% larger than the correct value for the case $\kappa = v$.

The paper[3] thus contains the prototype of the boundary-layer concept in that the conduction term is considered to be of the same order of magnitude as the convection term within the temperature boundary layer. Because of the linear behavior of temperature in the energy equation, it might have been a little easier to think of its boundary layer, rather than that relevant to velocity. Prandtl (1949) referred to this paper as "the first paper on free heat convection and at the same time the first on boundary layers! However, the dependence on x of the thickness and maximum velocity is not given in it" (translated). Calling it the first paper on free heat convection might be appropriate, but the first paper on boundary layers seems to be excessive praise. Prandtl's additional comment that the dependence on x of the thickness and maximum velocity is not given in the paper seems to be due to his oversight.

It appears to the writer that Prandtl did not notice this paper before publishing his boundary-layer theory (1905). If there were any opportunity for him to read the paper of Lorenz, he might have mentioned it in his paper, as is readily imaginable from his strict fairness in history, particularly in priority. His oversight just mentioned appears to support this conjecture. It is the writer's guess that Prandtl found the paper of Lorenz after the publication of his own paper and felt surprise at a thread of connection between the two papers.

Slow Acceptance

The second aspect that makes Prandtl's paper extraordinary is its very slow acceptance and growth. The statement has often been repeated that the paper occupied less than eight pages. In reply to Goldstein's question as to why he had kept it so short, Prandtl explained "that he had been given ten minutes for his lecture at the Congress and that, being still quite young, he had thought he could

---

[3] The writer is indebted to Prof. S. Corrsin and Prof. Y. Katto for this reference.

publish only what he had had time to say" (Goldstein 1969). The greater part of the paper was devoted to showing to the assembled mathematicians such items as the d'Alembert paradox, the Helmholtz vortex theorems, diagrams of streamlines involving separation, experimentally obtained photographs of flows past a projection and a circular cylinder, etc. As a result, the essentials of the boundary-layer theory were compressed into two and a half pages, largely descriptive and extremely curtailed in expression. It is quite certain that the paper was very difficult to understand at that time, making its spread rather sluggish.

In 1908 there appeared two papers on boundary layers, one by Blasius and the other by Boltze, both prepared as dissertations at Göttingen under Prandtl's guidance. Blasius applied Prandtl's theory to the detailed study of the flow along a flat plate placed parallel to a uniform stream, as well as of the flow around a circular cylinder that is started moving in a fluid at rest. Boltze investigated flow around a body of revolution, particularly a sphere. Subsequently, Prandtl (1910) applied the boundary-layer concept to the heat-transfer problem, and Hiemenz (1911), also in a Göttingen dissertation, carried out boundary-layer calculations with an experimentally determined pressure distribution on a circular cylinder. Töpfer (1912) refined the numerical computations of Blasius. Prandtl (1914) explained the change in flow pattern around a sphere on passing through the critical Reynolds number, which had been observed by Eiffel (1912), as due to transition of the flow in the boundary layer from laminar to turbulent.

Thus, in the first decade, there were seven papers on boundary layers, due to five authors, all at Göttingen. Through these papers the concept of boundary layers spread out of Göttingen, but only very slowly. Most of the papers were written in a more accessible and conventional form than Prandtl's original paper, but they seem to have escaped the attention they deserved. This may be demonstrated, for example, by reference to Lanchester's *Aerodynamics* (1907). On pages 50–51 of this book, Lanchester found that the frictional resistance of a flat plate would vary as $\rho v^{1/2} l^{1/2} U^{3/2}$, without knowing of Prandtl's result $1.1 \rho v^{1/2} l^{1/2} U^{3/2}$. He arrived at the result by comparing the frictional force with the loss of momentum in the boundary layer. Lanchester gave, in addition, an explanation of flow separation, less detailed than that of Prandtl, and also indications that the flow becomes turbulent at higher speeds. At any rate, this publication aroused the interest of Rayleigh (1911), who made a simple but less accurate estimate $2.26 \rho v^{1/2} l^{1/2} U^{3/2}$ for the frictional resistance on the basis of the analogous problem of an infinite flat plate that is started impulsively from rest, now referred to as the Rayleigh problem. No reference to Prandtl was made in this paper. A crude calculation was also made by Gümbel (1913) in an attempt to predict the frictional resistance of ships. He assumed the velocity distribution near a flat plate to be of the form $u = U\{1 - \exp[-y(U/2vx)^{1/2}]\}$ and obtained $2.83 \rho v^{1/2} l^{1/2} U^{3/2}$ for the frictional resistance. Reference was made to Blasius's result $1.33 \rho v^{1/2} l^{1/2} U^{3/2}$, which was rejected, however, since Froude's experimental data (1872, 1874) support Gümbel's formula. His comment that there is no information on velocity distribution in Blasius's paper makes one suspect that he did not fully understand the paper.

Mention should also be made of Zhukovskii's *Aérodynamique* (1916), the French edition of his lecture notes of 1911–1912 in Moscow. On pages 119–22 of this edition, Zhukovskii assumed that the fluid velocity is zero at the wall and rapidly increases until it becomes equal to the theoretical velocity of irrotational motion, the transition layer of fluid adjacent to the wall being thin and rotational. He then made a rough estimate of the thickness of the layer by assuming it to vary in inverse proportion to the theoretical velocity. In this connection no mention was made of Prandtl's boundary-layer theory. On page 198, however, there appears a reference to Prandtl's article on fluid motion in *Handwörterbuch der Naturwissenschaften* (1913). This article having contained, among other things, a brief but plain account of boundary layers, it is strange that it was referred to only in connection with vortex formation behind a bluff body, but not in connection with boundary-layer theory.

A blank of six years was caused by World War I in the record of publications on boundary layers. Then, Kármán (1921) proposed the momentum integral equation, obtained by integrating the momentum equation across the boundary layer, for approximate calculation of the development of boundary layers. K. Pohlhausen (1921) applied the method to several cases, using a polynomial approximation for the velocity distribution. E. Pohlhausen (1921) obtained the solution for forced convection in the boundary layer. Tollmien (1924) investigated the growth of the boundary layer on a circular cylinder impulsively set in rotation from rest. Burgers (1925) reported on experimental observations of the velocity distribution across the boundary layer on a flat plate, bringing to light the simultaneous presence of laminar and turbulent regions. The experiments, carried out by Burgers's student van der Hegge Zijnen (Burgers & van der Hegge Zijnen 1924) using a hot-wire anemometer, formed an important achievement in boundary-layer research, not only because it was the first experimental investigation on the subject but also because it was the first direct observation on the boundary layer itself. Up to that time, every experimental result was indirect, inferred from the overall aspects of the flow. For example, Prandtl's explanation (1914) of the critical Reynolds number of a sphere was based on the observation of a reduction in resistance at fairly low Reynolds numbers by inducing turbulence with a wire hoop fixed on a sphere. No direct observation was made on the boundary layer except smoke photographs of flow separation. Of course, to a limited extent measurements had been made before on the velocity of fluid in the neighborhood of the solid wall (Calvert 1893, Kempf 1913, Riabouchinskii 1914), but none of these had been systematic enough to afford a deeper understanding of boundary-layer flows.

Thus, in the second decade, the number of papers (six) was still about the same as in the first decade, but the interest in boundary layers had now spread out of Göttingen. With this momentum, boundary-layer theory became the subject of widespread attention and acceptance. A brief reference to boundary-layer theory appeared in Lamb's *Hydrodynamics* (1924). A suggestion was offered by Mises (1927) regarding the use of the stream function as one of the independent variables, so that the boundary-layer equation was reduced to a form analogous to the heat-conduction equation. In a remark on this paper, Prandtl (1928) expressed his pleasure in observing interest in boundary-layer theory spread outside of his group. He also stated on another occasion (1938) that he had used the same form in 1914 to apply the boundary-layer equation to flow through a two-dimensional channel, the boundary condition at the two walls being simply expressed in terms of the stream function. The result having been unpublished, however, he thought that "the priority in a usual sense should be due to Mises" (translated).

There are seemingly good reasons for the slow acceptance of the boundary-layer theory: favorable growth was hampered by the war; the first paper of Prandtl was so very short and published where no one could appreciate it; most of the practical requirements were more concerned with the gross aspects of the flow like the force and moment, rather than the local structure of the flow, etc. However, none of these reasons is convincing enough. It appears to the writer that the most essential reason is that Prandtl's idea was so much ahead of the times.

Sowing Seeds

The third aspect that makes Prandtl's paper extraordinary is the presentation of subjects to be pursued further, thus forming the source of some lines of subsequent developments in boundary-layer theory. The above-mentioned investigations due to Blasius (1908), Boltze (1908), and Hiemenz (1911) exemplified the outgrowth of the earlier developments. There are other subjects, however, that had to wait for solution until the arrival of new students.

One of the subjects posed in Prandtl's paper is concerned with the algebraic singularities affecting the numerical analysis, in which the velocity profiles are to be calculated starting with a given profile at the initial station, the pressure distribution having been prescribed. The singularities arise from the no-slip condition $u = 0$ at the wall. The problem was first taken up by Goldstein during his stay in Göttingen as a guest. Goldstein (1930) considered, among other things, the conditions to be satisfied by the initial profile for the absence of singularities. This provided a basis for devising a numerical step-by-step method of solution of the boundary-layer equations (Prandtl 1938, Görtler 1939, Hartree 1939).

Another subject posed is concerned with the possibility of controlling the boundary layer. Prandtl's paper contains an experimental demonstration of preventing separation by removing boundary-layer fluid by suction. It was only after World War I, however, that extensive experiments were carried out in Göttingen with the aim of practical utilization of boundary-layer control (Prandtl 1925a, 1927a; Ackeret 1925, 1926; Schrenk, 1928, 1931, 1935). The remaining subject is not concerned with the boundary layer itself, but is closely associated with it. By using the Helmholtz vortex theorem, Prandtl arrived at the result that in a region of closed streamlines in which the vorticity has been established by the action of very small viscosity, the vorticity should be uniform in a two-dimensional flow. The vorticity should be proportional to the radial distance from the axis of symmetry in an axially symmetric flow. After the lapse of half a century the problem was considered in detail by Batchelor (1956), but without reference to Prandtl. It is perfectly astonishing to find the seeds of subjects ranging from basic to practical problems sown in the soil of a single paper.

Genesis of Boundary-Layer Theory

Having received a degree at Munich in 1900 after writing a thesis on the buckling of a deep beam, Prandtl joined Maschinenfabrik Augsburg-Nürnberg as a mechanical engineer. His interest in fluid mechanics was awakened when he arranged a conical diffuser in a large air duct but failed to achieve the expected pressure recovery. The conical angle seems to have been a little too large, resulting in flow separation from the diffuser wall. At that time (1901), however, Prandtl had to leave Nürnberg to receive an appointment as professor of mechanics at the Technische Hochschule Hannover. The loss of pressure was of no serious concern to the engineering works, but the question as to why and how the flow separated from the wall occupied Prandtl's inquiring mind until, after three years, his concept of the boundary layer provided him with the answer. In 1904 he accepted Felix Klein's invitation to take charge of the newly established chair of applied mechanics at the University of Göttingen, which he held until his retirement in 1947, six years before he died.

It is interesting to learn that Prandtl's failure in diffuser design caused him to reflect seriously on the matters underlying the phenomenon. The accidental flow separation gave impetus to the concept of the boundary layer. On the occasion of his being elected an honorary member of the German Physical Society, Prandtl gave a talk entitled *My road to hydrodynamical theories* (Prandtl 1948). In response to Heisenberg's congratulatory address containing the statement that Prandtl had the ability to see through equations, without calculation, what solution they may possess, he said, "I should reply that I admit having no such ability, but I endeavor to gain as clear a conception as possible about the matters forming the basis of the problem and seek to understand the course of events. The equations do not come up until later, when I believe to have understood the problem: they are useful not only to produce quantitative information which cannot certainly be obtained by conception alone, but also to afford good means of adducing proofs for my conclusions, thus winning recognition from others" (translated). In the same talk he also formulated the heuristic principle of solution underlying the boundary-layer theory expressed by the following (translated): "When the complete problem appears hopeless in mathematics, it is advisable to examine what takes place if the relevant parameter of the problem is made to tend to zero. It is thereby assumed that the problem admits an exact solution when the parameter is set equal to zero from the beginning and that a simplified approximate solution is possible for very small values of the parameter. At the same time it is necessary to examine whether, in the limit as the parameter tends to zero, the solution tends to the solution for the case when the parameter is set equal to zero. The boundary conditions must be chosen in order that this is the case. As to the physical trustworthiness of the solution, the classical proposition '*Natura non facit saltus*' gives the guiding principle: in nature the parameter is possibly small, but not equal to zero. Thus, the first way is always the physically correct one."

Spread of Boundary-Layer Theory

As already mentioned, the boundary-layer theory spread very slowly but steadily from Göttingen to other groups within its country of origin and then to other countries of the world. The diffusion of the theory was considerably facilitated by the appearance in the third decade of Prandtl's book *Abriss der Strömungslehre* (1931), Tollmien's article in *Handbuch der Experimentalphysik* (1931), and Prandtl's article in *Aerodynamic Theory* (1935).

In the meanwhile, the concept of the boundary layer turned out to be remarkably fruitful, not only in forming the basis for approximate methods of calculation of practical utility but also in offering clarification of phenomena that were otherwise incomprehensible or at least obscure, thus exerting an enormous, far-reaching influence. It is no exaggeration to say that it paved the way for all modern developments in fluid mechanics. The concept, originally developed for laminar flow along a solid boundary, has been extended to the corresponding case of turbulent flow and also to boundary-free shear flows occurring in wakes and jets. Along with these extensions, the stability of laminar flow was examined as a possible key to understanding the origin of turbulence. According to Dryden's statistics (1955), there were in the third decade five to six papers per year on boundary layers.

The fourth decade produced Goldstein's *Modern Developments in Fluid Dynamics* (1938) and World War II. Goldstein's volumes received a widespread welcome as the most timely compendium of the existing knowledge, making an important contribution to the diffusion of boundary-layer theory. World War II apparently did not check the development, although it confined the spread of information largely to the country of origin. Many of the results of investigations completed during the war remained unknown in other countries until much later. Dryden's statistics indicate that there were in the fourth decade about fourteen papers per year on boundary layers. Before the impetus of the original idea was exhausted in dealing with a laminar boundary layer in incompressible fluids, some of the effort was turned to examining the effects of compressibility, in response to the requirements of high-speed flight. This trend continued to the subsequent decade, in which attention was further directed to real-gas effects. It may safely be said that the boundary-layer theory found its happy hunting ground in the field of aeronautical engineering. The expansion of aeronautical activity stimulated basic research on boundary layers. Along with these developments, the boundary-layer concept began to pervade other fields of engineering—mechanical engineering, chemical engineering, etc. On the other hand, studies of heat and mass transfer in moving fluids were greatly facilitated by the knowledge of boundary layers. These developments have created an almost exponential growth of interest in boundary-layer theory in recent years.

The remainder of this article is devoted to a brief historical review of the major branches of the subject in boundary-layer theory, more detailed for classical branches but less detailed for extended branches. Because of limitations of space and time, however, the period of coverage is restricted to about six decades after the birth of boundary-layer theory, that is, up to about the end of the 1960s. No attempt at an exhaustive survey is made, and the references are quoted sometimes merely by way of example.

DEVELOPMENTS IN MAJOR BRANCHES

Steady Two-Dimensional Laminar Boundary Layers

The form of similarity solution introduced by Prandtl (1905) and Blasius (1908) for flow on a flat plate was extended by Falkner & Skan (1930) to the case in which the free-stream velocity varies in proportion to x^m, representing irrotational flow in a corner formed by two plane boundaries meeting at an angle $\pi/(m+1)$. Subsequent studies by Hartree (1937) and Stewartson (1954) revealed nonuniqueness of the solution for negative values of m. The series solution initiated by Blasius (1908) and Hiemenz (1911) for flow past a blunt-nosed cylinder of arbitrary cross section was extended by Howarth (1934) and Görtler (1952, 1957). The series solution for a linearly retarded free stream was considered by Howarth (1938) and extended to the more general case by Tani (1949). Series solutions for flow past a parabolic cylinder (Van Dyke 1964b) and for flow past a blunt-nosed wedge (Chen et al 1969) are worthy of mention as rare examples provided with convergence considerations. The boundary-layer approximation was also applied by Goldstein (1933) to flow in a wake, and by Schlichting (1933a) to flow in a jet.

The approximate method of solution of Kármán and Pohlhausen (Kármán 1921, K. Pohlhausen 1921), based on the momentum integral equation and a quartic form of the velocity profile, was found to give good results in nonretarded flow but less satisfactory in the retarded region, as first noticed in Schubauer's experimental observations (1935) on flow past an elliptic cylinder. Almost immediately, the approximate method of Kármán & Millikan (1934), in which the boundary layer was divided into inner and outer regions with separate solutions, was applied by Millikan (1936) to Schubauer's ellipse with reasonable success. Attempts were subsequently made to secure improved accuracy of the method of Kármán and Pohlhausen by assuming a more adequate form of the velocity profile (Walz 1941, Mangler 1944, Timman 1949), or by using another integral relation in addition to the momentum integral equation (Wieghardt 1948, Loitsianskii 1949, Truckenbrodt 1952, Tani 1954). Along with these efforts an approximate method of integrating the momentum integral equation was suggested independently by Walz (1941), Hudimoto (1941), Tani (1941), and Thwaites (1949), yielding the relation now commonly referred to as the Thwaites formula. An improved version of the method of Kármán and Millikan was put forward by Stratford (1954).

The numerical solution of Hartree (1939) for a linearly retarded free stream suggested the presence of a singularity at the point of separation, where the wall shear stress vanishes. This led Goldstein (1948) to construct a singular solution containing an arbitrary constant in the neighborhood of separation. Stewartson (1958) reconsidered the problem and obtained the more general solution involving an infinite number of arbitrary constants. Terrill (1960) extended Stewartson's work to include suction and also gave a numerical solution for irrotational flow past a circular cylinder, which again suggested the presence of a singularity. It must be borne in mind that the singularity disappears when all the arbitrary constants are set equal to zero, while the only evidence of a singularity comes from the numerical investigations for prescribed pressure distributions. Without any evidence to the contrary, one may infer that the singular solution would be the most reasonable representation of which the boundary-layer equations are capable and that the solutions of the Navier-Stokes equations would exhibit an abrupt but regular approach to separation. Landau & Lifshitz (1959) gave a discussion on flow near separation by postulating that the normal component of velocity tends to infinity at the separation point.

Unsteady Two-Dimensional Laminar Boundary Layers

The growth of the boundary layer on a body set impulsively from rest into trans-lational motion, first studied by successive approximations (a series in time) by Blasius (1908), was extended by Goldstein & Rosenhead (1936) for a better estimate of the time required for separation, which occurs at the rear stagnation point for a circular cylinder. The method of analysis fails, however, in the case of a semi-infinite flat plate, transition from the time-dependent Rayleigh solution to the space-dependent Blasius solution occurring by way of an essential singularity (Stewartson 1951), which, of course, originates in the use of the boundary-layer approximation. Proudman & Johnson (1962) considered the flow near the rear stagnation point and showed that at large times there is an inner boundary layer of reversed flow.

The flow generated by the small-amplitude oscillation of a body in a fluid at rest was also studied by successive approximations (a series in amplitude), first by Rayleigh (1883) in connection with acoustic phenomena in the Kundt tube but without recourse to the boundary-layer concept, and later by Schlichting (1932) with a boundary-layer formulation. The important result that appears in the second approximation is the occurrence of a steady streaming in addition to the oscillatory flow components. Moore (1951) considered the case in which a semi-infinite flat plate moves with a gradually changing but arbitrary time-dependent velocity. Another problem of practical importance, in which the free-stream velocity exhibits a small-amplitude fluctuation in magnitude, was initiated by Lighthill (1954) and extended by Rott & Rosenzweig (1960) and Lam & Rott (1962). It was pointed out by Rott, Moore, and Sears (Rott 1956, Moore 1958) that the criterion of vanishing wall shear stress does not in general denote flow separation from the wall in the case of unsteady motions.

Three-Dimensional Laminar Boundary Layers

The extension of two-dimensional boundary-layer theory to flows with axial symmetry was considered first by Boltze (1908) and later by Millikan (1932). It was found independently by Stepanov (1947), Mangler (1948), and Hatanaka (1949) that the problem of an axially symmetric boundary layer on a body of revolution can be reduced to that of an equivalent two-dimensional flow past a cylinder. Glauert & Lighthill (1955) and Stewartson (1955) independently investi-gated the flow at large distances downstream on the outside of a long circular cylinder, where the thickness of the boundary layer is no longer small compared with the radius of the cylinder. The extension of boundary-layer calculation was made by Taylor (1950) and Cooke (1952) to include swirling motion and by Illingworth (1953) and Schlichting (1953) to include rotation of the body.

Flow past a yawed infinite cylinder was considered independently by Prandtl (1945b), Struminskii (1946), Jones (1947), and Sears (1948). This class of flows has the useful feature that the velocity components in planes normal to the generators of the cylinder can be determined independently of the velocity component parallel to the generators. Results of calculation illustrated the deviation of flow in the boundary layer from the direction of the free stream, a characteristic behavior of three-dimensional boundary layers. Formulation of the boundary-layer equations in curvilinear coordinates was given for flow over a general three-dimensional surface (Howarth 1951, Hayes 1951, Watson 1963). The approximate method of solution of Kármán and Pohlhausen was extended to three dimensions by Timman & Zaat (1955), Eichelbrenner & Oudart (1955b), and Cooke (1959). When the streamlines outside the boundary layer have small geodesic curvature, choice of the projections of those streamlines on the solid surface as a family of streamwise coordinate lines causes the velocity component in the crosswise direction to be small. This simplification leads to an equation for the streamwise velocity component that is analogous to that for axially symmetric flows and independent of the crosswise velocity component (Eichelbrenner & Oudart 1955b).

In two-dimensional flows separation occurs at, or very close to, the point where the wall shear stress vanishes, and if it is considered as a three-dimensional flow, there is a separation line of singularities. In truly three-dimensional flows, however, the wall shear stress has two components, and the concept of lines of wall shear stress or limiting streamlines (limits of streamlines as the wall is approached) is found to be useful. Both components of wall shear stress simultaneously vanish, in general, only at isolated singular points, which are either nodal or saddle points of the topographical pattern of limiting streamlines, while flow separation occurs along a line on which the parallel component of wall shear stress is not everywhere zero. Maskell (1955) and Eichelbrenner & Oudart (1955a) defined the separation line as the envelope of the limiting streamlines, but Lighthill (1963) proposed a more comprehensive definition of the separation line as a limiting streamline that issues from both sides of a saddle point of separation and, after embracing the body, disappears into a nodal point of separation.

Instability and Transition to Turbulence

The fact that the laminar-flow solutions of the Navier-Stokes equations are not observed at high Reynolds numbers brought out the question of the stability of flow, in particular, the question as to the existence of infinitesimal disturbances growing with time. Rayleigh (1880) examined the stability of a plane parallel flow by neglecting viscosity and showed that a necessary condition for instability is that the velocity profile has a point of inflection. Stability theory for viscous fluids was formulated by Orr (1907) and Sommerfeld (1909), but calculations (Hopf 1914) indicated complete stability when applied to plane Couette flow generated by parallel walls in relative motion. As regards the effect of viscosity, Prandtl (1921) pointed out its dual role in stabilizing by dissipating energy, but destabilizing by producing phase lags in a layer close to the wall, as illustrated by Tietjens's calculation (1925). Heisenberg (1924) showed that plane Poiseuille flow between parallel walls at rest becomes unstable at high Reynolds numbers, but the result was too incomplete to gain general acceptance. Tollmien (1929) considered the Blasius velocity profile near a flat plate and obtained the critical Reynolds number above which the flow becomes unstable to a traveling-wave type of disturbances in a certain frequency range. Schlichting (1933b) extended Tollmien's calculation to amplified disturbances. Squire (1933) reduced the problem of three-dimensional disturbances of a plane parallel flow to that of equivalent two-dimensional disturbances at a lower Reynolds number, enabling the theory to concentrate attention on two-dimensional disturbances when calculating the critical Reynolds number. Attempts at experimental verification of stability theory met with only little success (Prandtl 1933, Nikuradse 1933b, Schiller 1934). On the other hand, Dryden's experiments (1936) indicated that transition to turbulence in the flow near a flat plate originates in the turbulence of the free stream. Taylor (1936) postulated that transition results from momentary separation of the boundary layer caused by the fluctuating pressure gradient of the free-stream turbulence, and some of the conclusions on the overall aspects were confirmed by measurements of Dryden, Schubauer, Mock & Skramstad (1937) and Hall & Hislop (1938).

In spite of the notable achievement in surmounting mathematical difficulties, Tollmien's theory was disregarded for more than a decade until Schubauer & Skramstad (1943) observed transition preceded by slow oscillations, of the kind predicted by theory, in the boundary layer on a flat plate in a wind tunnel of very weak turbulence. The characteristics of the oscillations agreed so well with the predicted values that the theory was regarded as proven in every particular. Only a little later, Liepmann (1943) independently made a similar observation. It was clear that high levels of free-stream turbulence typical of earlier experiments had masked the existence of amplified waves. Lin (1945) improved Heisenberg's approach and obtained the boundary of neutral stability for plane Poiseuille and Blasius flows. Shen (1954) extended Lin's method of solution to amplified disturbances.

Squire's theorem does not hold for curved flows, where three-dimensional disturbances may grow due to the destabilizing effect of the centrifugal force. Disturbances take the form of cellular toroidal vortices in circular Couette flow between rotating cylinders when the rotation of the inner cylinder dominates, for which Taylor (1923) found extremely close agreement between theoretical prediction and experimental observation. Another example is provided by streamwise vortices produced in the boundary layer on a concave wall (Görtler 1940). Instability similar in form also occurs in a horizontal boundary layer heated from below (Jeffreys 1928), with buoyancy as the destabilizing agent.

Amplification of infinitesimal disturbances is but a prelude to the whole process of transition. Passage to the subsequent stage occurs as a result of disturbances of increased amplitude giving rise to nonlinear interactions. Stability theory including nonlinear effects, first stated by Landau (1944) and developed by Meksyn & Stuart (1951), Stuart (1958, 1960), and Watson (1960, 1962), was successfully applied to circular Couette flow for predicting the equilibrium of disturbances under supercritical conditions. On the other hand, there is experimental evidence for the boundary layer on a flat plate (Klebanoff and associates 1959, 1962) that the nonlinear effect manifests itself as a nearly periodic variation in the spanwise direction of the amplitude of the initially two-dimensional Tollmien waves. The nonlinear theory has not yet gone far enough to deal with three-dimensional disturbances in slowly growing boundary-layer flows, although the observed phenomena were fairly well accounted for by another form of nonlinear theory (Benney & Lin 1960; Benney 1961, 1964) based on some debatable assumptions. The spanwise variation of wave amplitude generates locally unstable velocity profiles possessing an inflection point (Kovasznay, Komoda & Vasudeva 1962), bringing about a rapid collapse into eddies, until eventually random oscillations characteristic of turbulence burst forth in small localized "spots" (Emmons 1951, Schubauer & Klebanoff 1955). The turbulent spots grow as they travel downstream, until they merge into a fully developed turbulent flow. The evolution leading to the formation of turbulent spots in boundary-layer and channel flows is rather abrupt compared with the gradual evolution observed during the transition process in circular Couette flow dominated by rotation of the inner cylinder (Taylor 1923, Coles 1965) and also in boundary-free shear flows in wakes and jets (Sato & Kuriki 1961, Browand 1966).

Since the Tollmien waves were first observed at very low free-stream turbulence levels, it had been thought that at higher turbulence levels transition occurs without the precedence of instability oscillations. The experiments of Bennett (1953) suggested, however, that the evolution leading to transition is not basically different at least up to moderately high turbulence levels. On the other hand, there has been no experimental evidence of momentary separation at or prior to transition, raising some doubt as to the validity of Taylor's postulate (1936). Observations of Tani & Sato (1956) and Klebanoff (1966) indicated that instability oscillations are also induced by the presence of a two-dimensional roughness element.

Boundary-Free Turbulent Shear Flows

The effect of turbulent fluctuations in causing apparent stresses to operate on the mean motion, vaguely anticipated by Saint-Venant (1843), was assumed by Boussinesq (1877) to be simply equivalent to an increase in viscosity, thus introducing the concept of eddy viscosity. Reynolds (1895) showed that the correlations between fluctuating velocity components give rise to apparent stresses, which now bear his name. Taylor (1915), and independently Prandtl (1925b), expressed the Reynolds shear stress in terms of the mean velocity gradient and the mixing length, which represents the mean distance traveled by lumps of fluid before losing their identities. It was tacitly assumed in Prandtl's formulation that the momentum is a transferable property, while the transfer of vorticity formed the basis of Taylor's theory.

The mixing-length approach was first applied to boundary-free shear flows in jets, wakes, etc., with the assumption that the mixing length is constant across the shear layer and proportional to its width (Tollmien 1926, Schlichting 1930). Both transfer theories yielded the same result for the mean velocity profile, but the vorticity-transfer theory predicted the mean temperature profile in the wake of a heated cylinder in better agreement with experiments (Taylor 1932). Later, Prandtl (1942) found that a more satisfactory description for mean velocity is provided by assuming eddy viscosity to be constant across the shear layer. In this formulation, however, one must for heated wakes and jets take the eddy diffusivity for heat greater than the eddy viscosity for momentum (Corrsin 1943).

A striking feature of boundary-free shear flows is that the region of shear is bounded by a relatively sharp but irregularly meandering interface that separates the turbulent motion possessing vorticity fluctuations from the surrounding irrotational motion. This phenomenon was first discovered by Corrsin[4] (1943), and was thoroughly investigated by Townsend (1949, 1950, 1956) and Corrsin & Kistler (1954). In particular, Townsend visualized a double structure of flow consisting of the main body of turbulence having relatively small eddies, loosely termed turbulent fluid and containing most of the turbulent energy, and a superposed system of slowly moving large eddies, responsible for distorting interface and entraining nonturbulent fluid. He advanced the hypothesis that large eddies are gaining energy from the mean flow at nearly the same rate as they are losing energy to the small eddies. Townsend postulated large eddies with streamwise elongation on the basis of his own measurements of velocity correlations in a two-dimensional wake, although the subsequent more comprehensive measurements of Grant (1958) suggested a pair of counter-rotating eddies with axes nearly normal to the center plane of the wake, and planes of circulation roughly normal to the maximum strain rate. Attempts were also made to interpret the motion of large eddies as due to the instability of turbulent fluid (Liepmann 1952, 1962; Landahl 1967). Measurements on pressure correlations in a turbulent jet (Mollo-Christensen 1967) revealed the coherent structure of large eddies, much more coherent than the chaotic randomness that had been thought to be the case.

Wall-Bounded Turbulent Shear Flows

For flow through a two-dimensional channel or a circular pipe, it was found experimentally that in the central region the velocity defect relative to the maximum value at the center depends only on the relative distance from the center for a given wall shear stress. This velocity-defect law, first enunciated by Darcy (1858), was interpreted by Kármán (1930) as suggesting that the mechanism of turbulence is almost independent of viscosity. By postulating that the turbulent fluctuations in the neighborhood of any two points are similar, Kármán derived the velocity profile expressed by a logarithmic function of the distance from the wall. On the other hand, dimensional arguments on the basis of Nikuradse's measurements (1932) led Prandtl (1932) to the law of the wall, in which the velocity in the wall region depends only on the shear stress at the wall, the distance from the wall, and the kinematic viscosity. Except very close to the wall, the velocity profile was found to be logarithmic. Shortly later, Prandtl (1933) showed that the assumption of the mixing length proportional to the distance from the wall yields the logarithmic velocity profile. It is important to note that the regions of validity of the velocity-defect law and the law of the wall overlap. Izakson (1937) and Millikan (1939) independently found that the logarithmic velocity profile is the direct outcome of the existence of a region of overlap, without need for any specific assumption on similarity or mixing length.

Adjacent to the wall the flow is principally viscous, forming a region called the viscous sublayer. The significance of the role of this layer in relation to heat transfer

[4] Corrsin's discovery was made in a subsonic turbulent jet. It is interesting to find that the irregular interface of a wake, clearly visible on the schlieren picture of a projectile in supersonic flight (for example, C. Cranz, 1927, *Lehrbuch der Ballistik*, Vol. 3, 2nd ed., Berlin), had remained unnoticed for so many years.

was noticed by Prandtl (1910) and Taylor (1916). Effects of wall roughness were also discussed with consideration for this layer (Nikuradse 1933a). Measurements of Laufer (1953) indicated that much of the turbulent energy is generated just outside the viscous sublayer. Einstein & Li (1956) visualized an inherently unsteady sublayer, periodically building up and disintegrating. The detailed mechanism involved, however, had not been made clear until Kline and associates (1959, 1967) visually observed the formation of low-speed streaks, which lift up and burst into ejection of low-momentum fluids into the fast-moving outer region.

For flow in boundary layers the velocity profile near the wall was found to be unaffected by the pressure gradient, following the law of the wall of the same form as for pipe flows (Ludwieg & Tillmann 1949). On the other hand, the flow in the outer region resembles more the boundary-free shear flows, and the similarity of the form of the velocity-defect law holds only for a particular type of pressure gradient (Rotta 1950, Clauser 1954). A breakthrough from a practical viewpoint was made by Coles (1956), who described the departure of the velocity profile from the law of the wall by a universal function, which has been called the law of the wake.

An approximate method of predicting boundary-layer growth was initiated by Kármán (1921) on the basis of the momentum integral equation and the velocity profile assumed by reference to pipe flows. The method was extended by taking account of pressure gradients and by employing additional equations (Buri 1931, Gruschwitz 1931, Doenhoff & Tetervin 1943, Head 1958, Rotta 1962, Walz 1966). The difficulty of extending these integral methods to wider classes of flows, coupled with the advent of high-speed computers, has turned general attention toward differential methods, in which the momentum equation is integrated numerically with an eddy viscosity or mixing-length hypothesis. Having recognized the conceptual weakness of the mixing-length formulation in which the eddy viscosity was equal to the product of the mixing length squared and the mean velocity gradient, Prandtl (1945a) made an improved proposal to take the eddy viscosity as the product of the mixing length and the root-mean-square velocity fluctuation, the latter of which was determined from the energy equation of fluctuating motion. This antedated by two decades the upsurge of interest in computing turbulent shear flows (Glushko 1965; Bradshaw, Ferriss & Atwell 1967; Nee & Kovasznay 1969). As noticed by Batchelor (1950), however, the eddy-viscosity hypothesis relating the shear stress directly to the local mean velocity gradient is physically sound provided there is energy equilibrium, a condition only roughly fulfilled in turbulent shear flows in the light of measurements by Townsend (1949), Laufer (1953), and Klebanoff (1954).

Boundary Layers in Compressible Fluids

The introduction of compressibility into boundary-layer theory was first stated by Busemann (1931). The density is now variable and related to pressure and temperature by the state equation of perfect gases, while the temperature is governed by the energy equation of the form simplified by the boundary-layer approximation. Calculations were made for flow on a flat plate by Busemann (1935), Kármán & Tsien (1938), Wada (1944), Crocco (1946), and Chapman & Rubesin (1949) by specifying the variation of viscosity with temperature but assuming Prandtl number as constant. The results showed a marked increase of the boundary-layer thickness and the temperature near the wall with increase of Mach number of the free-stream velocity. In the meanwhile, attempts were made to transform the equation for a compressible boundary layer into that for an incompressible boundary layer by confining attention to the flow of an idealized fluid, for which Prandtl number is unity and the viscosity is proportional to temperature, along a thermally insulated wall. Transformation of the normal coordinate was first introduced by Dorodnitsyn (1942) and independently by Howarth (1948) to correlate compressible and incompressible boundary layers in zero pressure gradient. This was followed by Illingworth (1949) and Stewartson (1949), who arrived independently at the transformation of both normal and streamwise coordinates for correlation in the more general case of nonzero pressure gradient. The transformation threw open the resources of incompressible boundary-layer theory to the idealized, but by no means unrepresentative, class of compressible flows. It proved useful also for relaxing the idealizing conditions when combined with the approximate method of solution of Kármán and Pohlhausen. Along this line Tani (1954) extended his solution for incompressible fluids to the compressible flow of a more representative fluid, for which the Prandtl number is slightly different from unity and the viscosity varies with temperature according to the Sutherland formula. Poots (1960) extended the solution to include heat transfer at the wall.

The stability of compressible boundary layers was first considered by Lees & Lin (1946) and followed by Lees (1947), Dunn & Lin (1955), Lees & Reshotko (1962), and Mack (1965). Two important results brought about by theory are that the boundary layer could be stabilized by sufficient cooling of the wall (Lees 1947) and that there could be more than one mode of instability, the mode of lowest frequency corresponding to the Tollmien wave for incompressible flows being less amplified than the higher mode at moderate supersonic speeds (Mack 1965).

In incompressible flows the pressure is determined by the velocity field so that the governing feature of turbulence is the fluctuating velocity field, or vorticity field. Compressibility brings in two more fields due to fluctuating pressure and temperature. Chu & Kovasznay (1958) considered small-amplitude fields in a homogeneous flow and found that interaction could be expected to second order in amplitude, the most interesting being the generation of a pressure field from vorticity-vorticity interaction and a vorticity field from temperature-pressure interaction. These correspond to sound generation by turbulence (Lighthill 1952) and vorticity generation by density gradient (Bjerknes 1898), respectively. However, it may be inferred (Morkovin 1962, Laufer 1968) from experimental results on turbulent boundary layers at moderate supersonic speeds that compressibility does not appear to add any substantial source of vorticity, suggesting that the basic mechanism differs little from that for incompressible flows. This afforded a basis for attempts at using a transformation to correlate compressible and incompressible turbulent boundary layers (Mager 1958, Coles 1964).

New phenomena are observed when the boundary layer interacts with the shock wave, which have no counterpart in incompressible flows. For example, when a shock wave impinges on a boundary layer, the pressure rise across the shock wave tends to be diffused in the boundary layer, making its effect felt some distance upstream of the point of impingement. If the shock wave is strong, the boundary layer separates, which in turn reacts upon the formation of the shock wave. The interaction is more spectacular when the boundary layer is laminar (Ackeret, Feldmann & Rott 1946; Liepmann 1946). The second example is provided by hypersonic flow over a flat plate with a sharp leading edge, where a falling pressure gradient is induced by the interaction of the thick boundary layer with the shock wave originating near the leading edge (Becker 1950, Lees & Probstein 1952, Lees 1953). The third example is offered by hypersonic flow near the stagnation point of a blunt-nosed body, where the boundary layer is influenced by vorticity and entropy gradients produced by the shock wave (Hayes & Probstein 1959). Besides these dynamical effects due to high Mach numbers, hypersonic considerations should include the real-gas effects associated with high temperatures, such as ionization, dissociation, and radiation.

Higher Approximations

We recall that Prandtl's boundary-layer theory yields the first approximation to the solution of the Navier-Stokes equations near the solid wall in the limit of small viscosity or large Reynolds number. The approximation having changed the type of the equation and reduced its order, difficulties can be expected to arise when attempts are made to improve on it. Prandtl himself (1935) suggested the possibility of improving the solution for flow on a flat plate by correcting for the effect of displacement thickness. Subsequently, various authors considered the effects of wall curvature, external vorticity, downstream disturbance, etc., in particular cases, bringing about more or less sporadic, but sometimes controversial, results (Van Dyke 1969). It was only in the 1950s that systematic studies were made by Lagerstrom and his associates to establish Prandtl's approximation as the basis of an asymptotic solution of the Navier-Stokes equations, leading to what is now known as the method of matched asymptotic expansions (Kaplun 1954, 1967; Lagerstrom & Cole 1955; Van Dyke 1962, 1964a). The basic idea is to construct two asymptotic expansions, outer and inner expansions, by iterating the Navier-Stokes equations about the inviscid solution and about the boundary-layer solution, respectively, and to match the two expansions in their overlap region of validity.

However, the inviscid solution is not unique for given boundary conditions, and it is difficult in general to select the relevant one that is the limit of the solution of the Navier-Stokes equations. For flow past a certain semi-infinite or streamlined body, one may expect that there is no separation and take the irrotational motion as the relevant inviscid solution. For flow past a bluff body involving separation, the relevant inviscid solution is unknown. Higher approximations have thus been found only for flows without separation. In such cases the first term of the outer expansion is the inviscid irrotational flow, from which the first term of the inner expansion is determined by Prandtl's approximation. The second term of the outer expansion is the irrotational flow due to an apparent source distribution representing the displacement effect of Prandtl's boundary layer. This then determines a correction to the boundary-layer solution, yielding the second term of the inner expansion. As a typical example one may cite the solution to the second approximation by Van Dyke (1964b) for flow past a parabolic cylinder. The wall shear stress was found to be reduced near the stagnation point by both displacement and curvature effects. Calculations were also carried out to higher approximation for flow over a semi-infinite flat plate (Imai 1957; Goldstein 1960; Libby & Fox 1963; Murray 1965, 1967). Contrary to Prandtl's expectations (1935), the second-order displacement effect vanishes and undetermined constants remain that depend on the details of flow near the leading edge, where the boundary-layer approximation fails. For a finite flat plate Kuo (1953) obtained a nonzero second-order displacement correction. Subsequently, however, a slightly more important correction was discovered by Stewartson (1969) and Messiter (1970), originating from a triple-deck structure near the trailing edge.

Even for moderately high Reynolds numbers the second-order correction to the boundary-layer solution is very small so that its calculation is mainly of theoretical interest. It is remarkable, however, to find that the concept of boundary-layer theory was extended and generalized to the method of matched asymptotic expansions, opening the way for treating singular perturbation problems for differential equations. Thus the ideas underlying the boundary-layer theory have been applied to sciences other than fluid mechanics and, in fluid mechanics, to problems other than those associated with small viscosity.

Flow with Separation

When separation occurs, one may not use the inviscid irrotational flow as a basis for setting up a uniformly valid solution, the situation being made even more difficult by the inevitable turbulence and large-scale unsteadiness resulting from instability at high Reynolds numbers. Thus, knowledge of flow with separation has been drawn mostly from experiments. One of the unknown elements is the relevant inviscid solution, although the free-streamline solution due to Helmholtz (1868) and Kirchhoff (1869) is still a likely candidate. When used for high but finite Reynolds numbers, however, the free-streamline solution predicts too small resistance, and various modifications have been suggested for a better description of flow around the body (Zhukovskii 1890, Riabouchinskii 1920, Gilbarg & Rock 1945, Roshko 1955, Woods 1955, Wu 1962).

Theoretical studies of interaction between boundary layer and free stream began first in supersonic flows, where the separated region is more or less localized so that its features depend only on the local properties of the flow, exhibiting what is called free interaction (Chapman, Kuehn & Larson 1957). Such a situation occurs, for example, when a shock wave impinges on the boundary layer. For this problem Lighthill (1953) divided the boundary layer into two layers, treating the outer as virtually inviscid and providing a pressure gradient, and the inner as virtually incompressible and producing changes in displacement thickness. A more detailed calculation along this line was carried out by Gadd (1957) to obtain the pressure distribution across the region of separation induced by the shock wave. Another method of approximate calculation based on the use of integral relations was developed by Lees & Reeves (1964). The method was also applied to the wake behind a bluff body or a backward-facing step (Reeves & Lees 1965). Korst (1956) and Chapman, Kuehn & Larson (1957) independently put forward a simple theory for flow past a concave wall with leading-edge separation by dividing the flow field into a recirculating region, in which the pressure is nearly constant, and a reattachment region, in which the total pressure is nearly constant along the dividing streamline.

In subsonic flows the problem is made rather difficult by the elliptic nature of the flow, although there are some classes of flows in which the separation is localized. In most subsonic flows past a body, however, separation occurs so catastrophically that the problem has so far not given way to theoretical treatment. In fact, there has been little new advance in the theory except Kármán's stability consideration (Kármán 1911, Kármán & Rubach 1912) of the vortex street in the wake of a two-dimensional bluff body. Particular mention should be made of the experimental investigations of Fage & Johansen (1928), Kovasznay (1949), Roshko (1953), Taneda (1959), and Gerrard (1966) as giving insight into the structure of flow downstream of a bluff body.

Literature Cited

Ackeret, J. 1925. *Z. Flugtechn. Motorluftschiff.* 16:44–52
Ackeret, J. 1926. *Z. Ver. Dtsch. Ing.* 70: 1153–58
Ackeret, J., Feldmann, F., Rott, N. 1946. *Mitt. Inst. Aerodyn. Zürich No. 10*
Bassett, A. B. 1888. *A Treatise on Hydrodynamics.* Cambridge: The University Press
Batchelor, G. K. 1950. *J. Aeronaut. Sci.* 17:441–45
Batchelor, G. K. 1956. *J. Fluid Mech.* 1: 177–90
Becker, J. V. 1950. *J. Appl. Phys.* 21:619–28
Bennett, H. W. 1953. *Rep. Kimberly-Clark Corp., Neenah, Wisconsin*
Benney, D. J. 1961. *J. Fluid Mech.* 10:209–36
Benney, D. J. 1964. *Phys. Fluids* 7:319–26
Benney, D. J., Lin, C. C. 1960. *Phys. Fluids* 4:656–57
Bjerknes, V. 1898. *Videnskabsselskabets Skrifter, Kristiania*
Blasius, H. 1908. *Z. Math. Phys.* 56:1–37
Boltze, E. 1908. Göttingen dissertation
Boussinesq, J. 1877. *Mém. Prés. Par Div. Sav. Acad. Sci. Paris* 23:46
Bradshaw, P., Ferriss, D. H., Atwell, N. P. 1967. *J. Fluid Mech.* 28:593–616
Browand, F. K. 1966. *J. Fluid Mech.* 26: 281–307
Burgers, J. M. 1925. *Proc. Int. Congr. Appl. Mech., 1st, Delft, 1924,* pp. 113–28
Burgers, J. M., van der Hegge Zijnen, B. G. 1924. *Verh. Akad. Wetensch. Amsterdam*

Ser. I 13: No. 3
Buri, A. 1931. Zürich dissertation
Busemann, A. 1931. In *Handbuch der Experimentalphysik,* ed. W. Wien, F. Harms, 4(I):341–460. Leipzig: Akademische
Busemann, A. 1935. *Z. Angew. Math. Mech.* 15:23–25
Calvert, G. A. 1893. *Trans. Inst. Nav. Archit.* 34:61–77
Chapman, D. R., Kuehn, D. M., Larson, H. K. 1957. *NACA Tech. Note No. 3869*
Chapman, D. R., Rubesin, M. W. 1949. *J. Aeronaut. Sci.* 16:547–65
Chen, K. K., Libby, P. A., Rott, N., Van Dyke, M. D. 1969. *Z. Angew. Math. Phys.* 20:919–27
Chu, B. T., Kovasznay, L. S. G. 1958. *J. Fluid Mech.* 3:494–514
Clauser, F. H. 1954. *J. Aeronaut. Sci.* 21: 91–108
Coles, D. 1956. *J. Fluid Mech.* 1:191–226
Coles, D. 1964. *Phys. Fluids* 7:1403–23
Coles, D. 1965. *J. Fluid Mech.* 21:385–425
Cooke, J. C. 1952. *J. Aeronaut. Sci.* 19:486–90
Cooke, J. C. 1959. *Aeronaut. Res. Coun., Rep. Mem. No. 3201*
Corrsin, S. 1943. *NACA Wartime Rep. W-94*
Corrsin, S., Kistler, A. L. 1954. *NACA Tech. Note No. 3133*
Crocco, L. 1946. *Monogr. Sci. Aeronaut. No. 3*
Darcy, H. 1858. *Mém. Savants Etrangers* 15:265–342
Doenhoff, A. E. von, Tetervin, N. 1943.

108   TANI

NACA Tech. Rep. No. 772
Dorodnitsyn, A. A. 1942. *Prikl. Mat. Mekh.* 6:449–85
Dryden, H. L. 1936. *NACA Tech. Rep. No. 562*
Dryden, H. L. 1955. *Science* 121:375–80
Dryden, H. L., Schubauer, G. B., Mock, W. C., Skramstad, H. K. 1937. *NACA Tech. Rep. No. 581*
Dunn, D. W., Lin, C. C. 1955. *J. Aeronaut. Sci.* 22:455–77
Eichelbrenner, E. A., Oudart, A. 1955a. *Publ. Off. Natl. Etude Rech. Aéron. No. 47*
Eichelbrenner, E. A., Oudart, A. 1955b. *Publ. Off. Natl. Etude Rech. Aéron. No. 76*
Eiffel, G. 1912. *C. R. Acad. Sci. Paris Ser. A* 155:1597–99
Einstein, H. A., Li, H. 1956. *Proc. Am. Soc. Civ. Eng.* 82: Paper No. 945
Emmons, H. W. 1951. *J. Aeronaut. Sci.* 18: 490–98
Fage, A., Johansen, F. C. 1928. *Philos. Mag. Ser.* 7 5:417–41
Falkner, V. M., Skan, S. W. 1930. *Aeronaut. Res. Coun., Rep. Mem. No. 1314*
Froude, W. 1869. *Rep. Br. Assoc. Adv. Sci.,* pp. 211–14
Froude, W. 1872. *Rep. Br. Assoc. Adv. Sci.,* pp. 118–24
Froude, W. 1874. *Rep. Br. Assoc. Adv. Sci.,* pp. 249–55
Gadd, G. E. 1957. *J. Aeronaut. Sci.* 24:759–71
Gerrard, J. H. 1966. *J. Fluid Mech.* 25: 401–13
Gilbarg, D., Rock, D. H. 1945. *Mem. US Nav. Ordnance Lab. No. 8718*
Glauert, M. B., Lighthill, M. J. 1955. *Proc. R. Soc. London Ser. A* 230:188–203
Glushko, G. S. 1965. *Bull. Acad. Sci. USSR, Mech. Ser. No. 4,* pp. 13–23
Goldstein, S. 1930. *Proc. Cambridge Philos. Soc.* 26:1–30
Goldstein, S. 1933. *Proc. R. Soc. London Ser. A* 142:545–62
Goldstein, S. 1938. *Modern Developments in Fluid Dynamics.* Oxford: Clarendon
Goldstein, S. 1948. *Q. J. Mech. Appl. Math.* 1:43–69
Goldstein, S. 1960. *Lectures on Fluid Mechanics.* New York: Interscience
Goldstein, S. 1969. *Ann. Rev. Fluid Mech.* 1:1–28
Goldstein, S., Rosenhead, L. 1936. *Proc. Cambridge Philos. Soc.* 32:392–401
Görtler, H. 1939. *Z. Angew. Math. Mech.* 19:129–40
Görtler, H. 1940. *Nachr. Ges. Wiss. Göttingen, Math.-phys. Kl.,* pp. 1–26
Görtler, H. 1952. *Z. Angew. Math. Mech.* 32:270–71

Görtler, H. 1957. *J. Math. Mech.* 6:1–66
Grant, H. L. 1958. *J. Fluid Mech.* 4:149–90
Gruschwitz, E. 1931. *Ing. Arch.* 2:321–46
Gümbel, L. 1913. *Jb. Schiffbautechn. Ges.* 14:393–509
Hall, A. A., Hislop, G. S. 1938. *Aeronaut. Res. Coun., Rep. Mem. No. 1843*
Hartree, D. R. 1937. *Proc. Cambridge Philos. Soc.* 33:223–39
Hartree, D. R. 1939. *Aeronaut. Res. Coun., Rep. Mem. No. 2426*
Hatanaka, H. 1949. *Rep. Inst. Sci. Technol., Tokyo Univ.* 3:115–17
Hayes, W. D. 1951. *Rep. US Nav. Ordnance Lab. No. 1313*
Hayes, W. D., Probstein, R. F. 1959. *Hypersonic Flow Theory.* New York: Academic
Head, M. R. 1958. *Aeronaut. Res. Coun., Rep. Mem. No. 3152*
Heisenberg, W. 1924. *Ann. Phys. Leipzig Ser.* 4 74:577–627
Helmholtz, H. v. 1868. *Monatsber. Königl. Akad. Wiss. Berlin,* pp. 215–28
Hiemenz, K. 1911. *Dinglers J.* 326:321–24, 344–48, 357–62, 372–76, 391–93, 407–10
Hopf, L. 1914. *Ann. Phys. Leipzig Ser.* 4 44:1–60
Howarth, L. 1934. *Aeronaut. Res. Coun., Rep. Mem. No. 1632*
Howarth, L. 1938. *Proc. R. Soc. London Ser. A* 164:547–79
Howarth, L. 1948. *Proc. R. Soc. London Ser. A* 194:16–42
Howarth, L. 1951. *Philos. Mag. Ser.* 7 42:239–43
Hudimoto, B. 1941. *J. Soc. Aeronaut. Sci. Jpn.* 8:279–82
Illingworth, C. R. 1949. *Proc. R. Soc. London Ser. A* 199:533–57
Illingworth, C. R. 1953. *Philos. Mag. Ser.* 7 44:389–403
Imai, I. 1957. *J. Aeronaut. Sci.* 24:155–56
Izakson, A. 1937. *Tech. Phys. USSR* 4:155–62
Jeffreys, H. 1928. *Proc. R. Soc. London Ser. A* 118:195–208
Jones, R. T. 1947. *NACA Tech. Rep. No. 884*
Kaplun, S. 1954. *Z. Angew. Math. Phys.* 5:111–35
Kaplun, S. 1967. *Fluid Mechanics and Singular Perturbations,* ed. P. A. Lagerstrom, L. N. Howard, C. S. Liu. New York: Academic
Kármán, Th. v. 1911. *Nachr. Ges. Wiss. Göttingen, Math.-phys. Kl.,* pp. 509–17
Kármán, Th. v. 1921. *Z. Angew. Math. Mech.* 1:233–52
Kármán, Th. v. 1930. *Nachr. Ges. Wiss. Göttingen, Math.-phys. Kl.,* pp. 58–76

Kármán, Th. v., Millikan, C. B. 1934. *NACA Tech. Rep. No. 504*

Kármán, Th. v., Rubach, H. 1912. *Phys. Z.* 13:49–59

Kármán, Th. v., Tsien, H. S. 1938. *J. Aeronaut. Sci.* 5:227–32

Kempf, G. 1913. *Jb. Schiffbautechn. Ges.* 14:504–5

Kirchhoff, G. 1869. *Crelle J. Math.* 70:289–98

Klebanoff, P. S. 1954. *NACA Tech. Note No. 3178*

Klebanoff, P. S. 1966. *Proc. Int. Congr. Appl. Mech., 11th, Munich, 1964*, pp. 803–5

Klebanoff, P. S., Tidstrom, K. D. 1959. *NASA Tech. Note D-195*

Klebanoff, P. S., Tidstrom, K. D., Sargent, L. M. 1962. *J. Fluid Mech.* 12:1–34

Kline, S. J., Reynolds, W. C., Schraub, F. A., Runstalder, P. W. 1967. *J. Fluid Mech.* 30:741–73

Kline, S. J., Runstalder, P. W. 1959. *Trans. ASME, J. Appl. Mech.* 26:166–70

Korst, H. H. 1956. *Trans. ASME, J. Appl. Mech.* 23:593–600

Kovasznay, L. S. G. 1949. *Proc. R. Soc. London Ser. A* 198:174–90

Kovasznay, L. S. G., Komoda, H., Vasudeva, B. R. 1962. *Proc. Heat Transfer Fluid Mech. Inst., Stanford, 1962*, pp. 1–26

Kuo, Y. H. 1953. *J. Math. Phys.* 32:83–101

Lagerstrom, P. A., Cole, J. D. 1955. *J. Ration. Mech. Anal.* 4:817–82

Lam, S. H., Rott, N. 1962. *Proc. Int. Congr. Appl. Mech., 10th, Stresa, 1960*, p. 239

Lamb, H. 1924. *Hydrodynamics*. 5th ed. Cambridge: The University Press

Lanchester, F. W. 1907. *Aerodynamics*. London: Constable

Landahl, M. T. 1967. *J. Fluid Mech.* 29:441–59

Landau, L. D. 1944. *Dokl. Akad. Nauk SSSR* 44:311–14

Landau, L. D., Lifshitz, E. M. 1959. *Fluid Mechanics*. Oxford: Pergamon

Laufer, J. 1953. *NACA Tech. Note No. 2953*

Laufer, J. 1968. In *Compressible Turbulent Boundary Layers*, ed. M. H. Bertram, pp. 1–13. *NASA SP-216*

Lees, L. 1947. *NACA Tech. Rep. No. 876*

Lees, L. 1953. *J. Aeronaut. Sci.* 20:143–45

Lees, L., Lin, C. C. 1946. *NACA Tech. Note No. 1115*

Lees, L., Probstein, R. F. 1952. *Rep. Dep. Aeronaut. Eng., Princeton Univ., No. 195*

Lees, L., Reeves, B. L. 1964. *J. Am. Inst. Aeronaut. Astron.* 2:1907–20

Lees, L., Reshotko, E. 1962. *J. Fluid Mech.* 12:555–90

Libby, P. A., Fox, H. 1963. *J. Fluid Mech.* 17:433–49

Liepmann, H. W. 1943. *NACA Wartime Rep. W-107*

Liepmann, H. W. 1946. *J. Aeronaut. Sci.* 13:623–37

Liepmann, H. W. 1952. *Z. Angew. Math. Phys.* 3:321–42, 407–26

Liepmann, H. W. 1962. In *Mécanique de la Turbulence*, ed. A. Favre, pp. 211–27. Paris: CNRS

Lighthill, M. J. 1952. *Proc. R. Soc. London Ser. A* 211:564–87

Lighthill, M. J. 1953. *Proc. R. Soc. London Ser. A* 217:478–507

Lighthill, M. J. 1954. *Proc. R. Soc. London Ser. A* 224:1–23

Lighthill, M. J. 1963. In *Laminar Boundary Layers*, ed. L. Rosenhead, pp. 1–45. Oxford: Clarendon

Lin, C. C. 1945. *Q. Appl. Math.* 3:117–42, 218–34, 277–301

Loitsianskii, L. G. 1949. *Prikl. Mat. Mekh.* 13:513–24

Loitsianskii, L. G. 1970. *Mechanics of Liquid and Gas*. Moscow: Nauka. (In Russian)

Lorenz, L. 1881. *Ann. Phys. Leipzig Ser. 2* 13:422–47, 582–606

Ludwieg, H., Tillmann, W. 1949. *Ing. Arch.* 17:288–99

Mack, L. M. 1965. In *Methods in Computational Physics*, ed. B. Alder, 4:247–99. New York: Academic

Mager, A. 1958. *J. Aeronaut. Sci.* 25:305–11

Mangler, W. 1944. *Z. Angew. Math. Mech.* 24:251–56

Mangler, W. 1948. *Z. Angew. Math. Mech.* 28:97–103

Maskell, E. C. 1955. *Aeronaut. Res. Coun., Rep. No. 18063*

Meksyn, D., Stuart, J. T. 1951. *Proc. R. Soc. London Ser. A* 208:517–26

Messiter, A. F. 1970. *SIAM J. Appl. Math.* 18:241–57

Millikan, C. B. 1932. *Trans. ASME, Appl. Mech. Sect.* 54:29–43

Millikan, C. B. 1936. *J. Aeronaut. Sci.* 3:91–94

Millikan, C. B. 1939. *Proc. Int. Congr. Appl. Mech., 5th, Cambridge, Mass., 1938*, pp. 386–92

Mises, R. v. 1927. *Z. Angew. Math. Mech.* 7:425–31

Mollo-Christensen, E. 1967. *Trans. ASME, J. Appl. Mech.* 34:1–7

Moore, F. K. 1951. *NACA Tech. Note No. 2471*

Moore, F. K. 1958. In *Grenzschichtforschung*, ed. H. Görtler, pp. 296–311. Berlin: Springer

Morkovin, M. V. 1962. In *Mécanique de la Turbulence*, ed. A. Favre, pp. 367–80.

Paris: CNRS
Murray, J. D. 1965. *J. Fluid Mech.* 21:337–44
Murray, J. D. 1967. *J. Math. Phys.* 46:1–20
Navier, C. L. M. H. 1823. *Mém. Acad. R. Sci. Paris* 6:389–416
Nee, V. M., Kovasznay, L. S. G. 1969. *Phys. Fluids* 12:473–84
Nikuradse, J. 1932. *Ver. Dtsch. Ing., Forschungsheft No. 356*
Nikuradse, J. 1933a. *Ver. Dtsch. Ing., Forschungsheft No. 361*
Nikuradse, J. 1933b. *Z. Angew. Math. Mech.* 13:174–76
Orr, W. M. F. 1907. *Proc. R. Irish Acad. Ser. A* 27:9–27, 69–138
Pohlhausen, E. 1921. *Z. Angew. Math. Mech.* 1:115–21
Pohlhausen, K. 1921. *Z. Angew. Math. Mech.* 1:252–68
Poisson, S. D. 1831. *J. Ec. Polytech. Paris* 13:139–66
Poots, G. 1960. *Q. J. Mech. Appl. Math.* 13:57–84
Prandtl, L. 1905. *Verh. Int. Math. Kongr., 3rd, Heidelberg, 1904,* pp. 484–91. Transl. 1928. *NACA Memo. No. 452*
Prandtl, L. 1910. *Phys. Z.* 11:1072–78
Prandtl, L. 1913. In *Handwörterbuch der Naturwissenschaften,* 4:101–40. Jena: Fischer
Prandtl, L. 1914. *Nachr. Ges. Wiss. Göttingen, Math.-phys. Kl.,* pp. 177–90
Prandtl, L. 1921. *Z. Angew. Math. Mech.* 1:431–36
Prandtl, L. 1925a. *Naturwissenschaften* 13:93–108
Prandtl, L. 1925b. *Z. Angew. Math. Mech.* 5:136–39
Prandtl, L. 1927a. In *Ergeb. Aerodyn. Versuchsanst. Göttingen,* 3:6–9. Munich and Berlin: Oldenbourg
Prandtl, L. 1927b. *J. R. Aeronaut. Soc.* 31:720–43
Prandtl, L. 1928. *Z. Angew. Math. Mech.* 8:249–51
Prandtl, L. 1931. *Abriss der Strömungslehre.* Braunschweig: Vieweg
Prandtl, L. 1932. In *Ergeb. Aerodyn. Versuchsanst. Göttingen,* 4:18–29. Munich and Berlin: Oldenbourg
Prandtl, L. 1933. *Z. Ver. Dtsch. Ing.* 77:105–14
Prandtl, L. 1935. In *Aerodynamic Theory,* ed. W. F. Durand, 3:34–209. Berlin: Springer
Prandtl, L. 1938. *Z. Angew. Math. Mech.* 18:77–82
Prandtl, L. 1942. *Z. Angew. Math. Mech.* 22:241–43
Prandtl, L. 1945a. *Nachr. Akad. Wiss.*

Göttingen, Math.-phys. Kl., pp. 6–19
Prandtl, L. 1945b. In *Festschrift zum 60. Geburtstage von A. Betz,* pp. 134–41. Göttingen
Prandtl, L. 1948. *Phys. Blätter* 4:89–92
Prandtl, L. 1949. *Führer durch die Strömungslehre.* 3rd ed. Braunschweig: Vieweg
Proudman, I., Johnson, K. 1962. *J. Fluid Mech.* 12:161–68
Rankine, W. J. M. 1864. *Trans. Inst. Nav. Archit.* 5:316–33
Rayleigh, Lord 1880. *Proc. London Math. Soc.* 11:57–70
Rayleigh, Lord 1883. *Philos. Trans. R. Soc. London Ser. A* 175:1–21
Rayleigh, Lord 1911. *Philos. Mag. Ser. 6* 21:697–711
Reeves, B. L., Lees, L. 1965. *J. Am. Inst. Aeronaut. Astron.* 3:2061–74
Reynolds, O. 1895. *Philos. Trans. R. Soc. London Ser. A* 186:123–64
Riabouchinskii, D. 1914. *Bull. Inst. Aérodyn. Koutchino* 5:51
Riabouchinskii, D. 1920. *Proc. London Math. Soc.* 19:206–15
Roshko, A. 1953. *NACA Tech. Note No. 2913*
Roshko, A. 1955. *J. Aeronaut. Sci.* 22:124–32
Rott, N. 1956. *Q. Appl. Math.* 13:444–51
Rott, N., Rosenzweig, M. L. 1960. *J. Aeronaut. Sci.* 27:741–47
Rotta, J. C. 1950. *Mitt. Max-Planck-Inst. Strömungsforsch. Göttingen No. 1*
Rotta, J. C. 1962. In *Progress in Aeronautical Sciences,* ed. D. Küchemann, 2:1–219. Oxford: Pergamon
Saint-Venant, B. 1843. *C. R. Acad. Sci. Paris Ser. A* 17:1240–42
Sato, H., Kuriki, K. 1961. *J. Fluid Mech.* 11:321–52
Schiller, L. 1934. *Z. Angew. Math. Mech.* 14:36–42
Schlichting, H. 1930. *Ing. Arch.* 1:533–71
Schlichting, H. 1932. *Phys. Z.* 33:327–35
Schlichting, H. 1933a. *Z. Angew. Math. Mech.* 13:260–63
Schlichting, H. 1933b. *Nachr. Ges. Wiss. Göttingen, Math.-phys. Kl.,* pp. 181–208
Schlichting, H. 1953. *Ing. Arch.* 21:227–44
Schmidt, E., Beckmann, W. 1930. *Tech. Mech. Thermodyn.* 1:341–49, 391–406
Schrenk, O. 1928. *Luftfahrtforschung* 2:49–62
Schrenk, O. 1931. *Z. Flugtechn. Motorluftschiff.* 22:259–64
Schrenk, O. 1935. *Luftfahrtforschung* 12:10–27
Schubauer, G. B. 1935. *NACA Tech. Rep. No. 527*
Schubauer, G. B., Klebanoff, P. S. 1955.

NACA Tech. Note No. 3489
Schubauer, G. B., Skramstad, H. K. 1943. *NACA Wartime Rep. W-8*
Sears, W. R. 1948. *J. Aeronaut. Sci.* 15: 49–52
Shen, S. F. 1954. *J. Aeronaut. Sci.* 21: 62–64
Sommerfeld, A. 1909. *Atti Congr. Int. Math., 4th, Rome, 1908*, pp. 116–24
Squire, H. B. 1933. *Proc. R. Soc. London Ser. A* 142: 621–28
Stepanov, E. I. 1947. *Prikl. Mat. Mekh.* 11: 203–04
Stewartson, K. 1949. *Proc. R. Soc. London Ser. A* 200: 84–100
Stewartson, K. 1951. *Q. J. Mech. Appl. Math.* 4: 182–98
Stewartson, K. 1954. *Proc. Cambridge Philos. Soc.* 50: 454–65
Stewartson, K. 1955. *Q. Appl. Math.* 13: 113–22
Stewartson, K. 1958. *Q. J. Mech. Appl. Math.* 11: 399–410
Stewartson, K. 1969. *Mathematika* 16: 106–21
Stokes, G. G. 1845. *Trans. Cambridge Philos. Soc.* 8: 287–305
Stokes, G. G. 1851. *Trans. Cambridge Philos. Soc.* 9(II): 8–106
Stratford, B. S. 1954. *Aeronaut. Res. Coun., Rep. Mem. No. 3002*
Struminskii, V. V. 1946. *Dokl. Akad. Nauk SSSR* 54: 765–68
Stuart, J. T. 1958. *J. Fluid Mech.* 4: 1–21
Stuart, J. T. 1960. *J. Fluid Mech.* 9: 353–70
Taneda, S. 1959. *J. Phys. Soc. Jpn.* 14: 843–48
Tani, I. 1941. *J. Aeronaut. Res. Inst., Tokyo Imp. Univ., No. 199*
Tani, I. 1949. *J. Phys. Soc. Jpn.* 4: 149–54
Tani, I. 1954. *J. Aeronaut. Sci.* 21: 487–95
Tani, I., Sato, H. 1956. *J. Phys. Soc. Jpn.* 11: 1284–91
Taylor, G. I. 1915. *Philos. Trans. R. Soc. London Ser. A* 215: 1–26
Taylor, G. I. 1916. *Aeronaut. Res. Coun., Rep. Mem. No. 272*
Taylor, G. I. 1923. *Philos. Trans. R. Soc. London Ser. A* 223: 289–343
Taylor, G. I. 1932. *Proc. R. Soc. London Ser. A* 135: 685–96
Taylor, G. I. 1936. *Proc. R. Soc. London Ser. A* 156: 307–17
Taylor, G. I. 1950. *Q. J. Mech. Appl. Math.* 3: 129–39
Terrill, R. M. 1960. *Philos. Trans. R. Soc. London Ser. A* 253: 55–100

Thwaites, B. 1949. *Aeronaut. Q.* 1: 245–80
Tietjens, O. G. 1925. *Z. Angew. Math. Mech.* 5: 200–17
Timman, R. 1949. *Rep. Trans. Natl. LuchtvLab. Amsterdam* 15: F29–45
Timman, R., Zaat, J. A. 1955. In *50 Jahre Grenzschichtforschung*, ed. H. Görtler, W. Tollmien, pp. 432–45. Braunschweig: Vieweg
Tollmien, W. 1924. Göttingen dissertation
Tollmien, W. 1926. *Z. Angew. Math. Mech.* 6: 468–78
Tollmien, W. 1929. *Nachr. Ges. Wiss. Göttingen, Math.-phys. Kl.*, pp. 21–44
Tollmien, W. 1931. In *Handbuch der Experimentalphysik*, ed. W. Wien, F. Harms, 4(I): 243–87. Leipzig: Akademische
Töpfer, C. 1912. *Z. Math. Phys.* 60: 397–98
Townsend, A. A. 1949. *Proc. R. Soc. London Ser. A* 197: 124–40
Townsend, A. A. 1950. *Philos. Mag. Ser. 7* 41: 890–906
Townsend, A. A. 1956. *The Structure of Turbulent Shear Flow.* Cambridge: The University Press
Truckenbrodt, E. 1952. *Ing. Arch.* 20: 211–28
Unwin, W. C. 1888. In *Encyclopaedia Britannica*, 12: 459–535, 9th ed.
Van Dyke, M. 1962. *J. Fluid Mech.* 14: 161–77
Van Dyke, M. 1964a. *Perturbation Methods in Fluid Mechanics.* New York and London: Academic
Van Dyke, M. 1964b. *J. Fluid Mech.* 19: 145–59
Van Dyke, M. 1969. *Ann. Rev. Fluid Mech.* 1: 265–92
Wada, K. 1944. *Rep. Aeronaut. Res. Inst., Tokyo Imp. Univ., No. 302*
Walz, A. 1941. *Ber. Lilienthal Ges. Luftfahrtf. No. 141*, pp. 8–12
Walz, A. 1966. *Strömungs- und Temperaturgrenzschichten.* Karlsruhe: Braun
Watson, J. 1960. *J. Fluid Mech.* 9: 371–89
Watson, J. 1962. *J. Fluid Mech.* 14: 211–21
Watson, J. 1963. In *Laminar Boundary Layers*, ed. L. Rosenhead, pp. 411–15. Oxford: Clarendon
Wieghardt, K. 1948. *Ing. Arch.* 16: 231–42
Woods, L. C. 1955. *Proc. R. Soc. London Ser. A* 227: 367–86
Wu, T. Y. 1962. *J. Fluid Mech.* 13: 161–81
Zhukovskii, N. E. 1890. *Math. Collect. Moscow Math. Soc.* 15: 121–278
Zhukovskii, N. E. 1916. *Aérodynamique.* Paris: Gauthier-Villars

Ann. Rev. Fluid Mech. 1977. 9: 113–44

INCOMPRESSIBLE BOUNDARY-LAYER SEPARATION

※8099

James C. Williams, III
Department of Mechanical and Aerospace Engineering, North Carolina State University, Raleigh, North Carolina 27607

INTRODUCTION

For high-Reynolds-number flow over bodies or in confined channels the effects of viscosity are generally limited to a thin layer, the boundary layer, adjacent to the bounding surface. When the imposed pressure gradient is adverse, however, the thickness of the viscous layer increases as momentum is consumed by both wall shear and pressure gradient, and at some point the viscous layer breaks away from the bounding surface. Downstream of this point (or line) of breakaway the original boundary-layer fluid passes over a region of recirculating flow. The point at which the thin boundary layer breaks away from the surface and which divides the region of downstream-directed flow, in which the viscous effects are quite limited in extent, from the region of recirculating flow is known as the separation point.[1] Two different types of post-separation behavior are known to exist. In some cases the original boundary layer passes over the region of recirculating fluid and reattaches to the body at some point downstream, trapping a bubble of recirculating fluid beneath it. The characteristic length of this separation bubble may be of the same order as the upstream boundary-layer thickness or many times the boundary-layer thickness. In other cases, the original boundary-layer fluid never reattaches to the body but passes downstream, mixing with recirculating fluid, to form a wake. For this wake-type of separation the characteristic dimension of the recirculating region is generally of the same order as the characteristic body dimension. In either case, the recirculating flow alters the effective body shape and hence the inviscid flow about the body.

[1] For two-dimensional flow over fixed walls, the point of vanishing shear coincides with the point of separation. As a result, the point of vanishing shear has been taken for years as a significant indicator of separation. For two-dimensional flow over moving walls, two-dimensional unsteady flow, and three-dimensional steady flow, the point or line of vanishing wall shear does not coincide with separation, and a general definition of separation cannot include vanishing wall shear as a characteristic of separation.

Separation is the controlling, if not dominant, feature of many fluid flows. For the flow over bluff bodies the location of separation determines the pressure drag on the body. For airfoils and blades at high angle of attack the conditions at separation dictate the circulation about the airfoil or blade and hence the lift. For flows in confined passages, such as diffusers, separation frequently drastically alters the flow field and hence the performance of the device.

In spite of its fundamental importance, the complete analytical or experimental description of separation at high Reynolds numbers remains one of the main unsolved problems of fluid mechanics. It is still not possible to calculate the entire flow field including both the boundary layer and the recirculating region when both the recirculating region and the Reynolds number are large. The extent to which the recirculating region alters the pressure distribution upstream of separation is therefore not known. When the flow field is such that separation is followed by a small bubble of recirculating fluid and reattachment of the main boundary layer, it is possible to integrate the steady laminar boundary-layer equations through the points of separation and reattachment if the wall shear or displacement thickness is prescribed, while a similar calculation is terminated by a singularity at separation if the pressure gradient is prescribed. In most practical cases, however, transition occurs after separation and accurate calculations are hampered by a lack of models for turbulent reverse flow. The model for unsteady separation has recently been verified analytically for the case where the separation point moves forward along the body, but the case where the separation point moves aft or oscillates is still in question. These are but a few of the problems that must be resolved before separation is completely understood.

An enlightening and thorough review of the problem of separation was given by Brown & Stewartson (1969). In the intervening eight years there has been considerable progress in understanding the problem of separation, although it may be safely said that the problem has not been solved at this point. In the present review an attempt is made to describe some of the progress made in the past eight years. Attention is limited to incompressible flows, although much of the present knowledge regarding incompressible separation is directly applicable to compressible flows. Two-dimensional steady laminar separation is considered initially and followed by a discussion of two-dimensional unsteady laminar separation and three-dimensional steady laminar separation. No attempt is made to review the theory of asymptotic solutions of the Navier-Stokes equations for the flow in the vicinity of separation. This theory, as applied to the problem of separation, has been reviewed recently in some detail by Stewartson (1974). The emphasis is on laminar boundary-layer separation. As the opportunity arises some comments are made regarding turbulent separation. Our understanding of turbulent separation is, however, so primitive that a complete and concise discussion of this problem is not possible at this time.

TWO-DIMENSIONAL STEADY SEPARATION

If one attempts to solve the boundary-layer equations for a prescribed pressure-gradient variation, up to and including separation, by any reasonable means that determines the entire structure of the boundary layer (this automatically excludes most integral techniques), one quickly finds that as the point of zero shear is approached it becomes impossible to complete the calculation. It was just such a problem encountered by Hartree that prompted Goldstein (1948) to postulate a singular behavior in the boundary-layer solution and to obtain a solution of the boundary-layer equations in the vicinity of the point of zero shear that is indeed singular. In the intervening 29 years much has been made of this singularity. It is clear, however, that such a singular behavior is not a physical property of the flow but is a characteristic of the solutions of the boundary-layer equations. From an engineering point of view then the singularity is an obstacle we must over-come and not a feature of the flow to be studied extensively.

Numerical Solutions of the Navier-Stokes Equations

In recent years, as a result of the development of high-speed large-capacity computers coupled with the development of sophisticated numerical techniques, it has been possible to obtain a limited number of solutions to the Navier-Stokes equations for flows that include separation. In spite of the fact that such solutions are currently limited to low or moderate Reynolds numbers they yield valuable clues to the nature of separation.

The flows for which exact (numerical) solutions to the Navier-Stokes equations have been obtained may be divided into two classes according to the size of the recirculating region downstream of separation. The first class is comprised of those flows in which the recirculating region is large, typically of the same order of magnitude as the characteristic body dimension. Solutions for the flow past a circular cylinder obtained by Dennis & Chang (1970), Son & Hanratty (1969), Keller & Takami (1966), for the flow past a sphere obtained by Jenson (1959), and for the flow past spheroids obtained by Masliyah & Epstein (1970) belong to this class. While these investigations shed considerable light on the nature of the recirculating region downstream of separation at low or moderate Reynolds numbers, they do not address the problem of separation directly. As expected, there is no indication of a singular behavior at separation in any of these solutions. In fact, this is one of the advantages of obtaining solutions to the full Navier-Stokes equations. The solution obtained by Leal (1973) for a special form of a linearly retarded flow over a plate also belongs to this class of flows. Solutions were obtained by Leal for relatively high Reynolds numbers, up to 800 based on plate length, and a special effort was made to obtain a fine resolution of the flow field in the immediate vicinity of the separation point. Again, there is no evidence of a singular behavior at separation in this solution. In fact, Leal's results indicate that in the vicinity of separation the skin friction varies linearly with distance from the separation point as predicted by Dean (1950) and not as the square root of this distance, as would be the case if a Goldstein-type singularity were to exist. Leal also found that in the Navier-Stokes solution the streamwise pressure gradient is less than that for the undisturbed potential flow and, as a result, the position of separation is downstream of that predicted by classical boundary-layer theory. More importantly, this result tends to substantiate the possibility, suggested by the solution of Catherall & Mangler (1966), that in any real flow the pressure distribution may adjust, by means of an interaction between the viscous layer and the external flow, to be that required for regular behavior of the viscous flow at separation.

The second class is comprised of those flows in which the recirculating region downstream of separation is small. The numerical solutions of Briley (1971) for a separation bubble that occurs after separation in a linearly retarded flow and of Ghia & Davis (1974) for the flow over a family of parabolas and rectangular slabs belong to this class.

Briley was motivated by the need to obtain solutions for high-Reynolds-number problems in which there are small, localized regions in which the boundary-layer assumptions are questionable. He therefore studied the case in which a separation bubble occurs as a result of a linearly retarded flow which is relieved to a constant free-stream velocity (and pressure) flow just downstream of the expected separation point. The resulting flow contains a small separation bubble of limited streamwise extent which is embedded within the boundary layer. The flow upstream of the bubble is given by the classical Howarth solution for a linearly retarded flow while the flow far downstream of the bubble is required to satisfy the boundary-layer equations for zero pressure gradient. The flow far above the wall is assumed irrotational with a constant x component of velocity while the usual no-slip and impermeable wall conditions are imposed at the wall. The boundary conditions are thus prescribed on the entire boundary of the domain in which the bubble is embedded, as required by the elliptic nature of the Navier-Stokes equations. The numerical solutions obtained showed no evidence of singular behavior at separation and reasonable distributions of both wall shear and displacement thickness were obtained.

Ghia & Davis obtained numerical solutions to the full Navier-Stokes equations for the flow past thick semi-infinite plates with vertical leading surfaces (blunt slabs). They formulate the problem in a new coordinate system obtained by a conformal transformation of the Cartesian coordinates. The steady-state solutions for the vorticity and the stream function are obtained as the asymptotic solutions, for large time, of the unsteady governing equations. An interesting, and important, feature of the solutions that are obtained is the fact that, for all the bodies studied, the solution far downstream is the well-known Blasius solution. Solutions obtained for relatively blunt bodies show the skin friction decreasing to a negative value and then rising to the proper Blasius value, indicating a separation with no hint of a singularity and a reattachment with a bubble of separated recirculating fluid entrapped within the viscous layer. Ghia & Davis carried their investigation further by obtaining solutions for the same geometry bodies using a simpler set of equations termed the "parabolized" Navier-Stokes equations. The parabolized Navier-Stokes equations are obtained by dropping the streamwise diffusion of vorticity terms from the complete Navier-Stokes equations. The results obtained are in very good agreement with the corresponding solutions of the complete equations. This result has some fascinating implications. First, it would seem to indicate that there is a set of equations, intermediate between the full Navier-Stokes equations and the boundary-layer equations, that may be used to obtain accurate solutions in the region where boundary-layer theory applies and in the reverse-flow region between separation and reattachment. Second, and perhaps more importantly, Ghia & Davis point out that parabolizing the Navier-Stokes equations is equivalent to making the boundary-layer assumptions but accounting for displacement effects. Thus, it may be possible (in fact, it is possible, as is shown below) to employ the boundary-layer equations in the regions mentioned above if proper accounting is made of the displacement effects.

Recently Ghia, Ghia & Tesch (1975) have compared the solutions obtained using the boundary-layer equations and several intermediate sets of model equations, which are between the boundary-layer equations and the full Navier-Stokes equations, with the numerical solutions of the Navier-Stokes equations for the same class of bodies as investigated by Ghia & Davis. Ghia, Ghia & Tesch conclude that "boundary-layer-type models can successfully predict separated flows if they take into account the interaction between the boundary layer and the external inviscid flow."

It would be improper to leave the impression that all problems that involve boundary-layer separation at high Reynolds numbers may now be solved by simply obtaining solutions to the full Navier-Stokes equations. To obtain accurate numerical solutions for the flow field requires grid sizes sufficiently small to resolve the physical nonuniformities within the flow field. At high Reynolds numbers these nonuniformities are usually confined to relatively small regions so that the requirement of small grid size usually leads to a total number of grid points that is prohibitively large from the standpoint of computer storage and computational time, even with the largest computers available at present.

Boundary-Layer Solutions

The very fact that in some cases the recirculating flow downstream of separation has the same order of thickness as the upstream boundary layer leads to the speculation that the boundary-layer equations might suffice in the analysis of these flows, because the basic boundary-layer assumption of a thin viscous layer is not violated in such flows. If the boundary-layer equations are to be used in the analysis of such flows two problems must be overcome. First, some way must be found around the separation singularity that occurs when the boundary-layer equations are solved for a prescribed pressure gradient. Second, some way must be found to account for the upstream influence of the flow in the recirculating region. Is it possible to solve these two problems and thus employ the boundary-layer equations in the solution of this type of flow? The emerging answer appears to be yes.

The problem of the separation singularity has been a particularly thorny one. It must be remembered that Goldstein (1948) was led to the consideration of the question of a singularity as a result of a set of careful numerical calculations performed by D. R. Hartree. To quote Goldstein, "As a result of his computations, Professor Hartree was convinced that there was a singularity in the solution at the position of separation, and I undertook to try and find some formula which would hold near this singularity and would help in finishing this computation." Thus,

Goldstein was looking for a solution singular at separation; and he found it. Stewartson (1958) reconsidered the same problem and again concluded that such a singular solution is possible. In recent years, literally thousands of numerical solutions of the boundary-layer equations in which the main stream pressure gradient was prescribed have encountered such a singularity, which effectively terminates the solution before separation is reached. Goldstein also recognized, however, that the possibility exists "that a singularity will always occur except for certain special pressure variations in the neighborhood of separation, and that, experimentally, whatever we may do, the pressure variations near separation will always be such that no singularity will occur."

Now it is well known that even when separation does not occur on the body the potential-flow pressure gradient is altered by the presence of the boundary layer. The boundary layer, in effect, alters the shape of the solid body by an amount proportional to the boundary-layer displacement thickness. Thus, it seems reasonable to assume that the pressure gradient could be altered in the vicinity of separation, where this displacement effect is certainly significant; perhaps altered sufficiently to eliminate the singularity. This idea led Catherall & Mangler (1966) to obtain solutions to the boundary-layer equations by a technique in which the displacement thickness was specified in place of an imposed pressure gradient. Using this technique they were able to integrate numerically the boundary-layer equations through the separation point and into a region of reverse flow and shallow separation bubbles without any evidence of singular behavior at the separation point. This type of calculation is termed an inverse calculation since the imposed pressure gradient is obtained as part of the solution rather than being prescribed. Catherall & Mangler did, however, encounter problems with convergence of their numerical solution in the region of reverse flow downstream of separation. This difficulty was probably due to the fact that they used the same finite-difference equation in both the forward flow and reverse flow regions. This is the second of the two problems mentioned above and is discussed later.

Klineberg & Steger (1974) were also able to obtain a numerical solution that is regular at separation by specifying a regular distribution of wall shear rather than a regular distribution of displacement thickness through separation and reattachment. Again this is an inverse solution in which the pressure gradient is obtained as part of the solution. In addition to obtaining these solutions, Klineberg & Steger investigated in detail the structure of the boundary layer at both separation and reattachment and the requirements on the external pressure gradient that are necessary for a regular solution at separation to exist. They found that when solutions are obtained for a prescribed imposed pressure gradient the shear distribution contains a saddle point singularity at separation that would account for the numerical difficulties encountered in these calculations. Further, the requirement on the computed external pressure distribution for the flow to pass smoothly through separation is:

$$\frac{dm}{dx} \geqq 0 \quad \text{at} \quad \tau_w = 0,$$

where m is the usual pressure-gradient parameter $(x/u_\delta)(du_\delta/dx)$, u_δ is the velocity at the edge of the boundary layer, x is the physical coordinate along the body surface, and τ_w is the wall shear. Such a constraint on the allowable pressure gradient is certainly consistent with the concept of an interaction between the boundary layer and the external flow fluid. Klineberg & Steger point out that Meksyn (1950) "has contended that a minimum in dm/dx was a necessary condition for regular separation."

One could argue that the displacement-thickness distribution prescribed in the inverse solution of Catherall & Mangler (1966) or the wall-shear distribution prescribed by Klineberg & Steger (1974) is somewhat arbitrary. A most convincing demonstration that the inverse techniques do pass through separation in a regular manner in realistic flow situations was given by Carter (1975), who recalculated the exact solution of the Navier-Stokes equations given by Briley (1971) using the boundary-layer equations and each of the inverse techniques (displacement thickness prescribed or wall shear prescribed). In one case, Carter used the displacement thickness obtained by Briley and calculated the wall-shear and external-velocity distribution (or pressure distribution); in the other case, Carter used the wall shear obtained by Briley and calculated the displacement thickness. The results obtained by Carter when the wall-shear distribution is prescribed and the displacement thickness is calculated are presented in Figure 1; the results obtained when the

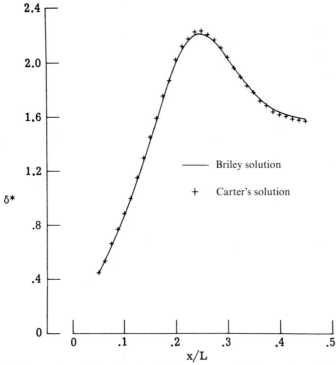

Figure 1 Comparison of Carter's calculated displacement thickness, for which the wall shear is prescribed, with Briley's exact solution. From Carter (1975).

displacement-thickness distribution is prescribed and the wall shear is calculated are presented in Figure 2. In each case, Carter's results are in excellent agreement with the solution of Briley. As an additional check, Carter also took as the prescribed condition the displacement thickness deduced by Klineberg & Steger (1974), who used a wall-shear-prescribed inverse technique, and calculated the corresponding wall-shear distribution. The wall shear deduced by Carter in this calculation is in excellent agreement with the wall shear prescribed by Klineberg & Steger.

It is clear from the above results that the problem of the separation singularity has been resolved. Regular solutions of the boundary-layer equations for separated flow are obtained when either the displacement-thickness distribution or the wall-shear distribution are prescribed. When the external-pressure distribution (or equivalently the external-velocity distribution) is prescribed, the solution of the boundary-layer equations leads to a singularity at separation.

The second of the problems mentioned above, that of accounting for the upstream influence in the region of reverse flow, has fortunately not been so difficult to resolve as the problem of the separation singularity was. The origin of this problem is demonstrated in the following formulation of the boundary-layer equations. Consider the boundary-layer equations for steady, incompressible, two-dimensional flow:

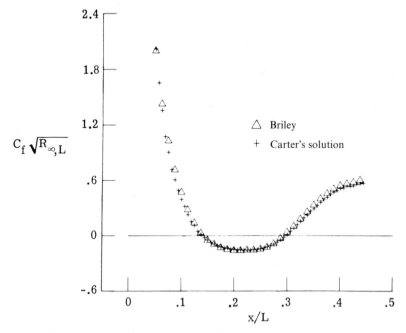

Figure 2   Comparison of Carter's calculated skin friction, for which the displacement thickness is prescribed, with Briley's exact solution. From Carter (1975).

$$\frac{\partial u}{\partial x} + \frac{\partial v}{\partial y} = 0, \tag{1}$$

$$u\frac{\partial u}{\partial x} + v\frac{\partial u}{\partial y} = u_\delta \frac{\partial u_\delta}{\partial x} + v\frac{\partial^2 u}{\partial y^2}, \tag{2}$$

where x and y are the spatial coordinates along and normal to the body surface, respectively, u and v are the corresponding velocity components, u_δ is the x component of velocity at the outer edge of the boundary layer, and v is the kinematic viscosity. A rather standard method of formulating these equations for solution is to introduce the stream function, ψ, defined by $u = \partial\psi/\partial y$, $v = -\partial\psi/\partial x$; to scale the x and y coordinates according to $\xi = x/l$, $\eta = y(u_\delta/vx)^{1/2}$, where ξ and η are the scaled x and y coordinates, respectively, and l is a characteristic dimension for the problem; and to introduce a scaled or normalized stream function, f, defined by

$$f(\xi, \eta) = \psi(x, y)/(u_\delta xv)^{1/2}.$$

The continuity equation is satisfied identically by the introduction of the stream function and the momentum equation becomes

$$f''' + (1 + M)ff''/2 + M(1 - f'^2) + \xi\left(f''\frac{\partial f}{\partial\xi} - f'\frac{\partial f'}{\partial\xi}\right) = 0. \tag{3}$$

Here M is the pressure-gradient parameter $M = (\xi/u_\delta)(du_\delta/d\xi)$ and primes denote differentiation with respect to η. For the purposes of the present demonstration it is convenient to rewrite Equation (3) in the form

$$f''' + \alpha_1 f'' + \alpha_2 f' + \alpha_3 = \alpha_4 \frac{\partial f'}{\partial\xi}, \tag{4}$$

where the α_i are the coefficients

$$\alpha_1 = (1 + M)f/2 + \xi\frac{\partial f}{\partial\xi}, \qquad \alpha_2 = -Mf',$$

$$\alpha_3 = M, \qquad\qquad\qquad \alpha_4 = \xi f'.$$

Equation (4) is the form of Equation (3) generally used as an intermediate step in arranging the boundary-layer equations for a numerical solution. More importantly, however, if f' is treated as the independent variable Equation (4) closely resembles the one-dimensional unsteady heat-conduction equation. In fact, if $\alpha_1 = \alpha_2 = \alpha_3 = 0$, Equation (4) has exactly the same form as the heat-conduction equation. The coefficient α_4 plays the same role in Equation (4) as the inverse of the thermal diffusivity $1/\alpha$ plays in the heat-conduction equation. In the heat-conduction equation, the thermal diffusivity (and therefore its inverse) is always positive, but in boundary-layer equations the coefficient α_4 will be negative in regions of reverse flow where f' is negative (provided the edge velocity is positive). If α_4 is always positive (no regions of reverse flow), Equation (4) is clearly parabolic and solutions are possible if appropriate boundary and initial conditions are prescribed. If a region of reverse flow is encountered in the problem, as occurs within a separation bubble, then Equation (4) is a parabolic equation of mixed type and the solution in the region of reverse flow requires a knowledge of the flow from downstream (in a global sense) as indicated by Brown & Stewartson (1969).

This problem has been solved by Klineberg & Steger (1974) and Carter (1975) by using backwards finite differencing in the regions of forward (downstream) flow but switching to forward differencing in the regions of reverse (upstream) flow. This allows for the proper flow of information in the region of reverse flow. Neither Klineberg & Steger nor Carter encountered any difficulty in integrating the boundary-layer equations in regions of reverse flow using this technique, provided the flow reattaches. Carter & Wornom (1975a) have shown, however, that a stable and accurate solution may be obtained in the region of reverse flow using a forward marching procedure if one neglects the streamwise vorticity-transport term in the boundary-layer form of the vorticity-transport equation, provided the reverse flow velocities are small.

An entirely different technique was employed by Klemp & Acrivos (1972) to insure the proper direction of the flow information in the reverse flow region. They divided the flow domain into two regions, an outer region in which the flow is always directed downstream and an inner region of reverse flow. In each region they used backwards differences with respect to the local flow direction, thus accounting for the downstream influence in the region of reverse flow. The interface between the two regions is, of course, not known at the outset. Klemp & Acrivos find this interface between the forward and reverse flow as part of the solution by iteration. The criterion used for determining the proper location of the interface is the continuity of shear across the interface. The specific problem solved, that of a finite stationary flat plate whose surface moves in a direction opposite the free stream at constant velocity, is unrealistic. As a result, singularities are present at the points of separation and reattachment because of discontinuities in the boundary condition at these points. Nevertheless, there is no a priori reason that this technique should not work in realistic flow problems, such as a separation bubble, if it were coupled with one of the techniques prescribed above for integrating through the separation point.

It now appears that solutions may be obtained within the framework of boundary-layer theory for flows in which laminar separation is followed by a shallow bubble of recirculating laminar flow and laminar reattachment. The problem of the separation singularity is eliminated by prescribing either the wall-shear or displacement thickness and the problem of integrating in regions of reverse flow may be handled by either using the appropriate differencing schemes in the region of reverse flow or integrating the equations in the appropriate direction locally. Several problems remain, however. First, although the separation singularity may be avoided by using inverse techniques, the required displacement thickness (or wall shear) to be used in the inverse techniques is not known a priori. The appropriate displacement thickness must be obtained as part of the overall problem from the interaction between the boundary layer and the inviscid flow. Calculations of laminar separation bubbles that include this viscous-inviscid interaction have been given by Carter & Wornom (1975b). Second, in most practical cases in which a separation bubble occurs the separation is laminar, but transition occurs within the bubble so that in the downstream portion of the bubble and at reattachment the boundary layer is transitional or turbulent. The techniques described above, which have proven successful in the case of wholly laminar separation bubbles, should also be applicable in this case where a portion of the bubble is transitional or turbulent. In fact, Briley & McDonald (1975) have made such a calculation, including both an approximate turbulence transport model and an approximate treatment of the interaction between the boundary layer and the inviscid flow. The results obtained are in good agreement with experimental measurements within such bubbles. The main difficulty in extending the laminar treatment to bubbles that include transitional or turbulent flow is the lack of experimentally verified turbulent transport models. Some attempts have been made to extend existing eddy-viscosity models for attached flow so that they might apply in regions of reverse flow [Alber (1971), Diewert (1975)], but the validity of these models remains to be verified.

Finally, a few words are in order with respect to the more practical problem of wholly turbulent separation. The calculation of a two-dimensional steady boundary layer requires the specification of a model for the momentum transport process. This model may be in the form of a turbulent-viscosity concept in which the turbulent or "eddy" viscosity is given by an algebraic equation or is obtained from a differential equation or in the form of one or more differential equations for such quantities as the turbulent kinetic energy or one or more of several length scales, or even the Reynolds stresses themselves [Launder & Spalding (1972)]. All of the turbulent transport models in use today have in common the fact that they are semi-empirical. As a result each model includes one or more empirical constants. In most cases these constants are evaluated so that the theory fits data for zero or mildly adverse pressure gradients. It is not surprising then that when applied to flows that lead to separation these theories yield only fair results [see for example the many calculations presented in Kline et al (1968)]. As separation is approached, some turbulent boundary-layer theories predict a singularity similar to that which occurs at laminar separation. This behavior deserves further investigation but because the singularity in laminar flow occurs at a point of zero shear at the wall, it may be that the same type of singularity exists at separation in the laminar sublayer of the turbulent boundary layer. In any event, it is clear that there is a need for more accurate models of the turbulent transport process near separation. In order to develop such models it will be necessary to enlarge the body of experimental data for turbulent boundary layers in the vicinity of separation.

UNSTEADY SEPARATION

For many years it was believed that unsteady laminar boundary-layer separation had the same characteristics as steady separation; that is, that unsteady separation was associated with the point of vanishing wall shear, the termination of the boundary layer, and the beginning of the wake or a bubble of "separated" fluid. Thus, many early investigations of the initial phases of unsteady flow over bodies tracked the point of zero shear and referred to this point as the separation point.

In 1956, however, Rott presented an analysis of the unsteady flow in the vicinity of a stagnation point and observed that for this flow, in which separation in the above sense is certainly not expected, the wall shear vanished with an accompanying flow reversal but without any hint of a singularity or a breakdown in the boundary-layer assumptions. Rott concluded that "evidently $\tau_w = 0$ does not necessarily mean separation for moving walls; in a system where the wall is at rest, the flow is not steady." This observation led a number of investigators to seek a generalized criterion for separation that would encompass both steady and unsteady flows.

In 1956 Sears proposed such a generalized model for separation, which postulates

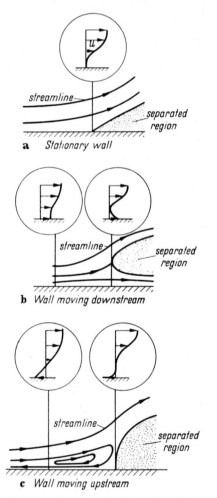

Figure 3 Moore's "expected streamlines and velocity profiles" for stationary and moving walls. From Moore (1958).

that the unsteady separation point is characterized by the simultaneous vanishing of the shear and the velocity at a point within the boundary layer *as seen by an observer moving with separation.* Moore (1958) independently arrived at the same model for unsteady separation on the basis of an investigation of steady flow over a moving wall. Moore was led to the conclusion that "for a slowly moving wall separation occurs when, at some point in the boundary layer, the profile velocity and shear simultaneously vanish." Figure 3 shows Moore's "expected streamlines and velocity profiles at separation for stationary wall and moving wall." On the basis of an intuitive relationship between steady flow over a moving wall and unsteady flow over a fixed wall, Moore was lead to postulate the velocity profiles for unsteady separation over a fixed wall that are shown in Figure 4. Moore also considered the possibility that a singularity of the Goldstein type occurs at the point where the velocity profile has simultaneously zero velocity and shear at a point above the moving wall. He concluded that such a singularity is in fact possible in the case of the downstream moving wall but for the upstream moving wall the results were inconclusive.

The investigations of Moore, Rott, and Sears have led to a model of unsteady separation, known as the Moore-Rott-Sears model, which postulates that the unsteady separation point is characterized by the simultaneous vanishing of the shear and velocity, in a singular fashion, at some point in the boundary layer, as

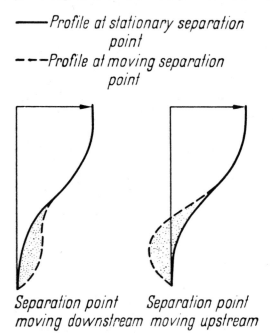

Figure 4   Moore's postulated velocity profiles for unsteady separation on a fixed wall. From Moore (1958).

seen by an observer moving with the separation point. Clearly such a model is a generalization of the earlier model of steady separation and includes steady separation as a special case.

It must be noted that the Moore-Rott-Sears model for the unsteady separation point has not been universally accepted. There are those who prefer to use the term "separation" to denote the point of vanishing wall shear and the term "break-down" or "breakaway" to denote the point where a singularity occurs, the boundary-layer thickness becomes unbounded, and boundary-layer theory breaks down. Only history will determine which terminology will finally be accepted. In the present review the Moore-Rott-Sears model is used and the term "separation" is understood to refer to the point where the boundary-layer equations break down due to a singularity in the solution and, as a result, the boundary-layer thickness becomes unbounded. Sears & Telionis (1975) have pointed out that there is some historical justification for this point of view. More importantly, however, the Moore-Rott-Sears model for unsteady separation contains the same elements, with the exception of vanishing shear at the wall, that have been taken as characteristics of steady separation for many years. In this respect, it is interesting to note that the separation singularity, which has proven such a problem in steady flow, has been instrumental in finding and identifying unsteady separation.

Steady Separation—Moving Walls

In recent years a number of attempts have been made to verify the Moore-Rott-Sears model for unsteady separation. Much of the interest in understanding unsteady separation arises from the importance of this phenomenon in dynamic stall or "lift hysteresis," which occurs in helicopter aerodynamics, and in rotating stall, which occurs in axial flow compressors.

The difficulty in either verifying this model or applying it in practice lies in the fact that the moving reference frame (moving with separation) is not, in general, known a priori. For this reason, early attempts to verify the model considered the related problem of steady separation over moving walls. The model, as applied to this problem, was verified experimentally by Vidal (1959) and Ludwig (1964), who investigated separation on a shrouded rotating cylinder in steady flow. Vidal showed that the separation was delayed (moved downstream) when the wall moved in the direction of the flow and was advanced (moved upstream) when the wall moved opposite the main flow. Ludwig made detailed velocity-profile measurements near separation on the cylinder and obtained velocity profiles that correspond to those postulated by Moore for both the upstream- and downstream-moving-wall cases. The model has been verified analytically by Telionis & Werle (1973) and Tsahalis & Telionis (1973), at least for the case of downstream moving walls. Telionis & Werle studied the case of flow over a parabola with a downstream moving wall and Tsahalis & Telionis studied the same problem but with the addition of blowing at the wall. In each of these studies it was shown that integration of the boundary-layer equations through a point of vanishing wall shear presents no special difficulties and that the solution is terminated in the vicinity of a station where the velocity and shear vanish in a singular fashion at a point within the boundary layer. This is taken as a verification of the Moore-Rott-Sears model for steady flow over downstream moving walls. The case of upstream moving walls presents special problems which are discussed later.

The relationship between unsteady boundary-layer separation over fixed walls and steady separation over moving walls has been, until quite recently, an intuitive one. Williams & Johnson (1974a) have shown that for the special case in which the external velocity distribution is a function of a linear combination of the spatial coordinate, x, and time, t, the unsteady boundary-layer flow over a stationary wall may be transformed into a steady boundary-layer flow over a wall moving with the speed of the unsteady separation point. Thus, the link between unsteady flow and flow over a moving wall was established in this case. A solution was obtained for a simple unsteady variation of Howarth's linearly retarded flow for which the unsteady velocity distribution in the external stream is given by $u_\delta = U_\infty(1 - Ax - Bt)$, where U_∞ is a constant velocity and A and B are constants. The results obtained show that the separation point is characterized by the simultaneous vanishing of both the shear and velocity at a point in the boundary layer in the flow "... seen by an observer moving with separation," and a singular behavior in the solution of the boundary-layer equations in the vicinity of this point. Again it should be noted that these results have been obtained only for the case where the separation point moves upstream or, correspondingly, when the wall moves downstream. In the case where the separation moves downstream, the corresponding steady-flow problem over a moving wall is one in which the wall moves upstream. In this case the u component of velocity is negative (upstream flow) near the wall and positive (downstream flow) in the upper part of the boundary layer, a situation that requires special techniques if a solution is to be obtained numerically.

Since it is now clear that there is a relationship between unsteady flow over fixed walls and steady flow over moving walls, it would appear that one way to at least partially verify the Moore-Rott-Sears model would be to seek similar solutions for moving-wall flows which yield the velocity profiles predicted by Moore. Danberg & Fansler (1975) have shown that velocity profiles corresponding to Moore's downstream-moving-wall model are indeed obtained as part of a family of similar velocity profiles. They also point out, however, that the profile postulated by Moore for separation over an upstream-moving wall is not a possible similar solution.

Tsahalis (1976), however, has obtained a solution for the case where the flow is steady with an adverse pressure gradient and the wall is moving in the direction opposite the main flow with a velocity proportional to the distance from the leading edge. The solution is obtained as an asymptotic solution, for large time, of the unsteady boundary-layer equations. The results obtained agree with the experimental results of Ludwig (1964) and indicate a singularity near the point of vanishing shear and velocity within the boundary layer, consistent with the Moore-Rott-Sears model. The assumed wall velocity, which depends on the distance from the leading edge, is somewhat awkward. Additional solutions for more realistic wall velocities, opposite the flow, would be helpful in understanding this problem of flow over upstream-moving walls.

Unsteady Separation—Fixed Wall

When one attempts to investigate unsteady separation by obtaining solutions of the unsteady boundary-layer equations, rather than investigating the related problem of steady flow over moving walls, three problems are encountered. First, the problem is one in three rather than two independent variables. While there are a variety of techniques available for solving boundary-layer problems in two dimensions, the techniques available for solving problems in three dimensions are quite limited. Second, since verification of the Moore-Rott-Sears model is expected, it is necessary to integrate the boundary-layer equations through the point of zero wall shear into regions of reverse flow for the case of an upstream-moving separation. It is generally acknowledged that integrations in regions of reverse flow are unstable, although, as is shown below, this is not necessarily the case. Third, the Moore-Rott-Sears model postulates that the separation point appears as a point of simultaneous vanishing of velocity and shear in a coordinate system moving with separation. The best way to identify this point is then in a moving coordinate system, but in a coordinate system whose motion is not known a priori.

The first solution of the full unsteady boundary-layer equations leading to separation was given by Telionis, Tsahalis & Werle (1973). They obtained a solution to the unsteady boundary-layer equations for an unsteady variation of Howarth's linearly retarded flow in which the adverse pressure gradient increases with time. The time-dependent pressure gradient was chosen so that at zero time it had one fixed value and thereafter increased with time to a new fixed value at a dimensionless time of 1.1. As a result of this choice steady separation occurred at $t = 0$ and $t = 1.1$. Between these two times, the separation point was determined by 1. a rapid increase in the number of iterations required to obtain convergence; and 2. a rapid increase in the normal or v component of velocity. As time increased from zero the separation point and the point of vanishing shear divided and moved forward along the plate with the separation point lagging the point of vanishing shear. At a dimensionless time $t = 1.1$ these two points again coalesced into a new steady-state separation point.

Telionis, Tsahalis & Werle employed a finite difference scheme that allowed them to solve the boundary-layer equations in two dimensions at each time step and special finite differencing techniques that allow for integration into the region of reverse flow beyond the point of vanishing shear. The fact that the number of iterations required for convergence of the solution increases rapidly and the vertical component of velocity apparently becomes unbounded in the vicinity of the point where the velocity has its maximum negative value is taken as evidence of the singularity postulated by Moore (1958) and Sears & Telionis (1975). It was not shown, however, that this point had zero velocity and shear in the coordinate system moving with separation, as postulated in the Moore-Rott-Sears model.

Additional verification of the Moore-Rott-Sears model, again only for the case of upstream-moving separation, has been given by Telionis & Tsahalis (1973a, 1973b). Telionis & Tsahalis (1973a) studied the case of a linearly retarded velocity distribution in which the adverse pressure gradient is impulsively increased by an increment at time $t = 0$. The boundary-layer equations were solved in the two spatial coordinates at each time step and "upwind differencing" was used in the region of the field where the velocity is reversed. The results indicate that the point of vanishing wall shear jumps abruptly to the leading edge of the plate at $t = 0+$ and then runs rapidly downstream, reverses course, and finally approaches a new steady-state location asymptotically. The separation point, on the other hand, remains relatively stationary at first and then moves upstream more rapidly and asymptotically approaches the new steady-state separation point. At time $t = 0-$ and at infinite time the separation point and the point of zero shear are of course coincident. Between these two times, the separation point lags behind the point of zero shear and the boundary layer downstream of the point of vanishing shear is thin and well behaved.

The unsteady boundary-layer development on a circular cylinder and on an elliptic cylinder at angle of attack impulsively set into motion has also been calculated by Telionis & Tsahalis (1973b). The numerical technique is essentially the same as that employed in their earlier work. The point of vanishing shear appears at the rear stagnation point at dimensionless time $t = 0.35$ and moves forward, with increasing time, asymptotically approaching $\theta = 104.5°$ at large time. From dimensionless time $t = 0.35$ until $t \cong 0.65$ the numerical integration is carried out beyond the point of vanishing shear to $\theta = 180°$ without any signs of a singularity. At $t \cong 0.65$ a singularity (separation) appears at approximately $\theta = 140°$ and thereafter moves forward along the cylinder, asymptotically approaching the point of zero shear at large time. The calculations for the elliptic cylinder show several regions of reverse flow that grow in magnitude with time and that appear prior to the appearance of any separation singularity. The calculations are terminated with the appearance of a separation singularity, which is interpreted as the initiation of a wake that would induce a circulation about the ellipse.

It should be noted that in the work of Telionis and his co-workers, discussed above, it is not shown directly that the separation singularity has all the properties postulated in the Moore-Rott-Sears model, that is, simultaneous vanishing of the shear and velocity in a coordinate system moving with separation. If however the solutions were transformed into a coordinate system moving with the separation singularity all the required properties would appear. The numerical methods employed by Telionis and his co-workers cannot, however, be applied to the case where the separation point moves rearward along the body. In this case the domain of integration is always expanding. If a solution is to be obtained at a time $t_1 + \Delta t$ (Figure 5) at an x station close to separation, say x_1, information would be required for the same x station at the previous time t_1. This point x_1, t_1 lies beyond the separation point at t_1 and thus no information is available to make the required calculations.

Additional verification of the Moore-Rott-Sears model for unsteady separation was given by Williams & Johnson (1974b, 1975). Again, however, the verification is only for cases in which the separation point moves upstream. Williams & Johnson employed the technique of semisimilar solutions, which scales the three independent variables into two, and worked in a coordinate system moving with separation. In the first of these papers Williams & Johnson studied an unsteady variation of Howarth's linearly retarded flow, and in the second they studied an unsteady variation of the Falkner-Skan velocity distributions, $u_\delta \sim x^m$. Since the equations of motion, after the semisimilar transformation, are of the same form as the nonsimilar, two-dimensional boundary-layer equations, it is possible to use rather standard implicit finite difference techniques to solve the equations. The numerical results obtained in the coordinate system moving with separation show that the point of simultaneous vanishing shear and velocity is approached in a singular fashion. Typical results obtained by Williams & Johnson (1974b) are shown in Figures 6 and 7. The velocity profiles in the coordinate system moving with separation are shown in Figure 6 for several values of ξ, the scaled (with time) x coordinate. The shear and velocity are seen to vanish simultaneously at a point just beyond $\xi = 0.221$. The corresponding velocity profiles in the body-fixed (stationary) coordinate system are shown in Figure 7.

It is instructive to consider the nature of the unsteady boundary layer within the framework of semisimilar solutions, for this formulation provides some insight into the problem of obtaining solutions for the case where separation moves downstream. In addition, this formulation indicates when it is possible to integrate into the region of reverse flow, beyond the point of vanishing wall shear, with conventional techniques. The method of semisimilar solutions is a mathematical technique for reducing the number of independent variables from three to two by an appropriate scaling of the x and y coordinates. In flows where separation is approached, this

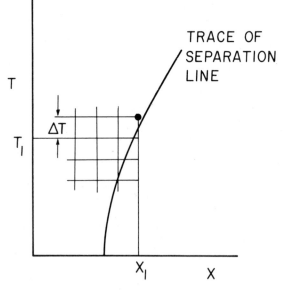

Figure 5   Finite-difference grid pattern in the $x-t$ plane in the case where separation moves rearward with time.

technique has a more important physical interpretation. It may be considered as a technique for stretching the physical coordinates x and y in such a manner that in the stretched x coordinate, ξ, separation always occurs at the same location. It is possible therefore to determine information on the motion of the separation point as the problem is being solved. Furthermore, in this new coordinate system the domain of integration is always fixed.

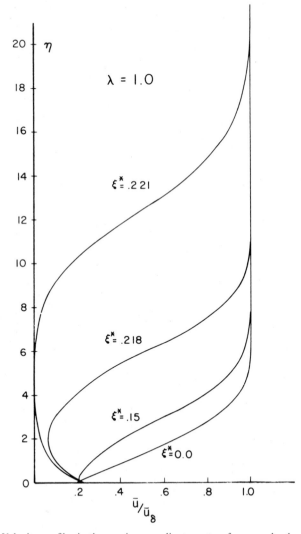

Figure 6 Velocity profiles in the moving coordinate system for several values of ξ. From Williams & Johnson (1974b).

The unsteady laminar boundary equations in two dimensions are:

$$\frac{\partial u}{\partial x} + \frac{\partial v}{\partial y} = 0, \tag{5}$$

$$\frac{\partial u}{\partial t} + u\frac{\partial u}{\partial x} + v\frac{\partial u}{\partial y} = \frac{\partial u_\delta}{\partial t} + u_\delta \frac{\partial u_\delta}{\partial x} + v\frac{\partial^2 u}{\partial y^2}, \tag{6}$$

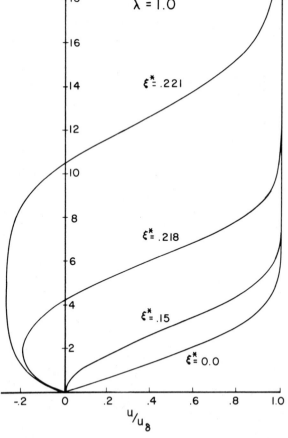

Figure 7 Velocity profiles in the fixed coordinate system for several values of ξ. From Williams & Johnson (1974b).

where x, y, u, v, u_δ, and v have been defined previously and t is time. The boundary conditions for the unsteady problem are:

$$u(x,0,t) = v(x,0,t) = 0, \qquad \lim_{y \to \infty} u(x,y,t) = u_\delta(x,t).$$

To obtain semisimilar solutions two new scaled coordinates defined by

$$\xi = \xi(x,t), \qquad \eta = y/v^{1/2}g(x,t)$$

and a nondimensional stream function, f, defined by

$$f(\xi,\eta) = \psi(x,y,t)/v^{1/2}g(x,t)u_\delta(x,t)$$

are introduced. Here $\xi(x,t)$ is the new scaled (with time) x coordinate and $g(x,t)$ is a scaling factor for the y coordinate. Both $\xi(x,t)$ and $g(x,t)$ are unknown at the outset. The continuity equation is satisfied identically by the introduction of the stream function, and the momentum equation, in terms of the new variables, becomes:

$$f''' + (d+e)ff'' + d(1-f'^2) + a(1-f') + b\eta f''/2 - c\frac{\partial f'}{\partial \xi}$$
$$+ h\left\{ f''\frac{\partial f}{\partial \xi} - f'\frac{\partial f'}{\partial \xi} \right\} = 0, \tag{7}$$

where the coefficients a, b, c, d, e, and h, which must be functions of ξ alone for semisimilar solutions to exist, are given by:

$$a(\xi) = (g^2/u_\delta)\,\partial u_\delta/\partial t, \qquad b(\xi) = \partial g^2/\partial t, \qquad c(\xi) = g^2\,\partial\xi/\partial t,$$

$$d(\xi) = g^2\,\partial u_\delta/\partial x, \qquad 2e(\xi) = u_\delta\,\partial g^2/\partial x, \qquad h(\xi) = u_\delta g^2\,\partial\xi/\partial x.$$

There are three additional equations relating these six coefficients. These equations insure that the second derivatives of u_δ, g, and ξ with respect to x and t are continuous. Equation (7) may be written in the same form as Equation (4), that is,

$$\frac{\partial^2 f'}{\partial\eta^2} + \alpha_1\frac{\partial f'}{\partial\eta} + \alpha_2 f' + \alpha_3 = \alpha_4\frac{\partial f'}{\partial\xi},$$

but now the α_i's are given by

$$\alpha_1 = (d+e)f + b\eta/2 + h\frac{\partial f}{\partial\xi}, \qquad \alpha_2 = -df' - a,$$

$$\alpha_3 = a + d, \qquad\qquad\qquad \alpha_4 = c + hf'.$$

Two interesting observations may be made regarding Equation (4). First, as is well known for the steady-flow case ($a = b = c = 0$) the coefficient α_4 is negative whenever there is reverse flow. Whenever α_4 is negative the numerical solutions of Equation (4) encounter difficulty unless some special technique is used to account for the upstream flow of information in the region of reverse flow. Second, the same type of difficulties arise when α_4 is negative even though there is no reverse flow. The best known example of this occurs in Stewartson's (1951) formulation of the problem of the unsteady boundary layer on a flat plate impulsively set into motion. In this case, $g = t^{1/2}$, $\xi = Ut/x$ (Stewartson uses the symbol τ in place of ξ), and $u_\delta = U = $ constant for $t \geq 0$, so that $a(\xi) = 0$, $b(\xi) = 1$, $c(\xi) = \xi$, $d(\xi) = 0$, $e(\xi) = 0$, and $h(\xi) = -\xi^2$. The corresponding coefficient α_4 is then $(\xi - \xi^2 f')$. In this problem f' is always positive but α_4 is negative, at least in part of the boundary layer, whenever $\xi > 1$. The physical explanation for the difficulties that occur for $\xi > 1$ is that it takes a finite time, $t = x/U$, for the influence of the leading edge to travel downstream to a given x station. Thus, for $\xi < 1$ the flow is unaffected by the leading edge and disturbances travel in the direction of increasing ξ. For $\xi > 1$ the local flow is influenced by the plate-leading edge and disturbances move in the direction of decreasing ξ. Hence, special techniques are required to solve the corresponding form of Equation (4) when $\xi > 1$ [Dennis (1972)]. The problem of the flat plate impulsively set into motion has been discussed at length by Stewartson (1964), Dennis (1972), and a number of other investigators. The important point for the present discussion is that it is not simply reverse flow that causes difficulty in obtaining solutions to Equation (4). These difficulties occur whenever α_4 is negative regardless of whether there is reverse flow or not.

This accounts for some interesting results that have been obtained using the method of semisimilar solutions. First, Williams & Johnson (1975) report that they were able to integrate Equation (7) into a region of reverse flow without any hint of an instability. Furthermore, the results obtained integrating into a region of reverse flow were in good agreement with results obtained, for the same flow, in a moving coordinate system where reverse flow is not a problem. The computation in which this result was obtained was one in which α_4 was always positive up to the point of separation. Wang (1975) has indicated, on the basis of the characteristics and subcharacteristics of the unsteady boundary-layer equations, that it should be possible to integrate the unsteady boundary-layer equations into regions of reverse flow. Wang's criteria for such integration and that given above for integration of the semisimilar equation into regions of reverse flow do not appear to be directly related. Second, Williams & Johnson (1974b) point out that in every case where they attempted to obtain a solution for a flow in which the separation point moved downstream, they encountered an instability that terminated the computation not far from the leading edge. In light of the above discussion, it is not difficult to determine the reason for this difficulty. In each of the cases attempted by Williams & Johnson for flows in which separation should move downstream, the coefficient $c(\xi)$ is negative so that near the wall the coefficient α_4 is negative. When this occurs, Equation (7) becomes a parabolic equation of the mixed type in which information flows in the direction of decreasing ξ. Since there is no downstream boundary condition to supply this information, it becomes impossible to complete the calculation. The physical reason for information travelling in the direction of decreasing ξ in this case is not altogether clear yet. Such a physical explanation would help in understanding this problem and might lead to methods that would lead to a solution of the problem of downstream-moving separation.

Neither the finite difference technique employed by Telionis and his coworkers nor the method of semisimilar solutions employed by Williams & Johnson has been able to yield solutions for the case of downstream-moving separation. Thus, the nature of downstream-moving separation is unknown and the interesting separation-velocity profile postulated by Moore (1958) (see Figure 4) for this case must be considered "lost" at this time.

Separation of Oscillating Boundary Layers

In the case where the imposed pressure gradient is adverse but oscillating in magnitude, one would expect that the separation point itself would oscillate, moving upstream during part of the cycle and downstream during part of the cycle. In view of the difficulties encountered in trying to determine the nature of downstream-moving separation, the prospects for an analytical treatment of this problem, using the techniques described previously, seem quite bleak. It would seem wise then to investigate this problem experimentally, at least initially.

Despard & Miller (1971) have performed a series of very careful experiments in which they measured the instantaneous velocity profiles in oscillating boundary layers subject to adverse pressure gradients. Their stated objective was "the determination of a rational definition of separation in oscillatory boundary-layer flows." The velocity profiles were measured on the rear portion of an airfoil-like body placed in a low-speed wind tunnel in which the free-stream velocity was varied periodically by a set of rotating shutter valves located downstream of the test section. The local pressure gradient on the body was varied by altering the contour of the upper surface of the wind-tunnel test section and boundary-layer velocity profiles were obtained using an array of hot-wire probes. Despard & Miller found that flow reversal within the boundary layer occurred without any unusual thickening of the boundary layer, which would be expected if separation occurred. On the other hand, they found that "at a point within half a boundary-layer thickness of the beginning of the region of wake formation, the entire cycle of profiles exhibited reverse flow or zero velocity gradient at the wall." This led them to propose a practical, empirical definition of separation as the farthest upstream point at which there is "zero velocity or reverse flow at some point in the velocity profile throughout the entire cycle of oscillation." Despard & Miller also note that for the flows studied, the "separation point" occurs upstream of the steady-flow separation point and that increasing the frequency of oscillation causes the "separation point" to move downstream toward the steady separation point. The "separation point" appears to be negligibly affected by variations in amplitude of the outer flow oscillations.

It should be noted that the type of flow studied by Despard & Miller is one in which the free-stream velocity varies periodically with time so that the impressed pressure gradient varies in magnitude but not in form during the oscillation. This type of pressure variation is far different from that which would occur, for example, on an airfoil oscillating in pitch. In this latter case, the pressure gradient at a given position varies both in magnitude and form. An experimental investigation, similar to that of Despard & Miller, but for the case of a pitching airfoil should shed considerable light on the nature of unsteady separation in this case.

Tsahalis & Telionis (1974) have studied several variations of the Howarth linearly retarded flow with a superimposed oscillatory component, using the same numerical techniques employed in their earlier work. It was necessary, however, to use a Taylor series extrapolation to avoid the problem of an ever-contracting domain of integration, which is characteristic of this type of solution. Nevertheless, the results obtained by Tsahalis & Telionis are essentially in agreement with the experimental results of Despard & Miller in that the point of separation, as determined by the appearance of a singularity, moves upstream during the transient portion of the flow and finally becomes stationary at a point upstream of the steady-state separation point. On the other hand, the results of Tsahalis & Telionis seem to indicate that, at least during part of each cycle, the point of vanishing shear is downstream of the "separation" singularity. This is at variance with the definition of separation given by Despard & Miller since in this definition all the velocity profiles just ahead of separation would have either negative or zero shear. Clearly, the problem of separation in oscillating boundary layers deserves further investigation.

THREE-DIMENSIONAL SEPARATION

In most practical flow situations the boundary layer is three-dimensional rather than two-dimensional. There are, of course, regions in many overall three-dimensional flows that may be treated quite satisfactorily using two-dimensional boundary-layer concepts or theory. On the inboard portions of high aspect ratio wings, for example, the flow is essentially two-dimensional. Sooner or later, however, one is confronted with regions in the flow field in which three-dimensional effects dominate the flow, as in the vicinity of the wing tips in the above example or on bodies of revolution at angle of attack or on low aspect ratio wings. In these regions of three-dimensional flow, separation can be quite complex. Smith (1975) and Peake, Rainbird & Atraghji (1972) have recently reviewed the phenomenon of three-dimensional separation in the field of aerodynamics, showing the complex structure of some separations on aircraft and missiles. Similar examples may be found in other branches of fluid mechanics.

Unfortunately, it is not possible as yet to predict accurately three-dimensional separation on many simple bodies, much less on complex configurations. This situation arises, in part at least, from the fact that our understanding of three-dimensional separation has often been obscured by the habit of thinking in terms of two-dimensional separation. As a result there are still unanswered questions regarding both separation and the flow leading up to separation. A second, and perhaps more important, reason for our inability to predict accurately three-dimensional separation is the complexity of the analytical problem due to the addition of another independent variable (the third spatial coordinate) and the corresponding dependent variable (the third velocity component). It has only been in recent years that it has been possible to solve the three-dimensional boundary-layer equations without such limiting assumptions as similarity, yawed infinite cylinders, or small cross flow.

Three-Dimensional Separation Criteria

A number of definitions for three-dimensional separation have been given in the literature. Moore (1956) has suggested that three-dimensional separation may be identified "by the existence of a bubble of fluid in the boundary not exchanging fluid with its surroundings." Stewartson (1964) defined the separation line as a "curve on the body dividing those points that are accessible to the streamline entering the zone of attachment from those points that are inaccessible from attachment." Eichelbrenner (1973) defines a separation region as "a region that is inaccessible to the viscous flow upstream" and the separation line on a solid wall as "a trace that is impervious to the wall streamlines." These definitions of separation and the separation line all have a common element, i.e. the flows in the separated and unseparated regions have different origins and the separation line divides these two regions of flow. Such definitions yield some insight into the nature of the separation phenomenon but they offer little help in determining the nature of the flow in the vicinity of the separation line. A beautifully simple demonstration of the nature of three-dimensional separation has been given by Lighthill (1963). Lighthill poses the question "how parallel to the surface are the neighboring streamlines." Clearly if these streamlines closely parallel the surface the viscous layer is thin and boundary-layer concepts are applicable, but if these streamlines diverge from the surface, separation is to be expected. The velocity along a streamline a small distance from the surface is given by

$$[\tau_x^2 + \tau_y^2]^{1/2} z,$$

where τ_x and τ_y are the wall-shearing stresses in the x and y directions respectively, x and y are the coordinates in the body surface, and z is the coordinate normal to the surface. If two of these streamlines lie a distance h apart, the volume rate through the area hz is

$$1/2 [\tau_x^2 + \tau_y^2]^{1/2} z^2 h = \varepsilon.$$

Since this volume flow rate, ε, is constant there are two mechanisms by which the streamlines can greatly increase their distance from the surface, that is, by which separation can occur. Separation will occur if both τ_x and τ_y approach zero simultaneously or if the streamlines run close together so that h approaches zero. Maskell (1955) reached the same conclusion, which is that there are two mechanisms for separation in three-dimensional flow, using the concept of "limiting streamlines." Limiting streamlines are those streamlines closest to the body surface, and their projections on the body surface have local slope

$$\frac{dx}{dy} = \frac{\tau_x}{\tau_y}.$$

Thus, the projections of the limiting streamlines coincide with the skin-friction lines on the body surface. These limiting streamlines are also often called "surface streamlines," although strictly speaking there are no streamlines on the body surface. The points at which τ_x and τ_y vanish simultaneously are "singular points"; hence, Maskell has defined the type of separation in which τ_x and τ_y vanish simultaneously as "singular separation." The second type of separation, in which the limiting streamlines run close together to form a line of separation, is defined by Maskell as "ordinary separation." This terminology is used throughout the remainder of this discussion. Maskell shows how the limiting streamlines may be used to construct a "skeleton structure of the viscous region" and his paper contains a number of examples of such skeleton structures depicting the type of separation that might be expected on various bodies. Maskell further concludes that the flow resulting from three-dimensional separation is composed of two basic elements: a bubble and/or a free vortex layer.

On a bluff axisymmetric body at zero angle of attack singular separation occurs at every point along the line of separation. But this is a very special case. For the same body at a slight angle of attack there are at least two points along the separation line at which singular separation occurs: the two points where the line of separation pierces the plane of symmetry. Away from these points on the separation line, however, the nature of separation is not clear. Brown & Stewartson (1969) illustrate two possibilities in Figures 3a and 3b of their review of separation. In one case the limiting streamlines actually meet the separation line at finite points so that the separation line is actually an envelope of the limiting streamlines. In the second case the limiting streamlines all converge on a nodal point of separation and hence only touch the separation line at this point. In addition to the question of whether or not the limiting streamlines meet the separation line at finite points, Brown & Stewartson raise the question of whether or not "the envelope of limiting streamlines coincides with a singularity in the solution to the three-dimensional boundary-layer equations." There is a third question which, although not directly related to the nature of the separation line, relates to the possibility of integrating the three-dimensional boundary-layer equations up to the separation line or at least close to it. This is the question of whether or not it is possible to integrate these equations into regions where one of the velocity components is reversed near the wall. Flow reversal of one of the velocity components often occurs just ahead of separation.

The Nature of Three-Dimensional Separation

As mentioned above, the projections of the limiting streamlines on the body surface coincide with skin-friction lines. It should be possible then to deduce some information regarding the behavior of the limiting streamlines near separation from photographs of the skin-friction lines made visible by experimental techniques such as the oil-flow technique. Such photographs abound in the literature, typical examples being given by Peake, Rainbird & Atraghji (1972) and Stetson (1972). These photographs tend to support the idea that the separation line is an envelope of the limiting streamlines.

Buckmaster (1972) has studied the three-dimensional boundary-layer equations close to a plane of symmetry and a curved separation line using a perturbation technique. He concludes that "the separation line is a skin-friction line and the other skin-friction lines meet this (tangentially) at finite points." He also points out that "the component of skin friction normal to the separation line vanishes like a square root." Thus, the separation line is apparently not only an envelope of the limiting streamlines but coincides with a singularity in the solution to the three-dimensional boundary-layer equations.

Unfortunately, the number of three-dimensional boundary-layer solutions that closely approach separation is quite limited. Many of the solutions that have been attempted are terminated at the onset of flow reversal of one of the velocity components. As will be seen later, the onset of flow reversal is not a critical or even physically significant condition, with respect to separation, as it is in steady two-dimensional flow.

Williams (1975) has obtained solutions to the three-dimensional boundary-layer equations, using the technique of semisimilar solutions, for several cases in which three-dimensional separation was approached. As in the application of this method to two-dimensional unsteady separation, the method of semisimilar solutions scales the three independent variables into two new variables. It may be viewed, in this application, as a scaling of the two surface coordinates, x and y, into a new coordinate, ξ, in such a way that separation occurs at a constant value of the scaled coordinate. This property of the transformation is quite useful since it yields at the outset information regarding the shape of the separation line. In all the cases studied by Williams the separation line was straight. As a result of this transformation the momentum equations have the character of the scaled momentum equation in the two-dimensional steady problems. Thus, the integration of the scaled momentum equations may be carried out using numerical methods that have proven successful in the two-dimensional problem. The numerical integration is carried out in the transformed coordinates by marching in the direction of increasing ξ. As a certain value of ξ is approached the number of iterations required for convergence increases rapidly until a point is reached where convergence cannot be obtained in a reasonable number of iterations. This is the same behavior as is observed in the two-dimensional steady case and, by analogy with that case, is taken as an indication of approaching a point of singular behavior in the boundary-layer equations, associated with separation. Since the shape of the separation line is known at the outset it is possible to determine the slope of the separation line. Williams noted that in all the cases studied except one, the slope of the limiting streamline approached very closely the slope of the separation line, indicating an ordinary separation in which the limiting streamlines run close together at separation. In the single exception the two components of skin friction appeared to be vanishing simultaneously, indicating singular separation at all points along the separation line. This line of singular separation seems, however, to be the exception rather than the rule.

An interesting calculation of the three-dimensional boundary-layer development in the plane of symmetry on a prolate spheroid at angle of attack has been given by Wang (1970). Such a solution, which is limited to the plane of symmetry, can, of course, offer only limited information regarding the nature of the separation line away from the plane of symmetry. However, Wang has used these results together with the results of a number of flow visualization studies to argue that a new type of separation[2] exists on this body at high angle of attack [Wang (1972)]. For this

---

[2] Actually, although both Stetson (1972) and Wang (1972) refer to this type of separation as "new," the beautiful flow visualization pictures obtained by Werlé (1962) appear to show this open type of separation on certain blunt bodies at high angle of attack.

new type of separation the separation line is apparently not closed and hence does not divide the body into two distinct regions, one accessible to streamlines entering the zone of attachment and one that is not accessible to these streamlines. Instead, streamlines from the attachment point that pass over the lee side of the body approach the separation line from one side while streamlines from the attachment point that pass over the windward side of the body approach the separation line from the other side. The separation line is an ordinary separation line in that the streamlines run close together near separation. As a result, a free vortex springs forth from the separation line. This free vortex is very similar to the free vortex that originates at the leading edge of a blunt delta wing. This new type of separation has also been demonstrated by Stetson (1972) for the case of a blunted cone at angle of attack in hypersonic flow. An interesting conceptional drawing of this type of separation is presented in Figure 18 of Stetson's paper.

Wang (1974) carried out a complete three-dimensional boundary-layer calculation for an ellipsoid of revolution of thickness ratio 1:4 at an angle of attack of 30°. Wang used a finite-difference numerical calculation in the meridional and circumferential coordinate system. These detailed calculations verify the existence of an open separation line, as postulated in Wang's earlier work, which lies very close, over most of the body, to the line of vanishing circumferential skin friction.

Geissler (1974) has calculated the laminar boundary-layer development on an ellipsoid of revolution at 15° angle of attack. The boundary-layer development is calculated using an implicit finite-difference technique and marching from the stagnation point along both the streamlines and the equipotential lines. An instability in the numerical calculation is taken as a signal of separation. This line of instability corresponds to an envelope of the limiting streamlines and hence is interpreted as a separation line of the free vortex type. Here again the separation line is open on the lee side of the body. The limiting streamlines from the windward side of the body pass around the body and approach the separation line from one side. Apparently the streamlines from the lee side approach the separation line from the opposite side, but it is not possible to continue the calculation into the region between the lee side line of symmetry and the separation line. This calculation may be taken as a verification of the new type of three-dimensional separation.

Integration into Regions with Flow Reversal

For three-dimensional flows there is no unique coordinate system that may be used with all bodies. Even on the same body, different coordinate systems may be used. In obtaining the boundary-layer solution on an ellipsoid of revolution, for example, Geissler (1974) used the streamline coordinate system while Wang (1974) used surface coordinates along the meridional and circumferential directions. While the pressure distribution on a three-dimensional body is independent of the coordinate system, the pressure gradients in the boundary-layer equations depend upon the coordinate system used. These imposed pressure gradients often lead to strong cross flows or secondary flows. Thus, depending upon the coordinate system used and the pressure distribution on the body, one may encounter regions on the body where flow reversal of one or the other velocity components occurs near the wall. Again, the question arises of whether or not it is possible to integrate numerically the boundary-layer equations into regions where one of the velocity profiles is partially reversed.

It has long been known that a disturbance at a point within the boundary layer is instantly transmitted up and down the normal to the body surface through that point and is convected downstream along all the streamlines passing through this line. This is Raetz's influence principle. Thus, for each point on the body, there is a zone of dependence and a zone of influence, in the shape of curvilinear wedges, one opening in the upstream direction and the other in the downstream direction. The zones of dependence and influence are bounded by the streamlines of maximum and minimum angles passing through the body normal at the point in question. Finite difference meshes that are constructed to be used in the numerical integration of the three-dimensional boundary-layer equations must account for the zones of dependence and influence so that the influence principle is not violated (Wang 1971). Wang (1974) has shown several ways in which three-dimensional meshes can be constructed to satisfy this requirement. Intuitively, one might expect that the question of whether or not integration into regions of reverse flow is permissible is related to the influence principle. This has been verified and the question answered by Wang (1975) who states, "In steady three-dimensional problems, so long as the zone of dependence is satisfied, flow involving either u- or v- profile reversal does not violate the initial-value idea and hence, can be calculated. The problem is still well posed." (Note that in this work Wang uses a rectangular coordinate system in which u and v are the velocity components parallel to the wall.)

Williams (1975) has obtained solutions that include regions of reverse flow using the method of semisimilar solutions. Here the requirements for integration into these regions are apparently quite different from those imposed by the influence principle. As applied to the three-dimensional boundary layer, the method of semisimilar solutions scales the three independent variables (the spatial coordinates) into two new variables. The continuity equation is satisfied identically by the introduction of a pair of stream functions. As a result of these operations each of the two momentum equations takes on a form that is very much like the transformed two-dimensional unsteady momentum equation [Equation (7)] and can further be rewritten in the form of Equation (4). Further, the coefficient α_4 [see Equation (4)] is the same in each of the momentum equations. The problem then is essentially a two-dimensional one (mathematically speaking) and, as in the two-dimensional unsteady case, whether or not integration into regions of reverse flow is permissible is determined by the sign of the coefficient α_4. Williams obtained several solutions in which reverse flow was encountered with no indication of convergence or stability difficulties in the numerical calculation. In each of these cases the coefficient α_4 was always positive. No cases in which α_4 became negative prior to separation were attempted.

CONCLUSIONS

In the past eight years, since the last review of separation appeared in Annual Review of Fluid Mechanics, significant progress has been made in both the understanding of the physics of separation and the analysis of boundary-layer flows leading to separation. Thus, it is now possible to integrate the two-dimensional laminar boundary-layer equations through a point of zero wall shear into a shallow separation bubble and through the point of reattachment again into the re-attached boundary layer. This has made possible the detailed study of the structure of such bubbles. Solutions have also been obtained for the laminar boundary layer ahead of an unsteady separation point when the separation point moves forward along the body. These solutions have verified the Moore-Rott-Sears model for separation in this case. Analytical solutions have also been obtained for the laminar boundary-layer development on several three-dimensional bodies, solutions that lead up to separation and, in one case at least, have indicated a new type of open three-dimensional separation line.

On the other hand, it is not yet possible to integrate the boundary-layer equations through the separation point when separation is followed by a large-scale recirculating (and in many cases unsteady) region. In this case, the separation point represents a boundary between two regions of vastly different scales and it is doubtful that this problem can be handled within the framework of boundary-layer theory alone. Sychev (1972) and Messiter (1975) have made the interesting suggestion that the theory of asymptotic solutions might be combined with free-streamline theory to provide a solution to this problem. Nor is it possible yet to obtain solutions to the unsteady boundary-layer equations in the vicinity of a downstream moving or oscillating separation point. Thus, the nature of downstream moving separation remains an enigma. There are, as yet, an insufficient number of solutions for the boundary layer on general three-dimensional bodies. As a result the complete picture of three-dimensional separation remains unclear. Finally, the full understanding of the extremely practical problem of turbulent boundary-layer separation, for steady two- or three-dimensional flow or for unsteady flow, remains clouded due to the poor understanding of the turbulent transport phenomenon in regions of small or negative shear. Hopefully, by the time of the next review of separation in Annual Review of Fluid Mechanics, many of these problems will be solved.

ACKNOWLEDGMENTS

The author is deeply indebted to Dr. James E. Carter of NASA Langley Research Center and Dr. W. R. Sears of the University of Arizona for reading portions of the manuscript and offering valuable criticisms and suggestions.

Literature Cited

Alber, I. E. 1971. Similar solutions for a family of separated turbulent boundary layers. *AIAA Pap. 71-203, AIAA Aerosp. Sci. Meet., 9th, New York, NY, Jan. 25–27*

Briley, W. R. 1971. A numerical study of laminar separation bubbles using the Navier-Stokes equations. *J. Fluid Mech.* 47:713–36

Briley, W. R., McDonald, H. 1975. Numerical prediction of incompressible separation bubbles. *J. Fluid Mech.* 69:631–56

Brown, S. N., Stewartson, K. 1969. Laminar separation. *Ann. Rev. Fluid Mech.* 1:45–72

Buckmaster, J. 1972. Perturbation technique for the study of three-dimensional separation. *Phys. Fluids.* 15:2106–2113

Carter, J. E. 1975. Inverse solutions for laminar boundary-layer flows with separation and attachment. *NASA TR-R447*, Washington, D.C., Nov. 1975

Carter, J. E., Wornom, S. F. 1975a. Forward marching procedure for separated boundary layer flows. *AIAA J.* 13:1101–3

Carter, J. E., Wornom, S. F. 1975b. Solutions for incompressible separated boundary-layers including viscous-inviscid interaction. *NASA SP347 Aerodynamic Analyses Requiring Advanced Computers, Part I*, pp. 125–150

Catherall, D., Mangler, K. W. 1966. The integration of the two-dimensional laminar boundary-layer equations past a point of vanishing skin friction. *J. Fluid Mech.* 26:163–82

Danberg, J. E., Fansler, K. S. 1975. Separation-like similarity solutions on two-dimensional moving walls. *AIAA J.* 13:110–12

Dean, W. R. 1950. Note on the motion of liquid near a position of separation. *Proc. Cambridge Philos. Soc.* 46:293–306

Dennis, S. C. R. 1972. The motion of a viscous fluid past an impulsively started semi-infinite flat plate. *J. Inst. Math. Appl.* 10:105–17

Dennis, S. C. R., Chang, G. 1970. Numerical solutions for steady flow past a circular cylinder at Reynolds numbers up to 100. *J. Fluid Mech.* 42:471–89

Despard, R. A., Miller, J. A. 1971. Separation in oscillating laminar boundary layer flows. *J. Fluid Mech.* 47:21–31

Diewert, G. S. 1975. Computation of separated transonic turbulent flows. *AIAA Pap. 75-829, AIAA Fluid Plasma Dyn. Conf., 8th, Hartford, Conn., June 16–18*

Eichelbrenner, E. A. 1973. Three-dimensional boundary layers. *Ann. Rev. Fluid Mech.* 5:339–60

Geissler, W. 1974. Three-dimensional laminar boundary layer over a body of revolution at incidence and with separation. *AIAA J.* 12:1743–45

Ghia, U., Davis, R. T. 1974. Navier-Stokes solutions for flow past a class of two-dimensional semi-infinite bodies. *AIAA J.* 12:1659–65

Ghia, K. N., Ghia, U., Tesch, W. A. 1975. Evaluation of several approximate models for laminar incompressible separation by comparison with complete Navier-Stokes solutions. *AGARD Conf. Proc. No. 168 Flow Sep.,* pp. 6-1 to 6-15

Goldstein, S. 1948. On laminar boundary-layer flow near a position of separation. *Q. J. Mech. Appl. Math.* 1:43–69

Jenson, V. G. 1959. Viscous flow round a

sphere at low Reynolds numbers (<40) *Proc. R. Soc. London Ser. A.* 249:346–66

Keller, H. B., Takami, H. 1966. "Numerical studies of steady viscous flow about cylinders," *Numerical Solutions of Nonlinear Differential Equations.* ed. D. Greenspan. New York: Wiley

Klemp, J. B., Acrivos, A. 1972. A method of integrating the boundary-layer equations through a region of reverse flow. *J. Fluid Mech.* 53:177–91

Kline, S. J., Morkovin, M. V., Sovran, G., Cockrell, D. J., eds. 1968. *Proc. Comput. Turbul. Boundary Layers, 1968, Air Force Off. Sci. Res. Intern. Flow Program Stanford Conf., Vol. 1. Methods, Predictions, Evaluation and Flow Structure.* Stanford Univ. 589 pp.

Klineberg, J. M., Steger, J. L. 1974. On laminar boundary-layer separation. *AIAA Pap. 74-94, AIAA Aerosp. Sci. Meet., 12th, Washington, D.C., Jan. 30–Feb. 1*

Launder, B. E., Spalding, D. B. 1972. *Mathematical Models of Turbulence.* London. Academic 169 pp.

Leal, L. G. 1973. Steady separated flow in a linearly decelerated free stream. *J. Fluid Mech.* 59:513–35

Lighthill, M. J. 1963. In *Laminar Boundary Layers,* ed. L. Rosenhead, pp. 1–113. London: Oxford. 687 pp.

Ludwig, G. R. 1964. An experimental investigation of laminar separation from a moving wall. *AIAA Pap. 64-6, AIAA Aerosp. Sci. Meet., New York, NY, Jan.* 20–22

Maskell, E. C. 1955. Flow separation in three-dimensions. *RAE Rep. Aero 2565,* Royal Aircraft Establishment, Farnborough, England

Masliyah, J. H., Epstein, N. 1970. Numerical study of steady flow past spheroids. *J. Fluid Mech.* 44:493–512

Meksyn, D. 1950. Integration of the laminar boundary-layer equation. I—Motion of an elliptic cylinder. Separation. II—Retarded flow along a semi-infinite plane. *Proc. R. Soc. London Ser. A.* 201:268–83

Messiter, A. F. 1975. Laminar separation—A local asymptotic flow description for constant pressure downstream. *AGARD Conf. Proc. No. 168 Flow Sep.* pp. 4-1 to 4-10

Moore, F. K. 1956. Three-dimensional boundary layer theory. *Adv. Appl. Mech.* 4:159–228

Moore, F. K. 1958. On the separation of the unsteady laminar boundary layer. *Boundary-Layer Research,* ed. H. G. Görtler, pp. 296–310. Berlin: Springer

Peake, D. J., Rainbird, W. J., Atraghji, E. G. 1972. Three-dimensional flow separation

144   WILLIAMS

on aircraft and missiles. *AIAA J.* 10: 567–80

Rott, N. 1956. Unsteady viscous flow in the vicinity of a stagnation point. *Q. Appl. Math.* 13:444–51

Sears, W. R. 1956. Some recent developments in airfoil theory. *J. Aerosp. Sci.* 23:490–499

Sears, W. R., Telionis, D. P. 1975. Boundary-layer separation in unsteady flow. *SIAM J. Appl. Math.* 28:215–35

Smith, J. H. B. 1975. A review of separation in steady, three-dimensional flow. *AGARD Conf. Proc. No. 168 Flow Sep.* pp. 31-1 to 31-17

Son, J. S., Hanratty, T. J. 1969. Numerical solution for flow around a cylinder at Reynolds numbers of 40, 200 and 500. *J. Fluid Mech.* 35:369–86

Stetson, K. F. 1972. Boundary layer separation on slender cones at angle of attack. *AIAA J.* 10:642–48

Stewartson, K. 1951. On the impulsive motion of a flat plate in a viscous fluid. *Q. J. Mech. Appl. Math.* 4:182–98

Stewartson, K. 1958. On Goldstein's theory of laminar separation. *Q. J. Mech. Appl. Math.* 11:399–410

Stewartson, K. 1964. *The Theory of Laminar Boundary-Layers in Compressible Fluids.* London: Oxford University Press. 199 pp.

Stewartson, K. 1974. Multistructured boundary layers on flat plates and related bodies. *Adv. in Appl. Mech.* 14:145–239

Sychev, V. V. 1972. Laminar separation. *Mekh. Zhid i Gaza.* No. 3: 47–59; *Fluid Dyn.* 7:407–417

Telionis, D. P., Tsahalis, D. Th., Werle, M. J. 1973. Numerical investigations of unsteady boundary-layer separation. *Phys. Fluids.* 16:968–73

Telionis, D. P., Tsahalis, D. Th. 1973a. The response of unsteady boundary-layer separation to impulsive changes of outer flow. *AIAA Pap. 73-684, AIAA 6th Fluid Plasma Dyn. Conf., 6th, Palm Springs, Calif., July 16–18*

Telionis, D. P., Tsahalis, D. Th. 1973b. Unsteady laminar separation over cylinders started impulsively from rest. *Report VPI 73-29*, Virginia Polytech. Inst. and State Univ., Blacksburg, Va.

Telionis, D. P., Werle, M. J. 1973. Boundary-layer separation from downstream moving boundaries. *J. Appl. Mech.* 40:369–74

Tsahalis, D. Th., Telionis, D. P. 1973. The effect of blowing on laminar separation. *J. Appl. Mech.* 40:1133–34

Tsahalis, D. Th., Telionis, D. P. 1974. Oscillating laminar boundary layers and unsteady separation. *AIAA J.* 12:1469–76

Tsahalis, D. T. 1976. Laminar boundary layer separation from an upstream moving wall. *AIAA Pap. 76-377. Presented at* AIAA 9th Fluid Plasma Dyn. Conf. San Diego, Calif., July 14–16

Vidal, R. J. 1959. Research on rotating stall in axial flow compressors; Part III—Experiments on laminar separation from a moving wall, *Wright Air Dev. Cent. Tech. Rep. 59-75*

Wang, K. C. 1970. Three-dimensional boundary layer near the plane of symmetry of a spheroid at incidence. *J. Fluid Mech.* 43:187–209

Wang, K. C. 1971. On the determination of the zones of influence and dependence for three-dimensional boundary layer equations. *J. Fluid Mech.* 48:397–404

Wang, K. C. 1972. Separating patterns of boundary layer over an inclined body of revolution. *AIAA J.* 10:1044–50

Wang, K. C. 1974. Boundary layer over a blunt body at high incidence with an open-type of separation. *Proc. R. Soc. London Ser. A.* 340:33–55

Wang, K. C. 1975. Aspects of "multitime initial-value problem" originating from boundary layer equations. *Phys. Fluids.* 18:951–55

Werlé, H. 1962. Separation on axisymmetric bodies at low speed. *Rech. Aéronaut.* 90: 3–14

Williams, J. C., III. 1975. Semi-similar solutions to the three-dimensional laminar boundary layer. *Appl. Sci. Res.* 31:161–86

Williams, J. C., III, Johnson, W. D. 1974a. Note on unsteady boundary-layer separation. *AIAA J.* 12:1427–29

Williams, J. C., III, Johnson, W. D. 1974b. Semi-similar solutions to unsteady boundary-layer flows including separation. *AIAA J.* 12:1388–93

Williams, J. C., III, Johnson, W. D. 1975. New solutions to the unsteady laminar boundary-layer equations including the approach to unsteady separation. *Unsteady Aerodynamics,* Proc. Symp., March 18–20, 1975, ed. R. B. Kinney. 261–82 pp.

Ann. Rev. Fluid Mech. 1977. 9:145–85

BUBBLE DYNAMICS ✕8100
AND CAVITATION

Milton S. Plesset
Department of Engineering Science, California Institute of Technology,
Pasadena, California 91125

Andrea Prosperetti
Istituto di Fisica, Università degli Studi, 20133 Milano, Italy

1 INTRODUCTION

The first analysis of a problem in cavitation and bubble dynamics was made by Rayleigh (1917), who solved the problem of the collapse of an empty cavity in a large mass of liquid. Rayleigh also considered in this same paper the problem of a gas-filled cavity under the assumption that the gas undergoes isothermal compression. His interest in these problems presumably arose from concern with cavitation and cavitation damage. With neglect of surface tension and liquid viscosity and with the assumption of liquid incompressibility, Rayleigh showed from the momentum equation that the bubble boundary $R(t)$ obeyed the relation

$$R\ddot{R} + \tfrac{3}{2}(\dot{R})^2 = \frac{p(R) - p_\infty}{\rho},$$ (1.1)

where ρ is the liquid density, p_∞ is the pressure in the liquid at a large distance from the bubble, and $p(R)$ is the pressure in the liquid at the bubble boundary. For this Rayleigh problem, $p(R)$ is also the pressure within the bubble. Incompressibility of the liquid means that the liquid velocity at a distance r from the bubble center is

$$u(r, t) = \frac{R^2}{r^2}\dot{R}.$$ (1.2)

The pressure in the liquid is readily found from the general Bernoulli equation to be

$$p(r, t) = p_\infty + \frac{R}{r}[p(R) - p_\infty] + \tfrac{1}{2}\rho\frac{R}{r}\dot{R}^2\left[1 - \left(\frac{R}{r}\right)^3\right].$$ (1.3)

While Rayleigh neglected surface tension and liquid viscosity and kept the pressure p_∞ constant, his dynamical equation (1.1) is easily extended to include these effects.

For a spherical bubble, viscosity affects only the boundary condition so that it becomes

$$p(R) = p_i - \frac{2\sigma}{R} - \frac{4\mu}{R}\dot{R}, \tag{1.4}$$

where now p_i is the pressure in the bubble and $p(R)$, as before, is the pressure in the liquid at the bubble boundary. The surface-tension constant and the coefficient of the liquid viscosity are σ and μ, respectively. By allowing p_∞ to be a function of time, one can use Equation (1.1) to describe the experimental observations on cavitation-bubble growth and collapse in a liquid flow (Plesset 1949). Other effects not considered by Rayleigh, such as the stability of the interface, the compressibility of the liquid, the effect of energy flow into or out of the bubble, and the physical conditions within the bubble, are described in the sections that follow.

We may write here a generalized Rayleigh equation for bubble dynamics, in view of Equation (1.4), as

$$R\ddot{R} + \tfrac{3}{2}(\dot{R})^2 = \frac{1}{\rho}\left\{p_i - p_\infty - \frac{2\sigma}{R} - \frac{4\mu}{R}\dot{R}\right\}, \tag{1.5}$$

where the pressure in the gas at the bubble wall, p_i, may be a function of the time, and the pressure at infinity, p_∞, may also be a function of the time. We may also, for future reference, define an equilibrium radius of a gas nucleus for given values of p_i and p_∞:

$$R_0 = \frac{2\sigma}{p_i - p_\infty}. \tag{1.6}$$

A bubble of this radius will clearly remain at rest if it is initially at rest, although it should be noted that this equilibrium is an unstable one.

The discussion of bubble dynamics divides itself in a natural way on the one hand into those situations in which the bubble interior consists for the most part of permanent gas and on the other hand into those situations in which the bubble interior is composed almost entirely of the vapor of the surrounding liquid. Vapor-bubble dynamics can be often simplified by a further subdivision into vapor-bubble dynamics in a subcooled liquid and into vapor-bubble dynamics in a superheated liquid (Plesset 1957). The subcooled-liquid case corresponds to that in which the vapor density is so small that latent heat flow does not affect the motion which is then controlled by the inertia of the liquid. In this sense the liquid may be said to be "cold," and the liquid is usually described as a cavitating liquid. The superheated liquid in a similar way may be described as one in which boiling phenomena occur. In boiling, the vapor-bubble dynamics is controlled by the latent heat flow rather than by the liquid inertia (Plesset 1969). Here the coupling of the energy equation with the momentum equation is essential. The case of the gas bubble also requires in most cases the simultaneous consideration of the momentum equation and the energy equation. A large body of literature has developed on this topic, which is considered in the following section.

2 GAS BUBBLES

We consider in this section the case in which the medium filling the cavity is essentially a permanent, noncondensable gas. We neglect all the effects associated with the vapor of the liquid which necessarily is present in the bubble together with the gas. Clearly this procedure is legitimate as long as the partial pressure of the vapor is small compared with the gas pressure.

Small-Amplitude Oscillations

The small-amplitude (linearized) oscillations of a permanent gas bubble in a liquid were first considered by Minnaert (1933), who showed that a substantial contribution to the sound emitted by running water comes from the free radial pulsations of entrained air bubbles. More recently the same type of problem has been considered in greater detail by several investigators including Devin (1959), Plesset & Hsieh (1960), Chapman & Plesset (1971), and Prosperetti (1976a).

In many instances, the case of small-amplitude forced radial oscillations arises when a bubble is immersed in a sound field of wavelength large compared with the bubble radius. Such a sound field may be introduced as an approximation in the Rayleigh equation by writing

$$p_\infty(t) = P_\infty(1 + \varepsilon \cos \omega t), \tag{2.1}$$

where P_∞ is the average ambient pressure, ω is the sound frequency, and ε is the dimensionless amplitude of the pressure perturbation which is supposed to be small. It is also assumed that the oscillations take place about the equilibrium radius R_0 defined by Equation (1.6), so that one may write

$$R = R_0[1 + x(t)], \tag{2.2}$$

with $x(t)$ a small quantity of order ε. Finally, one must introduce a suitable linearization of the internal pressure $p_i(t)$. Denoting by a dot differentiation with respect to time, we write

$$p_i(t) = p_{i,eq} + \frac{\partial p_i}{\partial x}\bigg|_{x=0,\dot{x}=0} x(t) + \frac{\partial p_i}{\partial \dot{x}}\bigg|_{x=0,\dot{x}=0} \dot{x}(t) + \cdots, \tag{2.3}$$

where $\partial p_i/\partial x|_{x=0,\dot{x}=0}x$ represents the component of the internal pressure perturbation in phase with the driving force, and $\partial p_i/\partial \dot{x}|_{x=0,\dot{x}=0}\dot{x}$ the component out of phase with it by $\pi/2$. Upon substitution of relations (2.1), (2.2), and (2.3) into Equation (1.5), one finds an equation of the harmonic-oscillator form

$$\ddot{x} + 2\beta\dot{x} + \omega_0^2 x = -\varepsilon\alpha e^{i\omega t}, \tag{2.4}$$

where the constant α is defined by $\alpha = P_\infty/\rho R_0^2$, and the damping constant β and the effective natural frequency ω_0 are given formally by

$$\beta = \frac{2\mu}{\rho R_0^2} - \frac{1}{2\rho}\frac{\partial p_i}{\partial \dot{x}}\bigg|_{x=0,\dot{x}=0} \tag{2.5}$$

and

$$\omega_0^2 = -\frac{1}{\rho}\frac{\partial p_i}{\partial x}\bigg|_{x=0,\dot{x}=0} - \frac{2\sigma}{\rho R_0^3}. \tag{2.6}$$

Further progress is clearly dependent upon the specification of the quantities appearing in Equation (2.3). A simple assumption is that the gas follows a polytropic law of compression with polytropic exponent κ

$$p_i = p_{i,eq}\left(\frac{R_0}{R}\right)^{3\kappa}. \tag{2.7}$$

If this assumption were strictly valid, one would find that energy dissipation arises only from liquid viscosity and compressibility (see below). With this assumption the natural frequency is

$$\omega_0^2 = 3\kappa\frac{p_{i,eq}}{\rho R_0^2} - \frac{2\sigma}{\rho R_0^3}. \tag{2.8}$$

If surface-tension effects are neglected, and if the pressure-volume relationship is taken to be adiabatic so that κ equals γ, the ratio of the specific heats of the gas, Equation (2.8) coincides with the result first obtained by Minnaert (1933).

Clearly, the accuracy of Equation (2.7) can be assessed only by considering the complete set of linearized conservation equations of mass, momentum, and energy both in the gas and in the liquid. An analysis of this type was first undertaken by Pfriem (1940; see also Devin 1959), who made use of a somewhat artificial Lagrangian formalism. An explicit treatment in terms of the conservation equations of continuum mechanics has been given by Plesset & Hsieh (1960) and more recently by Prosperetti (1976a). The case of the free oscillations has been analyzed by Chapman & Plesset (1971).

The results of these studies are two-fold. In the first place, it is found that the effective polytropic exponent κ exhibits a strong dependence on the driving sound frequency ω. For the case $\gamma = 7/5$ (diatomic gas) this dependence is illustrated in Figure 1 in terms of the dimensionless frequency $G_1 = MD_g\omega/\gamma R_g T_\infty$ and of another dimensionless parameter $G_2 = R_0^2\omega/D_g$. Here D_g denotes the thermal diffusivity of the gas, M its molecular weight, R_g the universal gas constant, and T_∞ the absolute temperature of the liquid at a distance from the bubble. The physical basis for the behavior depicted in Figure 1 can be clarified by noting that essentially three length scales are involved in the problem, namely, the bubble radius R_0, the wavelength of sound in the gas $\lambda_g = 2\pi(\gamma R_g T_\infty/M)^{1/2}/\omega$, and the thermal penetration depth in the gas, $L_{th} = (D_g\omega)^{1/2}$. The thermal penetration depth in the liquid is so small that in practice it can be taken to be zero with a negligible error. In terms of these three fundamental length scales, it is seen that, essentially, $G_1 \sim (L_{th}/\lambda_g)^2$ and $G_2 \sim (R_0/L_{th})^2$. If $(G_1 G_2)^{1/2}$ (i.e. R_0/λ_g) is small, the pressure within the bubble is spatially uniform, and one will observe an isothermal behavior, $\kappa \cong 1$ when $R_0 \ll L_{th}$ (i.e. G_2 is small); and one will observe an adiabatic behavior $\kappa \cong \gamma$ when $R_0 \gg L_{th}$ (i.e. G_2 is large). In the first case, the oscillations are too slow to maintain an appreciable temperature gradient in the bubble, whereas in the second case they are

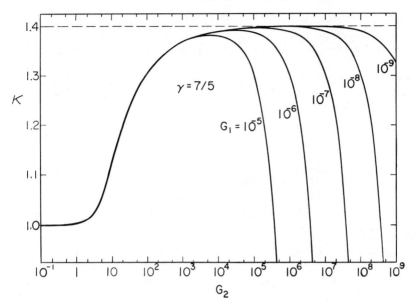

Figure 1   The effective polytropic exponent κ for the small-amplitude forced oscillations of a gas bubble containing a diatomic gas ($\gamma = 7/5$). The numbers labeling the curves denote different values of the dimensionless frequency $G_1 = MD_g\omega/\gamma R_g T_\infty$. The quantity G_2 is defined by $G_2 = \omega R_0^2/D_g$ (from Prosperetti 1976a).

so fast that most of the gas contained in the bubble is practically thermally insulated from the liquid. However, as was first pointed out by Plesset (1964) and by Plesset & Hsieh (1960), for frequencies so large that λ_g is of the order of R_0 or smaller [i.e. $(G_1G_2)^{1/2}$ is of order 1 or larger], pressure nonuniformities develop in the bubble and a polytropic pressure-volume relationship loses its thermodynamic meaning. As an associated effect, the polytropic exponent κ takes on values outside the range $1 < \kappa < \gamma$, and may become even negative (see Prosperetti 1976a). The interesting question of the thermodynamic behavior at high frequency of the bubble as a whole, however, is still meaningful. As discussed by Plesset (1964) and Prosperetti (1976a), one is led to the conclusion of an overall isothermal behavior, caused by the establishment in the interior of the bubble of a temperature distribution consisting of many standing waves of short wavelength.[1] In view of Equation (2.8), the frequency dependence of the polytropic exponent reflects itself in a frequency dependence of the effective natural frequency, in the sense that the pole of the oscillation amplitude determined from Equation (2.4) is a function of the driving frequency. The frequency at which resonance oscillations take place does remain well defined (see Prosperetti 1976a).

[1] Clearly, at very high frequencies one cannot consider the wavelength of the sound in the liquid large compared with the bubble radius. The bubble oscillations then will not be purely radial, but the physical content of this statement remains nevertheless applicable.

The second important result obtained by the analysis of the complete (linearized) fluid-mechanical problem concerns the damping of the radial oscillations. Figure 2 presents the results of Chapman & Plesset (1971) for the logarithmic decrement $\Lambda = 2\pi\beta/\omega_0$ of an air bubble in water. It is seen that, in the range 0.1 cm $> R_0 > 4 \times 10^{-4}$ cm, the thermal component, represented formally by the second term in Equation (2.5), dominates the viscous and acoustic contributions to the energy dissipation. The acoustic contribution can be computed by attributing a slight compressibility to the liquid, and the result is the addition of

$$\beta_{\text{acoustic}} = \tfrac{1}{2}\omega \frac{\omega R_0/c}{1+(\omega R_0/c)^2} \simeq \frac{1}{2}\left(\frac{\omega R_0}{c}\right)\omega \tag{2.9}$$

to the damping coefficient of Equation (2.5). This quantity determines the fraction of the work performed by the sound field on the bubble that is dissipated as sound waves radiated into the liquid. Results for the case of forced oscillations of an air bubble in water are shown in Figure 3 for two values of the equilibrium radius R_0 (the frequency corresponding to the resonant frequency of the bubble is indicated by an open circle in these figures). Except for extremely small bubbles for which viscosity is very important, the low-frequency damping is dominated by thermal effects, and the high-frequency damping by acoustic effects. The analytic expression for the thermal damping constant is rather involved and is not given here; the reader is referred to Prosperetti (1976a).

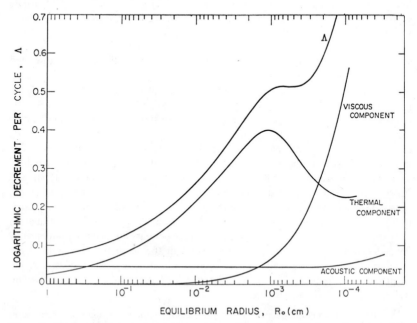

Figure 2   The logarithmic decrement for small-amplitude free oscillations of an air bubble in water as a function of the equilibrium radius R_0 (from Chapman & Plesset 1971).

As a final point, we may remark that Plesset & Hsieh (1960) have considered the small-amplitude oscillations as an initial-value problem and have shown that the solution does indeed approach the solution of the steady-state problem represented by Equation (2.4) as $t \rightarrow \infty$.

Considerable experimental work on gas-bubble oscillations has been reported in the literature. The data for the natural frequency can be said to agree well with the theory (Lauer 1951, Koger & Houghton 1968, Jensen 1974), whereas a large scatter appears in the data for the damping of the oscillations, as can be seen for example in Figure 2 of Koger & Houghton (1968). While this situation may certainly be ascribed to a large extent to the quality of the data (for discussions, see e.g. Devin 1959, Kapustina 1970, van Wijngaarden 1972), it appears that the situation has not yet been completely clarified. Some references to the original works are contained in the papers referenced above; in a recent work Ceschia & Iernetti (1974) made use of the subharmonic threshold to measure the damping of the oscillations and report acceptable agreement with theory for bubbles of different gases in water. An exception is hydrogen, for which anomalous behavior is reported. A different use of the subharmonic threshold, which appears capable of yielding very accurate data, has recently been proposed by Prosperetti (1976b).

Nonlinear Oscillations

With the assumption of polytropic behavior and with the neglect of thermal and acoustic dissipation, the dynamical Equation (1.5) takes the following form in an oscillating pressure field:

$$R\ddot{R} + \tfrac{3}{2}(\dot{R})^2 = \frac{1}{\rho}\left[p_{i,eq}\left(\frac{R_0}{R}\right)^{3\kappa} - P_\infty(1 - \eta \cos \omega t) - \frac{2\sigma}{R} - 4\mu \frac{1}{R}\dot{R} \right], \qquad (2.10)$$

where, in general, the dimensionless pressure amplitude η is not necessarily small. The numerical studies that can be found in the literature have shown the extreme richness of this equation which appears to lie beyond the capabilities of the available analytical techniques.

Among the earliest numerical work, the contribution of Noltingk & Neppiras (1950, 1951) should be mentioned. With the hypothesis of isothermal behavior ($\kappa = 1$) and with the neglect of viscous effects they were able to show the apparently explosive behavior of the solutions of Equation (2.10) which are sometimes found to consist of a very rapid growth followed by a violent collapse to very small values of the radius within a single period of the driving force. Basing themselves on these numerical results, they deduced some conclusions regarding the mechanism and the conditions for acoustic cavitation that have been used as a guide in much subsequent research on this subject.

Noltingk and Neppiras's work was somewhat limited by the capabilities of the computer at their disposal, but later Flynn (1964) and Borotnikova & Soloukin (1964) published several radius-versus-time curves that illustrate the complexity of the possible responses and their sensitivity to the parameters of the problem. Figure 4, from Borotnikova & Soloukin (1964), shows that the explosive behavior can be observed after several oscillations, as had already been conjectured by Willard (1953),

152   PLESSET & PROSPERETTI

both for bubbles driven below and above resonance. These examples give an indication of the importance of the initial conditions in the subsequent bubble motion, an area in which little research has been conducted.

The most extensive numerical investigation of Equation (2.10) is that undertaken by Lauterborn (1968, 1970a, 1970b, 1976), who has collected an impressive amount of results, still in part unpublished. Figure 5 from his 1976 paper illustrates the steady-state response $X_M = (R_{max} - R_0)/R_0$ (where R_{max} is the maximum value of the radius during the steady oscillations) as a function of the ratio ω/ω_0 for several values of the dimensionless pressure amplitude η. Here ω_0 is the resonant frequency for the linearized oscillations given by Equation (2.8). These calculations refer to an air bubble of radius $R_0 = 10^{-3}$ cm in water under a static pressure $P_\infty = 1$ bar. The complicated pattern of the resonances that gradually unfolds as the pressure amplitude is increased indicates very clearly the importance of nonlinear effects for large-amplitude bubble oscillations.

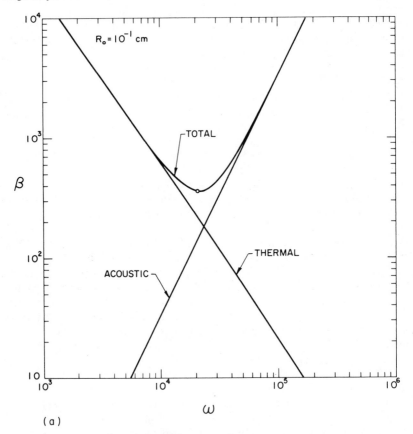

(a)

Figure 3   The damping coefficient β for the small-amplitude forced oscillations of

In Figure 5 the broken vertical lines represent the points at which the steady-state response is observed to jump from some low value to a higher one. This behavior is caused by an instability of the small-amplitude oscillation in a frequency region in which the response curve is really multivalued (see Figure 6). This behavior was found analytically by Prosperetti (1974, 1975, 1976b), who gave a description of small-amplitude forced oscillations including some nonlinear effects. Although his theory is quantitatively accurate only for relatively small values of $\eta[\eta \lesssim 0.25(1 + 2\sigma/R_0 P_\infty)]$ and can account for only a few of the resonances of the system, it gives an insight into the complicated effects of the initial conditions on the ensuing oscillatory motion. An example of his results is shown in Figure 7, which depicts the behavior of the first subharmonic amplitudes $u(t)$, $v(t)$ [with $u(t) \cos \frac{1}{2}\omega t - v(t) \sin \frac{1}{2}\omega t$, the sub-harmonic component in the oscillation] in the (u, v) plane for three different sets of initial conditions. The numbers along the curves represent time in units of ωt, and the dashed lines are the separatrices for the undamped forced oscillations (see

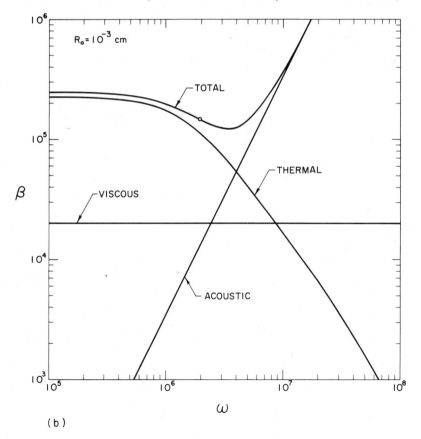

(b)

an air bubble in water as a function of the driving frequency (from Prosperetti 1976a).

Prosperetti 1975 for details). It is seen that it is not possible to express in a simple way the relation between the initial conditions and the value of the amplitudes in the steady-state oscillations (which is the point into which the curves spiral: the origin for curves *a* and *b*, and a nonzero value for curve *c*). One must expect that a much more complicated pattern of this type would be applicable to all the resonances of the complete Equation (2.10), particularly for large-amplitude oscillations for which a small change in the initial conditions can cause markedly different transients and steady-state motions. Such a behavior is indeed reported in the numerical studies by Lauterborn (1976).

In addition to furnishing a suitable example to illustrate the importance of initial conditions, subharmonic bubble oscillations are interesting per se in view of the apparent connection between a signal at half the frequency of the sound field and the development of cavitation in an acoustically irradiated liquid. A subharmonic component in the spectrum of a liquid undergoing acoustic cavitation was first reported by Esche (1952), and later by Negishi (1961). Subsequently De Santis et al (1967), Vaughan (1968), Neppiras (1969a,b), Mosse & Finch (1971), and Coakley (1971) established an apparent causal connection between the onset of cavitation and the appearance of the subharmonic signal. Neppiras (1969a,b), experimenting with single bubbles, showed that they were indeed capable of emitting at a frequency equal to one half of the driving frequency, and advanced the hypothesis that these

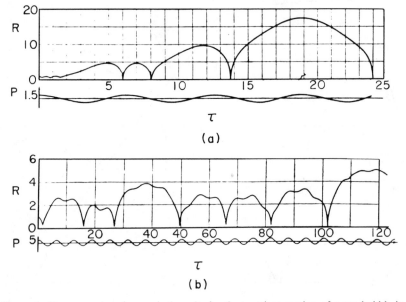

Figure 4 Two examples of numerical results for the transient motion of a gas bubble in an oscillating pressure field. For Figure 4*a* the conditions are $\eta = 1.5$, $\omega/\omega_0 = 0.154$; for Figure 4*b* $\eta = 5$, $\omega/\omega_0 = 1.54$. The ordinate scale is the dimensionless radius R/R_0; the abscissa scale is the dimensionless time ωt (from Borotnikova & Soloukin 1964).

subharmonically oscillating bubbles would evolve into transient cavities that would collapse and break up, and thus would produce the several phenomena associated with acoustic cavitation. Such a hypothesis encounters some difficulties which have been summarized by Prosperetti (1975). This author has also proposed an alternative explanation according to which the bubbles emitting at the subharmonic frequency do not participate actively in the process of cavitation, but act merely as monitors of its occurrence. Central to this explanation are the effects of the initial conditions for the oscillations. Before leaving this subject mention should be made of a work by Eller & Flynn (1969), in which the threshold for the instability of the purely harmonic motion in the subharmonic region is given. As discussed by Prosperetti (1976b), this threshold does not coincide with the true subharmonic threshold except in the immediate neighborhood of $\omega = 2\omega_0$.

A certain number of results concerning free oscillations of gas bubbles [described again by Equation (2.10) with $\eta = 0$] are also available. For the particular case of $\kappa = \frac{4}{3}$, and of vanishing dissipative and surface-tension effects, Childs (1973) has

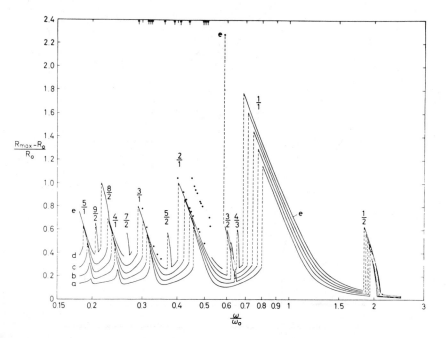

Figure 5 Response curves for the steady oscillations of an air bubble of equilibrium radius $R_0 = 10^{-3}$ cm in water as a function of the ratio of the driving frequency ω to the natural frequency for small oscillations ω_0. The fractions on the resonance peaks denote the order of the resonance [i.e. m/n indicates that m cycles of the oscillations take place during n cycles of the driving-pressure amplitude (from Lauterborn 1976)]. a is for $\eta = 0.4$, b is for $\eta = 0.5$, c is for $\eta = 0.6$, d is for $\eta = 0.7$, e is for $\eta = 0.8$. The dots and the arrows belong to curve e.

obtained an exact solution of Equation (2.10) in terms of elliptic integrals. Nonlinear effects on the small-amplitude motion have been analyzed by Prosperetti (1975), who included surface tension and dissipation. Lauterborn (1968, 1976) reported extensive numerical results on large-amplitude free oscillations including viscous dissipation.

It should be pointed out that in all the work on nonlinear oscillations mentioned above no detailed analysis of the gas contained in the bubble was attempted. There-

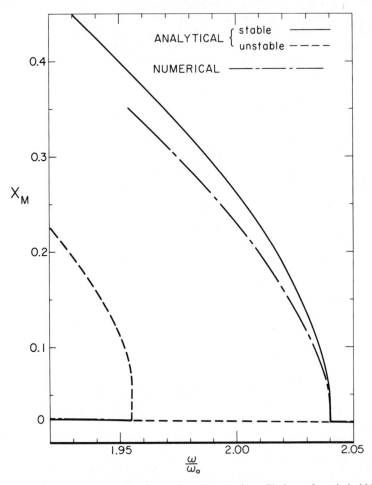

Figure 6   The response curve for the steady subharmonic oscillations of an air bubble of equilibrium radius $R_0 = 10^{-3}$ cm in water. The pressure amplitude is $\eta = 0.3$, and the ambient pressure $P_\infty = 1$ bar. Only viscous damping is taken into account, and the polytropic exponent is $\kappa = 1.33$. The numerical results shown have been obtained by Lauterborn (1974) (from Prosperetti 1974).

fore, all the results described here have been derived with the assumption of a polytropic relationship and sometimes with an assumption of an effective viscosity to account in an approximate way for the effect of the thermal and acoustic dissipation processes. Very recently Flynn (1975a) has derived a somewhat complex formulation of the general problem of cavitation dynamics that includes a detailed analysis of the gas behavior. This formulation has been applied by him to the study

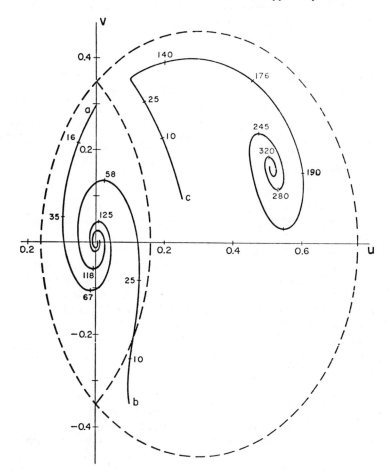

Figure 7 Phase-plane behavior of the transient subharmonic amplitudes u, v [where $u(t) \cos \frac{1}{2}\omega t - v(t) \sin \frac{1}{2}\omega t$ is the subharmonic component of the bubble oscillations] for different initial conditions. The driving frequency is $\omega = 1.8\omega_0$, the pressure amplitude $\eta = 0.4$, the ambient pressure $P_\infty = 1$ bar, and the polytropic exponent $\kappa = 1.33$. The numbers on the curves denote time in units of ωt. The steady oscillations will contain a subharmonic component for the case of the curve labeled c, but not for curves a or b (from Prosperetti 1975).

of large-amplitude free oscillations, and some interesting results on the energy dissipation of the cavity have been obtained (Flynn 1975b).

For large bubbles, such as those produced by underwater explosions, the most significant damping mechanism is acoustic energy radiation in the liquid, and therefore Equation (2.10) cannot be used. For this extreme case, Keller & Kolodner (1956) have obtained another equation that accounts for liquid compressibility in large-amplitude free oscillations. These results have recently been completed and extended to the case of plane and cylindrical bubbles by Epstein & Keller (1972). Other studies of compressibility effects in the motion of gas bubbles are discussed below.

Mass-Diffusion Effects

The mass-diffusion processes taking place across the bubble-liquid interface play a significant role in the behavior of gas bubbles because they may ultimately determine the presence or absence of bubbles in a liquid. The key to these processes is furnished by Henry's law, which establishes a connection between the partial pressure of a gas acting on a liquid surface, p_g, and the equilibrium (or saturation) concentration of gas in the liquid which we denote by c_s:

$$c_s = a p_g. \tag{2.11}$$

Here a is a constant characteristic of the particular gas-liquid combination and is primarily a function of temperature. Equation (2.11) is valid also at a nonplane interface.

If we consider first a situation in which the ambient pressure is fixed and equal to P_∞, then it is clear that unless the gas concentration c at the bubble surface satisfies Equation (2.11) (where we now interpret p_g as the gas pressure in the bubble and neglect for simplicity its vapor content), the bubble will not be in equilibrium, and it will either grow or shrink according to whether $c > c_s$ or $c < c_s$. The mathematical formulation for this process clearly consists of the diffusion equation in the liquid

$$\frac{\partial c}{\partial t} + \frac{R^2}{r^2} \dot{R} \frac{\partial c}{\partial r} = \frac{\alpha}{r^2} \frac{\partial}{\partial r} \left(r^2 \frac{\partial c}{\partial r} \right), \tag{2.12}$$

where α is the coefficient of diffusion of the gas in the liquid, subject to the boundary condition (2.11) at the (moving) bubble surface. Strictly speaking, the pressure p_g in (2.11) should be determined from the dynamical Rayleigh equation. The problem can be substantially simplified by observing that the bubble-wall velocities that are induced by mass diffusion alone are usually quite small in view of the relatively low value of α.[2] Indeed, a reasoning based on purely dimensional considerations suggests that

$$R\dot{R} = \alpha \frac{c_\infty - c_s}{\rho_g}, \tag{2.13}$$

---

[2] For air-water at 20°C, for example, $\alpha \simeq 2 \times 10^{-5}$ cm^2 sec^{-1}. An order-of-magnitude estimate for this quantity is given by the familiar result due to Einstein, $\alpha = 6\pi\mu a k_B T$ where k_B is the Boltzmann constant, and a is an equivalent radius of the gas molecule.

where c_∞ is the dissolved mass concentration at a large distance from the bubble and ρ_g is the density of the gas within the bubble. For typical values, one might take $(c_\infty - c_s)/\rho_g \sim 10^{-2}$, $R_0 \sim 10^{-3}$ cm; Equation (2.13) then gives an estimate of $\dot{R} \sim 10^{-4}$ cm sec^{-1} for the growth (or dissolution) velocity of an air bubble in water. These very low values allow one to disregard the inertial effects associated with the bubble motion and to neglect the convective term in Equation (2.12).

An approximate solution based on these simplifications was obtained some time ago by Epstein & Plesset (1950), who included surface-tension effects. Their result, with the neglect of the latter, takes the form

$$RR = \alpha \frac{c_\infty - c_s}{\rho_g}[1 + R(\pi\alpha t)^{-1/2}]. \tag{2.14}$$

The second term in this equation reflects the transient effect of the buildup of a diffusion boundary layer adjacent to the bubble surface that quickly becomes large compared with the bubble itself. Indeed, from (2.13) we find that, asymptotically, $R \sim [2\alpha(c_\infty - c_s)t/\rho_g]^{1/2}$ from which $R(\pi\alpha t)^{-1/2} \sim [2(c_\infty - c_s)/\pi\rho_g]^{1/2} \ll 1$. The physical meaning of the result of Equation (2.14) can be clarified by noting that if m_g denotes the gas content in the bubble so that $dm_g/dt = 4\pi R^2 \rho_g\, dR/dt$, elimination of dR/dt formally reduces it to the expression for the heat flux at the surface of a perfectly conducting sphere in an infinite medium (Carslaw & Jaeger 1959). Experimental evidence in favor of Equation (2.14) was reported by Krieger et al (1967), who proposed a method for the measurement of diffusion coefficients of gases in liquids based on the observation of bubble-dissolution rates. A systematic perturbation approach to the solution of the problem has been formulated by Duda & Vrentas (1969a,b), and an interesting treatment of the case of more than one gas has recently been given by Ward & Tucker (1975). Birkhoff et al (1958) have obtained a similarity solution for the complete problem including convection effects, and the same solution was obtained by Scriven (1959). Szekely and co-workers (1971, 1973) have given a detailed analysis that includes dynamic, viscous, and surface-kinetics effects.

The physical situation is quite different if the bubble is immersed in an oscillating pressure field, and it is readily seen that in this case it may grow even in an undersaturated solution. The mechanism giving rise to this effect may be explained in physical terms in the following way. Suppose that the amplitude of the forced oscillations is large enough so that the gas-liquid solution at the bubble surface becomes undersaturated during the compression half-cycle and supersaturated during the expansion half-cycle. Corresponding to these conditions there will be a mass exchange of alternating direction between the bubble and the liquid. If the interface were plane, the average flux over one oscillation would clearly be zero. However, as a consequence of the spherical geometry, on the average the surface area during the mass inflow is greater than during the mass outflow, so that a net increase in the mass of gas contained in the bubble results. In view of its second-order nature, this effect has been called "rectified mass diffusion" (Blake 1949). An additional consequence of the spherical geometry, however, combines with the area effect just mentioned, namely, the fact that during the expansion half-cycle the thickness of the diffusion layer adjacent to the bubble surface decreases, while it increases during the compression half-cycle. As a result concentration gradients (and hence mass-flow rates) during the mass inflow half-cycle are enhanced over those prevailing during the mass outflow part of the oscillation. A quantitative analysis of the convective effect shows it to be dominant, as was demonstrated by Hsieh & Plesset (1961), who gave the first rigorous treatment of the rectified diffusion process based on a linearized formulation of the problem. For the case in which the liquid is saturated in the absence of the bubble, neglecting surface-tension effects and assuming isothermal behavior, they find

$$R\dot{R} = \tfrac{2}{3}\alpha \frac{c_s}{\rho_g}\left(\frac{P_{\max}-P_0}{P_0}\right)^2,$$ (2.15)

where $P_{\max}$ and P_0 are the maximum and the average value of the bubble internal pressure, respectively.

If $P_{\max}$ is taken to be a constant, Equation (2.15) is readily integrated and for large times predicts a $t^{1/2}$ dependence of the bubble radius. The fact that in practice bubbles are not seen to grow indefinitely, as this result would imply, is a consequence of the instability of the spherical shape, which, as is discussed below, is an important effect for relatively large bubbles (Hsieh & Plesset 1961). A significant feature of Equation (2.15) is that it shows bubble growth by rectified diffusion to be a very slow process relative to the period of the sound fields commonly encountered. For instance, for an air bubble in water at 20°C at one atmosphere, with $(P_{\max}-P_0)/P_0 = 0.25$, Equation (2.15) predicts a doubling time that ranges from 1.1×10^6 sec for an initial radius of 10^{-1} cm to 1.1×10^2 sec for an initial radius of 10^{-3} cm.

When the liquid is not saturated, the bubble will eventually disappear if the mass flux caused by rectified mass diffusion is not sufficient to balance the loss of mass required by Henry's law. The value, η_{th}, of the pressure amplitude at which the two fluxes are equal corresponds to the threshold for the growth of the bubble by rectified diffusion, and was estimated very simply by Strasberg (1961) by equating Equation (2.13) and Equation (2.15). His result, which includes surface-tension effects, is

$$\eta_{th} = (\tfrac{3}{2})^{1/2}\left(1 + \frac{2\sigma}{R_0 P_0} - \frac{c_\infty}{c_s}\right)^{1/2}.$$ (2.16)

Here, and in the following, c_s is the saturation concentration at the average pressure of the sound field only for bubbles driven so much below resonance that inertial effects in their motion are insignificant. As was pointed out by Safar (1968), inertial effects introduce a correction in Equation (2.16) which for isothermal behavior becomes

$$\eta_{th} = (\tfrac{3}{2})^{1/2}\left(1 - \frac{\omega^2}{\omega_0^2}\right)\left(1 - \frac{c_\infty}{c_s} + \frac{2\sigma}{R_0 P_0}\right)^{1/2}\left(1 + \frac{4\sigma}{3R_0 P_0}\right)\left(1 + \frac{2\sigma}{R_0 P_0}\right)^{-1/2},$$ (2.17)

where ω_0 is given by Equation (2.8) with $\kappa = 1$.

A major approximation contained in the above results consists in the neglect of the nonlinear effects associated with the bubble oscillations. This aspect of the problem was analyzed by Eller & Flynn (1965), who made use of a method developed by Plesset & Zwick (1952) to describe the growth of vapor bubbles. While we return to this method in greater detail below, it may be mentioned here that its applicability is limited to those situations in which appreciable concentration gradients are present only in a layer adjacent to the bubble surface that is thin compared with the bubble radius. Clearly, for the threshold problem under consideration here, this condition takes the form $R_0 \gg (\alpha/\omega)^{1/2}$, and hence is usually met in practice. Eller and Flynn's result is

$$\frac{c_\infty}{c_s} - \left(1 + \frac{2\sigma}{R_0 P_0}\right) \frac{\langle R/R_0 \rangle}{\langle (R/R_0)^4 \rangle} = 0, \tag{2.18}$$

where the brackets denote the average of the enclosed quantity over one period of oscillation. The pressure amplitude of course enters in this equation through the averages, which can be computed either numerically or by an analytic perturbation scheme (Eller & Flynn 1965; Eller 1969, 1972, 1975). For the case in which $c_\infty = c_s$ and $\sigma/R_0 P_0 \ll 1$, Eller gives the result

$$\eta_{th} = \left(1 - \frac{\omega^2}{\omega_0^2}\right)\left(\frac{3\sigma}{R_0 P_0}\right)^{1/2}\left(1 - \frac{\omega^2}{8\omega_0^2}\right)^{-1/2}, \tag{2.19}$$

which is seen to coincide with Safar's modification of the Hsieh-Plesset theory except for the last factor, which is introduced by the nonlinear effects in the second-order approximation. A similar result has been derived from other techniques by Skinner (1970).

Both Equations (2.17) and (2.19) appear to predict a vanishing threshold at resonance. This feature of course is a consequence of the neglect of damping effects which have recently been analyzed by Eller (1975). He found a very pronounced minimum in the threshold curve at resonance and considered the effect of this behavior on the evolution in time of a population of bubbles with given initial radius.

Experimental observations of threshold values were first reported by Strasberg (1959, 1961) at 25 k Hz, and more recently by Eller at 26.6 k Hz (1969) and at 11 k Hz (1972) and by Gould (1974) at 20 k Hz. It is found that the data given by Strasberg at 25 k Hz and by Eller at 26.6 k Hz tend to fall somewhat below the theoretical predictions. Although the discrepancy is not serious for bubble radii greater than about 4×10^{-3} cm, a definite disagreement appears to exist for smaller radii. The data at 11 k Hz are in better agreement with theory, provided that the bubbles are taken to oscillate adiabatically rather than isothermally (Eller 1972). No data for bubbles smaller than 5×10^{-3} cm are reported, however. The data of Gould (1974) are also generally in good agreement with Equation (2.19) for a polytropic form of the pressure-volume relation. The smallest bubble observed by him had a radius of 5.5×10^{-3} cm.

A rigorous analysis of nonlinear effects in the theory of bubble growth by rectified diffusion is more complex than for the threshold calculation because now, in addition to the oscillatory boundary layer of thickness $(\alpha/\omega)^{1/2}$, one must take into account the layer in which the gas concentration is depleted because of the net mass diffusion into the bubble. It is easy to convince oneself that the thickness of this layer is of the order of the bubble radius or greater so that the applicability of Eller and Flynn's method is not immediately evident. The problem has recently been considered by Skinner (1972), who obtained a solution with the aid of multiple time- and length-scale expansions. Surprisingly, his results are practically equal to those previously derived by Eller (1969) by the same approach as that used for the threshold condition by Eller & Flynn (1965). A comparison with data (Eller 1969), however, shows that the bubble growth through resonance is much sharper than the theory predicts. Other large discrepancies between theory and experiment for bubble growth rates have also been reported by Eller (1969), who observed rates twenty times larger than the theoretical results. Recent experiments by Gould (1974) appear to substantiate Eller's conjecture that the disagreement is to be ascribed to acoustic streaming taking place in the vicinity of the bubble surface. He observed the appearance of surface oscillations on the bubble and of a streaming motion in its vicinity together with large increases in growth velocity above the theoretical values. Growth rates measured in the absence of streaming, however, tend to be smaller than the predicted values. Apparently, the connection between surface oscillations and streaming, although experimentally well documented (Kolb & Nyborg 1956, Elder 1959, Gould 1966), has not yet found a proper theoretical treatment. The studies of Davidson (1971) and of Davidson & Riley (1971) refer to the case in which the streaming is produced by an oscillatory motion of the center of a spherical bubble and is therefore not directly applicable to surface-oscillation-induced streaming.

Acoustic Cavitation and Applications

In the terminology of Blake (1949; see also Flynn 1964) we may distinguish between gaseous and vaporous acoustic cavitation. The first description refers to the phenomena associated with bubbles that have predominantly a noncondensable gas content; the second, to bubbles that have predominantly a vapor content. While we have discussed several features of the dynamics of gaseous cavities above, and we shall discuss the behavior of vapor cavities in the next section, it should be realized that at present it is not yet possible to relate all of the specific features of experimentally observed cavitation phenomena with the available experimental and theoretical work on single gas or vapor bubbles. Although it is generally agreed that cavitation damage, white noise, sonoluminescence, chemical reactions, and other features of cavitation are associated with violent bubble motion, at present not only a quantitative understanding of these phenomena is lacking but sometimes even the physical mechanisms through which they take place are obscure.

Violent collapse, as can be seen in Figure 4, is certainly a possible behavior for a gas bubble in an oscillating pressure field, and estimates of the temperatures and pressures reached by the bubble content toward the end of the collapse are of the order of several thousands of degrees Kelvin and of thousands of atmospheres, respectively (see e.g. Flynn 1964; 1975a,b). Unfortunately, during the later stages of the collapse the spherical shape of the bubble becomes highly unstable (see below), and the duration of the collapse itself is so short that adequate theoretical or experimental investigations appear very difficult. These same considerations can be applied also to vapor-bubble collapse in a highly subcooled liquid.

No less obscure is the beginning of a vapor- or gas-bubble life. This problem is essentially that of nucleation, and the uncertainties associated with it have a significant bearing not only on acoustic cavitation but, perhaps more importantly, on boiling heat transfer. Indeed, while it is observed that homogeneous-nucleation theory based on thermodynamic fluctuations describes very well the behavior of certain organic liquids, such as pentane, hexane, heptane, ether, and benzene (see e.g. Cole 1974, Blander & Katz 1975), the same conclusion cannot be drawn for such an important liquid as water for which the presence of impurities appears to play a determinant role. We cannot enter here into any detail on these matters, and we refer the reader to some pertinent recent papers (Apfel 1970, Gavrilov 1970, Holl 1970, Keller 1972). A useful summary of previous work is given by Flynn (1964).

Once nucleation has taken place, a gas bubble will grow by rectified diffusion over many periods of the driving pressure oscillations until it breaks up. The lifetime of a vapor bubble, however, will in general be much shorter, not more than a few acoustic cycles, and its growth will be determined more by dynamic effects than by a change in the mass content of the cavity. Some aspects of this process are dealt with in the following section.

The description of cavitation events is made even more difficult by the apparent importance of cooperative effects of many bubbles in several phases of the process (Willard 1953). Space limitations prevent us from entering into any detail here. The reader is referred to the works of Willard (1953), Strasberg (1959), Noltingk (1962), Flynn (1964), Coakley (1971), Hinsch, Bader & Lauterborn (1974), and Nyborg (1974). Work on sonoluminescence is described in Negishi (1961), Taylor & Jarman (1968, 1970), Margulis (1969), Saksena & Nyborg (1970), and Coakley & Sanders (1973).

Finally, for some industrial applications of acoustic cavitation the reader is referred to Neppiras (1965). A growing area of interest appears to be in biomedical applications (Rooney 1970, 1972; Nyborg 1974; Hill 1972).

3 CAVITATION BUBBLES AND VAPOR BUBBLES

In this section we consider the analysis of some features of the bubbles that are composed predominantly of vapor. As was suggested in a preceding section, these bubbles also usually originate from small gas nuclei. However, if a bubble grows to many times its initial size so rapidly that mass-diffusion effects are negligible, the gas content plays an unimportant role in the dynamics except possibly near the end of a violent collapse.

We are not concerned here with the problem of boiling heat transfer, which has recently been considered by Rohsenow (1971). Rather, we limit our discussion to the growth and collapse of vapor bubbles in unbounded liquids. In the presence of boundaries, the free surface loses its spherical symmetry. Some of these boundary effects are considered in the next section.

Preliminary Considerations

In the study of the dynamics of vapor bubbles it is helpful to make a distinction between two limiting cases according to the importance of thermal effects. The physical basis for this distinction rests on the combined effects of the strong temperature dependence of the equilibrium vapor density and of the rate of change of the equilibrium vapor pressure with temperature. To illustrate this point we may consider the case of a bubble in water that grows to a radius R in a time t. The amount of thermal energy required to fill the bubble with vapor in thermodynamic equilibrium with the water is $(4/3)\pi R^3 L \rho_v^e(T)$, where L is the latent heat of evaporation and $\rho_v^e(T)$ is the equilibrium vapor density at the water temperature T. This energy is made available through a drop ΔT in the temperature of a surrounding liquid layer of thickness of the order of the diffusion length given by $(Dt)^{1/2}$, where $D = k/\rho c_l$ is the thermal diffusivity in the liquid. The heat energy supplied is, therefore, $4\pi R^2 (D_l t)^{1/2} \rho c_l \Delta T$. When these two expressions are equated, one obtains an estimate of the temperature drop ΔT as

$$\Delta T \simeq \frac{1}{3} \frac{R \rho_v^e(T) L}{(D_l t)^{1/2} \rho c_l}. \tag{3.1}$$

For water at 15°C, with $R = 0.1$ cm and $t = 10^{-3}$ sec, $\rho_v = 1.3 \times 10^{-5}$ g cm^{-3}, one finds $\Delta T \sim 0.2$°C. This temperature drop causes a decrease in the vapor pressure of the order of 1%, and hence has an insignificant effect on the growth of the bubble. For water in the neighborhood of 100°C, however, the equilibrium vapor density is about 46 times its value at 15°C, and one finds $\Delta T \sim 13$°C, with a corresponding decrease in vapor pressure of roughly 50%. Clearly, the bubble-growth dynamics will be drastically altered with a longer growth time to the same radius. Thermal effects rather than inertial effects now dominate the growth process. We refer to the first case as "cavitation" bubbles and to the second one as "boiling" or "vapor" bubbles. A similar difference in behavior will be observed in the collapse process. For a cavitation bubble the internal pressure will remain practically constant until the latest stages of the collapse, while for a boiling bubble a much greater effect from the vapor pressure will be observed.

The Dynamics of a Cavitation Bubble

With the neglect of viscous effects and by use of the identity

$$\frac{1}{2} \frac{1}{R^2 \dot{R}} \frac{d}{dt} (R^3 \dot{R}^2) = R \ddot{R} + \tfrac{3}{2} \dot{R}^2, \tag{3.2}$$

Equation (1.5) can be integrated once if the ambient pressure is taken to be independent of time and if, in the light of the above discussion, the internal pressure $p_i = p_v$ is also regarded as a constant. The result is

$$\dot{R}^2 = \left(\frac{R_0}{R}\right)^3 \dot{R}_0^2 + \frac{2}{3} \frac{p_v - p_\infty}{\rho} \left[1 - \left(\frac{R_0}{R}\right)^3\right] - \frac{2\sigma}{\rho R} \left[1 - \left(\frac{R_0}{R}\right)^2\right], \tag{3.3}$$

where the subscript zero denotes the initial conditions for the growth. If $p_v > p_\infty$, it is seen that for $R \gg R_0$ the velocity is approximately equal to its asymptotic value

$$R = \left(\frac{2}{3}\frac{p_v - p_\infty}{\rho}\right)^{1/2}, \tag{3.4}$$

which can also be obtained by equating the kinetic energy of the flow to the work performed by the pressure forces. It is shown below that Equation (3.4) gives also the growth velocity of a boiling bubble at high superheats or low ambient pressures in the early and intermediate stages. If $p_v < p_\infty$ but an initial impulse is imparted to the bubble wall, Equation (3.3) predicts that the bubble would reach a maximum radius that, with the neglect of surface tension, is given by

$$R = [1 + \tfrac{3}{2}\rho \ \dot{R}_0^2(p_\infty - p_v)^{-1}]R_0. \tag{3.5}$$

Under the same assumptions one can obtain an expression similar to (3.3), valid for the collapse of the bubble starting from some initial radius R_i. If the initial velocity is taken to vanish, one has

$$\dot{R}^2 = \frac{2}{3}\frac{p_\infty - p_v}{\rho}\left[\left(\frac{R_0}{R}\right)^3 - 1\right] + \frac{2\sigma}{\rho R}\left[\left(\frac{R_i}{R}\right)^2 - 1\right]. \tag{3.6}$$

This equation would predict that the velocity approaches infinity as $R^{-3/2}$ as $R \to 0$, which is unacceptable. Therefore, one must conclude that the approximations under which Equation (3.6) is obtained will eventually break down during the collapse process. We return to this point below. Let us observe here that, in the absence of surface-tension effects, one can compute from Equation (3.6) the time t_0 required for complete collapse of the cavity:

$$t_0 = \frac{\Gamma(5/6)}{\Gamma(1/3)}\left[\frac{3\pi\rho}{2(p_\infty - p_v)}\right]^{1/2} R_i \simeq 0.915 \left(\frac{\rho}{p_\infty - p_v}\right)^{1/2} R_i, \tag{3.7}$$

which is the result of Rayleigh (1917). An extension to the collapse of a closed cavity of arbitrary shape has been given by Miles (1966), who, for oblate and prolate spheroids, finds a correction to (3.7) of the order of the fourth power of the eccentricity (the quantity R_i is defined as the radius of a sphere of equivalent volume for this problem).

Let us now return to the question of the dynamics of the bubble in the later stages of the collapse. An important feature of this problem that should be kept in mind in the following discussion is that the spherical configuration of the bubble surface is unstable during the collapse, so that analyses based on the assumption of spherical symmetry cannot be rigorously correct. Nevertheless, this assumption furnishes an order-of-magnitude estimate of the several quantities involved, and it gives a qualitative picture of the phenomenon. It is in this spirit that the following considerations should be interpreted. The most obvious reason for the nonphysical behavior of (3.6) for $R \ll R_i$ is the neglect of liquid compressibility which becomes important as soon as bubble-wall velocities become comparable with the speed of sound in the liquid. A very successful modification of the Rayleigh equation that takes into account liquid compressibility was obtained by Gilmore (1952; see also Plesset 1969) on the basis of the Kirkwood-Bethe (1942) approximation. This approximation consists in assuming that the quantity $r(h+\frac{1}{2}u^2)$, with u the liquid velocity and h the enthalpy, is propagated unaltered along the outgoing characteristics, $dr = (u+c)\,dt$; that is,

$$\frac{D}{Dt}[r(h+\tfrac{1}{2}u^2)] + c\frac{\partial}{\partial r}[r(h+\tfrac{1}{2}u^2)] = 0. \tag{3.8}$$

In this equation $c = c(p)$ denotes the speed of sound in the liquid, and $D/Dt = \partial/\partial t + u\partial/\partial r$ is the material derivative. The derivatives with respect to r in the second term of Equation (3.8) can be eliminated with the aid of the momentum equation, $Du/Dt = -\partial h/\partial r$, and of the continuity equation,

$$\frac{\partial u}{\partial r} = -\frac{1}{c^2}\frac{Dh}{Dt} - \frac{2u}{r}.$$

The result then leads to the following equation for the bubble wall

$$\left(1 - \frac{1}{C}\dot{R}\right)R\ddot{R} + \tfrac{3}{2}(\dot{R})^2\left(1 - \frac{1}{3C}\dot{R}\right) = H\left(1 + \frac{1}{C}\dot{R}\right) + \left(1 - \frac{1}{C}\dot{R}\right)\frac{R}{C}\dot{H}. \tag{3.9}$$

H and C denote here the values of the quantities h and c at the bubble wall. If surface tension and viscosity are omitted from the boundary conditions and the bubble internal pressure p_v is taken to be a constant, Equation (3.9) can be integrated analytically once to obtain

$$\log\frac{R}{R_i} = -2\int_{\dot{R}_0}^{\dot{R}}\frac{U(U-C)\,dU}{U^3 - 3CU^2 + 2HU + 2HC}. \tag{3.10}$$

The bubble-wall velocity, $\dot{R}$, has been written as U in this equation. The initial conditions correspond to a bubble of radius R_i that at $t=0$ undergoes a step increase in the ambient pressure. The initial "release" velocity $\dot{R}_0$ for this situation is easily computed from (3.9) and found to be given by $\dot{R}_0 \simeq (p_v - p_\infty)/\rho_\infty c_\infty$, where ρ_∞ and c_∞ are the values of the liquid density and sound speed at large distance from the bubble. In the approximation $|H| \ll C^2$, Equation (3.10) gives the following correction to Equation (3.6):

$$\left(\frac{R_i}{R}\right)^3 = \left(1 - \frac{1}{3}\frac{\dot{R}}{C}\right)^4\left(1 + \frac{3}{2}\frac{\rho_\infty}{p_\infty - p_v}\dot{R}^2\right), \tag{3.11}$$

from which we find $\dot{R} \propto R^{-1/2}$ as $R \to 0$, in contrast to the incompressible approximation that was shown to give $\dot{R} \propto R^{-3/2}$.

The accuracy of Equation (3.9) can be assessed through a comparison with the full solution of the problem which can be obtained only numerically. Such an analysis was conducted by Hickling & Plesset (1964), who considered both empty cavities (i.e. essentially $p_\infty - p_v$ = constant) and cavities with a very small gaseous content that undergo isentropic or isothermal compression. An example of their results is shown in Figure 8, which shows that Equation (3.9) is accurate up to bubble-wall Mach numbers of the order of 5. Their analysis included also the details of the liquid flow, with the computation continued past the rebound point of the gas bubble up to the formation of the shock wave in the liquid. They found

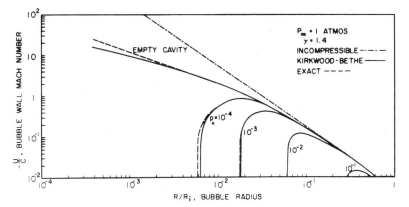

Figure 8   The bubble-wall Mach number as a function of the dimensionless radius R/R_i for a gas bubble collapsing in a compressible liquid (from Hickling & Plesset 1964).

that the peak pressure has an approximate $1/r$ dependence during the rebound, with pressures in the cavity of the order of tens of kilobars. Results of the same order of magnitude were obtained by Flynn (1975b). While this estimate of the internal pressure (and hence of the pressure distribution in the neighboring liquid) may be off by as much as an order of magnitude, it appears likely that the $1/r$ dependence would not be greatly affected by a more accurate description of the latest stages of the collapse. It may be of interest to observe that the Kirkwood-Bethe approximation gives good results for collapse up to relatively large Mach numbers since the collapse motion leads to an expansion wave in the liquid. For bubble growth at very large velocities, the same accuracy could not be expected since this motion would lead to the propagation of a shock wave into the liquid.

Following Benjamin (1958), Jahsman (1968) has given an analytic treatment of the collapse of a gas bubble in a compressible liquid based on a perturbation expansion in terms of a small parameter that is essentially the liquid compressibility. His results show that the Kirkwood-Bethe assumption is verified for the zero and first-order terms of the expansion, but breaks down in higher orders. In this connection mention should be made also of a work by Hunter (1960), who, guided by a numerical investigation, developed a similarity solution for the collapse of an empty cavity in the neighborhood of the collapse point in the region where $c \gg c_\infty$; he found that in the latest stages of the collapse $\dot{R} \propto R^{-0.8}$. It should be remarked that the early work on this subject, summarized by Cole (1948, see also Trilling 1952), was motivated by interest in underwater explosions. Most of the later contributions discussed above were undertaken to investigate the mechanism of cavitation damage, since it was conjectured that the very strong pressure pulses radiated by a collapsing cavity might be responsible for it. The current understanding of cavitation damage, however, is quite different, as is discussed in the following section.

In addition to the neglect of compressibility effects, a second obvious reason for the failure of Equation (3.6) to model in an acceptable way the physical behavior of a collapsing bubble is the neglect of the variation of p_v that is brought about not only by the increasing importance of even small quantities of noncondensible gas in the cavity as $R \rightarrow 0$ but also by the fact that condensation of the vapor cannot keep up with the bubble-wall motion when its velocity becomes of the order of the speed of sound in the vapor. Some remarks on this aspect of the problem are made below.

A quantitative experimental confirmation of the collapse time predicted by Equation (3.7) was given by Lauterborn (1972a), who realized a very close physical approximation to the Rayleigh empty-bubble model by focusing a giant pulse of a Q-switched ruby laser on an interior point of a liquid mass (Lauterborn 1972b). He reported an experimental value of $t_0 = 297 \times 10^{-6}$ sec to be compared with a theoretical value of 300×10^{-6} sec for a 0.378-cm bubble in water. It should be remarked that the final stage of the collapse is so rapid that Equation (3.7) is negligibly affected by the liquid compressibility, the behavior of the bubble content, or the deviation from spherical shape discussed below.

The first investigation of the behavior of a cavitation bubble in a time-dependent pressure field was carried out some time ago by the integration of the Rayleigh equation with constant p_i for a bubble entrained in a liquid flowing past a submerged object (Plesset 1949). For the ambient pressure $p_\infty(t)$ he made use of the experimentally determined pressure distribution along the object in the absence of the bubble at the point occupied by the bubble at time t, and he obtained a very good agreement with the data. Another interesting study of cavitation-bubble dynamics in a time-varying pressure field has been made by Hsieh (1970), who obtained approximate analytic expressions for the maximum radius that a cavity would attain in a viscous liquid subject to a transient pressure pulse. From his results he concluded that the empirical relation $\overline{P_{th} \propto \mu^{0.2}}$, obtained by Bull (1956) with a stress-wave technique, between liquid viscosity and threshold pressure for cavitation inception appears to be due more to a coincidental distribution of nuclei in the different liquids tested than to other more basic features of the cavitation process.

The Growth of Vapor Bubbles

As mentioned above, a vapor bubble will designate a bubble in the dynamics of which thermal effects play a dominant role. It is supposed that the nucleus from which the bubble will eventually grow is a small spherical cavity of radius R_0 in a liquid at uniform temperature T_∞. At equilibrium, the internal pressure in the nucleus will be the equilibrium vapor pressure corresponding to the liquid temperature, $p_v^e(T_\infty)$, and the equilibrium radius is

$$R_0 = \frac{2\sigma}{p_v^e(T_\infty) - p_\infty} \tag{3.12}$$

where as usual p_∞ denotes the ambient pressure, which is here taken to be constant. The pressure p_∞ corresponds to a well-determined equilibrium temperature, a boiling temperature, which we denote by T_b. Equation (3.12) implies $T_\infty > T_b$; the difference $\Delta T = T_\infty - T_b$ is termed the liquid superheat. This simple model for the vapor nucleus is certainly highly idealized, but it can be shown that initial conditions do not affect significantly the growth of the bubble.

The process of bubble growth in superheated liquids can readily be described in physical terms as follows. When the equilibrium situation depicted by (3.12) is disturbed, the bubble starts to grow very slowly under the restraining effect of surface tension. If the initial superheat is sufficiently large, the growth velocity will eventually reach the asymptotic value for a cavitation bubble, as given by (3.4), before the rate of vapor inflow (which is proportional to R^2) is so large as to produce a substantial cooling of the surrounding liquid. At this point, both inertial and thermal effects limit the growth rate. The growth rate then begins to decrease making inertial effects less and less important until the radius has grown so large that the growth process is limited only by the rate at which heat can be supplied to the bubble wall. The velocity of growth during this asymptotic stage is readily estimated by noting that the bubble internal pressure will have decreased nearly to p_∞, so that the bubble surface temperature is T_b. The heat flow into the bubble from the liquid is then approximately $4\pi R^2 k_l \Delta T/(D_l t)^{1/2}$, where D_l and k_l are the liquid thermal diffusivity and conductivity, respectively, and t is the time from inception of the growth. This heat flux will be balanced by the absorption of the latent heat necessary to vaporize the liquid into the bubble, which is given by $4\pi R^2 L \rho_v^e(T_b)\dot{R}$ where L is the latent heat and $\rho_v^e(T_b)$ is the equilibrium vapor density corresponding to the boiling temperature. Equating the above quantities, we find

$$\dot{R} = \left(\frac{3}{\pi}\right)^{1/2} \frac{k_l}{L\rho_v^e(T_b)} \frac{T_\infty - T_b}{(D_l t)^{1/2}}. \tag{3.13}$$

The factor $(3/\pi)^{1/2}$ has been inserted in (3.13) so as to give the result that is obtained in the asymptotic analysis of Plesset & Zwick (1954). Integration of Equation (3.13) leads to $R \propto t^{1/2}$, a widely used result in bubble-growth theory. It should be emphasized that Equation (3.13) expresses only an asymptotic relationship that is valid only for times large enough so that the growth velocity predicted by it is smaller than the inertia-controlled value (3.4). Indeed, this remark enables one to make a distinction between a cavitation and a boiling bubble since the characteristic growth rate for cavitation-bubble growth is (3.4) and the characteristic growth rate for the boiling-bubble growth is (3.13) (in this connection, see also Brennen 1973).

The complete mathematical formulation of the problem of spherical-bubble growth is similar to the one indicated above for the mass-diffusion problem inasmuch as one has a partial differential equation for the energy (or temperature) in the liquid,

$$\frac{\partial T}{\partial t} + \frac{R^2}{r^2}\dot{R}\frac{\partial T}{\partial r} = \frac{D_l}{r^2}\frac{\partial}{\partial r}\left(r^2\frac{\partial T}{\partial r}\right), \tag{3.14}$$

that is coupled to the Rayleigh equation. For the time being we assume that the internal pressure in the bubble is given by the equilibrium vapor pressure corresponding to the bubble surface temperature as determined by (3.14). Some comments on nonequilibrium effects are made below. The initial and boundary conditions for Equation (3.14) are

$$T(r,0) = T_\infty,$$ (3.15a)

$$T(r,t) \to T_\infty \quad \text{as } r \to \infty,$$ (3.15b)

$$-4\pi R^2 k_l \frac{\partial T}{\partial r} = L\frac{d}{dt}(\tfrac{4}{3}\pi\rho_v^e R^3), \quad \text{at } r = R(t),$$ (3.15c)

where the last equation is an expression of the heat balance at the bubble boundary.

A general solution to the problem posed by (3.14) and (3.15) was obtained by Plesset & Zwick (1952) under the assumption that appreciable temperature gradients are established only in a thin layer surrounding the cavity radius. Their result for the bubble surface temperature $T_s(t)$ is

$$T_s(t) = T_\infty - \left(\frac{D_l}{\pi}\right)^{1/2}\int_0^t \frac{R^2(\theta)\dfrac{\partial T}{\partial r}(R,\theta)}{\left[\displaystyle\int_\theta^t R^4(\lambda)\,d\lambda\right]^{1/2}}\,d\theta,$$ (3.16)

where $\partial T(R,t)/\partial r$ is an arbitrarily specified temperature gradient at $r = R(t)$. Upon substitution of (3.15c) for this quantity and with neglect of the temperature dependence of L, ρ_v, k_l, and with the approximation of the pressure-temperature relation by a straight line, Plesset & Zwick (1954) obtained the following formulation of the problem:

$$p^{7/3}\frac{d^2p}{du^2} + \tfrac{7}{6}p^{4/3}\left(\frac{dp}{du}\right)^2 = 3\left[1 - \mu\int_0^u (u-v)^{-1/2}\frac{dp}{dv}(v)\,dv - \frac{2\sigma}{p^{1/3}}\right],$$ (3.17)

where $p = R^3/R_0^3$ is the normalized volume and u is a new dimensionless time scale defined as

$$u(t) = \frac{\alpha}{R_0^4}\int_0^t R^4(\theta)\,d\theta.$$ (3.18)

The constants α and μ are defined by

$$\alpha = \{[p_v^e(T_\infty)-p_\infty]/\rho_l\}^{1/2}/R_0,$$

$$\mu = \frac{1}{3}\left(\frac{2\sigma D_l}{\pi}\right)^{1/2}\rho_v^e(T_b).\frac{L}{k_l}(T_\infty-T_b)^{-1}\{\rho_l[p_v^e(T_\infty)-p_\infty]\}^{-1/4}.$$ (3.19)

From (3.17) and (3.18) they then proceeded to obtain various asymptotic expansions for the radius-time dependence, the leading term of which, for large times, is given by (3.13) (see also Plesset & Zwick 1955). The experimental data of Dergarabedian (1953, 1960) obtained for low superheats (up to about 6°C) in water, carbon tetrachloride, benzene, and other organic liquids agree very well with the results of Plesset and Zwick, but for these cases the asymptotic stage is reached so fast that essentially only the asymptotic results (3.13) could be tested with the data. The same can be said of the work of Niino et al (1973), who obtained superheats of up to

14°C. A close experimental simulation with high superheats of the theoretical conditions to which the complete Plesset-Zwick theory applies is rather difficult because large thermal gradients in the liquid can be easily generated, and further because observation of the bubble in the inertia-controlled phase of the growth, when the radius is small and the growth velocity large, is difficult. The best support of the theory (at least as a mathematical approximation to a more complete set of equations) can be found in comparing its predictions with the results of some extensive numerical computations performed by Dalle Donne & Ferranti (1975), who solved the complete energy equation, Equation (3.14), simultaneously with the Rayleigh equation. Their investigation was primarily motivated by doubts about the validity of the thin-thermal-layer assumption for liquids of high thermal conductivity such as liquid sodium. A detailed comparison of the Plesset-Zwick theory with Dalle Donne and Ferranti's results is available elsewhere (Plesset & Prosperetti 1976, 1977). In summary it can be said that the agreement with Equation (3.15) is extremely good up to liquid-sodium superheats of 300°C provided that the liquid and vapor properties are evaluated at the boiling temperature as indicated in Equations (3.19). An example of this comparison is given in Figure 9. The physical reason for this result is simply that at very high superheats inertial effects play a

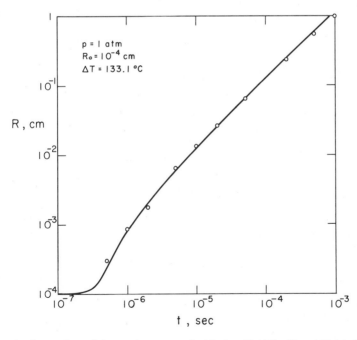

Figure 9 Comparison of the results computed with the aid of the Plesset-Zwick formulation [Equation (3.17), full line] with those obtained by Dalle Donne & Ferranti (1975) by means of numerical integration of the complete equation (3.14) (*open circles*) for the growth of a vapor bubble in superheated sodium (from Plesset & Prosperetti 1976).

dominant role essentially until the bubble surface has cooled to the boiling temperature.

Plesset & Prosperetti (1976) have also shown that for $p \gg 1$ Equation (3.17) admits a scaling such that a universal bubble-growth law for any liquid and any superheat is obtained in the form

$$S = S(\tau)$$

where

$$S = \mu^2 R/R_0, \qquad \tau = \alpha\mu^2 t. \tag{3.20}$$

Several examples of growth velocities $\dot{R}(t)$ scaled according to (3.20) (i.e. graphs of $dS/d\tau$ versus τ) for different values of the parameter μ are shown in Figure 10. The dots on the curves mark the point where $R/R_0 = 10$; it is seen that approximately from this point on, all the curves fall on a single line which demonstrates the validity of the scaling law (3.20). Similar results are obtained for the radius. In the dimensionless variable of Equation (3.20) the inertia-controlled growth velocity (3.4) has the constant value $dS/d\tau = (2/3)^{1/2}$, and the thermally controlled one given by (3.13) is $dS/d\tau = \pi^{-1}(3\tau)^{-1/2}$. This curve is also shown in Figure 10, and it is apparent that it describes a substantial part of the growth only for relatively small subcooling (large values of μ). It is also shown by Plesset & Prosperetti (1976) that the analytic expression for bubble growth obtained by Mikic, Rohsenow & Griffith (1970) can be written in terms of the scaled variables S, τ provided that the pressure-temperature relationship is approximated not by a tangent but by a chord (see Theofanous & Patel 1976). This expression is

$$S = \frac{2}{\pi^2} (\tfrac{2}{3})^{3/2}[(\tfrac{1}{2}\pi^2\tau + 1)^{3/2} - (\tfrac{1}{2}\pi^2\tau)^{3/2} - 1], \tag{3.21}$$

and its derivative is also plotted in Figure 10. It is seen that Equation (3.21) is a good approximation except in the region where inertial and thermal effects are of comparable importance.

In addition to the studies mentioned, other attempts have been made at deducing a scaling law for vapor-bubble growth. We note here those of Birkhoff et al (1958) and of Scriven (1959), who showed that with the assumption $R = \beta t^{1/2}$, the energy equation (3.14) admits a similarity variable that reduces it to an ordinary differential equation that can be solved in closed form. Insertion of the result into the boundary condition (3.15c) determines then the constant β as a function of the several parameters of the problem. For low superheats, it is found that the value of β coincides with the one obtained upon integration of (3.13). The limitations of this approach should be clear from the discussion given here. In particular, it is seen from Figure 10 that a $t^{1/2}$ dependence of the radius cannot account for dynamical effects, so that the applicability of such an analysis must be restricted to low superheats. In the problem of mass diffusion, however, which has the same mathematical structure, but for which inertial effects are in general much less pronounced, the similarity approach is quite appropriate. A very useful feature of it for the application to the mass-diffusion problem is that the diffusion equation is solved exactly without any assumption concerning the thickness of the boundary layer.

Vapor-Bubble Collapse

The theoretical modeling of the collapse of vapor bubbles under conditions in which thermal effects play a significant role is more difficult than the analysis of their growth, and no entirely satisfactory theoretical results are available. The principal difficulty here is that, unlike the growth case, the thickness of the liquid layer in which substantial temperature gradients are present cannot in general be taken to be small compared with the bubble radius for all times. An approach that makes use of the Plesset-Zwick result (3.16) for the surface temperature has been followed by Florschuetz & Chao (1965), but the applicability of that expression to the collapse case is open to doubt, especially for those cases in which large thermal effects result in small collapse velocities. Among the notable features of the results of Florschuetz and Chao is a nonmonotonic decrease in the bubble radius when both inertial and thermal effects are important in the collapse. The experimental evidence in favor of this point appears to be rather unclear for the collapse of spherical bubbles in unbounded liquids (Levenspiel 1959, Akiyama 1965, Hewitt & Parker 1968), possibly because shape deformations during the collapse introduce errors in the precise determination of the radius. It may be mentioned, however, that no such behavior was observed by Gunther (1951) for the collapse of hemispherical bubbles on a heated solid wall. Florschuetz and Chao also reported experiments performed with the aid of a free-fall tower to minimize the effects of translational motion and initial buoyancy-induced deformations. They observed that the shape of all of the bubbles deviated from the spherical during the collapse. While for small subcoolings ($\Delta T_i \lesssim 13°C$) they appeared to oscillate in the lowest mode between prolate and oblate spheroidal configurations, for larger subcoolings much more marked deviations occurred with the formation of jets and large deformations beginning early in the collapse process.

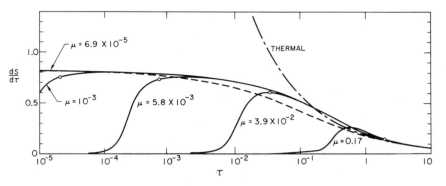

Figure 10 Dimensionless growth velocity versus dimensionless time for vapor-bubble growth, for several values of the parameter μ, which shows the scaled growth behavior. The open circles denote the points where the bubble radius equals ten times its initial value. The dashed line is the Mikic-Rohsenow-Griffith (1970) result (3.21), and the line labeled "thermal" is the asymptotic result of Plesset & Zwick (1954), Equation (3.13) (from Plesset & Prosperetti 1976).

Bubble collapses at high subcoolings (up to 60°C) in water have been obtained by Board & Kimpton (1974), whose data however appear to be affected by strong wall effects. Very recently Delmas & Angelino (1976) have studied the collapse of rising bubbles with subcoolings ΔT_i between 7°C and 42°C in water. They reported two different collapse modes, one for large bubbles and high subcoolings, for which inertial effects are important, the other for smaller bubbles and lower subcoolings. The bubbles belonging to the first class appear to occupy a well-defined region in the $(R_i, \Delta T_i)$-plane. They exhibit large collapse velocities and rapid ($< 1.4 \times 10^{-4}$ sec) fragmentation upon attainment of the minimum radius. The initial conditions giving rise to the second collapse mode are less predictable. The bubbles present an irregular surface and selected condensation sites. Fragmentation takes place earlier than in the other case and the smaller bubbles to which it gives rise continue to condense. These results are interesting, but it is difficult to estimate the effect of the velocity of rise of the bubbles on which the authors did not make any comment.

Other Topics in the Dynamics of Vapor Bubbles

In recent years research in the acoustic cavitation process in cryogenic liquids (see e.g. Mosse & Finch 1971, Neppiras & Finch 1972, Akulichev 1974) and in the development of ultrasonic bubble chambers (Akulichev et al 1970; Brown et al 1970; Hilke 1973; Shestakov & Tkachev 1973, 1975) have prompted the theoretical investigation of the behavior of vapor bubbles in oscillating pressure fields (Finch & Neppiras 1973, Wang 1973). In these studies, the energy equation in the liquid and the Rayleigh equation have been solved under the assumption of small-amplitude oscillations, allowing for the evaporation-condensation coupling expressed by Equation (3.15c). It has been found that when the liquid temperature is sufficiently close to the saturation temperature, there are two values of the bubble radius that give rise to resonant oscillations for a fixed driving frequency. One of them corresponds to that found for a permanent gas bubble, while the other one is smaller, approximately by an order of magnitude for the case of water at one atmosphere and at frequencies of the order of 10^4 Hz. The resonance peaks are broader than for gas bubbles, and they merge to give a large oscillation amplitude for values of the radius spanning one or two orders of magnitude. In addition to this effect of the evaporation of the liquid, Wang (1973) has also considered the simultaneous diffusion of a permanent gas into the bubble.

In these studies, the thermodynamic state of the vapor in the bubble has been assumed to correspond to conditions of thermodynamic equilibrium at the instantaneous bubble-wall temperature. In a certain sense, this assumption corresponds to the validity of Henry's law, Equation (2.11), in the case of rectified diffusion of mass into the bubble. Therefore, one may expect a rectified heat-transfer effect into the bubble by means of which the average vapor content of the bubble increases with time. Such a process has indeed been demonstrated theoretically (Wang 1974), but no experimental verification of the theory is available. It should be noticed that this effect would be of first (rather than second) order in the oscillation amplitude if thermodynamic nonequilibrium effects for the vapor were accounted for. Whether this increased order of magnitude would give rise to a contribution bigger than the rectified effect depends strongly on the accommodation coefficients for evaporation and condensation. This aspect of the problem has not as yet been investigated.

More generally, the question of effects associated with a state of thermodynamic nonequilibrium between liquid and vapor is of course an important one in the dynamics of vapor bubbles. An estimate of the difference between the equilibrium vapor density ρ_v^e and the actual vapor density ρ_v is readily obtained by a mass balance over the bubble

$$\frac{d}{dt}(\tfrac{4}{3}\pi R^3 \rho_v) = 4\pi R^2 J,$$ (3.22)

where J is the net mass flux at the bubble wall which, on the basis of the elementary kinetic theory of gases, can be written as

$$J = \alpha \left(\frac{R_g T}{2\pi M}\right)^{1/2} (\rho_v^e - \rho_v).$$ (3.23)

Here R_g is the gas constant, M is the molecular weight of the vapor, and α is the accommodation coefficient for evaporation, taken to be equal to that for condensation. Continuity of temperature across the bubble surface has also been assumed in (3.23). Substitution of (3.23) into (3.22) and neglect of the term involving $d\rho_v/dt$ gives then (Plesset 1964):

$$\frac{\rho_v^e - \rho_v}{\rho_v^e} = \frac{\dot{R}}{\alpha(R_g T/2\pi M)^{1/2} + \dot{R}}.$$ (3.24)

In view of the fact that the sound speed in the vapor is given by $c = (\gamma R_g T/M)^{1/2}$, this equation shows that the nonequilibrium correction is of the order of the Mach number of the bubble-wall motion in the vapor whenever $\alpha \simeq 1$, and therefore negligible except perhaps near the end of violent collapses. Much more significant effects may be expected, however, for small values of α. Nonequilibrium analyses of vapor-bubble growth have been given by Bornhorst & Hatsopoulos (1967) and by Theofanous et al (1969a). As one might expect, very small values of α give rise to decreased growth rates, but the results corresponding to $\alpha = 1$ are in very good agreement with the equilibrium behavior. Theofanous et al (1969b) have treated in a similar way the case of the collapse of vapor bubbles, although their assumption of a thin thermal layer may be questionable in this case for the reasons already stated.

In addition to the neglect of thermodynamic equilibrium, the formulation of vapor-bubble dynamics given here has been simplified in other respects. For instance, the equation of continuity at the bubble surface gives

$$\rho_l(u_l - \dot{R}) = \rho_v(u_v - \dot{R})$$ (3.25)

(where u_l and u_v are the liquid and vapor velocities at the liquid-vapor interface, respectively), from which

$$\dot{R} = \frac{\rho_l}{\rho_l - \rho_v} u_l + \frac{\rho_v}{\rho_l - \rho_v} u_v.$$ (3.26)

Clearly, setting $u_l = \dot{R}$, as was implicitly done in the derivation of the Rayleigh equation, is legitimate only if $\rho_v \ll \rho_l$. While this relation is normally satisfied by liquids at temperatures sufficiently below the critical point, its applicability in some special situations (such as cryogenic liquids) should not be automatically taken for granted. Similarly, the condition of continuity of normal stresses, Equation (1.4), and of conservation of energy at the interface, Equation (3.15c), in principle should contain other terms of order ρ_v/ρ_l. For a more detailed exposition of these matters the reader is referred to Scriven (1959) and Hsieh (1965). The latter author gives also a useful unified treatment of several mathematical aspects of the whole subject of bubble dynamics.

4   THE DYNAMICS OF NONSPHERICAL BUBBLES

It is obvious that as soon as the property of spherical symmetry is lost, the analysis of the various aspects of bubble dynamics becomes exceedingly complex both from the theoretical and the experimental standpoint (Hsieh 1972b). This situation is unfortunate, in view of the practical importance of the effects associated with deviations from the spherical shape and in view of the multitude of factors that promote such deviations. Among these one may mention the inherent dynamical instability of contracting bubbles, the proximity of solid boundaries or of free surfaces, and buoyancy effects. Such important phenomena as bubble breakup or coalescence and cavitation damage are dominated by these effects which are also found to increase heat and mass-transfer rates.

The first problem that is encountered in the dynamics of nonspherical bubbles is that of the specification of the bubble shape. An obvious possibility, which is used in practically all the work described below, is the use of an expansion in terms of spherical harmonics $Y_n^m(\theta, \varphi)$,

$$S(r, \theta, \varphi, t) = r - R(t) - \sum_{n,m} a_{nm}(t) Y_n^m(\theta, \varphi), \qquad (4.1)$$

where $S(r, \theta, \varphi, t) = 0$ is the equation of the bubble surface, $R(t)$ is the instantaneous average radius, and $a_{nm}(t)$ is the amplitude of the spherical harmonic component of order n and degree m. It should be emphasized, however, that (4.1) is not the only possible choice, nor is it the most convenient for all problems.

A basic result in the dynamics of nonspherical bubbles has been obtained by Plesset (1954), who solved the equation of motion of an incompressible, inviscid, unbounded fluid with a free surface given by (4.1) in the small-amplitude approximation $|a_{nm}|/R \ll 1$. In this approximation the equations for R and the a_{nm}'s are uncoupled, the first one being just the Rayleigh equation. The equations for the a_{nm}'s do not contain the index m (which accordingly will be dropped in the following) and are found to be

$$\ddot{a}_n + 3(\dot{R}/R)\dot{a}_n + (n-1)[(n+1)(n+2)\sigma/\rho_l R^3 - \ddot{R}/R]a_n = 0. \qquad (4.2)$$

If the surface-tension term is neglected, this equation reduces to one also given by Birkhoff (1954). It is easily seen that $n = 1$ corresponds to a translation of the bubble center. In this case Equation (4.2) gives just $R^3 \dot{a}_1 = $ constant, which expresses the constancy of the total liquid momentum which is proportional to the bubble volume times its translational velocity. If all the other a's vanish this happens of course to be an exact result independent of the small-amplitude approximation. This situation, however, is of little interest to us here, and we shall restrict our considerations to the case $n \geq 2$ in the following.

The introduction of viscous effects complicates the matter substantially, giving to Equation (5.2) an integro-differential structure (Prosperetti 1976c) except in the case where the viscous-diffusion length is small compared with the bubble radius, for which it becomes

$$\ddot{a}_n + [3\dot{R}/R + 2(n+2)(2n+1)\nu_l/R^2]\dot{a}_n$$

$$+ (n-1)[(n+1)(n+2)\sigma/\rho_l R^3 + 2(n+2)\nu_l \dot{R}/R^3 - \ddot{R}/R]a_n = 0, \qquad (4.3)$$

ν_l denoting the liquid kinematic viscosity. Both Plesset (1954) and Prosperetti (1976c) have also considered the dynamical effects of the fluid contained in the bubble. The expressions given above are the limiting forms obtained when its density and viscosity vanish.

The Stability of Spherical Growth and Collapse

Even with the neglect of the surface-tension term (which has always a stabilizing effect, but which is important only for very small bubble radii), the qualitative behavior of the solution of Equation (4.2) is not readily understood a priori. Indeed, the stability characteristics do not depend only on the acceleration of the interface, as in the familiar plane case of the Rayleigh-Taylor instability, but also on its velocity. This situation is clearly a consequence of the geometry, since the divergence of the streamlines during bubble growth has a stabilizing effect, while the reverse occurs during collapse. In the limiting case of very large acceleration and of small velocity, such as take place for instance during the early stages of the growth of an underwater-explosion bubble, Equation (4.2) predicts an instability of the spherical shape, which is indeed known to exist (Cole 1948, illustration facing p. 247). Another example of the occurrence of this situation is furnished by the large-amplitude radial pulsations of a permanent gas bubble, for which the radial acceleration becomes very large and positive near the position of minimum radius, with the consequent possibility of large deformations and breakup. A similar situation might be expected to take place also in the early stages of the growth of a cavitation or boiling bubble, but here the duration of this stage of large acceleration and small velocity is so short that the instability does not have the time to develop.

The quantitative investigation of the stability for a bubble expanding or collapsing under a fixed pressure difference $P = p_i - p_\infty$ (cavitation bubble) has been carried out by Plesset & Mitchell (1956), who were able to obtain closed-form solutions of Equation (4.2). The qualitative features of their results can be readily understood with the aid of Equations (3.4) and (3.6). For the growth case we have $\dot{R}^2 \simeq (2/3)P/\rho$, from which one readily obtains $a_n \to $ constant as $R \to \infty$. Hence, although on the basis of the plane Rayleigh-Taylor case one would expect an unstable behavior during the growth, in practice no significant deviations from the spherical shape take place because the acceleration is large only for a small fraction of the process, and its destabilizing influence is very effectively counterbalanced by the stretching of the bubble surface produced by the divergence of the streamlines. It may also be remarked that, in spite of the factor $(n-1)$ multiplying the radial acceleration in Equation (4.2), high growth rates for instabilities of large order n do not take place because of viscous effects [cf Equation (4.3)].

For the study of the collapse of a cavitation bubble it is expedient to set $b_n = (R_i/R)^{3/2}a_n$, so that Equation (4.2) takes the form

$$\ddot{b}_n - [(3/4)(\dot{R}/R)^2 + (n+\tfrac{1}{2})\ddot{R}/R - (n-1)(n+1)(n+2)\sigma/\rho_l R^3]b_n = 0. \tag{4.4}$$

Now, from Equation (3.6) we have during collapse that $\dot{R}^2 \simeq -(2/3)(P/\rho_l)(R_i/R)^3$ from which the coefficient of b_n in this equation is seen to have the form $-nc^2 R^{-5}$, where c is a constant. An approximation of the W.K.B. type gives then

$$a_n \simeq R^{-1/4} \exp\left(\pm icn^{1/2} \int R^{-5/2}\, dt \right),$$

whence it is seen that a_n increases proportionally to $R^{-1/4}$ and oscillates with increasing frequency as $R \to 0$ (Plesset & Mitchell 1956; Birkhoff 1954, 1956). The decisive role played by the sinklike converging nature of the flow in the development of this instability is apparent from the second term of Equation (4.2).

Clearly, Equation (4.4) implies the breakup of the bubble when the amplitude of the oscillations becomes of the order of R. The results of the linearized approximation, however, cannot be expected to be quantitatively valid up to this point. A fully nonlinear numerical calculation of the shape of an empty bubble during the collapse has been carried out by Chapman & Plesset (1972). They have obtained local bubble-wall velocities of the order of hundreds of meters per second at the points where the breakup of the bubble takes place. They have also shown that the linearized result is surprisingly accurate for most of the duration of the collapse. The greatest error in the description of the bubble shape is caused by the growth of harmonic components of different orders from those initially present, which is induced by the nonlinear couplings and which cannot be predicted on the basis of the linear theory.

The instability of the spherical shape of collapsing vapor bubbles at high sub-coolings is well documented in the literature (see e.g. Florschuetz & Chao 1965, Figure 11; and Lauterborn 1974). However, no reduction of the data in terms of spherical harmonic components has yet been attempted.

Bubble Collapse in the Vicinity of a Rigid Boundary

We have seen that, in principle, deviations from the spherical shape for a bubble in an unbounded liquid can take place only through the amplification of preexisting small perturbations. No such initial perturbation, however, is necessary for the occurrence of deformations of a bubble near a boundary, because the asymmetry of the flow that the boundary itself introduces is sufficient to give rise to highly distorted bubble shapes. It is possible to get an insight into these effects by observing that, with the neglect of viscosity, the potential problem can be solved by the method of images that replaces the rigid boundary by an image bubble equal to the real one and located symmetrically to it with respect to the boundary. From this observation one is led to expect that the portion of the bubble farther from the wall acquires a greater velocity than the one near the wall, because there the velocity of the flow induced by the image bubble adds to the collapse velocity, while in the second case it decreases the collapse velocity. This asymmetry leads to the formation of a high-velocity jet directed toward the wall, while the overall characteristics of the sinklike flow of the image attracts the bubble toward the wall (Lauterborn & Bolle 1975).

That jet formation during bubble collapse could be responsible for cavitation damage was suggested as early as 1944 by Kornfeld & Suvorov, but their conjecture was not explored further, and the accepted explanation remained that of the large pressures associated with the latest stages of bubble collapse that had originally been put forward by Rayleigh (1917). The first experimental demonstration of jet formation was obtained by Naudé & Ellis (1961), and definitive conclusions about the mechanism of cavitation damage was later reached by Benjamin & Ellis (1966). This latter paper contains also an extensive summary of the history of the subject and of related work. Additional considerations on the magnitude of the stresses induced by jet impingement can be found in Plesset & Chapman (1971) and in Plesset (1974). Direct measurement of jet velocities is rather difficult. Operating at low ambient pressure, Benjamin & Ellis (1966) obtained velocities of 10 m sec^{-1}, which scale to velocities an order of magnitude greater at an ambient pressure of one atmosphere. The value of 120 m sec^{-1} is reported by Kling & Hammitt (1972) and Lauterborn & Bolle (1975), and is consistent with the numerical results of Plesset & Chapman (1971).

As might be expected, an analytical treatment of the problem of bubble collapse in the vicinity of a rigid boundary is very difficult. An early perturbation approach is that of Rattray (1951), who was successful in predicting the possibility of jet formation and also in predicting the elongation of the bubble in a direction perpendicular to the boundary in the early stages of the collapse. The first complete theoretical analysis of the collapse of an empty cavity in the neighborhood of a rigid wall was obtained numerically by Plesset & Chapman (1971), whose results have been experimentally confirmed by Lauterborn & Bolle (1975). Figure 11, from the work of these authors, shows a comparison of the theoretical results (*continuous lines*) with the experimental ones (*open circles*). The comparison is quite satisfactory, particularly in view of the difficulty in establishing a correspondence between theoretical and experimental initial conditions. The case of a bubble collapsing attached to a wall has also been worked out numerically by Plesset & Chapman (1971).

Surface Oscillations

Before considering some of the phenomena associated with the interplay between radial motion and surface deformations of permanent gas bubbles, we discuss the shape oscillations of a bubble of fixed radius R_0. The governing equation for this

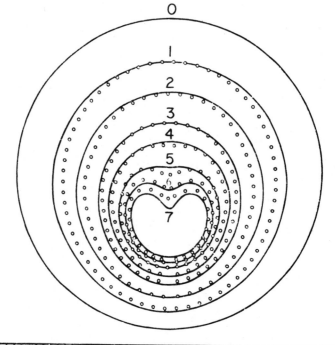

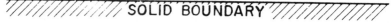

Figure 11   Comparison of the experimental results of Lauterborn & Bolle (1975, open circles) with the theoretical ones of Plesset & Chapman (1971) for the collapse of a cavitation bubble in the vicinity of a rigid boundary (from Lauterborn & Bolle 1975). Bubble shapes corresponding to later values of time than those shown here are given by Plesset & Chapman (1971).

situation is readily obtained from (4.2) or (4.3), and one finds the following expression for the natural frequency of the nth mode

$$\omega_{0n}^2 = (n-1)(n+1)(n+2)\sigma/R_0^3\rho_l. \tag{4.5}$$

The familiar dispersion relation for capillary waves on a flat liquid surface is readily obtained from this equation by observing that the wavelength λ is given by $\lambda = 2\pi R_0/n$ and by letting R_0 and n tend to infinity with λ held fixed. For small viscosity, one obtains from (4.3) the viscous damping constant β_n of the oscillations as $\beta_n = (n+2)(2n+1)\nu_l/R_0^2$ (cf Lamb 1932, p. 475, p. 641).

An understanding of the general behavior of the surface deformations for a bubble undergoing radial pulsations can be obtained in the limit of small-amplitude oscillations by writing $R = R_0(1+\delta \sin \omega t)$ and then retaining only linear terms in δ (Hsieh & Plesset 1961; Hsieh 1972a, 1974a). Equation (4.4) takes then the form of a Mathieu equation,

$$\ddot{b}_n+(\omega_{0n}^2+\alpha_n \sin \omega t)b_n = 0, \tag{4.6}$$

with $\alpha_n = [(n+1/2)\omega^2 - 3\omega_{0n}^2]\delta$ and ω_{0n}^2 given by (4.5). It is well known from the theory of the Mathieu equation that, depending on the magnitudes of the constants α_n and ω_{0n}^2, the solutions of (4.6) may have the form of modulated oscillations the amplitude of which grows exponentially with time. Although Equation (4.5) itself is not valid unless $|b_n| \ll R_0$, this behavior does indicate that parametric excitation of the instability of the spherical shape is possible, and describes the early stages of its development. It may be observed that the most easily excited surface oscillation is the one corresponding to $n = 2$, the threshold of which has a minimum in the vicinity of $\omega = 2\omega_0$. The acoustic emission from a bubble executing surface oscillations in this mode would contain a subharmonic component, and indeed this possibility for the origin of the subharmonic component in acoustically cavitating liquids has also been considered. However, the observed intensity of the signal is incompatible with this hypothesis because, as pointed out by Strasberg (1956), bubbles executing shape oscillations are inefficient sound sources because of the rapid decrease of the velocity potential with distance from the bubble center.

Approximations to Equation (4.4) of a higher order in δ than (4.6) have been carried out by Eller & Crum (1970) in an attempt to explain the observed instability of the position of a pulsating bubble in a sound field first reported by Benjamin and Strasberg (see Benjamin 1964). Although the connection between the onset of surface distortions and the observed erratic translatory motion of the bubble is not obvious, the instability thresholds computed from (4.4) do exhibit a fair agreement with the experimental data for the onset of the bubble translatory motion. More direct experimental observations on the onset of surface oscillations are not very readily obtained in low-viscosity liquids for several reasons. Gould (1974) reported some results that have a large scatter but show agreement with theory in the general trend. Storm (1974) has studied large bubbles trapped in a gel, but the interpretation of his observations is somewhat obscured by the complex rheological characteristics of the suspending medium. Results of studies with large bubbles trapped on a plate have also been reported (Gould 1966, Howkins 1965, Blue 1967), but here a difficulty is encountered in accounting for the presence of the rigid boundary in the theory (in this connection see also Strasberg 1953).

Numerical results for stability thresholds for freely oscillating gas bubbles have been obtained by Strube (1971), who solved simultaneously the Rayleigh equation for a freely pulsating gas bubble and Equation (4.4). He presents results on maximum bubble deformations (which frequently appear shortly after the instant of minimum radius) and gives a chart of the stability boundaries. No work of this type is available for forced oscillations. Experiments on the distortion of a translating gas bubble subject to a pressure step were reported by Smulders & Van Leeuwen (1974) and compared with a theoretical analysis by Hermans (1973). Several results drawn from the theory are confirmed on a qualitative basis although a quantitative comparison was not attempted by the investigators.

As the last topic of this article, we mention a variational approach to nonspherical bubble dynamics formulated by Hsieh (1974b). The difficulties encountered in the accurate description of the shape of nonspherical bubbles are such that there certainly is a strong case for attempting a variational attack. The results obtained so far, however, are rather limited, and it appears that additional research is necessary before the variational formulation can fulfill its promises of effectiveness and relative simplicity.

5 CONCLUDING REMARKS

It is clear that, in view of the very large number of papers on the subject, even the present list of references is quite incomplete. Further, in view of the fact that the present article is the first one on this subject to appear in the *Annual Review of Fluid Mechanics,* we felt that a relatively detailed exposition of the most important established results was in order even at the expense of omitting some more recent developments such as the dynamics of bubbles in non-Newtonian liquids. We have also tried to cover the fundamental physical aspects of the problems with little or no mention of practical developments. Thus we have been very concise in our coverage of the literature on bubble dynamics in connection with boiling heat transfer. We have also given only a summary of the processes and results for acoustic cavitation and flow cavitation. Finally, we have neglected entirely the phenomena related to the translatory motion of bubbles that have recently been considered elsewhere by Harper (1972).

Literature Cited

Akiyama, M. 1965. *Bull. Jpn. Soc. Mech. Eng.* 8:683–94
Akulichev, V. A. 1974. *Sov. Phys. Acoust.* 20:94–95
Akulichev, V. A. et al. 1970. *Sov. Phys. Acoust.* 15:439–43
Apfel, R. E. 1970. *J. Acoust. Soc. Am.* 48:1179–86
Benjamin, T. B. 1958. *Symp. Naval Hydrodyn., 2nd, Washington, D.C.,* pp. 207–33
Benjamin, T. B. 1964. *Cavitation in Real Liquids,* ed. R. Davies, pp. 164–80. New York: Elsevier
Benjamin, T. B., Ellis, A. T. 1966. *Philos. Trans. R. Soc. London Ser. A* 260:221–40
Birkhoff, G. 1954. *Q. Appl. Math.* 12:306–9
Birkhoff, G. 1956. *Q. Appl. Math.* 13:451–53
Birkhoff, G., Margulies, R. S., Horning, W. A. 1958. *Phys. Fluids* 1:201–4
Blake, F. G. 1949. *Harvard Univ. Acoust. Res. Lab., Tech. Mem. No. 12*
Blander, M., Katz, J. L. 1975. *AIChE J.* 21:833–48
Blue, J. E. 1967. *J. Acoust. Soc. Am.* 41:369–72
Board, S. J., Kimpton, A. D. 1974. *Chem. Eng. Sci.* 29:363–71
Bornhorst, W. J., Hatsopoulos, G. N. 1967. *J. Appl. Mech.* 34:847–53

Borotnikova, M. I., Soloukin, R. I. 1964. *Sov. Phys. Acoust.* 10:28–32
Brennen, C. 1973. *J. Fluids Eng.* 95:533–41
Brown, R. C. A., Harigel, G., Hilke, H. J. 1970. *Nucl. Instrum. Methods* 82:327–30
Bull, T. H. 1956. *Br. J. Appl. Phys.* 7:416–18
Carslaw, M. S., Jaeger, J. C. 1959. *Conduction of Heat in Solids,* 2nd ed. Oxford: Clarendon
Ceschia, M., Iernetti, G. 1974. *J. Acoust. Soc. Am.* 56:369–73
Chapman, R. B., Plesset, M. S. 1971. *J. Basic Eng.* 93:373–76
Chapman, R. B., Plesset, M. S. 1972. *J. Basic Eng.* 94:142–45
Childs, D. R. 1973. *Int. J. Non-Linear Mech.* 8:371–79
Coakley, W. T. 1971. *J. Acoust. Soc. Am.* 49:792–801
Coakley, W. T., Sanders, M. F. 1973. *J. Sound Vib.* 28:73–85
Cole, R. 1974. *Adv. Heat Transfer* 10:86–166
Cole, R. H. 1948. *Underwater Explosions.* Princeton, NJ: Princeton Univ. Press. Reprinted by Dover Publications, New York
Dalle Donne, M., Ferranti, M. P. 1975. *Int.*

J. Heat Mass Transfer 18:477–93
Davidson, B. J. 1971. *J. Sound Vib.* 17:261–70
Davidson, B. J., Riley, N. 1971. *J. Sound Vib.* 15:217–33
Delmas, H., Angelino, H. 1976. *Int. J. Heat Mass Transfer* 19:118–21
Dergarabedian, P. 1953. *J. Appl. Mech.* 75:537–45
Dergarabedian, P. 1960. *J. Fluid Mech.* 9:39–48
De Santis, P., Sette, D., Wanderlingh, F. 1967. *J. Acoust. Soc. Am.* 42:514–16
Devin, C. 1959. *J. Acoust. Soc. Am.* 31:1654–67
Duda, J. L., Vrentas, J. S. 1969a. *Chem. Eng. Sci.* 24:461–70
Duda, J. L., Vrentas, J. S. 1969b. *AIChE J.* 15:351–56
Elder, S. A. 1959. *J. Acoust. Soc. Am.* 31:54–64
Eller, A. 1969. *J. Acoust. Soc. Am.* 46:1246–50
Eller, A. 1972. *J. Acoust. Soc. Am.* 52:1447–49
Eller, A. 1975. *J. Acoust. Soc. Am.* 57:1374–78
Eller, A., Crum, L. A. 1970. *J. Acoust. Soc. Am.* 47:762–67
Eller, A., Flynn, H. G. 1965. *J. Acoust. Soc. Am.* 37:493–503
Eller, A., Flynn, H. G. 1969. *J. Acoust. Soc. Am.* 46:722–27
Epstein, D., Keller, J. B. 1972. *J. Acoust. Soc. Am.* 52:975–80
Epstein, P., Plesset, M. S. 1950. *J. Chem. Phys.* 18:1505–9
Esche, R. 1952. *Acustica* 2:(AB)208–18
Finch, R. D., Neppiras, E. A. 1973. *J. Acoust. Soc. Am.* 53:1402–10
Florschuetz, L. W., Chao, B. T. 1965. *J. Heat Transfer* 87:209–20
Flynn, H. G. 1964. *Physics of Acoustic Cavitation in Liquids. Physical Acoustics,* ed. W. P. Mason, Vol. 1B:58:172. New York: Academic
Flynn, H. G. 1975a. *J. Acoust. Soc. Am.* 57:1379–96
Flynn, H. G. 1975b. *J. Acoust. Soc. Am.* 58:1160–70
Gavrilov, L. R. 1970. *Sov. Phys. Acoust.* 15:285–95
Gilmore, F. R. 1952. *Calif. Inst. Technol. Eng. Div. Rep. 26-4, Pasadena, Calif.*
Gould, R. K. 1966. *J. Acoust. Soc. Am.* 40:219–25
Gould, R. K. 1974. *J. Acoust. Soc. Am.* 56:1740–46
Gunther, F. C. 1951. *Trans. ASME* 73:115–23
Harper, J. F. 1972. *Adv. Appl. Mech.* 12:59–129
Hermans, W. A. 1973. PhD thesis. Tech. Univ. Eindhoven, Netherlands
Hewitt, H. C., Parker, J. D. 1968. *J. Heat Transfer* 90:22–26
Hickling, R., Plesset, M. S. 1964. *Phys. Fluids* 7:7–14
Hilke, H. 1973. *Ultrasonics* 11:51–52
Hill, C. R. 1972. *J. Acoust. Soc. Am.* 52:667–72
Hinsch, K., Bader, F., Lauterborn, W. 1974. In *Finite-Amplitude Wave Effects in Fluids,* ed. L. Bjørnø, pp. 240–44. Guilford: I.P.C. Sci. Technol. Press
Holl, J. W. 1970. *J. Basic Eng.* 92:681–88
Howkins, S. D. 1965. *J. Acoust. Soc. Am.* 37:504–8
Hsieh, D. Y. 1965. *J. Basic Eng.* 87:991–1005
Hsieh, D. Y. 1970. *J. Basic Eng.* 92:815–18
Hsieh, D. Y. 1972a. *J. Acoust. Soc. Am.* 52:151
Hsieh, D. Y. 1972b. *J. Basic Eng.* 94:655–65
Hsieh, D. Y. 1974a. *J. Acoust. Soc. Am.* 56:392–93
Hsieh, D. Y. 1974b. In *Finite-Amplitude Wave Effects in Fluids,* ed. L. Bjørnø, pp. 220–26. Guilford: I.P.C. Sci. Technol. Press
Hsieh, D. Y., Plesset, M. S. 1961. *J. Acoust. Soc. Am.* 33:206–15
Hunter, C. 1960. *J. Fluid Mech.* 8:241–63
Jahsman, W. E. 1968. *J. Appl. Mech.* 35:579–87
Jensen, F. B. 1974. *J. Fluids Eng.* 96:389–93
Kapustina, O. A. 1970. *Sov. Phys. Acoust.* 15:427–38
Keller, A. 1972. *J. Basic Eng.* 94:917–25
Keller, J. B., Kolodner, I. I. 1956. *J. Appl. Phys.* 22:1152–61
Kirkwood, J. G., Bethe, H. A. 1942. *OSRD Rep. No. 588*
Kling, C. L., Hammitt, F. G. 1972. *J. Basic Eng.* 94:825–33
Koger, H. G., Houghton, G. 1968. *J. Acoust. Soc. Am.* 43:571–75
Kolb, J., Nyborg, W. L. 1956. *J. Acoust. Soc. Am.* 28:1237–42
Kornfeld, M., Suvorov, L. 1944. *J. Appl. Phys.* 15:495–506
Krieger, I. M., Mulholland, G. W., Dickey, C. S. 1967. *J. Phys. Chem.* 71:1123–29
Lamb, H. 1932. *Hydrodynamics.* Cambridge: The University Press. Reprinted 1945 by Dover Publications, New York
Lauer, H. 1951. *Acustica* 1:AB 12–24
Lauterborn, W. 1968. *Acustica* 20:14–20
Lauterborn, W. 1970a. *Acustica* 22:238–39
Lauterborn, W. 1970b. *Acustica* 23:73–81
Lauterborn, W. 1972a. *Xème Congrès International de Cinématographie Ultra-Rapide,*

pp. 206–9. Paris: Assoc. Natl. Rech. Technique

Lauterborn, W. 1972b. *Appl. Phys. Lett.* 21: 27–29

Lauterborn, W. 1974. In *Finite-Amplitude Wave Effects in Fluids,* ed. L. Bjørnø, pp. 195–202. Guilford: I.P.C. Sci. Technol. Press

Lauterborn, W. 1976. *J. Acoust. Soc. Am.* 59: 283–93

Lauterborn, W., Bolle, H. 1975. *J. Fluid Mech.* 72: 391–99

Levenspiel, O. 1959. *Ind. Eng. Chem.* 51: 787–90

Margulis, M. A. 1969. *Sov. Phys. Acoust.* 15: 135–51

Mikic, B. B., Rohsenow, W. M., Griffith, P. 1970. *Int. J. Heat Mass Transfer* 13: 657–66

Miles, J. W. 1966. *J. Fluid Mech.* 25: 743–60

Minnaert, M. 1933. *Philos. Mag.* 16: 235–48

Mosse, A., Finch, R. D. 1971. *J. Acoust. Soc. Am.* 49: 156–65

Naudé, C. F., Ellis, A. T. 1961. *J. Basic Eng.* 83: 648–56

Negishi, K. 1961. *J. Phys. Soc. Jpn.* 16: 1450–65

Neppiras, E. A. 1965. *Ultrasonics* 3: 9–17

Neppiras, E. A. 1969a. *J. Sound Vib.* 10: 176–86

Neppiras, E. A. 1969b. *J. Acoust. Soc. Am.* 46: 587–601

Neppiras, E. A., Finch, R. D. 1972. *J. Acoust. Soc. Am.* 52: 335–43

Niino, M., Toda, S., Egusa, T. 1973. *Heat Transfer Jpn. Res.* 2(4): 26–36

Noltingk, B. E. 1962. *Encyclopedia of Physics,* ed. S. Flugge, Vol. XI/2. *Acoustics* II: 258–86

Noltingk, B. E., Neppiras, E. A. 1950. *Proc. Phys. Soc. London. Sect. B* 63: 647–85

Noltingk, B. E., Neppiras, E. A. 1951. *Proc. Phys. Soc. London. Sect. B* 64: 1032–38

Nyborg, W. L. 1974. In *Finite-Amplitude Wave Effects in Liquids,* ed. L. Bjørnø, pp. 245–51. Guilford: I.P.C. Sci. Technol. Press

Pfriem, H. 1940. *Akust. Zh.* 5: 202–12

Plesset, M. S. 1949. *J. Appl. Mech.* 16: 277–82

Plesset, M. S. 1954. *J. Appl. Phys.* 25: 96–98

Plesset, M. S. 1957. *Symp. Naval Hydrodyn., 1st, Washington D.C., 1956,* pp. 297–323

Plesset, M. S. 1964. Bubble dynamics, Chap. I. In *Cavitation in Real Liquids,* ed. R. Davies. Amsterdam: Elsevier

Plesset, M. S. 1969. *Topics in Ocean Engineering,* ed. C. I. Bretschneider, Vol. 1, pp. 85–95. Houston: Gulf Publ. Co.

Plesset, M. S. 1974. In *Finite-Amplitude Wave Effects in Fluids,* ed. L. Bjørnø, pp.

203–9. Guilford: I.P.C. Sci. Technol. Press

Plesset, M. S., Chapman, R. B. 1971. *J. Fluid Mech.* 47: 283–90

Plesset, M. S., Hsieh, D. Y. 1960. *Phys. Fluids* 3: 882–92

Plesset, M. S., Mitchell, T. P. 1956. *Q. Appl. Math.* 13: 419–30

Plesset, M. S., Prosperetti, A. 1976. *J. Fluid Mech.* In press

Plesset, M. S., Prosperetti, A. 1977. *Int. J. Heat Mass Transfer.* In press

Plesset, M. S., Zwick, S. A. 1952. *J. Appl. Phys.* 23: 95–98

Plesset, M. S., Zwick, S. A. 1954. *J. Appl. Phys.* 25: 493–500

Plesset, M. S., Zwick, S. A. 1955. *J. Math. Phys.* 33: 308–30

Prosperetti, A. 1974. *J. Acoust. Soc. Am.* 56: 878–85

Prosperetti, A. 1975. *J. Acoust. Soc. Am.* 57: 810–21

Prosperetti, A. 1976a. Thermal effects and damping mechanisms in the forced radial oscillations of gas bubbles in liquids. *J. Acoust. Soc. Am.* In press

Prosperetti, A. 1976b. Application of the subharmonic threshold to the measurement of the damping of oscillating gas bubbles. *J. Acoust. Soc. Am.* In press

Prosperetti, A. 1976c. *Q. Appl. Math.* In press

Rattray, M. 1951. *Perturbation effects in cavitation bubble dynamics.* PhD thesis. Calif. Inst. Technol., Pasadena, Calif.

Rayleigh, Lord. 1917. *Philos. Mag.* 34: 94–98

Rohsenow, W. M. 1971. *Ann. Rev. Fluid Mech.* 3: 211–36

Rooney, J. A. 1970. *Science* 169: 869–71

Rooney, J. A. 1972. *J. Acoust. Soc. Am.* 52: 1718–24

Safar, M. H. 1968. *J. Acoust. Soc. Am.* 43: 1188–89

Saksena, T. K., Nyborg, W. L. 1970. *J. Chem. Phys.* 53: 1722–34

Scriven, L. E. 1959. *Chem. Eng. Sci.* 10: 1–13

Shestakov, V. D., Tkachev, L. G. 1973. *Sov. Phys. Acoust.* 19: 169–72

Shestakov, V. D., Tkachev, L. G. 1975. *Int. J. Heat Mass Transfer* 18: 685–87

Skinner, L. A. 1970. *J. Acoust. Soc. Am.* 47: 327–31

Skinner, L. A. 1972. *J. Acoust. Soc. Am.* 51: 378–82

Smulders, P. T., Van Leeuwen, H. J. W. 1974. In *Finite-Amplitude Wave Effects in Fluids,* ed. L. Bjørnø, pp. 227–33. Guilford: I.P.C. Sci. Technol. Press

Storm, D. L. 1974. In *Finite-Amplitude Wave Effects in Fluids,* ed. L. Bjørnø, pp. 234–39. Guilford: I.P.C. Sci. Technol. Press

Strasberg, M. 1953. *J. Acoust. Soc. Am.* 25: 536–37

Strasberg, M. 1956. *J. Acoust. Soc. Am.* 28: 20–26

Strasberg, M. 1959. *J. Acoust. Soc. Am.* 31: 163–76

Strasberg, M. 1961. *J. Acoust. Soc. Am.* 33: 359

Strube, H. W. 1971. *Acustica* 25: 289–303

Szekely, J., Fang, S. D. 1973. *Chem. Eng. Sci.* 28: 2127–40

Szekely, J., Martins, G. P. 1971. *Chem. Eng. Sci.* 26: 147–60

Taylor, K. J., Jarman, P. D. 1968. *Br. J. Appl. Phys.* (*J. Phys. D*) 1: 653–55

Taylor, K. J., Jarman, P. D. 1970. *Aust. J. Phys.* 23: 319–34

Theofanus, T. G., Biasi, L., Isbin, H. S., Fauske, H. K. 1969a. *Chem. Eng. Sci.* 24: 885–97

Theofanus, T. G., Biasi, L., Isbin, H. S., Fauske, H. K. 1969b. Paper presented at *Natl. Heat Transfer Conf., 11th, Minneapolis, August 1969*

Theofanus, T. G., Patel, P. D. 1976. *Int. J. Heat Mass Transfer* 19(4): 425–29

Trilling, L. 1952. *J. Appl. Phys.* 23: 14–17

Van Wijngaarden, L. 1972. *Ann. Rev. Fluid Mech.* 4: 369–96

Vaughan, P. W. 1968. *J. Sound Vib.* 7: 236–46

Wang, T. 1973. *J. Acoust. Soc. Am.* 56: 1131–43

Wang, T. 1974. *Phys. Fluids* 17: 1121–26

Ward, C. A., Tucker, A. S. 1975. *J. Appl. Phys.* 46: 233–38

Willard, G. W. 1953. *J. Acoust. Soc. Am.* 25: 669–86

Ann. Rev. Fluid Mech. 1977. 9 : 187–214
Copyright © 1977 by Annual Reviews Inc. All rights reserved

UNDERWATER EXPLOSIONS   ✕8101

Maurice Holt
Department of Mechanical Engineering, University of California,
Berkeley, California 94720

1 INTRODUCTION

The behavior of underwater explosions has been systematically investigated during the past thirty-five years, starting with many independent investigations undertaken during the second world war. The early work carried out in England, Canada, and the United States is surveyed thoroughly in Cole (1965). Investigations carried out at the same time in the Soviet Union (with emphasis on point-source explosions) are described in Sedov (1959). An excellent survey of later work extending up to the mid-fifties is given by Snay (1956). No connected account of later work has appeared since but the writer understands that Dr. Snay is now preparing a book covering his unique knowledge of this extensive field.

This brief survey begins with a review of the formation and early growth of explosions detonated in water, regarded as an unlimited medium. Then the effect of gravity is considered, especially concerning migration of gas bubbles created by detonation of chemical explosives. Finally, a section is devoted to the very important influence of the ocean surface on underwater and near-surface explosions.

2 FORMATION OF SPHERICAL EXPLOSIONS

We first consider explosions in a uniform unlimited medium, which, since they result from release of a large amount of energy from a source of small dimensions, can be treated as spherical. A distinction must be made between nuclear explosions, which can be regarded as due to the instantaneous release of energy at a point, and chemical explosions, which result from the detonation of a charge of finite dimensions consisting of explosive such as TNT or PETN. We shall assume that this charge is always spherical.

Figure 1 shows the principal boundaries in the disturbance field resulting from the detonation of a spherical charge of chemical explosive such as PETN. The explosion is initiated at the center of the charge and a uniform detonation wave OO' converts the solid explosive in the shaded region $OO'M$ into gaseous products in region d. The detonation velocity is constant because the Chapman-Jouguet condition is satisfied. When the detonation wave reaches the outer shell of the charge it continues to propagate in the outer undisturbed medium as a spherical blast

HOLT

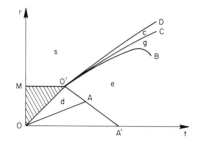

Figure 1   Principal boundaries in the spherical-explosion problem.

wave $O'D$. The gases released by the explosion are contained within a sphere with path $O'C$; this is the gas bubble. At the end of the detonation phase the highly compressed gaseous products expand and escape into the surrounding medium. This is achieved through a centered expansion with center O', and this is initially a centered simple wave with head $O'A$ and tail tangential to $O'B$. It is shown in Holt (1956) and Berry & Holt (1954) that the centered-wave region e and the initially uniform region g overlap in the sense that the characteristic line defining the tail of the expansion e crosses the characteristic line of the same family originating at O' in region g. Physically this means that the gas in the centered-wave region overexpands and has to be recompressed to satisfy the condition of continuity of pressure at the interface $O'C$. As a consequence the tail of the centered wave $O'B$ immediately develops into a second spherical shock wave.

   The second shock is characteristic of all spherical explosions from chemical charges but its behavior in air is different from that in water, as illustrated in Figures 2 and 3. For an air explosion the boundaries $O'D$, $O'C$, and $O'B$ are initially very close to each other. The gas sphere grows rapidly and the second shock follows the bubble boundary for some time before coming to rest and reversing direction. In water the boundaries are less crowded together. The bubble velocity is much

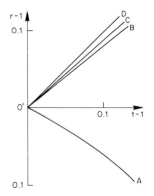

Figure 2   Chemical explosion in air. Principal boundaries.

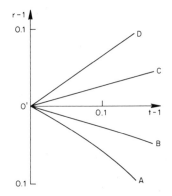

Figure 3 Chemical explosion in water. Principal boundaries.

smaller than that of the main blast wave, while the second shock immediately moves inwards towards the center of the gas bubble.

Another distinctive feature of explosions in water is that the gas bubble pulsates between a series of maximum and minimum values of bubble radius. In the early stages of the explosion the bubble expands until the mean pressure inside it falls below the pressure in the compressed water surrounding it. It then comes to rest and contracts until its mean pressure exceeds that in the surrounding water, at which time it again expands outwards. During the contracting phase compressive waves are sent inwards, giving rise to a converging spherical shock wave which is reflected at the center as an outgoing shock. A new shock is formed during each contraction so that multiple shocks are characteristic of underwater explosions.

In nuclear explosions the release of energy occurs so rapidly that the detonation phase can be ignored. The products of the explosive consist of a limited number of radioactive particles that are diffused in the region disturbed by the main spherical blast wave. In underwater nuclear bursts a bubble is formed, but it consists of steam or even dissociated water resulting from the radiative heating of the undisturbed ocean. It is bounded by a change-of-phase surface rather than a gas-water interface. Nevertheless it pulsates in a manner similar to that of a gas bubble from a spherical explosion. Although no second shock accompanies the main blast wave on initiation of a nuclear explosion, several shocks are generated by the vapor-bubble oscillations.

3   METHODS OF CALCULATING SPHERICAL EXPLOSION DISTURBANCES

Disturbances due to a spherical explosion in an unlimited uniform medium are governed by the equations of unsteady motion of a compressible fluid with spherical symmetry. We consider the equations in Eulerian form. These are hyperbolic and must be solved subject to appropriate initial data. In the case of a chemical explosive these data are obtained by first calculating the detonation process. This provides initial values of the unknowns on a minus characteristic line, $O'AA'$ in Figure 1. In addition, we must construct initial data on a positive characteristic traversing the whole field of disturbance, represented by zones e, g, and c, very close to O', the point representing the end of the detonation phase.

If perfect-gas behavior is assumed in each region the equations of motion are

$$\frac{\partial v}{\partial t} + v\frac{\partial v}{\partial r} + \frac{1}{\rho}\frac{\partial p}{\partial r} = 0,$$

$$\frac{\partial \rho}{\partial t} + \frac{\partial \rho v}{\partial r} + (v-1)\frac{\rho v}{r} = 0, \tag{3.1}$$

$$\frac{\partial}{\partial t}\left(\frac{p}{\rho^\gamma}\right) + v\frac{\partial}{\partial r}\left(\frac{p}{\rho^\gamma}\right) = 0,$$

where v is the fluid velocity, p the pressure, ρ the density, and γ the specific heat ratio. The independent variables t, the time, and r, the space coordinate, are measured from an origin representing the initiation of the explosion. The factor $v-1$ takes values 0, 1, 2 for plane, cylindrical, and spherical symmetry, respectively.

In the detonation region, provided that the Chapman-Jouguet condition is satisfied, no fundamental length or time influences the motion and all the unknowns are functions of a single similarity variable r/t. Equations (3.1) then reduce to ordinary differential equations which are solved subject to boundary conditions behind the detonation wave OO' and the line $O'A$ representing complete expansion of the detonation products to vacuum conditions. Both end points are singularities of the equations, and solutions in their neighborhood must be represented by series expansions. The detonation solution was obtained independently by Taylor (1950), Döring & Burkhardt (1949), Zel'dovich (1942), Sedov (1956), and Yavorskaya (1956). Berger & Holt (1959) refined the procedure near the singular points.

To construct initial data on a positive characteristic extending from $O'A$ to $O'D$ close to O', the equations of motion are solved in series of the form

$$u = u_0(\theta) + u_1(\theta)\xi + u_2(\theta)\xi^2 + \cdots,$$

where

$$r = 1 - \xi^2 \sin\theta,$$
$$t = 1 - \xi^2 \cos\theta,$$

so that ξ is $O(t-1)^{1/2}$, and θ is an angular coordinate based on O'. Different series are used in each of the regions e, g, and c and these are matched across boundaries to satisfy shock or contact-discontinuity conditions. The lowest-order terms are the same as those in one-dimensional flow in a shock tube, with a centered wave in region e and initially uniform conditions in regions g and c. The higher-order coefficients are determined from sets of ordinary differential equations. The expansions have to be modified near the boundary $O'B$, which is a singular characteristic, and the variable ξ must be strained by Lighthill's technique (1949). The coefficients are calculated up to the second-order terms. Full details are given in Holt (1956) and Berry & Holt (1954).

Once data have been constructed on these two characteristics the further development of the disturbance is calculated by applying the method of characteristics to Equations (3.1). The calculation for air is given in Berry, Butler & Holt (1955), while that for sea water is described in Berger & Holt (1962). In the latter work the equation of state due to Jones (1948) is used in the detonation region instead of the perfect-gas equation and the Tait equation is applied in water (this ensures that the density does not vanish with the pressure).

Friedman (1961) observed some simplifying features in the spherical-explosion problem and obtained a good approximate solution to the problem in quasi-analytical form. The detonation of a plane slab of explosive can be determined analytically, of course, provided perfect-gas behavior is assumed, since all unknowns can be expressed in terms of constant boundary speeds and Riemann invariants. Friedman assumes that the effect of spherical (or cylindrical) symmetry can be determined from small-perturbation solutions of the one-dimensional equations that can be represented in closed form. He uses Whitham's rule (Whitham 1958) to determine the motion of the main blast wave and a modified version of this due to Chisnell (1957) to calculate the path of the second shock. Friedman applies his theory to the spherical-shock-tube problem in which a pressurized sphere is burst instantaneously in an infinite medium at time $t = 0$. This problem was solved numerically by Boyer et al (1958) and investigated experimentally by Boyer (1960). Friedman's results compare well with those from the full numerical calculations for early times after initiation. Chan, Holt & Welsh (1968) subsequently applied Friedman's solution to an explosion from a pressurized sphere in water.

Brode (1958, 1959) calculated spherical explosions in air resulting from chemical charges, as well as from pressurized spheres, using the Lagrangian form of the equations of motion. This has the advantage of following fluid particles so that interfaces can be immediately identified at each time. Their numerical technique is based on that of von Neumann & Richtmyer (1950) for solving the point-source explosion and uses artificial viscosity to represent motion in the region of shocks. This is expensive in computer time and produces fuzzy shocks. On the other hand, shocks do not have to be followed deliberately and are automatically accounted for.

The main effects of nuclear explosions, such as blast-wave-strength variation with distance and pressure profile at a given location, can be determined from similarity solutions of Equations (3.1). These were obtained independently by Taylor (1950) and Sedov (1945a,b). Similarity solutions are valid because the pressure ahead of the main blast wave can be neglected compared with that behind in an intense explosion, so that only two physical constants with independent dimensions play a role, namely, the energy released by the explosion and the density of the undisturbed medium.

To follow Sedov's analysis we rearrange the two constants in the problem so that one contains the mass while the other is independent of mass with dimensions including length to the first power. Denoting the constants by a and b, we write

dimensions of $a = [a] = ML^k T^s$,

$$[b] = LT^{-\delta}.$$

Then the pressure is determined by a functional relationship

$$p = f(r, t, a, b).$$

According to the Pi theorem, if we take r, t, and a as basic parameters, the dimensionless form of this is

$$\frac{r^{k+1}t^{s+2}}{a}p = f\left(1, 1, 1, \frac{b}{rt^{-\delta}}\right).$$

This shows that the dimensionless pressure is a function of a single similarity variable

$$\lambda = \frac{r}{bt^\delta}.$$

The same is true of the other unknowns. Now the two constants in the point-explosion problem are the energy E and the undisturbed density ρ_1. Then take $a = E$ so that $[a] = ML^2T^{-2}$ and $k = 2$, $s = -2$. Also take $b_1 = E/\rho_1$ so that $[b_1] = L^5T^{-2}$. Now choose $b = b_1^{1/5}$ and consequently $[b] = LT^{-2/5}$. Thus in the point-explosion problem $\delta = \frac{2}{5}$.

If dimensionless variables corresponding to v, p, and ρ are denoted by V, P, and R, respectively, Sedov shows that the property of conservation of energy in the disturbed fluid contained by the main nuclear blast wave leads to the existence of an energy integral. This reduces to the simple form

$$\frac{\gamma P}{R} = \frac{(\gamma - 1)V^2(V - \delta)}{2(\delta/\gamma - V)}. \tag{3.2}$$

When Equation (3.2) is combined with two quadratures derived from the similarity forms of Equation (3.1) we obtain the distribution of pressure, density, and velocity behind the main blast wave. The pressure distribution for an air blast is shown in Figure 4.

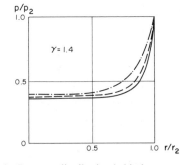

Figure 4   Pressure distribution behind a strong shock.
——————— spherical symmetry;
- - - - - - cylindrical symmetry;
- · - · - · - plane symmetry.

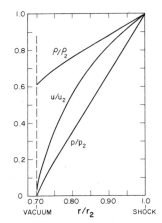

Figure 5   Density, velocity, and pressure distributions behind a strong shock in water.

The application of this similarity analysis to a point-source explosion in water was carried out by Collins & Holt (1968). The Tait equation of state

$$\frac{p}{p_0} = \left(\frac{\rho}{\rho_0}\right)^n - 1,$$

with $n = 7$, was used instead of the perfect-gas equation for air explosions. Figure 5 shows variation of unknowns with distance behind the shock, expressed as ratios to shock values. It is evident that the pressure decays very rapidly behind the shock and a cavitating core is formed around the center of the explosion.

To calculate the later development of a nuclear explosion, when the ambient pressure is no longer negligible compared with the shock pressure, the von Neumann-Richtmyer technique can be applied. At intermediate times perturbations of the similarity solutions can be calculated; such techniques were developed by Sedov (1959) and Sakurai (1965).

4   BUBBLE MOTION

Almost as important as the motion of the main blast wave from an underwater explosion is that of the bubble containing detonation gas products or water vapor. In the case of nuclear explosions this bubble will contain most of the radioactive debris and it is important to trace its progress. Furthermore, bubble motion displaces a great deal of water and frequently produces surface waves of very large amplitude.

The simplest theories of underwater bubble motion neglect both gravitational effects and compressibility and are based on solutions of the equations governing unsteady incompressible flow with spherical symmetry. The radial velocity is then inversely proportional to radius squared and integration of the momentum equation from the bubble surface ($r = a$) to large distances leads to a relation

$$t = \left(\frac{3\rho_0}{2P_0}\right)^{1/2} \int_{a_0}^{a} \frac{da}{[(a_m/a)^3 - 1]^{1/2}}, \tag{4.1}$$

where P_0 is the hydrostatic pressure and a_0 is the initial bubble radius, if the internal energy of the gas is neglected. This result is due to Herring (1941) and the radius-time plot is shown in Figure 6, together with experimental measurements made by Ewing & Crary (1941).

Herring (1941) and Taylor (1941) extended this theory to take account of gravity. If the bubble center has an upward vertical velocity U, the disturbance created by the oscillating bubble can be represented by a velocity potential

$$\phi = \frac{a^2}{r}\left(\frac{da}{dt}\right) + \frac{1}{2}\frac{a^3}{r^2} U \cos \theta. \tag{4.2}$$

Using Equation (4.2) we can set up an energy equation in the form

$$2\pi\rho_0 a^3 \left(\frac{da}{dt}\right)^2 + \frac{\pi}{3}\rho_0 a^3 U^2 + \frac{4\pi}{3}\rho_0 a^3 gz = Y - E(a), \tag{4.3}$$

where z is the height of the bubble center, Y is the energy released by the explosion, and E is the internal energy of the bubble. Since $U = dz/dt$, Equation (4.3) has two unknowns, a and z. The second equation relating a and z is obtained by equating the impulse of the buoyant force on the gas sphere to the vertical momentum acquired by the surrounding water and reduces to

$$U = -\frac{dz}{dt} = \frac{2g}{a^3}\int_0^t a^3\,dt. \tag{4.4}$$

This was first derived by Herring.

The combined oscillation and migration of the bubble derived from Equations (4.3) and (4.4) are shown in Figure 7. These are based on calculations by G. I. Taylor in which U was assumed to be small and $E(a)$ was neglected.

Both of these simple theories are based on the assumptions that the bubble is spherical and that the disturbed water has a superimposed radial and upward

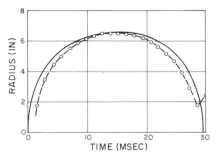

Figure 6   Early stages of bubble growth.

motion. Observations of bubble motion both in small-scale laboratory tests and full-scale explosions have shown that the actual bubble behavior is much more complicated. During the first bubble contraction phase the buoyancy effect is dominant and distorts the spherical form of the bubble significantly. The underside of the bubble becomes re-entrant during the final contraction period so that the bubble as a whole is shaped like a kidney, as shown in Figure 8. When this shape moves upwards the flow over the bubble boundary separates near the rear concave section and a symmetrical, recirculating core is formed that develops into a vortex ring. The combined bubble and vortex ring then migrate together towards the ocean surface.

The conjecture that a vortex ring is attached to the gas bubble, first proposed by Snay (1960), is supported by experimental evidence, both direct and indirect. Figure

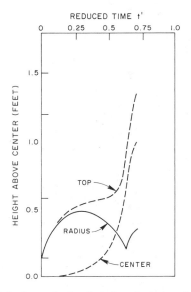

Figure 7 Pulsation and migration of a spherical-explosion bubble.

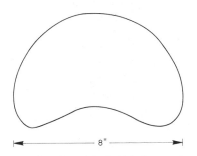

Figure 8 Distortion of gas bubble due to migration.

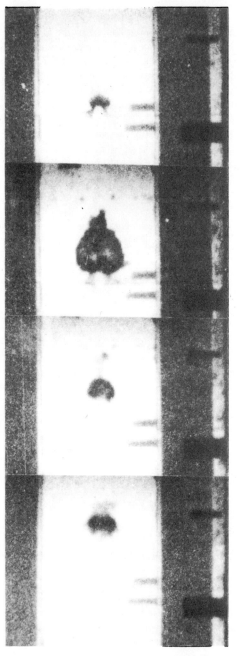

e. t = 0.068 sec, First Minimum

f. t = 0.081 sec, Second Maximum

g. t = 0.100 sec, Second Minimum

h. t = 0.129 sec

Figure 9   Formation of spherical vortex in a migrating bubble.

9 shows photographs of the vapor bubble generated by an exploding wire and released in an explosion tank partly filled with water. The development of the kidney shape is clearly shown both at the first and second bubble minima. In Operation Wigwam measurements were made of the disturbance generated by a deep underwater explosion from a chemical charge. In spite of the long travel time required for the gas bubble to reach the surface the magnitude of the surface disturbance generated indicated that the bubble arrived there with a large amount of energy. In fact, the energy is much larger than the calculated value based on the simple spherical-bubble theory. However, if a vortex ring is combined with the bubble a substantial amount of kinetic energy acquired during the early stages of the explosion is stored in the rotational component conveyed by the ring and little of this is dissipated by radiation during the oscillations and migration of the bubble.

The feasibility of the vortex-ring theory has been examined by Holt & Heiskell (1966) by comparing the energy of a single pulsating bubble with that of a bubble combined with a vortex ring. In the combined motion the bubble is treated as a Hill's Spherical Vortex.

The motion of the water outside the bubble does not depend on the details of the interior bubble motion and is influenced only by the motion of the bubble boundary. The present theory of bubble migration, which gives an expression for the translational velocity and period of bubble oscillation, is based on an energy balance of the water outside the bubble together with a momentum equation for the bubble as a whole. It is reasonable to conclude that the existing migration theory is un-affected by the model of vortex motion inside the bubble. Therefore, in making calculations the existing values of translational velocity and bubble periods may be used.

The bubble energy consists of internal energy, kinetic energy, and potential energy. The internal energy is proportional to the bubble surface pressure. This pressure depends on the bubble radius and translational velocity and is therefore independent of interior bubble motion. The potential energy is also independent of the behavior inside the bubble, since it also depends on the bubble radius and mean density. Hence any difference in bubble energies for its two interior-motion models will be revealed by a comparison of kinetic energies alone.

To modify the Taylor-Herring theory we may still use Equations (4.3) and (4.4) provided that we modify the contribution to the kinetic energy from the vortex ring.

Now in Equations (4.3) and (4.4) the only term that could possibly depend on the interior bubble motion is $E(a)$. This is proportional to the pressure at the surface of the bubble. Therefore, the value of the pressure depends only on the radius of the bubble and its upward velocity and is in no way influenced by the details of the motion inside the bubble.

From these properties it may be concluded that the migrating characteristics of the bubble are independent of the mode of motion in the interior of the bubble. Whether the bubble moves as a radially pulsating sphere or as an oscillating spherical vortex will have no influence on the velocity of upward translation and

the periodic motion of the bubble boundary. The only difference in the two modes
of motion will be in the energies of the fluid inside the bubble.

The kinetic energy of the bubble is calculated on the assumption that it moves
as an oscillating spherical vortex. The expression for the potential of a spherical
vortex of constant radius moving in a uniform stream is given in Lamb (1932).
Following the work of Turner (1964) and Levine (1959) it is assumed that the same
expressions apply instantaneously when the spherical vortex is oscillating, provided
the radius a is a function of time.

The expressions for the radial and the traverse velocity components inside a
spherical vortex of radius a moving with velocity U are

$$v_r = \tfrac{3}{2}U\left(\frac{1-r^2}{a^2}\right)\cos\theta,$$

$$v_\theta = \tfrac{3}{2}U\left(1 - \frac{2r^2}{a^2}\right)\sin\theta,$$

(4.5)

where r is the radius measured from its center and θ is the angle measured from
the upward vertical.

The kinetic energy of the spherical vortex is

$$T_v = \int_0^a\int_0^\pi 2\pi\rho r^2[(U\cos\theta+v_r)^2+(v_\theta-U\sin\theta)^2]\sin\theta\,dr\,d\theta,$$

(4.6)

$$T_v = \frac{46}{21}\pi U^2 a^3 = 2.18\pi U^2 a^3.$$

(4.7)

For comparison the kinetic energy of the bubble is calculated on the assumption
that it simply pulsates in the radial direction, with vortex-ring motion completely
absent.

It is assumed that the density is uniform throughout the bubble at any given
time but is a function of time. Then the equation of continuity is

$$\frac{dp}{dt} + \frac{dv_r}{dr} + \frac{2\rho v_r}{r} = 0,$$

or

(4.8)

$$\frac{dv_r}{dr} + \frac{2v_r}{r} = -K,$$

where $k = (d/dt)\log\rho$, and v_r is the radial velocity. The integral of Equation (4.8) is

$$v_r = -\tfrac{1}{3}Kr$$

(4.9)

(since $v_r = 0$ when $r = 0$). Let V be the velocity of the bubble boundary $= da/dt$.
Then Equation (4.9) holds at the boundary so that

$$v_r = \frac{V}{a}r.$$

(4.10)

The kinetic energy of the pulsating bubble is

$$T_p = \int_0^a \int_0^\pi 2\pi\rho r^2 (U^2 + v_r^2 + 2Uv_r \cos\theta) \sin\theta \, dr \, d\theta. \tag{4.11}$$

This reduces to

$$T_p = 4\pi\rho a^3 \left[\tfrac{1}{3} U^2 + \tfrac{4}{5} a^2 \omega^2 \left(1 - \frac{a^2}{a_{max}^2} \right) \right], \tag{4.12}$$

where a = radius at time t, a_{max} = maximum radius, $\omega = \pi/[\text{bubble period (in sec)}]$ and $V = a_{max} \cos^2 \omega t$.

From Equations (4.7) and (4.12) we obtain the ratio of the kinetic energy of the bubble when moving as a spherical vortex to that of a simple pulsating bubble as

$$\frac{T_v}{T_p} = \frac{2.18\pi\rho U^2 a^3}{4\pi\rho a^3 [\tfrac{1}{3} U^2 + 4a^2\omega^2 (1 - a^2)/a_{max}]}. \tag{4.13}$$

If $a = a_{max}$, we then have

$$\frac{T_v}{T_p} = \frac{2.18\pi U^2 a^3}{4\pi a^3 (\tfrac{1}{3} U^2)},$$

$$\frac{T_v}{T_p} = \frac{2.18}{1.33} = 1.64. \tag{4.14}$$

Thus the kinetic energy of the spherical vortex exceeds that of the pulsating bubble by up to 64%. The increase will be slightly smaller between maxima and minima of bubble oscillations. This result is in qualitative agreement with observation.

5 SURFACE EFFECTS

Underwater explosions not only cause significant blast-wave effects both below and above the ocean surface, they are also responsible for the propagation of large-amplitude surface waves, which under certain conditions can cause enormous damage to shorelines and harbors.

This general subject is treated in a book just published by Le Méhauté (1976), so the present survey is confined to special topics familiar to the writer.

Figure 10 shows the variation of amplitude of a surface wave generated by an underwater or near-surface explosion with depth of charge. Two peaks in the amplitude can be observed. The lower one corresponds to a charge depth equal to about eight charge radii. In this case the explosion is completely submerged and the maximum occurs because an explosion at this depth will deliver the greatest kinetic energy on arrival at the free surface, this being the condition when the gas bubble reaches its first maximum just below the free surface. The depth corresponding to this maximum is called the lower critical depth.

The other maximum gives an amplitude that is larger by an order of magnitude than amplitudes at neighboring charge depths and occurs when the charge center is approximately one half charge radius below the surface. This location is called the upper critical depth.

The lower-critical-depth effect can be satisfactorily explained using estimates of maximum bubble energy based on spherical-explosion calculations and bubble-migration data. The upper-critical-depth effect is very difficult to investigate theoretically because the flow field resulting from a near-surface explosion is never spherically symmetric showing strong dependence on the angular coordinate measured from the charge center. Moreover, the blast wave from such an explosion interacts immediately with the free surface. We describe two investigations connected with the latter topic. In the first the disturbance created by a point-source explosion located exactly on the ocean surface is calculated. In the second the reflection and transmission of an underwater spherical blast wave at the ocean surface is considered.

6   POINT-SOURCE EXPLOSION AT THE OCEAN SURFACE

The problem of a point-source explosion released from a position on an ocean surface, initially undisturbed, was systematically investigated by Collins & Holt (1968). Several particular aspects of the problem were considered in the USSR, notably by Grib et al (1960), Zaslavskii (1962, 1963), Bezhanov (1962), and Shurshalov (1969).

If the pressure ahead of the blast wave is neglected, similarity properties may be used since the energy released and the undisturbed densities in air and water (the latter have the same dimensions) are the only physical parameters of significance. The flow is axially symmetric and the independent variables are a similarity variable of the form r/t^{δ} (as in point-source spherical explosions) and the angle θ between the upward vertical through the origin (taken at the source of the explosion) and a radius vector.

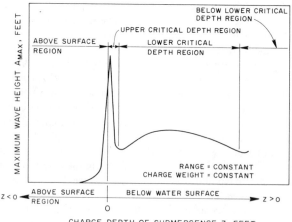

Figure 10   Variation of surface-wave amplitude with depth of explosive charge.

The coordinate system is shown in Figure 11. The dimensionless unknowns used in the similarity analysis are U, V, P, and R where

$$u = \frac{r}{t} U(\lambda, \theta), \qquad v = \frac{r}{t} V(\lambda, \theta), \qquad \rho = \rho_1 R(\lambda, \theta), \qquad p = \rho_1 \frac{r^2}{t^2} p(\lambda, \theta),$$

where (u, v) are, respectively, the (r, θ) velocity components, p is the pressure, and ρ the density.

The governing equations, reduced in terms of the similarity variables λ, θ are

$$\lambda \left[(\delta - U) \frac{\partial U}{\partial \lambda} - \frac{1}{R} \frac{\partial P}{\partial \lambda} \right] - V \frac{\partial U}{\partial \theta} = U^2 - V^2 - U - \frac{P}{R}(k+1),$$

$$\lambda \left[(\delta - U) \frac{\partial V}{\partial \lambda} \right] - \frac{1}{R} \frac{\partial P}{\partial \theta} - V \frac{\partial V}{\partial \theta} = 2UV - V,$$

$$\lambda \left[-\frac{\partial U}{\partial \lambda} + (\delta - U) \frac{1}{R} \frac{\partial R}{\partial \lambda} \right] - \frac{1}{R} \frac{\partial RV}{\partial \theta} = -s - kU + V \cot \theta, \tag{6.1}$$

$$\lambda(\delta - U) \left[\frac{1}{P} \frac{\partial P}{\partial \lambda} - \frac{\gamma}{R} \frac{\partial R}{\partial \lambda} \right] - \frac{V}{P} \frac{\partial P}{\partial \theta} + \frac{\gamma V}{R} \frac{\partial R}{\partial \theta} = -s(1 - \gamma) - 2 - [k(1 - \gamma) + 1 - 3\gamma]U,$$

where $s = 0$, $k = -3$, and $\delta = 2/5$.

The point explosion at O sends blast waves into both air and water but, since the shock speeds will differ considerably, the air disturbance can be assumed to be completely independent. Moreover, we treat the atmosphere as a vacuum.

We now consider the interaction between a curved shock wave originating at O with a plane free surface OB, which initially divides an ocean at rest from a vacuum. To analyze this we transfer polar coordinates from O to B as shown in Figure 11, and with Equations (6.1) in terms of r, ϕ instead of λ, θ. We then seek solutions of these equations in the neighborhood of B, using series expansions of the form

$$U(r, \phi) = U_0(\phi) + rU_1(\phi) \dots,$$

$$U(r, \phi) = V_0(\phi) + rV_1(\phi) \dots,$$

$$P(r, \phi) = P_0(\phi) + rP_1(\phi) \dots,$$

$$R(r, \phi) = R_0(\phi) + rR_1(\phi) \dots.$$

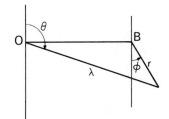

Figure 11   Coordinate system in surface-explosion analysis.

The zero terms satisfy the equations

$$AU_0' - \frac{\cos\phi}{R_0}P_0' = 0,$$

$$AV_0' + \frac{\sin\phi}{R_0}P_0' = 0,$$

$$U_0'\cos\phi - \frac{A}{R_0}R_0' - V_0'\sin\phi = 0,$$

$$\frac{P_0'}{P_0} - \frac{\gamma}{R_0}R_0' = 0,$$

(6.2)

where

$$A = (\delta - U_0)\cos\phi + V_0\sin\phi.$$

Equations (6.2) have two solutions, the first corresponding to constant conditions $U_0' = V_0' = P_0' = R_0' = 0$, and the second given by

$$A^2 = \gamma P_0/R_0,$$

$$\frac{dU_0}{d\phi} = \frac{2\cos\phi}{(\gamma+1)}\{V_0\cos\phi - (\delta - U_0)\sin\phi\},$$

$$\frac{dV_0}{d\phi} = \tan\phi\frac{dU_0}{d\phi},$$

(6.3)

$$P_0 = KR_0^\gamma,$$

which represents a centered expansion fan analogous to the Prandtl-Meyer expansion.

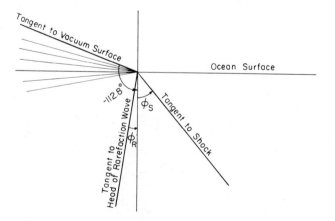

Figure 12   Underwater shock interaction with ocean surface.

The shock-surface interaction pattern is shown in Figure 12. The leading boundary is the shock wave and its angle of inclination with the free surface is determined from a criterion proposed by Zaslavskii (1963). He argues that as the incident shock wave travels along the surface from the origin of the explosion O its angle of inclination, α, to the vertical will gradually increase. When α is small, expansive disturbances sent from surface points behind the shock overtake the shock and weaken it, causing α to increase. When α reaches a critical value α^* the shock point on the surface moves with the same speed as the head of the rarefaction wave following it. If α exceeds this value, expansion waves behind the shock can no longer influence it and the shock strength would tend to increase, causing the angle α to be reduced again. The condition $\alpha = \alpha^*$ proposed by Zaslavskii therefore represents an equilibrium condition under which regular surface-shock-rarefaction matching can be maintained.

The conditions behind the shock are

$$U_0 = \frac{2}{\gamma+1}\cos^2\alpha^*, \qquad V_0 = U_0\tan\alpha^*,$$

$$P_0 = \frac{2\delta^2}{\gamma+1}\cos^2\alpha^*, \qquad R_0 = \frac{\gamma+1}{\gamma-1}.$$

applied at $\phi = \alpha^*$.

These conditions determine the values of U_0, V_0, P_0, and R_0 at the head of the rarefaction wave. We can then integrate Equations (6.3) up to the tail of the expansion fan corresponding to the vacuum surface $P_0 = 0$. Figures 13 and 14 show the variation of P_0 and U_0 through the fan. The disturbed ocean surface (coinciding with the vacuum surface) is inclined at an angle 22.8° above the undisturbed surface so that a lip of disturbed water is formed behind the incident shock.

In an attempt to construct the shape of the disturbed ocean surface in the neighborhood of the interaction point the above analysis was recently extended to the first-order terms represented by $U_1(\phi)$, $V_1(\phi)$, $P_1(\phi)$, and $R_1(\phi)$ (Falade & Holt 1976).

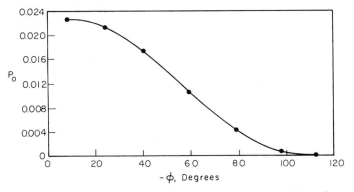

Figure 13 Pressure variations near shock-free surface-interaction point.

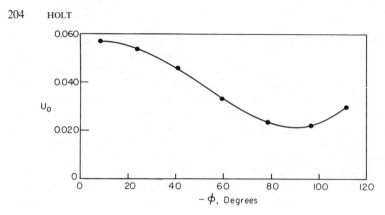

Figure 14   Velocity variation near shock-free surface-interaction point.

The first-order coefficients satisfy a system of nonlinear first-order ordinary differential equations

$$AX' = f,$$

where X is a column vector $\{U_1, V_1, P_1, R_1\}$, f is a column vector depending on the first- and zero-order coefficients, and A is a 4×4 matrix in the unknowns and similarity variables. This system is singular at the head of the rarefaction region and throughout that region. To be able to integrate the equations up to the head of the rarefaction wave a correct choice must be made for the value of the shock curvature. If the shock angle is written $\beta = \alpha + \eta r$ then the correct solution corresponds to $\eta = 48.55$.

The first-order equations reduce to three ordinary differential equations and an algebraic condition required to guarantee existence of any solution. The system can be solved provided series solutions are used in the neighborhood of the head of the rarefaction wave. The first-order coefficients P_1 and U_1 are shown in Figures 15 and 16. Combining the zero- and first-order terms, the shape of the disturbed ocean surface can be determined. It is evident that a high, isolated spray region is formed by the explosion and this does not in itself have a strong influence on surface-wave generation.

7   INTERACTION BETWEEN AN UNDERWATER BLAST WAVE AND THE OCEAN SURFACE

The interaction between a blast wave generated by a completely submerged explosion and the ocean surface has been analyzed by Ballhaus & Holt (1974). The analysis is intended to describe the initial stages of this interaction. If h is the depth of the explosion center and a_{ow} is the speed of sound in undisturbed water, the analysis is valid for times that are small compared with h/a_{ow} and at distances from the center of reflection that are small compared with h.

At time $t = 0$, the explosion produces a spherical shock wave that begins to

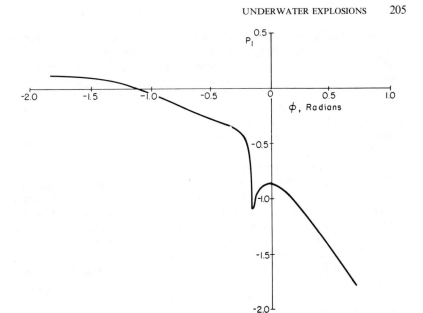

Figure 15   First-order pressure coefficient in shock-free surface-interaction analysis.

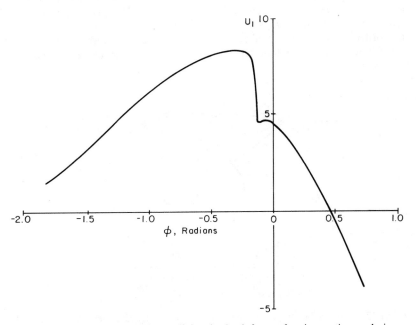

Figure 16   First-order velocity coefficient in shock-free surface-interaction analysis.

propagate radially outward. The flow field behind the shock wave is spherically symmetric. Hence, the flow variables (pressure p, density ρ, speed of sound a, and radial velocity V) are all expressed in terms of the independent variables R and t, where R is the distance from a point in the flow field to the explosion center. At time $t = t_0$, the underwater shock wave begins to interact with the ocean surface, thereby upsetting the spherical symmetry of the flow. For times greater than t_0 the flow is axially symmetric, and the flow variables are expressed in terms of the independent variables r, z, and t, where r and z are distances in the radial and axial directions, respectively.

After the underwater shock wave reaches the ocean surface, it is partially transmitted into the atmosphere and partially reflected back into the water region in the form of an expansion wave. The initial strengths of the transmitted shock wave and reflected expansion wave are equivalent to those that would result from the interaction between a plane shock wave and a plane water-air contact front.

The solution is sought in the form of Taylor-series expansions for the flow variables in terms of small values of r, z, and $(t - t_0)$. Different series expansions must be found for the three flow regions: 1. the air region behind the transmitted shock wave, 2. the water region between the reflected expansion wave and the ocean surface, and 3. the water region inside the expansion wave. In the expansion region flow variables are expanded in series for small r and $(t - t_0)$ along the surfaces $z = f(r, t, V_0)$, where f is also expanded for small r and $(t - t_0)$. The coefficients of the series expansions for the flow variables are functions of N_0, the parameter of the family of surfaces. The requirements that the flow variables satisfy the equations of motion and that the surfaces satisfy certain wave relations provide a system of first-order, ordinary differential equations. Expanding the known spherically symmetric flow behind the underwater shock wave in Taylor series for the flow variables

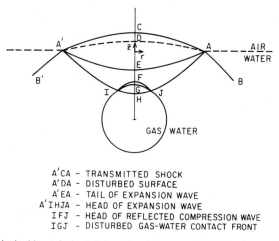

A'CA – TRANSMITTED SHOCK
A'DA – DISTURBED SURFACE
A'EA – TAIL OF EXPANSION WAVE
A'IHJA – HEAD OF EXPANSION WAVE
IFJ – HEAD OF REFLECTED COMPRESSION WAVE
IGJ – DISTURBED GAS-WATER CONTACT FRONT

Figure 17   Principal boundaries in interaction between spherical underwater shock and the ocean surface.

along the leading characteristic surface of the expansion wave then provides the boundary conditions necessary to integrate the differential equations. Integration of the equations provides the coefficients of the flow variable expansions and the surfaces $z = f(r, t)$.

In the water region between the reflected expansion wave and the ocean surface, the flow variables are expanded in Taylor series with constant coefficients for small r, z, and $(t - t_0)$. The series coefficients are determined by satisfying the equations of motion, matching with the solution for the expansion region at the tail of the expansion wave, and requiring that pressure and normal velocity be continuous across the ocean surface.

In the air region disturbed by the transmitted shock wave, the flow variables are also expanded in similar Taylor series with constant coefficients. The series co-efficients are found in a manner similar to that developed by Vasil'ev (1960), who obtained the solution for the flow behind a blast wave reflected from a plane rigid wall. The transmitted shock wave is written in the form $z = f_{sh}(r, t)$, where f is expanded for small r and $(t - t_0)$. The Hugoniot relations are successively differen-tiated along the shock-wave surface f_{sh} in two directions. Taking the limit as r, $z \to 0$, and $t \to t_0$ gives a set of algebraic equations relating the series coefficients of the flow variables to those of the surfaces f_{sh}. Taking the limit as r, $z \to 0$, and $t \to t_0$ in the equations of motion and requiring that pressure and normal velocity be continuous across the ocean surface provides the remaining algebraic equations needed to solve for the coefficients of the series expansions of f_{sh} and the flow variables. Taking account of the fact that the ocean surface moves with the local material velocity gives the coefficients for a series expansion for the shape of the ocean surface $z = f_s(r, t)$.

The paths of the important boundaries in the t, z plane on the axis of symmetry are shown in Figure 17. Figure 18 shows the positions of transmitted shock, the

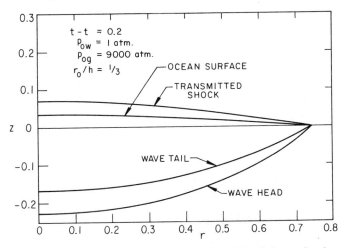

Figure 18   Calculated boundaries in underwater spherical shock-free surface interaction.

disturbed ocean surface, and the head and tail of the reflected expansion wave, a short time after contact between the incident wave and the ocean surface. These results were evaluated for an explosion from a pressurized gas sphere centered at a depth three times the sphere radius and at an initial pressure of 9000 atm. Figure 19 shows the pressure-time curve for this disturbance at a point just below the original ocean surface on the axis of symmetry.

8 COMPARISON OF SURFACE WAVES GENERATED BY SURFACE AND SUBMERGED EXPLOSIONS

These interaction analyses can be combined with blast-wave calculations to determine the nature of surface waves generated by point-source explosions placed firstly on the ocean surface and secondly at a point completely below the surface. The surface-wave calculations are carried out using the method of Kranzer & Keller (1959), which requires as initial data the impulse resulting from the pressure field generated by the explosion applied on some horizontal plane below the surface. It is evident that the surface explosion will produce the larger surface wave since the applied impulse in this case has no negative contribution, while that from a submerged explosion will be reduced by the contribution from the reflected expansion wave at the ocean surface. Detailed calculations of these effects are in progress at the present time.

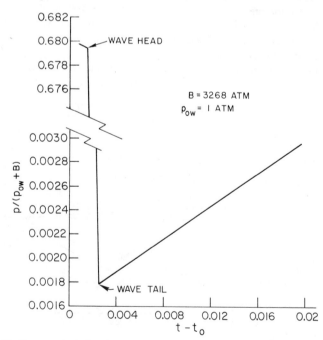

Figure 19   Pressure variation due to spherical underwater shock-free surface interaction.

Very complete calculations of explosions and the surface waves they produce have been carried out by Bjork & Gittings (1972) for a series of four TNT spheres and a surface nuclear burst. The TNT charge weights range from 1 to 385 lb while the nuclear burst is equivalent to 1.1×10^{10} lb of TNT. Each phase of the explosion is modeled as realistically as possible. For the TNT detonation a Jones-Wilkins-Lee equation of state is used (Lee 1968) in a one-dimensional Lagrangian code. The blast-wave disturbance to air and water is calculated by a two-dimensional Eulerian code (three independent variables) with an empirical and complicated equation of state for water. Finally, the data from these calculations are fed into a Kranzer-Keller analysis to determine the surface-wave forms at some distance from the explosion source.

The TNT calculations are all made for charges placed close to the upper critical depth and establish that wave heights comparable to those observed experimentally are calculated. The evidence of the existence of an equivalent upper critical depth

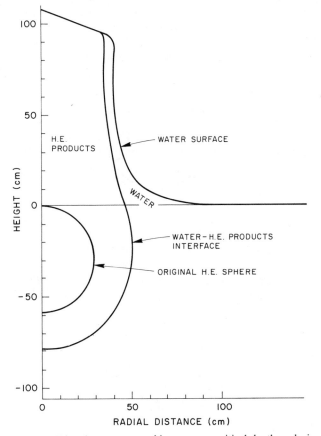

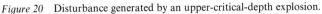

Figure 20   Disturbance generated by an upper-critical-depth explosion.

for nuclear explosions is not conclusive, and this question needs further investigation. The Bjork-Gittings calculations do not cover disturbances from thoroughly submerged explosions and therefore do not finally prove the existence of the upper critical depth effect. Nevertheless, the calculations supply remarkably detailed information about surface explosions. Of particular interest is Figure 20, which shows the shape of the bubble and disturbed ocean surface shortly after initiation. As shown by the analysis of the point-explosion–ocean-surface interaction, the water projected above the ocean surface is confined to a thin cylindrical shell and is unlikely to make a significant contribution to large-amplitude surface waves. These must result from the disturbance below the surface, especially the cavity caused by the blast wave.

9   THE IMPULSE THEORY OF UPPER CRITICAL DEPTH EFFECT

Sakurai (1972) has offered a very convincing explanation of the upper-critical-depth phenomenon based on the impulse rather than the energy properties of a surface or near-surface explosion.

If an explosion is completely submerged the cavity produced in the surrounding

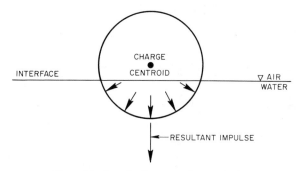

Figure 21   Impulse from a surface explosion.

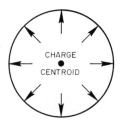

Figure 22   Impulse from a submerged explosion.

ocean is a result of the gas bubble, which migrates upwards and oscillates simultaneously and only disturbs the ocean surface in the later stages of its motion. In the case of a surface explosion there is no gas bubble in the underwater sense since explosive products vent into the atmosphere immediately and the cavity in the ocean is caused by the direct action of blast-wave pressure rather than displacement from a gas bubble. Moreover, the surface explosion causes a net downward impulse to be applied to the ocean (Figure 21), while the impulse from a completely submerged explosion is distributed symmetrically in all directions (Figure 22), producing zero net effect on the ocean.

As Sakurai points out, in the ideal case of a point (or line) detonation at the ocean surface, the early development of the surface deformation can be described by a similarity solution (Sedov 1959, Deribas & Pokhozhaev 1962). If impulse is the key characteristic parameter then similarity analysis shows that the cavity expands like $t^{1/4}$ for a point source or $t^{1/3}$ for a line source, where t is the time. Films of cavities produced by model surface explosions verify the $t^{1/3}$ dependence for exploding wires (Deribas & Pokhozhaev 1962) and t^m with $0.27 < m < 0.30$ for short exploding wires (approximating point sources).

Sakurai calculates the energy applied to the water as a result of direct impact of the water shock and uses this to determine the cavity radius generated by the explosion of a TNT charge. For a small TNT charge weighing 3.5×10^{-3} lb the observed cavity radius is 17 in compared with a radius of 20 in determined from the impulse theory. Sakurai also shows that calculated wave heights are close to those observed for similar charges. This is seen in Figure 23.

10   A NUMERICAL MODEL OF THE UPPER CRITICAL DEPTH PROBLEM

A simple model of an explosion centered at the upper critical depth was investigated recently by Hall & Holt (1976). The explosion results from the rupture of a sphere

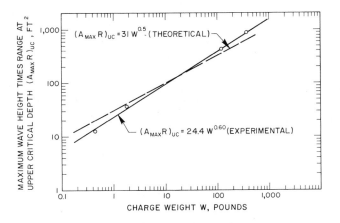

Figure 23   Wave heights due to near-surface explosions.

containing pressurized air with its center located at a depth below the surface equal to half the sphere radius. After rupture the pressurized air initiates a blast wave that propagates into the ocean. At the upper end of the ruptured sphere air escapes into the atmosphere. This situation is modeled by the interaction of an expanding sphere compressing the water below the ocean surface with a vertical column of air expanding vertically upwards. The compression is treated as spherical while the escape process is treated one-dimensionally. This model is consistent with the results obtained by Bjork & Gittings (1972). Each phase of the disturbance is determined numerically by a technique developed by Godunov [see Alalykin et al (1970)] for calculating unsteady motion with plane, cylindrical, or spherical symmetry. This method uses a moving network tied to the important boundaries in the problem under investigation. For example, in the basic piston problem (compression of a gas by a piston moving with constant speed) the network extends from the piston on the left to the shock on the right.

The model is illustrated in Figure 24. The pressure profile and cavity shape at a

PHYSICAL PICTURE

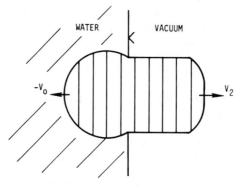

X-T DIAGRAM

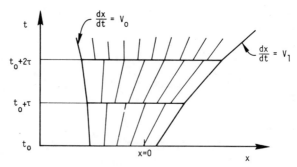

Figure 24   Model of surface explosion.

CELL RAD AT T = 0.600

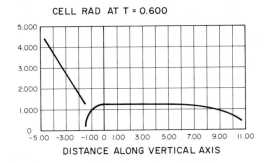

DISTANCE ALONG VERTICAL AXIS

Figure 25   Pressure profile and cavity shape from surface explosion.

time shortly after rupture are shown in Figure 25. The results are used to determine the impulse applied on a plane below the surface providing initial data for surface-wave calculations using the Kranzer-Keller technique.

ACKNOWLEDGMENTS

The preparation of this survey was supported by the Office of Naval Research, Field Projects Branch, under Contract N 00014-75-C-0151 monitored by Mr. Jacob L. Warner, Scientific Officer. The author wishes to thank Mr. Warner, and also Mr. John W. Pritchett of Systems, Science and Software, for assistance with many of the references.

Literature Cited

Alalykin, G. B., Godunov, S. K., Kireeva, I. L., Pliner, L. A. 1970. *Solutions of One-dimensional Problems in Gas Dynamics in Moving Networks.* Moscow: Nauka

Ballhaus, W. F. Jr., Holt, M. 1974. Interaction between the ocean surface and underwater spherical blast waves. *Phys. Fluids* 17: 1068–79

Berger, S. A., Holt, M. 1959. Spherical explosions in sea water. *Proc. Midwestern Conf. Fluid Mech., 6th, Univ. of Texas,* pp. 118–39

Berger, S. A., Holt, M. 1962. The implosive phase of a spherical explosion in sea water. *Phys. Fluids* 5: 426–31

Berry, F. J., Holt, M. 1954. The initial propagation of spherical blast from certain explosives. *Proc. R. Soc. London Ser. A* 227: 258–80

Berry, F. J., Butler, D. S., Holt, M. 1955. The early development of spherical blast from a particular charge. *Proc. R. Soc. London Ser. A* 227: 258–70

Bezhanov, K. A. 1962. The interaction between a shock wave and the free surface of a liquid. *USSR Comput. Math. Math. Phys.* 1: 142–59

Bjork, R. L., Gittings, M. L. 1972. Wave generation by shallow underwater explosions. *Syst. Sci. Software Rep. DNA 2949Z 3SR-1008*

Boyer, D. W., Brode, H. L., Glass, I. I., Hall, J. G. 1958. Blast from a Pressurized Sphere. *Univ. of Toronto, Inst. Aerophys., UTIA Rep. No. 43*

Boyer, D. W. 1960. An experimental study of the explosion generated by a pressurized sphere. *J. Fluid Mech.* 9: 401–29

Brode, H. L. 1958. A calculation of the blast wave from a spherical charge of TNT. *Rand Corp. Rep. P-975*

Brode, H. L. 1959. Blast wave from a spherical charge. *Phys. Fluids* 2: 217–29

Chan, B. C., Holt, M., Welsh, R. L. 1968. Explosions due to pressurized spheres at the ocean surface. *Phys. Fluids* 11: 714–22

Chisnell, R. F. 1957. The motion of a shock wave in a channel with application to cylindrical and spherical shock waves. *J. Fluid Mech.* 2: 286–98

Cole, R. H. 1965. *Underwater Explosions.* New York: Dover

Collins, R., Holt, M. 1968. Intense explosions

at the ocean surface. *Phys. Fluids* 11:701–13

Deribas, A. A., Pokhozhaev, S. I. 1962. Formulation of the problem of a strong explosion on the surface of a fluid. *Dokl. Akad. Nauk SSSR* 144:524–26

Döring, W., Burkhardt, G. 1949. Contributions to the Theory of Detonation. *Wright Field Tech. Rep. No. F-TS-1227-1A* (Brown Univ. transl.)

Ewing, M., Crary, A. 1941. Multiple impulses from underwater explosions. *Woods Hole Oceanogr. Inst. Ann. Rep.* (See Cole 1965)

Falade, A., Holt, M. 1976. On the reflection of a strong underwater shock wave at the ocean surface. Submitted to *Phys. Fluids*

Friedman, M. P. 1961. A simplified analysis of spherical and cylindrical blast waves. *J. Fluid Mech.* 11:1–15

Grib, A. A., Ryzhov, S. A., Khristianovich, S. A. 1960. The theory of short waves. *Zh. Prik. Mat. i Tekh. Fiz.* 1960: No. 1, 63–74

Hall, R. M., Holt, M. 1976. Numerical solutions of the upper critical depth problem. *AIAA J.* 14:191–98

Herring, C. 1941. Theory of the pulsating of the gas bubble produced by an underwater explosion. (See Cole 1965, p. 288)

Holt, M. 1956. The initial behavior of a spherical explosion. I. Theoretical analysis; II. Application to PETN charges in air and water. *Proc. R. Soc. London Ser. A* 234:89–109, 110–15

Holt, M., Heiskell, R. H. 1966. Vortex motion as related to migrated steam bubbles from underwater nuclear explosions. *US Naval Radiol. Def. Lab. Curr. Rep.*

Jones, H. 1948. In *Proc. Symp. Combust. Flame Explos. Phenom., 3rd,* pp. 590–94

Kranzer, H. G., Keller, J. B. 1959. Water waves produced by explosions. *J. Appl. Phys.* 30:398–407

Lamb, H. 1932. *Hydrodynamics.* Cambridge: Univ. Press

Lee, E. L. et al. 1968. Adiabatic expansion of high explosive detonation products. *Lawrence Radiat. Lab. Rep. UCRL-50422*

Le Méhauté, B. 1976. *An Introduction to Hydrodynamics and Water Waves.* New York: Springer

Levine, J. 1959. Spherical vortex theory of bubble like motion in cumulus clouds. *J. Meteorol.* 16:653–62

Lighthill, M. J. 1949. A technique for rendering approximate solutions to physical problems uniformly valid. *Philos. Mag. Ser. 7* 40:1179–1201

Sakurai, A. 1965. Blast wave phenomena. *Basic Developments in Fluid Dynamics,* ed. M. Holt. New York: Academic

Sakurai, A. 1972. Upper critical depth generation. *US Army Eng. Waterways Exp. Stn. Res. Rep. N-72-1*

Sedov, L. I. 1945a. On certain unsteady compressible fluid motions. *Prik. Mat. Mekh.* 9:293–311

Sedov, L. I. 1945b. On unsteady motions of a compressible fluid. *C.R. (Dokl.) Acad. Sci. URSS* 47:91–93

Sedov, L. I. 1956. On the motion of a gas in stellar bursts. *Dokl. Akad. Nauk SSSR* 111:780–82

Sedov, L. I. 1959. *Similarity and Dimensional Methods in Mechanics* (Engl. transl. ed. M. Holt). New York: Academic

Shurshalov, L. V. 1969. On the problem of an intense explosion at the boundary of a half space containing a perfect gas. *Prik. Mat. Mekh.* 33:358–63

Snay, Hans G. 1956. Hydrodynamics of underwater explosions. *Proc. Symp. Naval Hydrodyn. 1st,* Washington, D.C., pp. 325–52

Snay, H. G. 1960. The hydrodynamic background of radiological effects of underwater nuclear explosions. *US Naval Radiol. Def. Lab. Rev. Lect. No. 103, vol. 2,* pp. 1–52

Taylor, G. I. 1941. Underwater explosion research. *Br. Minist. Supply, David Taylor Model Basin Rep. TMB 510*

Taylor, G. I. 1950. The dynamics of the combustion products behind plane and spherical detonation fronts in explosives. *Proc. R. Soc. London Ser. A* 200:235–47

Turner, J. S. 1964. The flow into an expanding spherical vortex. *J. Fluid Mech.* 18:195–208

Vasil'ev, M. M. 1960. On the reflection of a spherical shock wave at a plane surface. *Vychisl. Mat.* 1960: No. 6, 89–99

von Neumann, J., Richtmyer, R. D. 1950. A method for the numerical calculation of hydrodynamic shocks. *J. Appl. Phys.* 21:232–37

Whitham, G. B. 1958. On the propagation of shock waves through regions of non-uniform area of flow. *J. Fluid Mech.* 4:337–60

Yavorskaya, I. M. 1956. Solution of some problems on detonation in a medium with variable density. *Dokl. Akad. Nauk SSSR* 111:783–86

Zaslavskii, B. I. 1962. Certain solutions of the equations of "short waves." *Zh. Prik. Mekh. i Tekh. Fiz.* 1962: No. 1, 34–38

Zaslavskii, B. I. 1963. On the reflection of a spherical shock wave in water at a free surface. *Zh. Prik. Mekh. i Tekh. Fiz.* 1963: No. 6, 50–58

Zel'dovich, B. I. 1942. On the distribution of pressure and velocity in the products of a detonation. *Zh. Eksper. Teoret. Fiz.* 1942:389–406

Ann. Rev. Fluid Mech. 1977. 9:215-28
Copyright © 1977 by Annual Reviews Inc. All rights reserved.

HYDRODYNAMICS OF   ⋇8102
THE UNIVERSE

Ya. B. Zel'dovich

Institute of Applied Mathematics, Academy of Sciences, Moscow, USSR

1 INTRODUCTION: RELATIVISTIC VERSUS NEWTONIAN APPROACH

Immediately after the mathematical formulation of the Newtonian laws of dynamics and gravitation and their successful application to celestial mechanics, the next goal of cosmological application of these laws was put forward. As a matter of fact this goal was achieved only in the middle of the twentieth century. Ironically one had to await Einstein's discovery of general relativity (1917), which is the relativistic generalization of the theory of gravitation. Thereafter the cosmological solutions of general relativity equations were found describing our Universe [Friedmann 1922, 1924; for observational confirmation see Hubble (1929)]. As a result the concept of an evolving and expanding universe was firmly established. The theoretical papers in which the possibility of the Newtonian approach was shown [McCrea & Milne (1934) and others] had only methodical interest; they did not give substantially new results and remained rather unnoticed by the general reader.

The very possibility of treating infinite space filled with matter by Newtonian theory seemed doubtful. The gravitational potential is infinite, and the gravitational force cannot be calculated unambiguously but depends on the limiting process by which we pass from a finite to an infinite distribution of matter.

These features of the problem, called the "gravitational paradox," seemed to preclude the possibility of a Newtonian approach. But the general theory of relativity was built in accord with the correspondence principle just so that it contains Newtonian theory as a limiting case. Therefore one must not wonder that the *local* properties of the Universe are described by Newtonian theory (compare Peebles 1965). The gravitational paradox is solved not by means of mathematical complications, but by exact logical analysis of the concept of uniformity. This approach to the unperturbed Friedmann-Robertson-Walker picture of the Universe is exhibited in Section 2.

The classical hydrodynamical approach proved to be of utmost importance in the problem of the formation of galaxies. One has to analyze the departure of the actual Universe from the idealized strictly homogeneous state. It is a general belief, several centuries old, that gravitational instability is the ultimate cause of matter clustering and inhomogeneous density distribution. The following sections deal with gravitational instability. First the idealized case of a static universe is treated following Jeans (1902) in Section 3.

But we know that the Jeans background of a static universe does not describe the actual situation in our Universe. It disagrees with observation and moreover it is not in accord with theory. The perturbations in the expanding universe were treated first by Lifshitz (1946) in a beautiful fully relativistic manner. He obtained correct results, but their correspondence to the classical Jeans results was well understood only one decade later by Bonnor (1957).

The Newtonian treatment of small perturbation theory is given in Section 4 for an expanding universe.

The recent work of our group concerning the extension of the theory to perturbations of finite amplitude is treated briefly in Section 5. Simple arguments are given showing that the ultimate phase of perturbation growth is the formation of shock waves. Matter compressed by shock waves forms rather flat clumps, sometimes called pancakes. Most of the foregoing is written as examples of formal solutions of hydrodynamic equations with gravitational interaction. A brief Section 6 establishes the connection of the theory to observational evidence.

This article has two goals: (a) to show the importance of modern concepts of homogeneity, self-similarity, and group theory in hydrodynamics. The very important example of the Universe as a whole shows how these concepts are broadening the field of hydrodynamic application; and (b) to push specialists in hydrodynamics into the fascinating field of astronomical, astrophysical, and cosmological research. A detailed treatment of the problem as a part of cosmological research will appear soon in English (see Zel'dovich & Novikov 1975). Excellent modern treatises have been written by Peebles (1971) and Weinberg (1972).

2   ISOTROPIC AND HOMOGENEOUS SOLUTION, GROUP-THEORY DEFINITION OF HOMOGENEITY, FRIEDMANN SOLUTION

Let us write the equations of gravitation and fluid motion for an ideal gas with a definite equation of state $p = p(\rho, S)$:

$$\Delta\varphi = 4\pi G\rho, \tag{2.1}$$

$$\frac{\partial\rho}{\partial t} = -\nabla(\rho\mathbf{u}), \tag{2.2}$$

$$\frac{\partial s}{\partial t} = -(\mathbf{u}\nabla)s, \tag{2.3}$$

$$\frac{\partial\mathbf{u}}{\partial t} + (\mathbf{u}\nabla)\mathbf{u} = -\nabla\varphi - \frac{1}{\rho}\nabla p. \tag{2.4}$$

We are asking for a solution of the form

$$\rho = \rho(t), \qquad p = p(t), \qquad s = \text{const},$$

$$\mathbf{u} = H\mathbf{r}, \qquad \varphi = \frac{2\pi G}{3}\rho(t)r^2. \qquad (2.5)$$

It is easy to put (2.5) into (2.1–2.4) and to verify that the spatial distributions in (2.5) satisfy the equations exactly, and the system of partial differential equations is reduced to two ordinary equations for time dependent $\rho(t)$ and $H(t)$:

$$\frac{d\rho}{dt} = -3H\rho, \qquad \frac{dH}{dt} + H^2 = -\frac{4\pi G\rho}{3}. \qquad (2.6)$$

But we shall not hurry to solve (2.6). Let us first analyze the essence of the assumptions (2.5). Do they correspond to the idea of a homogeneous universe? What exactly is the homogeneity? Giordano Bruno preached that there are many— an infinite number—of suns and planets and types of human society; he was an advocate, even a champion, of statistical homogeneity of the Universe. The Inquisition was strongly opposed and Bruno was burned alive. Really, the homo- geneity concept was exciting strong feelings in medieval society, much more violent than the controversy between Sir Fred Hoyle, author of the steady-state Universe and, say, Allan Sandage or John Archibald Wheeler in our day.

Let us return to our assumptions (2.5). We take density and entropy independent of the spatial coordinates.[1] Shifting the origin we do not change the density and entropy. This is homogeneity. But what about the velocity? At first glance Hubble's law describes a situation with matter at rest at the origin ($\mathbf{r} = 0$), but moving everywhere else with all velocity vectors pointing from a common point. Is the origin really distinguished from other points? Does the Hubble law violate homogeneity?

To obtain the right answer one must remember the principle of Galileo. The measurement of velocity is always relative. We can add a constant vector to all velocities and nothing will change. Therefore the answer to our question about Hubble's law is that homogeneity is not disturbed: we can shift the origin of coordinates in space, but this transformation must be accompanied by a correspond- ing Galilean transformation, i.e. a corresponding shift of the origin in velocity space. Substituting

$$\mathbf{r}' = \mathbf{r} - \mathbf{r}_0, \qquad \mathbf{u}' = \mathbf{u} - H\mathbf{r}_0 \qquad (2.7)$$

into (2.5), we see that

$$\mathbf{u}' = H\mathbf{r}'. \qquad (2.8)$$

A new observer located at the point $\mathbf{r} = \mathbf{r}_0$ will measure the same pattern of movement as the old observer in the former origin $r = 0$. This is the new, modern

---

[1] Smoothing over a volume containing many galaxies is assumed, of course. Discussion of local inhomogeneity is postponed to subsequent sections.

operational or pragmatic definition of homogeneity: the possibility of going to another place and observing the same picture from everywhere.[2]

Along these lines it is not difficult to understand the situation with gravitational potential. At first glance $\varphi \sim r^2$ singles out the origin by two properties:

$$r = 0, \qquad \varphi = 0$$

and

$$r = 0, \qquad \nabla \varphi = 0.$$

If the origin is shifted without change of potential, $\mathbf{r}' = \mathbf{r} - \mathbf{r}_0$, the old φ expressed in terms of the new $\mathbf{r}'$ is

$$\varphi = \frac{2\pi}{3} G \rho r'^2 + \frac{4\pi G}{3} \rho(\mathbf{r}'\mathbf{r}_0) + \frac{2\pi G}{3} \rho r_0^2 \tag{2.9}$$

and both properties are lost. But actually this does not destroy the homogeneity. It is well-known for a potential to be defined up to a constant. Therefore the last $(\sim r_0^2)$ term is easily discarded.

This is a property shared by the electrostatic and the gravitational potential. But the special case of the gravitational potential has one extra property: if one system of coordinates is moving relative to the other with acceleration $\mathbf{g}$, the situation is equivalent to an additional gravitational potential equal to $(\mathbf{gr})$, i.e. linearly dependent on the coordinates. The famous "experiment" of Einstein with a falling elevator cabin is just an example of this theorem: in the accelerated frame the additional potential cancels the Earth's gravitation (linear in the coordinates in the small cabin volume) and weightlessness is achieved.

The conclusion is that by going to an accelerated frame and making appropriate transformation of the potential we can achieve

$$\varphi' = \frac{2\pi G}{3} \rho r'^2, \tag{2.10}$$

which means that the homogeneity proof is completed!

The success of the classical treatment of cosmology by the set of equations (2.1–2.6) is based on a new, refined understanding of the homogeneity. Of course one must have also the moral courage to work with a potential that is infinite (instead of zero) at infinite distance.

---

[2] To a hydrodynamicist it is important to know that the Hubble flow is not the only one with such a property. One generalization is the tensor linear law $u_i = H_{ik} x_k$ with nine space-independent functions of time $H_{ik}(t)$. It contains vortex $(H_{ik} - H_{ki})$ motion and anisotropic dilatation. Hubble flow is the particular case $H_{ik} = H(t) \delta_{ik}$ of the tensor law. The general property is the absence of a characteristic space scale and velocity scale, the dimensions of H_{ik} being t^{-1}.

There is also another type of homogeneous flow, $v_x = w \cos kz$, $v_y = w \sin kz$. The shift δz along the z-axis must be accompanied by a rotation around this axis $\delta \varphi = k \, \delta z$ in order to show the homogeneity. This is the same type as in a plane circular polarized electromagnetic wave or in the Bianchi VII type of metric (Lukash 1974).

Isotropy is also tacitly assumed: the Poisson equation would be solved also by $\alpha x^2 + \beta y^2 + \gamma z^2$ with $\alpha + \beta + \gamma = 1$ instead of r^2 but then isotropy (the equivalence of different directions) would be lost. The loss of potential isotropy would spoil also the Hubble expansion isotropy.

The ideas of homogeneity and isotropy of the Universe are now results of careful astronomical (optical, radio, and x-ray) observations. They are no more the simplest a priori assumptions of a theoretician, as was the situation in 1922–1924 when Friedmann obtained his solution.

Returning to technical problems, we see that the system (2.6) is easily solved. After eliminating t we obtain

$$\frac{dH^2}{d\rho} = \frac{2}{3}\frac{H^2}{\rho} + \frac{8\pi G}{9}, \qquad H^2 = \frac{8\pi G\rho}{3} + b\rho^{2/3}, \tag{2.11}$$

where b is the constant of integration. Using present values of the Hubble constant and the density we obtain b from

$$H_0^2 = \frac{8\pi G}{3}\rho_0 + b\rho_0^{2/3},$$

$$b = \frac{H_0^2}{\rho_0^{2/3}}(1 - \Omega), \qquad \Omega = \rho/\rho_c, \tag{2.12}$$

$$\rho_c = 3H_0^2/8\pi G.$$

Having solved the equation for $H(\rho)$, we can now easily find t as a function of ρ. A very useful quantity is $a = \exp \int H\, dt$, which is the linear scale. The distance between any two given material points is proportional to a.

Now an easy analysis shows that the most important property of the solution depends on the sign of b, which in turn depends on the dimensionless density Ω. Ω is the ratio of the actual density ρ_0 to the so-called critical density ρ_c formed from H_0. If $\Omega < 1$ so that $b > 0$, then H never changes its sign, and the expansion continues forever. If $\Omega > 1$, then b is negative and at a given density $\rho = (3|b|/8\pi G)^3$ the Hubble H vanishes, and the expansion stops and changes to a compression. The formulas are especially simple in the case $\Omega = 1$, $b = 0$:

$$\rho = \frac{1}{6\pi G t^2}, \qquad H = \frac{2}{3t}, \qquad a = \text{const } t^{2/3}. \tag{2.13}$$

One could obtain the foregoing results by considering simply a finite sphere of radius a with constant density and Hubble's velocity distribution inside, $\mathbf{u} = H\mathbf{r} = (\mathbf{r}/a)(da/dt)$. The corresponding equation of motion would be

$$\frac{d^2a}{dt^2} = -\frac{GM}{a^2}, \qquad M = \frac{4\pi}{3}\rho a^3 = \text{const.}$$

We preferred a more sophisticated approach used also by Peebles (1969) in order to show explicitly that we do not fear contemplating an infinite distribution of matter, nor do we need to cut an artificial sphere out of the infinite uniform matter.

The concepts of self-similarity and uniformity and their intimate relation with group theoretical ideas on the shift of the origin of coordinates, Galilean transformation, etc., are no less important for a hydrodynamicist than the facts about the Universe.

3 PERTURBATION THEORY OF A STATIC UNIVERSE

We study the same equations (2.1 to 2.4) but insert small perturbations into the static model:

$$\rho = \rho_0[1 + \delta_k \exp(i\mathbf{kr})],$$

$$\mathbf{u} = \mathbf{u}_k \exp(i\mathbf{kr}),$$

$$s = s_0,$$

$$\varphi = \varphi_k \exp(i\mathbf{kr}), \tag{3.1}$$

$$p = p_0 + c^2\rho_0\,\delta_k \exp(i\mathbf{kr}),$$

$$c^2 = (\partial p/\partial\rho)_{s=s_0}.$$

The quantities with index k are considered small, and second-order quantities are neglected. The set of equations for the first-order quantities is

$$\dot{\delta}_k = -i(\mathbf{ku}_k), \qquad k^2\varphi_k = -4\pi G\rho_0\,\delta_k,$$

$$\dot{\mathbf{u}}_k = -i\mathbf{k}(\varphi_k + c^2\,\delta_k). \tag{3.2}$$

There is a trivial solution for vortex motion $\mathbf{u}_{k^1}$, $\mathbf{u}_{k^2} \perp \mathbf{k}$; in this case $\delta_k = \varphi_k = 0$, $\mathbf{u}_{k^1} = \text{const}$, $\mathbf{u}_{k^2} = \text{const}$.

The interesting results correspond to density perturbations coupled with the longitudinal velocity components $\mathbf{u}_{k^3} = v_k\mathbf{k}/|\mathbf{k}|$. The equations are ($k = |\mathbf{k}|$)

$$\dot{\delta}_k = -iv_k k,$$

$$\dot{v}_k = -ik(-4\pi G\rho_0/k^2 + c^2)\,\delta_k. \tag{3.3}$$

These linear equations whose coefficients do not depend on time have exponential solutions

$$\delta_k = \delta_{0k}\exp(\gamma t), \qquad v_k = v_{0k}\exp(\gamma t). \tag{3.4}$$

The dispersion equation for γ is finally found to be

$$\gamma = \pm\sqrt{4\pi G\rho_0 - c^2 k^2}. \tag{3.5}$$

This result was obtained by Jeans in 1902. The important qualitative features and asymptotics of this solution are:

1. When the waves are long enough (k small), the $\gamma_{1,2}$ are real, one of them being positive, and there is an exponentially growing mode of perturbation. This is

the mathematical description of *gravitational instability* of uniformly distributed matter.

2. For large k one has $\gamma = \pm ikc$. This result describes short acoustic waves with negligible (in the limit $k \to \infty$) gravity influence.

3. There is a definite k, and corresponding wavelength, corresponding to $\gamma_{1,2} = 0$, which separates the instability and acoustic regions. They are called the Jeans wave vector and wavelength, given by

$$k_J = \sqrt{4\pi G\rho_0}/c, \quad \lambda_J = 2\pi/k_J = 2\pi c/\sqrt{4\pi G\rho_0}. \tag{3.6}$$

4. For $k \ll k_J$, $\lambda \gg \lambda_J$ there is asymptotically no dispersion; the dependence of γ on k vanishes:

$$\gamma_{\lim} = \sqrt{4\pi G\rho_0}. \tag{3.7}$$

In the linear theory the principle of superposition of solutions holds:

$$\rho = \rho_0\left[1 + \sum \delta_k \exp\left(\gamma_k t + i\mathbf{k}\mathbf{r}\right)\right]. \tag{3.8}$$

For a smooth initial distribution with all $|\mathbf{k}|$ much less than k_J the exponents of all Fourier components are the same. As a result the solution is factorized

$$\rho = \rho_0\left[1 + \delta_+(\mathbf{r}) \exp\left(t\sqrt{4\pi G\rho_0}\right) + \delta_-(\mathbf{r}) \exp\left(-t\sqrt{4\pi G\rho_0}\right)\right]. \tag{3.9}$$

If the initial ($t = 0$) functions δ_+ and δ_- are of the same order of magnitude (and so are the vortex and potential parts of the initial velocity), the limiting form for sufficiently large t is

$$\mu = \rho_0\left[1 + \delta_+(\mathbf{r}) \exp\left(t\sqrt{4\pi G\rho_0}\right)\right],$$

$$\operatorname{div}\mathbf{u} = \sqrt{4\pi G\rho_0}\,\delta_+(\mathbf{r}) \exp\left(t\sqrt{4\pi G\rho_0}\right), \quad \operatorname{rot}\mathbf{u} \ll \operatorname{div}\mathbf{u}. \tag{3.10}$$

Once more remember that these results are valid for smooth functions or for low pressure and small sound velocity. The growth of a perturbation with a constant form of space distribution $\delta \sim \delta_+(\mathbf{r})$ is characteristic.

Assume that somewhere an excess of density occurs, for example $\delta_+ \sim \exp(-r^2/r_0^2)$. It attracts more matter and growth. But the induced flow corresponds to $\varphi \sim 1/r$ for $r \gg r_0$, the acceleration and the velocity being proportional to $-\mathbf{r}/r^3$. One can conclude that the density and mass are growing due to inflow from infinity rather than by depletion of the near neighborhood, as is seen from the fact that $\operatorname{div}\mathbf{u} \sim \operatorname{div}(\mathbf{r}/r^3) \equiv 0$. This is characteristic for the long-range gravitational forces.

At least the major flaw of Jeans's treatment must be pointed out. The unperturbed solution $\rho = $ const, $\mathbf{u} = 0$ (independent of the space coordinates and time) is not real, as we have already seen in Section 2. We calculated the perturbations of potential φ_k assuming that the unperturbed potential was $\varphi_0 = $ const, but this is not a solution of the unperturbed Poisson equation $\Delta\varphi_0 = 4\pi G\rho_0$. We give the correct treatment in the following section. Anticipating its results, one must wonder to what high extent the inherently wrong results of Jeans come near to the exact ones.

4 PERTURBATION THEORY OF AN EXPANDING UNIVERSE

The correct treatment of the perturbation problem consists in perturbing the solution given in Section 2.

$$\rho = \rho_0(t)[1 + \delta_k(t) \exp{(i\mathbf{kr})}],$$

$$\mathbf{u} = H(t)\mathbf{r} + \mathbf{u}_k \exp{(i\mathbf{kr})}, \tag{4.1}$$

$$\varphi = \frac{2\pi G}{3}\rho_0 r^2 + \varphi_k \exp{(i\mathbf{kr})}$$

with the functions $\rho_0(t)$ and $H(t)$ satisfying Equations (2.6).

The solutions could be found in this form only for varying wave vector

$$\frac{d\mathbf{k}}{dt} = -H\mathbf{k}. \tag{4.2}$$

Physically speaking this means that the wavelength of a given mode of perturbation is increasing during expansion, proportional to the scale $a(t)$ of the unperturbed solution. One can introduce the Lagrangian coordinates $\boldsymbol{\sigma}(\xi, \eta, \zeta)$ of the unperturbed solution

$$\mathbf{r} = a(t)\boldsymbol{\sigma}. \tag{4.3}$$

Then the exponent in the space dependence of the perturbation can be written

$$\exp{(i\mathbf{kr})} = \exp{(i\boldsymbol{\kappa\sigma})}, \qquad \boldsymbol{\kappa} = a(t)\mathbf{k}. \tag{4.4}$$

The statement given above means that the independent solutions have constant Lagrangian-borne wave vector $\boldsymbol{\kappa}$, but varying Eulerian wave vector $\mathbf{k}$.

We omit the algebra. The final results are:

1. The vortex movement with $\mathbf{u}_k \perp \mathbf{k}$ is again independent of density perturbations. During expansion the vortex velocity decreases in inverse proportion to the scale. This is in accord with conservation of the circulation $\int \mathbf{u}\, d\mathbf{l}$ around a closed line expanding with the whole Universe.

2. The short waves behave like acoustic waves. Their frequency changes proportional to $c(t)/a(t)$; their amplitude changes corresponding to the law of adiabatic invariance $(E_{\text{acoust}} \sim \omega)$.

3. There is a definite wavelength at which the oscillatory time dependence of perturbations vanishes. This wavelength is numerically very close to the momentary value of the Jeans wavelength (see the preceding section).

4. The long waves exhibit gravitational instability. In the limit of $k^2 \ll 4\pi G\rho_0/c^2$ the dispersion vanishes, i.e. the form of $\delta_k(t)$ no longer depends on k.

We give the results for the simplest case of the unperturbed "flat" universe[3] corresponding to a parabolic velocity of expansion.

[3] The name "flat" comes from general-relativistic ideas about Riemannian curved space. We do not use these ideas here.

This solution was given above (2.13). There are two modes of density perturbations

$$\delta_+ \sim t^{2/3}, \qquad u_+ \sim t^{1/3};$$
$$\delta_- \sim t^{-1}, \qquad u_- \sim t^{-4/3}. \tag{4.5}$$

Again the absence of dispersion in the long-wave limit allows us to write the answer in factorized form (with account taken of wave-vector change):

$$\delta = t^{2/3}\,\delta_+(\mathbf{r}/a) + t^{-1}\,\delta_-(\mathbf{r}/a),$$
$$\mathbf{u} = t^{1/3}\mathbf{u}_+(\mathbf{r}/a) + t^{-4/3}\mathbf{u}_-(\mathbf{r}/a) + t^{-2/3}\mathbf{u}_{\text{vort}}(\mathbf{r}/a), \tag{4.6}$$

where

$$\operatorname{rot}\mathbf{u}_+ = \operatorname{rot}\mathbf{u}_- = 0, \qquad \operatorname{div}\mathbf{u}_+ = -\frac{2}{3}\frac{\delta_+}{t}, \qquad \operatorname{div}\mathbf{u}_- = \frac{\delta_-}{t}, \qquad \operatorname{div}\mathbf{u}_{\text{vort}} = 0.$$

If all the modes are of the same order of magnitude at some initial moment t_0, then asymptotically for $t \gg t_0$ the only surviving mode corresponds to the growing solution. Again due to superposition the result can be written in the factorized form

$$\rho = \frac{1}{6\pi G t^2}\left[1 + t^{2/3}\,\delta_+(\mathbf{r}/a)\right],$$
$$\mathbf{u} = \frac{2}{3}\frac{\mathbf{r}}{t} + t^{1/3}\mathbf{u}_+(\mathbf{r}/a), \tag{4.7}$$
$$a = \text{const } t^{2/3}$$

with arbitrary δ_+ but u_+ adjusted to δ_+.

The most important change when compared with Jeans' theory is the power law $t^{2/3}$ of (4.7) instead of the exponential one $\exp(\gamma t)$ of (3.10).

When first discovered by means of difficult general-relativistic calculations, the power law appeared to be a drastic change; it seemed that relativity leads to vanishing of instability. Actually the change is very small! In Jeans' theory $\gamma = \pm\sqrt{4\pi G\rho_0}$. In the expanding universe $\rho_0 = 1/6\pi G t^2$ is no longer constant. The very natural generalization is to take $\gamma = \gamma(t)$ according to Jeans and to use $\int \gamma\, dt$. This leads to $\gamma = (1/t)\sqrt{\tfrac{2}{3}}\ln t$ and $\int \gamma\, dt = \sqrt{\tfrac{2}{3}}\ln t$.

But obviously $\exp\left(\pm\sqrt{\tfrac{2}{3}}\ln t\right) = t^{\pm\sqrt{\tfrac{2}{3}}}$. The power law is just what one expects from Jeans' theory applied to the expanding universe! The growth of perturbations at every given moment in the expanding universe is different from Jeans' predictions only inasmuch as $\tfrac{2}{3} = 0.67$ is different from $\sqrt{\tfrac{2}{3}} = 0.85$.

What really matters is the finite time of growth and change of physical properties in the expanding universe (see Section 6) as compared with the static, infinitely old Jeans background.

5   APPROXIMATE NONLINEAR SOLUTION AND ITS QUALITATIVE FEATURES

We proceed to construct an approximate solution describing the behavior of the gas when the density contrasts are of order unity and the linear treatment is no longer valid. The most obvious difficulty with the formulas given above is that $\delta = (\rho - \bar{\rho})/\bar{\rho}$ must be negative in some places and positive in others (in order that the average over all space $\delta = 0$). The growth of δ will lead to $\delta < -1$ somewhere with $\rho < 0$, which has no meaning. Clearly it is preferable to take another quantity instead of density as the basis of approximate formulation of the results; specifically we propose the consideration of *displacements*, which means using Lagrangian coordinates. The advantage of using them is seen also from the results of the preceding section: the wavelength of a single perturbation mode is stretched during expansion and the general solution is given in terms of $\mathbf{r}/a$, which is the Lagrangian coordinate of the unperturbed background. Having guessed an approximate solution in Lagrangian coordinates, we calculate the density by exact formulas. This procedure guarantees that no paradox arises.

The chances of finding an adequate approximation are better when a narrow class of problems is solved. We consider the case of smooth initial perturbations, with all characteristic lengths much greater than the Jeans wavelength. This means that dispersion is neglected. In the differential equations it is equivalent to fully neglecting the pressure.

All movements are supersonic, $\mathrm{Ma} = \infty$, because $c^2 \equiv 0$. Clearly the hydrodynamics of flows with $\mathrm{Ma} = \infty$ degenerates into the motion of independent particles in a common self-consistent gravitational field.

The second specification of the problem is that we consider an initial situation with perturbations of all types, but all of amplitude of the same small order of magnitude. Thereafter the growing mode is increasing; the damped mode and the vortex mode are decreasing during expansion.

The situation we are interested in, with large amplitude of perturbations, occurs late. In this period all modes except the growing one have already vanished. We ask for the approximate solution describing the unperturbed universe with growing mode only superimposed.

All together the arguments given above lead to the following formulation:

$$\mathbf{r}(\boldsymbol{\sigma}, t) = a(t)\boldsymbol{\sigma} + b(t)\boldsymbol{\psi}(\boldsymbol{\sigma}). \tag{5.1}$$

Here $\boldsymbol{\sigma}$ is the Lagrangian coordinate attached to particles of the gas and $\mathbf{r}(\boldsymbol{\sigma}, t)$ the Eulerian coordinate, i.e. the actual position in space of a given particle ($\boldsymbol{\sigma}$ is the label) at the moment t.

The first term describes the Hubble expansion: with $\psi = 0$ one would obtain

$$\mathbf{u} = \left.\frac{\partial \mathbf{r}}{\partial t}\right|_{\sigma} = \frac{da}{dt}\boldsymbol{\sigma} = \frac{1}{a}\frac{da}{dt}\mathbf{r} = H\mathbf{r} \tag{5.2}$$

with

$$H = \frac{1}{a}\frac{da}{dt}, \qquad a = \exp \int H\,dt.$$

The second term describes the growing perturbations. The corresponding motion under the influence of gravitation is potential, so that we can specify

$$\psi = \text{grad}_\sigma\,\mu(\sigma).$$

The function $b(t)$ is taken from the linear theory of perturbation as the law of growth of small perturbations of long wavelength. It is growing more rapidly than the scale a:

$$\frac{d\ln b}{dt} > H = \frac{d\ln a}{dt}.$$

In the case of the simple "flat" (parabolic) universe obeying (2.13)

$$\mathbf{r} = t^{2/3}\boldsymbol{\sigma} + t^{4/3}\,\text{grad}_\sigma\,\mu(\sigma). \tag{5.3}$$

The distribution of density is given by the inverse of the volume element in $\mathbf{r}$ (Eulerian) space corresponding to a given element in σ space:

$$d\vartheta = d^3 r = \frac{D(x, y, z)}{D(\xi, \eta, \zeta)}\,d^3\sigma. \tag{5.4}$$

An easy calculation gives the density

$$\rho = \text{const}\left| t^{2/3}\,\delta_{ik} + t^{4/3}\,\frac{\partial^2\mu}{\partial\sigma_i\,\partial\sigma_k}\right|^{-1} \tag{5.5}$$

with $\sigma_1 = \xi, \sigma_2 = \eta, \sigma_3 = \zeta$. Having in mind that in the unperturbed case

$$(\mu = 0) \qquad \rho = \bar\rho = \frac{1}{6\pi G t^2},$$

we can write

$$\rho = \frac{\bar\rho}{(1 + \alpha t^{2/3})(1 + \beta t^{2/3})(1 + \gamma t^{2/3})} \tag{5.6}$$

where α, β, γ are the three principal values[4] of the tensor

$$\frac{\partial^2\mu}{\partial\sigma_i\,\partial\sigma_k}.$$

[4] This means that by choosing the appropriate orientation of the coordinate axis (which is not the same for different particles) we shall obtain in these axes

$$\frac{\partial^2\mu}{\partial\xi^2} = \alpha, \qquad \frac{\partial^2\mu}{\partial\eta^2} = \beta, \qquad \frac{\partial^2\mu}{\partial\zeta^2} = \gamma,$$

and mixed derivatives are zero.

One can check the precision of the approximate solution. Doroshkevich, Riabenkiy & Shandarin (1973) proposed the following procedure: take the solution (5.1) or (5.3). The resulting displacement leads through the continuity equation to the density distribution (5.6). But we can ask what is the gravitational potential needed to induce the displacements described by (5.1) or (5.3). It is given by the equation of motion

$$\frac{d^2\mathbf{r}}{dt^2} = -\operatorname{grad}_r \varphi'. \tag{5.7}$$

Further we ask what is the density distribution ρ' needed to induce the gravitational potential φ': by Poisson's law it is given by

$$\rho' = \frac{1}{4\pi G}\Delta\varphi' = -\frac{1}{4\pi G}\operatorname{div}_r\left(\frac{d^2\mathbf{r}}{dt^2}\right). \tag{5.8}$$

Should the ρ given by (5.5) and the ρ' given by (5.8) coincide exactly, one could deduce that the initial form (5.1) or (5.3) is an exact solution, with exact self-consistent motion, density, and gravitational potential.

The calculations give (again for the case $\Omega = 1$)

$$\rho' = \bar{\rho}\,\frac{1+(\alpha\beta+\alpha\gamma+\beta\gamma)t^{4/3}+\alpha\beta\gamma t^2}{(1+\alpha t^{2/3})(1+\beta t^{2/3})(1+\gamma t^{2/3})}.$$

The solution is exact in the unidimensional case $\beta = \gamma = 0$, $\alpha \neq 0$. In the general case the difference between ρ and ρ' is of second and third orders as compared with $\alpha t^{2/3}$, $\beta t^{2/3}$, $\gamma t^{2/3}$. This was obvious, because the approximate theory was modeled so as to include the linear approximation.

Most important are the common properties of ρ and ρ': if one of α, β, γ is negative (say α), then the density becomes infinite at a definite moment t_c when $1 + \alpha t_c^{2/3} = 0$. The denominators in ρ and ρ' are the same, which means that the approximate theory has a finite relative error when $\rho \to \infty$; the qualitative picture is undoubtedly true.

The point is that density is infinite due to *one* of the denominators. This means unilateral compression. It is a general property of a smooth initial distribution of particle velocities that in the general nondegenerate case the intersection of trajectories takes place on so-called caustic surfaces.

A more thorough investigation of the overall picture goes as follows: given the initial distribution of perturbations at an early date when they are small, one must find $\mu(\sigma)$, i.e. the space distribution of displacements corresponding to the growing mode. Furthermore the quadratic form $\partial^2\mu/\partial\sigma_i\,\partial\sigma_k$ and its principal values α, β, γ are to be found.

These α, β, γ are functions of the particle coordinates. One can find the particle σ_m where the smallest of them, say α, has the greatest negative minimum $\alpha_m < 0$. This is the particle where the density first goes to infinity. Subsequently the pressure cannot be neglected in the compressed gas. New particles are impinging on already compressed gas and are brought to rest by receding shock waves.

On the other hand, new infinities are building up in adjacent particles with

$|\alpha| < |\alpha_m|$ as time goes on. The overall picture is the formation of a rather flat[5] region of compressed gas, with caustics on the equatorial line and shock waves on top and bottom.

Further elaboration of the picture, in both hydrodynamic and statistical theory, is difficult, with complicated numerical and analytical calculations, and it seems not to be in line with the *Annual Review of Fluid Mechanics*.

6 OUTLINE OF THE ASTRONOMICAL SITUATION

A very brief account should be given of the results of astronomical investigations and their relation to the initial statements and the results of the theory outlined above.

The general picture of an expanding universe is now established firmly. The relative velocity of distant galaxies is measured by the Doppler red shift of spectral lines; the Hubble constant is of the order of 60^{+40}_{-20} kilometers per second per megaparsec. The distance unit megaparsec is 3.08×10^{24} cm, so that in time units $H = 2 \times 10^{-18}$ sec$^{-1} = (1.7 \cdot 10^{10}$ yr$)^{-1}$ is most probable.

This estimate is roughly confirmed by the age of the stars. Theories of a static or steady universe, as well as theories explaining the red shift by the aging of photons, are strongly disfavored it seems to me.

Statistical counts of distant objects, as well as the investigation of the 2.7°K blackbody radio radiation, confirm strongly the isotropic character of the expansion and, indirectly, the uniformity of the Universe on a scale greater than, say, 100 or 200 megaparsecs. The region investigated is very much greater, up to 3000 megaparsecs for galaxies, 5000 for radio sources, and 6000 for blackbody radiation. The overall average density of the Universe is still under violent discussion; there are estimates from $2 \cdot 10^{-31}$ g cm^{-3} up to 10^{-29} g cm^{-3}, corresponding to $0.03 \leq \Omega \leq 2$.

The blackbody radiation extrapolated in the past leads to the picture of a hot big bang.[6]

The hydrodynamic theory outlined above applies to the period when the temperature falls below 3000°K, and protons and electrons combine into hydrogen atoms. The atoms do not interact practically with radiation. We have to do with two independent interpenetrating gases: free radiation and atoms. The atomic gas pressure is 10^8 times less than the radiation pressure. It is the atomic gas that we study, on the background of an unperturbed uniform radiation field.

The atomic gas is one of the heirs of the fully ionized plasma that existed earlier, at temperatures higher than 4000°K. During this phase the radiative viscosity damped short-wave common perturbations of radiation and matter density (corresponding to masses less than 10^{13} solar masses). We use this information to begin

[5] Its flatness is only asymptotic. So long as the perturbation is smooth, all characteristic quantities including the α-value and the direction of the α-axis are changing slowly inside the correlation length, which describes the smoothness of the perturbation function.

[6] See corresponding articles in the *Annual Review of Astronomy and Astrophysics*.

the atomic gasdynamics with a smooth velocity distribution, which enables us to neglect pressure. The initial perturbations are of the order of 10^{-3} (dimensionless). The formation of shock waves and compressed regions probably occurred when the scale of the Universe was 5 or 10 times smaller than now, and the Universe was 5–15 times younger (time elapsed from the big bang).

A characteristic feature of the approximate theory is the occurrence of shock waves and, during subsequent evolution, of turbulence induced by shock waves and heat-transfer processes. Thus we hope to explain the rotation of galaxies in spite of the potential motion during the perturbation growth.

There are claims that the superclusters of galaxies are flat, perhaps because the initial clouds of compressed gas were flat. This is not proven yet.

There are other types of theories of galaxy formation, of which the best known operates with initial perturbations of vortex type, giving density contrast by nonlinear interaction. Initial small-scale quiet proton density perturbations on an unperturbed radiation background are also possible and in fashion now. Again this article is not an appropriate place to discuss the comparative advantages of various astronomical theories. Our goal is to show how astronomy uses hydrodynamics and how welcome hydrodynamicists would be in the astronomical community.

ACKNOWLEDGMENTS

Special thanks are due to Prof. Fiszdon of the Polish Academy of Sciences. He invited me to make a report on the *Hydrodynamics of the Universe* at the Biennial Symposium on Fluid Mechanics in Bialowieza, Poland. A short account of the report is published in the proceedings of that symposium (Zel'dovich 1976). Prof. Van Dyke, present at the symposium, asked me for a more comprehensive review; I am grateful for his encouragement. Last but not least, my colleagues Doroshkevich, Novikov, Shandarin, and Sunyaev have done much to elucidate the overall picture presented above.

Literature Cited

Bonnor, W. B. 1957. *MNRAS* 117:104
Doroshkevich, A. G., Riabenkiy, N. S., Shandarin, S. F. 1973. *Astrofizika* 9:258
Einstein, A. 1917. Kosmologische Betrachtungen zur allgemeinen Relativitätstheorie. *Sitzungsber. Berlin Akad.*
Friedmann, A. A. 1922. *Z. Phys.* 11:377; 1924. *Z. Phys.* 21:326
Hubble, E. P. 1929. *Proc. Natl. Acad. Sci. USA* 15:168
Jeans, J. H. 1902. *Philos. Trans.* 129:44
Lifshitz, E. M. 1946. *JETP* 16:587
Lukash, V. N. 1974. *JETP* 67:1594
McCrea, W., Milne, E. 1934. *Q. J. Math.*

5:73
Peebles, P. J. E. 1965. *Am. J. Phys.* 33:106
Peebles, P. J. E. 1969. *Ap. J.* 155:393
Peebles, P. J. E. 1971. *Physical Cosmology.* Princeton, NJ: Princeton Univ. Press
Weinberg, S. 1972. *Gravitation and Cosmology.* New York: Wiley
Zel'dovich, Ya. B., Novikov, I. D. 1975. *Structure and Evolution of the Universe.* Moscow: Nauka (in Russian); English transl. 1977. ed. G. Steigman. Chicago Univ. Press. In preparation
Zel'dovich, Ya. B. 1976. *Fluid Dyn. Trans.* 8:279

Ann. Rev. Fluid Mech. 1977. 9: 229–74

PULMONARY FLUID DYNAMICS

✸8103

T. J. Pedley

Department of Applied Mathematics and Theoretical Physics, University of Cambridge, Cambridge, England

1 INTRODUCTION

1.1 *Physiological Background*

The main functions of respiration are the supply of oxygen to the tissues of the body and the removal of carbon dioxide from them. In all but the smallest animals this is achieved in two stages: stage I—the transfer of gases between the surrounding medium and the blood, and stage II—the circulation of blood to the tissues. Large quantities of O_2 and CO_2 can be transported by the blood because of the presence of hemoglobin, with which these gases can combine rapidly and reversibly: whole blood can normally take up well over 50 times as much oxygen as can be dissolved in the same volume of water. In this article I deal solely with stage I, the transport between the surrounding medium and the blood. In water-breathing animals the organs that effect this transport are the gills; in air-breathing animals they are the lungs. I discuss only the latter here and restrict myself further to the lungs of mammals, especially man. [The respiratory systems of other classes of animals are extremely interesting; for example, respiratory systems of birds are constructed very differently from those of mammals and raise some intriguing fluid-mechanical problems that are as yet unsolved; see Lighthill (1975, section 11.3) and references therein; see also Brackenbury (1972) and Duncker (1974).] I also ignore the complicated chemical mechanisms of O_2 uptake and CO_2 release by the blood, and merely describe the delivery of O_2 to, and the removal of CO_2 from, the membrane that separates the air from the blood.

At the end of a normal, quiet expiration, the air in the lungs is at rest, the tissues of the lung are in elastic equilibrium, and the respiratory muscles are virtually inactive; the volume of air in the lungs (called the functional residual capacity, or FRC) is roughly half of the total lung capacity (TLC), which in a man 1.7 m in height is about 6×10^{-3} m^3. On inspiration, the inspiratory muscles (principally the diaphragm in quiet breathing) contract and cause the volume of the thorax to increase. The lung, whose outer (pleural) surface is separated from the inner surface of the thorax by a thin, discontinuous space containing fluid, therefore also expands, and air is sucked into it through the nose and mouth. The pressure drop between the outside air and the pleural space is balanced by the elastic recoil forces of the lung itself (tending to restore volume to FRC) and by the viscous forces that resist the flow of air into the airways. Expiration, during quiet breathing, is achieved by switching off the inspiratory muscles, so that the lung volume is reduced by the elastic recoil, and air is driven out against the viscous resistance. The recoil forces can be assisted by the expiratory muscles if expiration has to be accelerated or if lung volumes below FRC are required. The elastic component of the pressure drop between the mouth and pleural space can be understood through a structural examination of the lung tissue (and of the thin layer of surface-active fluid that lines the small air spaces). The viscous component can be understood only by means of a fluid-mechanical analysis of airflow in the airways. The former is not within the scope of this review (see, instead, West & Matthews 1972, Hoppin & Hilde-brandt 1977, Horn & Davis 1975), but the latter is its principal subject.

The gas-exchange membranes, whose total surface area in man is about 70 m^2, form the walls of about 3×10^8 small sacs, the alveoli, whose diameters range between 75 and 300 μm, and between which the blood flows in thin sheets about 7 μm in thickness [the pulmonary circulation is another fascinating fluid-mechanical phenomenon: see Fung & Sobin (1969) and Fishman & Hecht (1969)]. Most of the air space consists of alveoli. The atmosphere is connected to the alveoli by a complex diverging system of branching tubes, the bronchial tree, which is described in more detail below. The volume of air contained in the airways (as opposed to the alveolar region) at FRC is about 0.15×10^{-3} m^3, while the total gas volume in the lung is about 3×10^{-3} m^3; the volume inspired per breath during quiet breathing is about 0.45×10^{-3} m^3 [all numbers quoted are for a "typical" man, about 40 years old and about 1.7 m in height; for an appreciation of the wide range of values to be found in a normal population, see Cotes (1965)]. The average partial pressure of O_2 (P_{O_2}) in alveolar gas is the same as that in arterial blood, about 13.3 kN m^{-2} (100 mm Hg), while that in moist fresh air is 19.8 kN m^{-2}, and that in mixed venous blood is 5.3 kN m^{-2}. The corresponding values for CO_2 (P_{CO_2}) are 5.3 kN m^{-2} (arterial and mean alveolar), 6.4 kN m^{-2} (mixed venous), 0.02 kN m^{-2} (moist fresh air). (Most of the air both inside and outside the lungs consists of nitrogen, whose partial pressure both in the alveoli and in moist air is about 75 kN m^{-2}.) It should be emphasized that the values quoted for alveolar P_{O_2} and P_{CO_2} are average values; not only does the composition of alveolar gas vary during a breath (a swing of 0.5–0.6 kN m^{-2} is often quoted for P_{O_2}, and 0.4 kN m^{-2} for P_{CO_2}) but at any given time there must be gradients of gas partial pressures across the alveoli in order for a net flux to take place. Further-more, conditions vary between alveoli in different regions of the lung. In order to understand the process of gas transfer it is necessary to understand both how fresh air is mixed with gas flowing in the airways during inspiration (and expiration) and how gas in the airways is mixed with alveolar gas. This represents a complicated convection-diffusion problem, and before it can be solved it is again necessary to describe and understand the flow itself.

It is important to understand the transport in the lung of respiratory gases other than O_2 and CO_2. Anesthetic gases are obviously very important. Many others are also used for experimental and diagnostic purposes: CO, which is readily taken up by the blood, is used to test the efficacy of pulmonary gas exchange (Comroe 1974, pp. 164–66); radioactive xenon, ^{133}Xe, which is not readily taken up by the blood, is used to observe directly the distribution of inspired gas (Dollfuss, Milic-Emili & Bates 1967); and other physiologically inert gases (He, Ar, Ne, SF_6) are used to assess different factors influencing gas mixing (Farhi 1969). Gas mixtures based on SF_6 (density 5.0 times that of air) and He (density 0.14 times that of air) are also used to model the mechanical loads experienced by divers, breathing air at several times atmospheric pressure, and by high altitude fliers, breathing at very low pressures. Divers also experience greatly enhanced respiratory losses of heat and water vapor, which are presumably associated with increased mixing resulting from greater Reynolds numbers at higher gas densities (Linderoth et al 1975). Pollutants in the atmosphere form another class of substances whose mixing and deposition in the lung need to be understood for medical purposes. As well as noxious gases and vapors, these often take the form of solid particles or liquid droplets, which can be regarded as dense gases of very small diffusivity. Their deposition in the airways can result in damage to the mucociliary clearance mechanism, which means that more pollutants tend to accumulate in the small airways and alveoli, where they cause further damage to the gas-exchange system (see Brain, Proctor & Reid 1977). On the other hand, locally acting drugs are often supplied to patients with lung disease (especially asthma) in the form of inhalants carried into the respiratory tract as an aerosol. Very little systematic information on the pattern of deposition or the effectiveness of these aerosol treatments is yet available, and this forms an active area of clinical respiratory research.

A knowledge of the airflow pattern in normal lungs is sufficient for an analysis of gas and particle mixing and deposition. It is also adequate for an understanding of the viscous resistance to breathing in normal subjects. However, the main application of respiratory physiology is to respiratory disease, and the understanding of the mechanics of normal lungs is only a prelude to an investigation of diseased lungs. At least three different types of chronic disease that obstruct the mechanical function of the lung are recognized: (*a*) asthma, in which the muscle in the walls of some bronchi constricts, blocking the airways and increasing their resistance; (*b*) bronchitis, in which the bronchi are blocked or narrowed by inflammation or some other mechanism; and (*c*) emphysema, in which the walls of some of the alveoli and/or small airways are damaged, exhibit altered elastic properties, and no longer contribute to gas exchange. However, it is possible to distinguish clearly between these diseases only at post mortem; the functional differences are often very hard to distinguish. One of the important goals of clinical respiratory physiologists is to devise a discriminating diagnostic test, for example by isolating the aspect of lung mechanics that is affected; no generally agreed upon test is available, however. Furthermore, no test is yet available that can clearly diagnose lung disease at an early stage, before symptoms appear, except in subjects who have been frequently tested throughout their lives. Much of this difficulty, of course, is due to the inherent variability between subjects, but it is still the hope (and belief) of many workers in the field that a test can be found, on the basis of an improved understanding of the mechanics of normal and diseased lungs. This article is concerned with the fluid mechanics of the normal lung, for which we are developing a reasonable understanding. The detailed fluid mechanics of the diseased lung is very poorly understood, however, at least at any level deeper than "narrowed bronchi lead to increased airway resistance." This is an area in which fluid dynamicists can make useful contributions as long as they base their work firmly on the real anatomical features of the diseased lung.

The reader who wants to learn more about respiratory physiology in health and disease should read both a basic textbook [see, for example, Comroe (1974) or West (1974)] and some of the up-to-date reviews [see Macklem (1971); Brain, Proctor & Reid (1977); West,(1977); Macklem & Permutt (1977); and Widdicombe (1974), for example].

1.2   Anatomy

It is convenient to divide the human respiratory tract into three regions: (a) the upper airways, comprising nose, mouth, and all airspaces above and including the larynx; (b) the tracheobronchial tree (or conducting airways), comprising the trachea

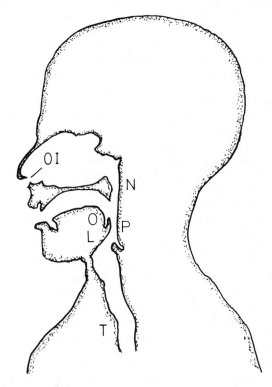

Figure 1   Lateral silhouette of the upper airways. OI = ostium internum, N = nasopharynx, O = oropharynx, P = pharynx, L = larynx, T = trachea.

(which originates just below the larynx) and the repeatedly bifurcating bronchi to which it leads, down to airways about 0.05 cm in diameter; and (c) the gas-exchanging region of the lung, comprising the alveoli and terminal airways (alveolar ducts and alveolar sacs) whose walls consist of alveolar openings.

UPPER AIRWAYS   In normal, quiet breathing, inspired air moves through the nose, nasopharynx, pharynx, and larynx (Figure 1). When nasal obstruction occurs (through congestion of the nasal blood vessels, constriction of the palatal muscles, or blockage with mucus), or when a high air flow rate is required (as in exercise), breathing occurs largely through the lower-resistance pathway of the mouth, oropharynx, and larynx. The anatomy of all these airways is complex, and very little precise data are available. The nostrils lead into two narrow, collapsible constrictions (the ostia interna, each with a cross-sectional area of about 0.32 cm²) that open into the main convoluted nasal passages; these are also narrow (their width does not normally exceed 1 mm) but have a considerable cross-sectional area because of their great depth. These converge and bend into the nasopharynx, which leads into the larynx as shown in Figure 1 [this description is culled from Swift & Proctor (1977), from which further details may be derived; see also Proctor (1964)]. Since the shape and dimensions of the oral cavity, depending as they do on the position of the tongue and soft palate and on the angle at which the jaw is held, defy any exact description, it is important (as physiologists are aware) to standardize conditions in these airways when airway-resistance measurements are made.

More is known about the larynx, which has the smallest cross-sectional area of any part of the upper airways used during mouth breathing, and hence is likely to make an important contribution to their resistance. The laryngeal aperture (or glottis) is not circular, but has the form of a narrow triangle, wider at the back. However neither its shape nor its cross-sectional area is constant, varying both with the flow rate through it (in either direction) and with lung volume. As either quantity increases, the laryngeal area increases and the resistance to airflow decreases (Stanescu et al 1972). Panting also produces a wide opening of the glottis, although the flow rates and lung volume may be relatively low. Control of the laryngeal aperture is a reflex process.

CONDUCTING AIRWAYS   The trachea is a roughly circular tube, about 1.8 cm in diameter and about 12 cm in length, that conducts all inspired air from the larynx to the bifurcation at which the right and left main bronchi originate. These again divide (after about 2.2 and 5.0 cm respectively) into lobar bronchi that themselves divide, and so on. There are about 20 generations of branching between the trachea and the gas-exchanging regions of the lung. Most branchings occur dichotomously (each parent airway giving rise to two daughters), but the two daughter tubes are not usually identical. The first detailed measurements of the lengths and diameters of airways were made by Weibel (1963), who measured all the airways of the first six generations, as well as samples of all other generations, in five normal human lungs that were inflated to 75% of TLC. From this data he constructed a symmetric model of the bronchial tree, in which all of the airways of any generation are assumed to be identical. The lengths and diameters of some of the airways of this model are given in Table 1, together with average velocities ($\bar{u}$) and Reynolds numbers (Re) at two overall flow rates (denoted by $\dot{V}$, equal to the rate of change of lung volume V): $\dot{V} = 0.5 \times 10^{-3}$ m^3 sec^{-1} corresponds to fairly quiet breathing, while $\dot{V} = 2 \times 10^{-3}$ m^3 sec^{-1} corresponds to fairly vigorous breathing. The increase in velocity down to generation 3 reflects a slight decrease in overall cross-sectional area with distance from the trachea (area 2.5 cm^2) to that generation. The rapid fall in velocity thereafter reflects a rapid expansion of cross-sectional area, which is about 940 cm^2 at generation 19. Although it ignores the asymmetry of the real bronchial tree, this model is the one most commonly used in analyses of pressure drop and gas mixing in the lung. This is partly because no systematic asymmetry has been identified (apart from the greater lengths of the pathways to the bottom of the lung than of those to the top); hence a model in which the random asymmetries are averaged out seems reasonable. The main reason, however, is that any accurate asymmetric model is extremely difficult to use. For example, the prediction of pressures in the bronchi, given a uniform alveolar pressure, would require the solution of about 2^{20} simultaneous equations, even if the pressure drop associated with an asymmetric junction were known (which it is not).

A full description of an asymmetric lung has been given by Horsfield & Cumming (1968a,b), who measured the lengths and diameters of every airway of diameter 0.06 cm or more in one human lung. They also numbered the airways upward from those of diameter 0.06 cm (order 0) to the trachea (order 25), using the techniques developed by geographers in the description of river systems. Because of the asymmetry, some order-0 airways will join airways of order greater than 1, and so on. They found some pathways with only 8 branches between the trachea and order-0 airways, and some with 24. This ordering scheme makes it possible (although very laborious) to construct a complete model of the bronchial tree, as far as the relative arrangement of the airways is concerned. However, predictions of the pressure drop (for example) would still be inaccurate because (a) the angle of branching and (b) the vertical position of each branch are not reported. The former is needed for a fluid-

Table 1   Airway dimensions and velocities

Generation	Diameter (cm)	Length (cm)	$\dot{V} = 0.5 \times 10^{-3}$ m^3 sec^{-1} $\bar{u}$ (cm sec^{-1})	Re	$\dot{V} = 2.0 \times 10^{-3}$ m^3 sec^{-1} $\bar{u}$ (cm sec^{-1})	Re
Trachea	1.80	12.0	197	2325	790	9300
1	1.22	4.76	215	1719	859	6876
2	0.83	1.90	235	1281	941	5124
3	0.56	0.76	250	921	1002	3684
4	0.45	1.27	202	594	809	2376
5	0.35	1.07	161	369	643	1476
10	0.13	0.46	38	32	151	127
15	0.066	0.20	4.4	1.9	17.8	7.6
20	0.045	0.083	0.3	0.09	1.2	0.37

mechanical analysis; the latter is required because of the vertical gradient in pleural pressure that is present in vivo, which means that the driving pressure depends on the vertical position of the airway in question.

Much of this article is concerned with the flow patterns induced in airways by the presence of junctions. Therefore, in addition to airway lengths and diameters, the detailed geometry of individual junctions is important; in particular, we need to know the angles of branching, the sharpness of the flow divider, the radii of curvature of the tube walls at the junction, the way in which the parent tube changes shape as it approaches the bifurcation, whether the tubes are themselves curved or straight, and whether their walls are rough or smooth. Very little data are available; I present what is available as a description of a "typical" junction, which is assumed to be symmetric, is depicted in Figure 2, and is used for the fluid-mechanical studies presented below. (a) The diameter ratio (d_2/d_1) is 0.78, corresponding to an area ratio (both daughters to parent) of 1.2, which is typical of airways beyond the first four generations (as we have seen, the area falls for the first three). (b) The length-to-

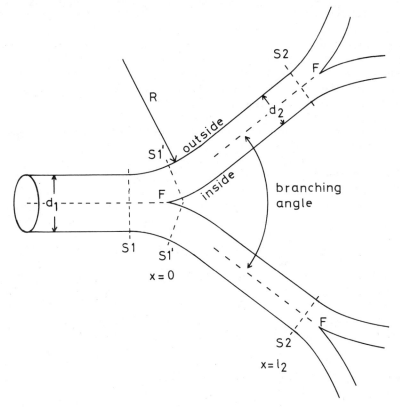

Figure 2   Diagram of a "typical" symmetric bifurcation, with a second generation downstream. F = flow dividers. S1, S1', and S2 are the stations referred to in text.

diameter ratio of a branch is typically 3.5. (c) The angle of branching is 70°. Horsfield & Cumming (1967) measured the branching angles of over 200 bifurcations in a single lung cast and found that the mean angle varies from 64° in airways of diameter greater than 0.4 cm to 100° for airways of diameter less than 0.1 cm. (d) The flow divider is usually sharp, at least in larger airways. (e) The radius of curvature of the outer wall of a junction (R, Figure 2) is a very variable quantity, lying between 1 and 30 times the radius ($\frac{1}{2}d_1$) of the parent tube; an average value of between 5 and 10 seems reasonable. (f) As it approaches the junction, the parent tube first becomes elliptical without change of cross-sectional area, before changing both shape and area as the daughter tubes emerge. Note that the cross-sectional area at the junction is not uniquely defined, since it depends on whether the plane of measurement is perpendicular to the axis of the parent tube or to that of a daughter tube. (g) Daughter tubes are initially curved and of constant area, straightening when the branching angle has been achieved. [The data in d–g above are all taken from Horsfield et al (1971).] (h) The walls of most airways appear smooth, although rings of cartilage in the trachea and main bronchi may cause slight indentations. However, these are too small to affect the flow (Schroter & Sudlow 1969), and in any case all airways are normally lined with a layer of mucus, which tends to make them even smoother. This mucus may be moved along by the action of the under-lying cilia, but the velocities are normally small compared with that of the mean airflow (Sackner, Rosen & Wanner 1973, Blake 1975).

   Finally, we should recall that, like all biological tissues, the walls of airways have elastic and viscoelastic properties and contain both passive elements and some smooth muscle. The lengths and diameters of airways increase as lung volume is increased, varying approximately as the cube root of lung volume (Hughes, Hoppin & Mead 1972), which is consistent with the simple hypothesis that lung tissue as a whole expands and contracts almost isotropically, at least during quiet breathing. The implications of this for flow and pressure drop in the airways are examined below and are found to be predictable. However, a different situation can arise during forced expiration, when some airways appear to collapse, changing their volume much more rapidly than the lung as a whole. In these circumstances the distensibility of the airways is reduced by the action of the surrounding lung tissue, which tends to hold them open; the degree of reduction depends on lung volume and is difficult to quantify; only recently has such work begun (Hughes et al 1974; Hyatt, Rodarte & Wilson 1975). The fluid mechanics of forced expiration is discussed in Section 2.2.

   Measurements of airway distensibility have normally been made in excised dog lungs. In vivo, there is likely to be some contraction (or "tone") in the bronchial smooth muscle, which should affect both the caliber and the distensibility of airways. The effect of this can be investigated either by measuring airway dimensions in open-chested (but living) animals, with and without blocking of the nerves that cause the muscle action, or by measuring physiological quantities such as airway resistance, with and without a blocking drug. However, the results of such experiments are not clear-cut and are by no means completely understood (see Macklem 1971).

   Many physiological experiments are performed on animals, especially dogs, whose anatomy is different from man's: for example, the branches of canine airways are systematically asymmetric, with many equal smaller tubes coming successively off a single large tube. Caution must be used in extrapolating the results of such "model" studies to man.

GAS-EXCHANGING REGION   This region is distinguished from the conducting airways by the presence of alveoli in the walls of air passages. According to Weibel's model, this region comprises generations 17 and above. There are three generations of respiratory bronchioles, airways of diameter 0.14 cm or less with an increasing number of alveoli on their walls as the generation number increases. These are followed by several generations of alveolar ducts, whose walls consist entirely of alveolar openings. Terminal alveolar ducts are called alveolar sacs. Horsfield & Cumming's model gives the same picture, except that the diameters of respiratory bronchioles are said to vary between 0.06 and 0.04 cm. The authors agree that there is a great increase in total cross-sectional area at this level (from 300 cm^2 at generation 17 to 11800 cm^2 at generation 23, according to Weibel). Thus the mean velocity of airflow falls very rapidly in this region, and even at the highest flow rates the Reynolds number will not exceed 1 in the smallest airways (although it may not be very much smaller). Furthermore, gas transport by molecular diffusion will be at least as important as that by convection. The detailed geometry of alveoli is still the subject of investigation (Weibel & Gil 1977); they are thought to be polyhedral cups, in close contact, arranged rather like the cells of a honeycomb. However, since models of airflow in terminal regions are still rather crude, we need consider their geometry no further, except to remark that there are direct communications between alveoli, in the form of holes about 50 μm in diameter (the pores of Kohn) in their walls. In various species the ventilation of some regions may be achieved solely through these pores (Macklem 1971).

2   FLOW IN BRANCHED TUBES

2.1   *Inspiratory Flow*

SINGLE BIFURCATION: EXPERIMENTS   Since a direct, in vivo measurement of air velocities and pressures is virtually impossible, our knowledge is largely based on experiments in model airways and other tubes, as well as on theoretical deduction. We begin by examining inspiratory flow in a single symmetrical bifurcation, typical of the conducting airways, as depicted in Figure 2 and described above. Although this is already a simplified model of a pulmonary bifurcation, we cannot proceed without further simplifications. We therefore take the flow to be laminar and steady (that is, we assume that flow in the lung is quasi-steady, with negligible local accelerations) and the tube walls to be rigid. The validity of these assumptions is discussed in subsequent sections. The air is assumed to be incompressible, because airspeed is always much smaller than sound speed. We also suppose that the flow in the parent tube, upstream of the bifurcation, is axisymmetric and is either Poiseuille flow or partially developed laminar entrance flow; the Reynolds number is taken to be large, of the order of 500. Of course, the parent tube of one junction is in general the daughter of another, so the flow in it is far from symmetric. The one exception is the trachea, which receives the air via the usually constricted larynx, where an asymmetric turbulent jet is probably formed. This spreads rapidly, so that upstream of the first bifurcation the flow will in general be turbulent across the whole cross section of the trachea. Both turbulent flow and the flow through several bifurcations are considered later.

It is not difficult to construct a qualitative picture of the flow pattern to be expected in the daughter tube (Pedley, Schroter & Sudlow 1971). First, the flow is split into two streams, so that a new boundary layer is formed on the inside wall, with maximum axial velocity just outside the boundary layer (Figure 3). Second, the flow into each daughter tube turns a corner, so that secondary motions are set up as in a uniform curved tube: the faster-moving fluid moves toward the inside wall of the junction (the outside of the bend), tending to keep the point of maximum axial velocity close to that wall, and the slower-moving fluid near the walls moves toward the outside. In addition, depending on the sharpness of the corner in the outer wall (i.e. on $2R/d_1$; see Figure 2), there may be a region of separated flow. Apart from growth of the boundary layer, and the possible closure of a separation bubble, this pattern will not be greatly modified in a distance of 3.5 diameters (which is the normal length of an airway).

Two types of experiments have been carried out in models of single bifurcations: flow visualization (using smoke as a tracer in airflow, or dye in water flow) and measurement of velocity profiles (using hot-wire anemometry or similar techniques).

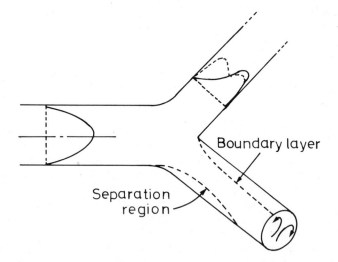

Figure 3 Qualitative picture of flow downstream of a single junction with Poiseuille flow in the parent tube. Direction of secondary motions, new boundary layer, and separation region are indicated in the lower branch; velocity profiles in the plane of the junction (———), and in the normal plane (----------) are indicated in the upper branch.

All of the flow-visualization studies have confirmed the above qualitative picture, that fluid particles move downstream in the daughter tubes in two helical paths, for various branching angles, area ratios, and Reynolds numbers (in the approximate range 100–1400). Schroter & Sudlow (1969) used a "typical" bifurcation, as described above, except that (a) $2R/d_1$ was too small ($=1$), (b) the parent tube did not become elliptical before the junction, and (c) the daughter tubes were not initially curved (all of the other flow-visualization studies have suffered from the same defects). Schroter and Sudlow found that the secondary flows were strong enough to complete one cycle within three diameters of the junction. This indicates that the secondary velocities could be as high as 50% of the average axial velocities. However, the velocity measurements of Olson (1971) showed that the maximum value is about 30% when $2R/d_1 = 7$ and is lower still for more gradually curved tubes (at the given branching angle of 70°). This suggests that the unnaturally sharp curvature of the outer wall is responsible for such large secondary velocities. Schroter and Sudlow also observed separation and reversed flow at the outer wall of the junction, although this does not imply a region of dead water, because the secondary motions sweep new fluid into the separation bubble from the sides (Brech & Bellhouse 1973). Zeller, Talukder & Lorenz (1970) further confirmed that if the flow rates in the two daughter tubes are unequal, separation readily occurs at the outer wall of the tube with the smaller flow rate. Olson (1971) did not observe separation of his more gradually curved models, despite having the same area expansion of 1.2 as Schroter and Sudlow. Experiments on a bifurcation with a blunt flow divider suggest that even there the influence of the bifurcation on flow in the parent tube is insignificant more than one diameter upstream of the flow divider (Pedley, Schroter & Sudlow 1977; Pacome 1975).

Finally, a number of authors have observed instability and transition to turbulence first in the daughter and later in the parent tube of a bifurcation, at a parent-tube Reynolds number (Re_c) well below the transition Reynolds number in a long straight tube: Pacome (1975) gave $Re_c = 680$ for a 70° bifurcation, although Schroter and Sudlow still had laminar flow at $Re = 700$, and Olson (1971) reported $Re_c > 1300$ for his more smoothly curved models. Both Stehbens (1959) and Clarke, Jones & Oliver (1972) reported that Re_c falls as the branching angle is increased, but in neither case is it clear whether the authors distinguished clearly between complicated three-dimensional laminar flows and true turbulence.

The best measurements of velocity downstream of a bifurcation have been made by Olson (1971); they remain largely unpublished, although a few have been reported by Pedley, Schroter & Sudlow (1977). Olson's measurements are the most reliable for the following reasons:

1. His models were very carefully made and conform closely to the "ideal" bifurcation described above, including the gradual change in shape of the parent tube and the initial curvature of the daughter tubes (with $2R/d_1$ equal to 7 or more).

2. He made careful measurements *both* with conventional hot wires, which can be interpreted as measurements of axial velocity, u, as long as the component of transverse velocity normal to the wire is much less than u (which in most cases

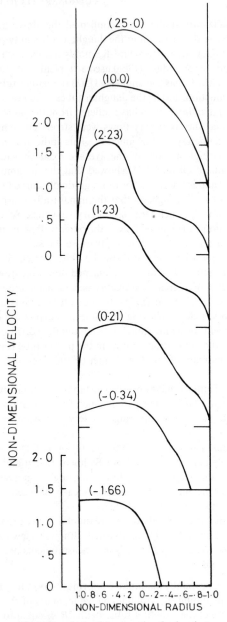

Figure 4   Velocity profiles in the plane of the junction for inspiratory flow through a single bifurcation. Flat entry profile; Re = 530; flow divider is on *left*. Figures in brackets are the numbers of tube diameters from the flow divider. Redrawn from Olson (1971), by courtesy of the author.

was the case), *and* with a specially designed probe consisting of two wires, of which one is given a pulse of heat and the other records the temperature downstream. Rotating the probe about the pulsed wire until the response of the receiver wire is at a maximum gives both the magnitude and the direction of the velocity component in the plane perpendicular to the wire. Repeating the measurement at the same point in a perpendicular plane gives the complete velocity vector. Using this procedure at many points, Olson was able to map out profiles of the transverse-velocity components v and w, as well as those of u.
3. Finally, Olson took measurements at more points in a given cross section and at more cross sections than any other worker. However, because the experimental procedure was so laborious, he reported measurements of secondary velocities at only two Reynolds numbers, in one model bifurcation, and with a flat entry profile a few diameters upstream.

Typical velocities measured by Olson in the daughter tubes of the bifurcation are presented in Figures 4–6. Figure 4 shows the development of the axial-velocity profile in the plane of the bifurcation from a station 1.66 daughter-tube diameters upstream of the flow divider to many diameters downstream, at a daughter-tube Reynolds number Re_2 of 530. As expected, the maximum velocity occurs close to the inner wall of the bifurcation, although near the flow divider there is evidence

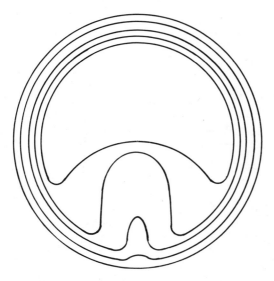

FLOW DIVIDER

Figure 5 Contour plot of axial velocity in the daughter tube at 2.2 diameters from the junction. Inspiratory flow; flat entry profile; $Re = 660$. Contours are at steps of 0.4 non-dimensional velocity (they have been slightly smoothed in reproduction). Redrawn from Olson (1971), by courtesy of the author.

of a slight skewing of the profile away from the inner wall, outside the boundary layer. This presumably reflects the inviscid development of the flat profile upstream as it rounds the bend; inviscid flow in a curved tube has the form of a potential vortex, centered on the center of curvature. Similar results were found at other

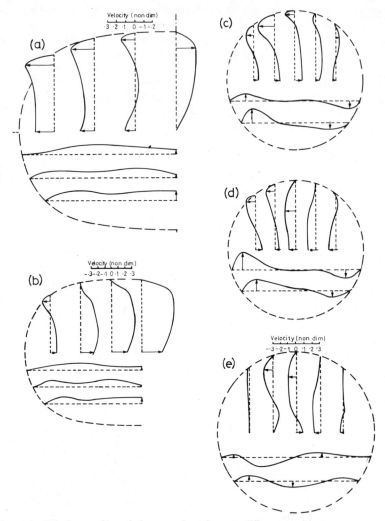

Figure 6 Velocity profiles of the secondary flow at different positions upstream and downstream of the flow divider (*right*); flat entry profile; Re = 935. (*a*) 0.34 *d* upstream of the flow divider (*d* = diameter of the daughter tube); the plane of profiles is perpendicular to the axis of the parent tube. (*b*) 0.34 *d* upstream; the plane of profiles is perpendicular to the axis of the daughter tube. (*c*) 0.21 *d* downstream. (*d*) 2.23 *d* downstream. (*e*) 5.0 *d* downstream. Redrawn from Olson (1971), by courtesy of the author.

values of Re_2 (in the range 300 to 1500). Olson found no separation from the outside wall. He also remarked that a parabolic entry profile made little difference to the results, at least after 2 or 3 diameters.

In Figure 5 a contour map of longitudinal velocity is plotted, measured at 2.2 diameters downstream of the flow divider and at $Re_2 = 660$. The main features are that (a) the region of high shear on the inside wall (the boundary layer) extends more than half way around the tube and (b) the contours have a winged appearance, demonstrating that a velocity profile in the plane perpendicular to the bifurcation will be M-shaped, with a significant dip in the middle (cf Figure 3).

In Figure 6 we reproduce Olson's measurements of the profiles of the secondary-velocity components at a somewhat higher Reynolds number ($Re_2 = 935$). They are shown at four stations, three downstream and one 0.34 d_2 upstream of the flow divider. At this upstream station, the profile is plotted first in a plane perpendicular to the parent-tube axis, and then in a plane perpendicular to the daughter-tube axis. The difference is striking. Whereas in the first case the secondary motion is seen to be primarily toward the outside of the *junction*, in response to the change of tube shape (except on the minor axis where the effect of the imminent flow divider is strong), in the second case it corresponds to a strong sideways motion toward the outside of the *bend* (the flow divider). It is this aspect of the secondary flow that presents itself first to the daughter tubes, and it can be seen to persist downstream of the junction. At 0.21 diameters it is still evident, with a boundary layer developing at the walls in which the secondary flow back to the outside wall is to be seen, as expected from the flow-visualization studies. Farther downstream, however, at 2.23 diameters, the situation has become very confusing again; here almost all the secondary motion is directed toward the outer wall of the junction, and there is no obvious explanation for this. Even farther downstream, the effect is less strong. It is as if the flow pattern varied sinuously with distance downstream, as has been reported for turbulent flow in pipe bends of large angle (Rowe 1970). However, no such variation is obvious in the axial profiles.

The picture of the secondary motions in the elliptical transition region was confirmed by Olson from measurements in a tube that gradually became elliptical (with constant area), but which had no bifurcation at the other end. An example of the measurements, at a (parent-tube) Reynolds number of 1620, is shown in Figure 7a. The similarity between this and Figure 6a is obvious. For comparison, the corresponding plot with a parabolic entry profile is shown in Figure 7b; the secondary motions are far less strong.

A number of authors have made less accurate measurements in less carefully made bifurcations. Among the most important (because of the use subsequently made of them; see Section 3) were by Schroter & Sudlow (1969). They made hot-wire measurements of (longitudinal) velocity profiles in the plane of the junction and the plane perpendicular to it, usually with a parabolic inlet profile. They too found peak velocities near the inside wall and an M-shaped profile in the plane perpendicular to the junction, and qualitatively their results differed from Olson's only in the absence of the skew outside the boundary layer on the inside wall (because the inlet profile was not flat) and in the presence of a minimum in the profile in the

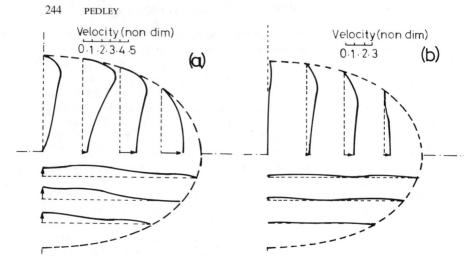

Figure 7 Velocity profiles of the secondary flow in a tube that is initially circular and becomes increasingly elliptical with constant area. Re = 1617; (*a*) flat entry profile; (*b*) parabolic entry profile. Redrawn from Olson (1971), by courtesy of the author.

plane of the junction, with a second maximum near the outside wall. They assumed this to be associated with the separation demonstrated by their flow-visualization experiments, since the hot-wire anemometer gives a positive signal whatever the direction of flow. Qualitatively similar results have been obtained by Schreck & Mockros (1970), Brech & Bellhouse (1973), and Berger, Calvet & Jacquemin (1972). The measurements of Berger, Calvet, and Jacquemin were made in the first generation of a branched-tube network containing several generations, with the later generations still attached. It is not clear, however, what effect this might have had on the measured profiles.

SINGLE BIFURCATION: THEORY   Attention is still restricted to steady laminar flow at high Reynolds number. The geometrical configuration of even a single bifurcation is so complicated that a complete analytical theory is out of the question, and we must await the development of suitable numerical methods before we can investigate all the details of the flow. [The flow in two-dimensional bifurcations has been analyzed numerically, but since such a model inevitably excludes all the interesting phenomena and gives no clue to the structure of the three-dimensional flow even in an idealized example (Smith 1976a), it is not worth describing the results. One hopes that the methods can soon be extended to three-dimensional bifurcations.] Theoretical insight into the development of the secondary motions and the distortion of the axial-velocity profile will be possible only with the help of crude simplifying assumptions, which must always be assessed critically. One such simplification is to neglect the effect of viscosity on the gross changes in the flow that take place in the region of the junction itself, on the grounds that viscosity has no time to influence the flow pattern except in the boundary layers. This is likely to be reasonable for a short distance beyond the junction, as long as separation does not occur, and will break down where secondary motions cause fluid originating near the wall to penetrate into the core from the outside of the bend. The development of the boundary layers, both on the flow divider in the daughter tubes and upstream of the junction in the parent, may be analyzed separately.

Olson's observations on the flow in the elliptic transition region suggest that the nonviscous core problem can also be divided into two stages. First, one should compute the secondary motions (and the distortion of the axial velocity profile) caused by streamline curvature in the increasingly elliptic parent tube upstream of the flow divider. This neglects any blocking effect that the presence of the flow divider may have on the core flow just upstream, in the hope [based on the flow-visualization studies reviewed above, but perhaps not justified in the light of the work of Smith (1976b,c,f)] that this is confined to a small region, especially when the flow divider is sharp. Second, one should calculate the effect that the initial curvature of the daughter tube will have on the already distorted flow entering it.

A tentative start has been made to the first stage by Sobey (1976), who has calculated the distortion of a weakly sheared axisymmetric flow in a tube which is circular (radius a_0) for $x < 0$ and which has slowly varying (but not necessarily small) ellipticity downstream $(x > 0)$. He solved the problem using classical secondary-flow methods (reviewed by Horlock & Lakshminarayana 1973), that involve the following stages:

1. Calculate the potential flow in the slowly varying tube as a power series in ε $(= a_0/l$, where l is the length scale for longitudinal variations). The leading term is a uniform stream with velocity equal to U_0/ab, where U_0 is the average velocity in $x < 0$ and where $a_0a(x)$ and $a_0b(x)$ are the major and minor semi-axes of the elliptical cross section, respectively. The next term in the velocity potential, of order ε, can easily be calculated and is determined uniquely by the condition that volume flux is uniform along the tube. The error term is $O(\varepsilon^3)$.
2. Let the transverse vorticity in the upstream flow be $O(\delta U_0/a_0)$, where $\delta = o(\varepsilon)$; then the streamlines of the $O(\varepsilon)$ potential flow will be a first approximation to the actual streamlines. Assuming that vortex lines are carried along the potential-flow streamlines, one can compute a first approximation to the disturbance vorticity field. This includes longitudinal components that are associated with secondary-velocity components of $O(U_0\varepsilon\delta)$, as well as an $O(U_0\delta)$ perturbation to the axial velocity.
3. These result in further distortions to the axial-velocity profile of $O(U_0\delta^2)$, which can also be computed without too much difficulty.

This approximation procedure is self-consistent for values of x less than $O(a_0/\varepsilon\delta)$. Its greatest weakness is that it cannot treat large, $O(U_0/a_0)$, upstream vorticity, and therefore any results can have at most qualitative relevance to real pipe flows, except perhaps those (as in some of Olson's experiments) with almost flat entry profiles and thin boundary layers.

Sobey's results can be summarized quite simply. When the entry profile is perfectly flat, the secondary-velocity profiles are also flat, and these motions are stronger farther away from the center of the tube. The contours of constant axial velocity are ellipses that remain similar to the tube itself. When there is a slight shear, the secondary velocities are greater near the wall of the tube, and the contours of constant axial velocity are distorted, so that the axial velocity is increased near the ends of the major axis and decreased near the ends of the minor axis. The predicted secondary profiles agree qualitatively with those of Figure 7a, except of course in the boundary layer on the walls.

An extremely crude estimate of the flow in the (straight) daughter tubes of a symmetrical bifurcation has been made by Scherer (1972), who ignored any perturbation to the flow in the parent tube upstream of the flow divider. He also assumed that the flow in the daughter tubes does not change with distance along them and then calculated the simplest possible flow pattern consistent (a) with this assumption, (b) with the nonviscous equations of motion, and (c) with the fact that the flow must be symmetric about the plane containing the axes of all tubes in the bifurcation. The flow entering from the parent tube is accounted for by arranging that the axial and transverse velocity components on the axis of the daughter tube at its entrance are the same as would occur at that point if the parent tube continued uninterrupted. He worked out the details for a flat entry profile of velocity U_0, and his results are in rough qualitative agreement with Schroter and Sudlow's profiles, including the M-shaped profile in the plane normal to that of the junction.

Before further progress can be made on the core-flow problem, a theory will have to be developed to analyze the rapid inviscid distortion, over a short length of tube, of a flow with large transverse vorticity. It is difficult to see how any perturbation theory (such as Sobey's) can suffice.

Progress has also been made with the boundary-layer analysis, although this is necessarily restricted to idealized situations in which the core flow is not grossly distorted by secondary vorticity. Smith (1976a) examined the effect of dividing the flow into two without making it simultaneously turn a corner by considering steady flow in an infinitely long, straight circular pipe of radius a (whose axis is the line $y = z = 0$), which is split by a semi-infinite plane $z = 0$ in $x > 0$. Poiseuille flow, $u = 2\bar{u}(1 - y^2/a^2 - z^2/a^2)$, is present as $x \rightarrow -\infty$. Smith analyzed the flow on a long, boundary-layer length scale, $x/a = O(\mathrm{Re})$, but concentrated attention on small values of $x/(a\,\mathrm{Re})$. A Blasius-type boundary layer is set up on the splitter plate, corresponding to the z-dependent outer velocity $u = 2\bar{u}(1 - z^2/a^2)$. This generates a z-dependent normal velocity v that causes both a displacement of the inviscid core flow and a transverse velocity w inside the boundary layer and in the core. The disturbance to the core flow follows both kinematically and because a favorable pressure gradient $\{p = -P_1[x/(a\,\mathrm{Re})]^{-1/2}\}$ is set up. A double boundary layer is then formed on the outer, curved wall: the outer, inviscid layer of thickness $\propto [x/(a\,\mathrm{Re})^{1/4}]$ reduces the normal velocity to zero, and the inner, viscous one of thickness $\propto [x/(a\,\mathrm{Re})^{3/8}]$ reduces the tangential velocity to zero. No complete solution is computed, but the analysis appears to be self-consistent and seems to have a solution whatever the value of the constant P_1. This suggests that P_1 is determined by the short, $x = O(a)$, length-scale solution in the neighborhood of $x = 0$, and that it involves a complete solution of the three-dimensional Navier-

Stokes equations, as for a bifurcation of finite angle. Smith also showed that the corresponding two-dimensional problem does not suffer from this indeterminacy on the long length scale and is quite irrelevant from the three-dimensional point of view.

In other papers, Smith (1976b–f), showing an impressive analytical mastery of modern boundary-layer techniques, has been concerned with the flow past slowly varying obstructions (including branchings) in otherwise parallel channels or pipes. The flow depends on the height, slope, length, and symmetry of the obstruction and on whether it is situated in a region where the flow would otherwise be effectively fully developed or one where it would be of the entry-flow type. Only the latter is likely to be relevant in the lung; but most of Smith's papers concern fully developed oncoming flow (except for Smith 1976b). It is therefore not appropriate to list all of his results, and we merely comment on three types of situation that can be distinguished. (a) When the obstruction is extremely slender, the flow is everywhere described by the boundary-layer equations with lateral length scale a (an appropriate transverse scale for the undisturbed flow) and longitudinal length scale aRe. (b) When the obstruction is not quite so slender, the flow over it represents an interaction between the pressure gradient induced by the inviscid displacement of the core flow and the viscous boundary layer that must be interposed between the displaced core flow and the wall. There is no influence on the upstream flow, and the calculation of the downstream flow can be continued past points of (regular) flow separation, for example on the downstream side of an indentation. (c) When the obstruction is sufficiently large or abrupt, there is a significant effect on the upstream flow, especially in the boundary layers, and this is the aspect of greatest interest because upstream flow separation can occur. Furthermore a novel type of singularity can arise, at least in fully developed two-dimensional channel flows that are asymmetrically disturbed at $x = 0$ (Smith 1976c). Separation takes place on one wall far upstream of the disturbance $[X = (x/a\,\text{Re}^{1/7}) \approx -0.49]$, while on the other wall the boundary layer becomes progressively thinner, and a singularity (on the long length scale) occurs at $X = 0$. This means that on the short length scale $x = O(a)$ the flow over the obstacle has to be calculated with upstream conditions quite different from Poiseuille flow.

We should remain clearly aware that there is still a wide gulf between any of these theoretical flow patterns and those observed in model bronchial bifurcations.

SEVERAL BIFURCATIONS   In the lung, the flow in the parent tube of a bifurcation is not a unidirectional laminar flow, but has itself emerged from an earlier bifurcation (tracheal flow apart). Thus all of the complicated experimental and theoretical flow patterns outlined above are no more than first approximations to reality. There are no theories for flow in a network of more than one bifurcation, and few experiments, the only ones known to the author being by Schroter & Sudlow (1969); Berger, Calvet & Jacquemin (1972); and Pacome (1975). The first authors measured profiles in a two-generation model, with one of the second-generation bifurcations being in the plane of the first and the other perpendicular to it. All of the bifurcations were geometrically similar. They measured velocity profiles in the same two planes as before at several distances from the flow divider and at three Reynolds numbers in each case. The second authors used a two-generation model with all bifurcations in the same plane and with a different area ratio at each generation, as given by the first two generations of Weibel's model (Table 1). Pacome extended this model to five generations, also using Weibel's data and also keeping the first three generations coplanar, but for geometrical reasons the last two were not all coplanar.

All of these authors used their velocity-profile measurements in an estimate of pressure drop (see below), and the only profiles presented are in the daughters of the second bifurcation. When this is coplanar with the first, but not otherwise, the velocity fields are symmetric about the plane of the junctions, as expected. The slopes of the velocity profiles near the wall are again high at least halfway around the tube from the flow divider (note, however, that extrapolation of a measured velocity profile to zero is a notoriously inaccurate way of obtaining the shear rate at the wall). However, the main impression is of a very complicated flow, varying significantly with distance downstream because of the redistribution of axial momentum by the secondary motions. Such disturbed (but still laminar) flow is to be anticipated almost everywhere in the lung.

PRESSURE DROP   In order to predict the pressure drop across a complicated network of branched tubes like the lung at a given volume flow rate, we must first know the pressure drop across a single tube of the network. Direct measurement of the pressure in a single bifurcation does not give useful results because the pressure differences are very small, the pressure varies as much within the cross section of the tube as it does across a junction (Jaffrin & Hennessey 1972), and the contribution to the pressure drop of kinetic-energy changes is large. We must therefore estimate the pressure drop by using the measured velocity profiles, and check the estimates by measuring the pressure drop across a model with several generations. Furthermore, we cannot use a momentum equation to estimate the pressure drop, because the walls of the tubes are not parallel, and so the flow divider, for example, exerts a longitudinal component of force on the flow in the parent tube (and in any case wall shear stress is extremely difficult to measure accurately). It is therefore necessary to use an energy equation.

We refer to a junction and its daughter tubes as a single unit (between stations S1 and S2 in Figure 2). The energy equation for the fluid within the unit can then be written:

$$(\hat{p}_1 + \tfrac{1}{2}\rho\hat{q}_1^2) - (\hat{p}_2 + \tfrac{1}{2}\rho\hat{q}_2^2) = D/Q, \tag{1}$$

where Q is the volume flow rate through the unit (we reserve $\dot{V}$ for flow rate into the lung as a whole), D is the total rate of viscous energy dissipation in the unit, p is the pressure, q is the total fluid velocity, and subscripts 1 and 2 refer to stations S1 and S2. The symbol "^" refers to the average of a quantity, weighted by the longitudinal velocity component across the cross section of the tube. Thus

$$\hat{p} = \frac{1}{Q}\int pu\,dA, \tag{2}$$

etc. (Pedley, Schroter & Sudlow 1977). Note that if either p or u is uniform across the tube, then $\hat{p}$ is equal to $\bar{p}$, the conventional average pressure. Also, if we write

$$\hat{q}^2 = \beta\bar{u}^2, \tag{3}$$

where $\bar{u}$ is the average velocity, we note that $\beta = 1$ for parallel flow with a flat profile and $\beta = 2$ for Poiseuille flow. For complicated three-dimensional flows such as this, the only rational meaning that can be given to the words "pressure drop" is that derived from Equations (1) and (2).

In applying (1) to a bifurcation, two more simplifying assumptions are made. (a) We neglect the viscous dissipation at the junction itself, i.e. between stations S1 and S1', where the flow is thought to be largely determined by nonviscous mechanisms; this is reasonable since most of the dissipation takes place in the boundary layer in the daughter tube, where the velocity gradient is high. (b) We neglect the contributions to $\hat{q}^2$ and to D of the transverse components of velocity v and w; this is very crude, leading to a possible underestimate of at least 10% in D (when v and w reach 30% of $\bar{u}$), but no more accurate yet still manageable alternative has been suggested.

Pedley, Schroter & Sudlow (1970a) used rather crude interpolation procedures to estimate the dissipation per unit length (dD/dx) at each station in each daughter tube and at each Reynolds number for which profile measurements were made. The results were expressed as the ratio Y of the actual dissipation rate to that for Poiseuille flow at the same Reynolds number in the same tube. D was calculated from dD/dx, and its ratio Z to the corresponding Poiseuille-flow value was reported. The results showed considerable scatter, but two general conclusions could be drawn: Y (i.e. dD/dx) decreased on average with increased distance (x) from the flow divider; both Y and Z decreased (on average) as Re decreased. Because most of the energy dissipation is expected to take place in the boundary layer growing on the flow divider, these authors proposed a model in which Y and Z would depend on Re, x, and the tube length l and diameter d in the same way as for entry flow in a straight tube. According to this model, Y and Z in any tube would be given by

$$Y = \tfrac{1}{2}\gamma\left(\frac{d}{x}\,\mathrm{Re}\right)^{1/2}, \qquad Z = \gamma\left(\frac{d}{l}\,\mathrm{Re}\right)^{1/2}, \tag{4}$$

where the dimensionless constant γ is independent of Re, x, l, and d. These equations can be used to deduce values of γ from the values of Y and Z derived from the measured velocity profiles. Pedley, Schroter & Sudlow (1970a) found no systematic dependence of the values of γ obtained in this way on either Re or x/d (l/d was constant in the experiments), and therefore they regarded the entry-flow model as confirmed. The only systematic variation was that the values of γ obtained from second-generation tubes were uniformly smaller than those obtained from the first, by 25% on average. This indicates that greater complexity in the flow modifies the boundary layer on the flow divider. Nevertheless, this error was within two standard deviations, and the authors chose to use their overall mean value of $\gamma = 0.33$ for predictions of energy dissipation in the lung.

The values of β (Equation 3) obtained from the measured profiles at a distance

6 cm from the flow divider (close to station S2, Figure 2) showed *no* systematic variation with Re or with generation number, and the average value of 1.7 was used in all predictions.

The above results concerning dissipation suggest that in any system of symmetric branched tubes, with the same branching angle and area ratio and with Reynolds numbers in the same range, D will be proportional to $Q^{5/2}$, and the loss of total head, averaged as in (1) and (2), will be proportional to $Q^{3/2}$. Berger, Calvet & Jacquemin (1972) obtained conflicting results by a similar method. They found that a good fit to their data is given by $D \propto Q^2$, so that the loss of total head is directly proportional to Q. Such a linear relationship is inherently very unlikely, although the constant of proportionality was much greater than in Poiseuille flow, and as yet it has not proved possible to explain the discrepancy. There were slight differences in the setup, in that the area ratios in their model were taken directly from Weibel's data instead of assuming a uniform average value, and there was always at least one generation of branching downstream of the tube in which measurements were made, whereas Schroter and Sudlow made measurements in the last tube of a network. Berger et al attribute all of the discrepancy to this, although no reason for the effect on dissipation is given. There were also differences in practice, the most important of which was that Berger et al took measurements only at one station (station S2) in each daughter tube, asserting without justification (or truth) that the flow would be independent of x. This invalidates their claim that the total head loss for the whole tube (and hence the whole network) is proportional to Q, but does not explain the fact that, at station S2, the ratio Y is independent of Q according to them, but is proportional to $Q^{1/2}$ according to Pedley et al. It would seem that a third, independent set of experiments is required.

It is interesting to note that a student of Calvet, Pacome (1975), has extended the measurements of Berger et al to five generations, but he measured pressure drops as well as velocity profiles (the lateral variations in pressure are small compared with the overall pressure drop in a network of this size). His most important result was that when the kinetic energy terms ($\frac{1}{2}\beta\rho\bar{u}^2$) in (1) are correctly accounted for, the total head loss (D/Q) is accurately proportional to $Q^{3/2}$. This vindicates the model of Pedley, Schroter & Sudlow (1970a), even if doubts remain about the estimation of dissipation.

Berger et al also disagreed with Pedley et al in the value to be taken for the kinetic energy factor β. They found a value, lying between 1.44 and 1.55 in generation 1 and between 1.09 and 1.29 in generation 2, which was in general lowest for the highest Reynolds numbers: the parent-tube Reynolds number took several values between 420 and 2800, at the highest of which the flow was turbulent. Douglass (1973), on the other hand, found values of β around 2.0 in generation 1 and around 1.5 in generation 2 for turbulent flow at Re $> 10^4$, with no systematic dependence on Re.

TURBULENCE: FLOW IN THE UPPER AIRWAYS   Virtually all of the above results concern laminar flow in branched tubes. However, in rapid breathing the value of Re in the trachea may rise above 10^4 (Table 1), especially if high-density gases are inspired. Furthermore during mouth breathing the flow enters the trachea via a constriction (the larynx) and therefore normally takes the form of a turbulent jet, which spreads to fill the tube. Owen (1969) has estimated that a tracheal Reynolds number of 3000 (i.e. a flow rate of about 0.67×10^{-3} m^3 sec^{-1}) is required for the boundary layer as well as the core to be turbulent. He also estimated the rate at which turbulence in the trachea would decay, as it passes through subsequent generations in which Re is below the critical value for a long straight tube. He concluded that for a tracheal Re of 3100 the intensity of turbulence would have diminished to a quarter of its tracheal value by generation 3 (Re = 1230). Olson, Iliff & Sudlow (1972) found random eddies in the core of the trachea and major branches of a cast of the upper and central airways (from mouth to generation 4) at all flow rates above 0.2×10^{-3} m^3 sec^{-1} (tracheal Re = 950). Thus at all but the quietest rates of breathing we may expect the flow to be turbulent for several generations of branching. Since most of the pressure drop is predicted to occur in the first few generations (Section 3), the effect of turbulence on it should be examined.

Pedley, Schroter & Sudlow (1971) argued that the presence of turbulence will not affect the pressure drop significantly for Re $< 10^4$, as long as the boundary layer on the flow divider remains laminar and hence of approximately the same thickness as in laminar flow. If this is the case, the mean velocity profile downstream of the junction will be similar to that in laminar flow, and the turbulent intensity will be similar to that in fully developed pipe flow and, therefore, will not contribute significantly to the viscous dissipation. However, the observations that the critical Reynolds number for branched tubes is less than that in straight pipes suggest that the boundary layer is implicated in transition to turbulence and probably does not remain laminar. Thus it will be thicker than in laminar flow, and the dissipation associated with the mean velocity profile will be less; the total dissipation will also be less unless there is a sharp increase in turbulent intensity. This suggests that Pedley et al's predictions of dissipation are, if anything, an overestimate (although in purely laminar flow they are thought to be an underestimate).

Douglass & Munson (1974) have measured velocity and pressure profiles in a two-generation coplanar model (similar to Schroter & Sudlow's, but with 60° branching angles) at parent-tube Reynolds numbers between 1.1×10^4 and 8.5×10^4. The flow was fully turbulent in all cases. They found mean velocity profiles that were qualitatively similar to laminar ones, except that the profile in the plane perpendicular to the junction was not markedly M-shaped, resembling rather the profile in fully developed pipe flow. Instead of using the measured velocities to estimate dissipation (D), they used Equation (1) to calculate D directly from the measured pressures and velocities. They found that D increased in proportion to Re^3 (compared with $Re^{5/2}$ for the laminar-entry-flow model), which is similar to the relation for turbulent flow in a rough tube. However, the absolute values of D remained below those predicted for laminar flow by Equation (4), by a factor of between 0.12 and 0.75, confirming the above expectation.

The only experiments known to the author in which velocity profiles have been measured at the intermediate values of tracheal Reynolds number at which the lung normally functions ($10^3 < \text{Re} < 10^4$) are by Berger, Calvet & Jacquemin (1972), discussed above, and by Olson, Iliff & Sudlow (1972) made in airway casts (for mouth breathing). Those by the latter authors are also the only known model experiments in which airway asymmetry has been present, both in the anatomical details and in the flow rates, which the authors adjusted so that the flow rate in any branch was proportional to the volume of lung ventilated by that branch. Their mean velocity profiles showed predictable asymmetric features, such as the maximum velocity in the trachea being near the posterior wall since the laryngeal aperture is wider at the back, and a slight rotation of the flow in the trachea, resulting presumably from the fact that the right lung of the model received nearly 20% more of the flow than the left. By the level of generation 2, however, the profiles began to resemble those in symmetric bifurcations. These authors made a rough estimate of energy dissipation and also found it to be below the laminar entry-flow predictions. It is clear that many more experiments on turbulent flow in branched tubes at values of Re between 10^3 and 10^4 will be needed before its influence on pressure drop can be understood. Dekker (1961) has also done experiments on a cast to ascertain the critical tracheal Reynolds number for transition to turbulence. He obtained a critical Re of about 500 for inspiratory flow with an open larynx.

I can find no reports of velocity measurements (in vivo or in casts) upstream of the larynx during mouth breathing, perhaps because of the extreme variability of the geometry. Swift & Proctor (1977) reported measurements in a cast of those upper airways used during nose breathing. The main feature seems to be a very high velocity at the constriction (the ostium internum; Figure 1) about 1.5 cm from the nostril. The velocity recorded here during inspiratory flow at 0.42×10^{-3} m^3 sec^{-1} [the critical flow rate for turbulence in the trachea during mouth breathing according to Olson, Iliff & Sudlow (1972)] was about 18 m sec^{-1}. This is about 6 times the maximum velocity in the bronchi at the same flow rate assuming the maximum to be twice the mean (Table 1). As the driving-pressure difference increases, the flow rate through the nose does not increase indefinitely but reaches a plateau of about 10^{-3} m^3 sec^{-1}. This is a result of the collapse of the walls of the nostrils when the pressure at the ostium internum becomes much less than atmospheric because of the high air velocities (cf sniffing). This should be compared with the mechanics of forced expiration discussed in Section 2 below. Swift and Proctor also reported a turbulent jet downstream of the ostium internum and suggested that this is the main source of turbulence in the trachea during nose breathing, when the larynx is normally not constricted. Note that physiological experiments are almost invariably conducted on subjects breathing through their mouths.

VERY SMALL AIRWAYS   In the terminal airway units (airways and alveoli with diameters less than 0.05 cm), Re is normally below 1, or at most close to 1 (Table 1). Thus inertia is unlikely to be important, and pressure drop will be proportional to flow rate, although it will still be somewhat greater than for Poiseuille flow in each tube, because the entrance length is about 1.5 diameters even when Re $\ll$ 1 (Lew & Fung 1969). The contribution of the terminal units to overall pressure drop is expected to be very small (Section 3), but the flow in these units is important for gas and particle transport (Section 4).

Steady low Reynolds number flow in model terminal units has been studied theoretically by Davidson & Fitz-Gerald (1972). There are no qualitative surprises; the results confirm the absence of separation and show how the gas entering an expanding region will achieve a closer proximity to the walls near the entrance than to those far away, with implications for mass transport.

The unsteady character of the flow may also be important in terminal regions, although inertia is unimportant there, if on expiration the alveolar wall does not retrace exactly the path it took on inspiration. In such a case the flow as a whole will also not be reversible, so that some of the gas that was not originally in the alveolus is left there after the breath. This may have a significant effect on gas exchange, although the full implications have not yet been resolved. The potential irreversibility of alveolar flow was elegantly explained by E. J. Watson in an appendix to a paper by Cinkotai (1974), who demonstrated the phenomenon during flow-visualization studies on a large-scale model of a pulsating alveolar sac (at low Re).

UNSTEADY FLOW   The preceding paragraph is the first in which the oscillatory character of the breathing cycle has been recognized: both the applied pressure gradient and the volume flow rate vary with time, changing direction twice every cycle. Furthermore the lengths and diameters of the airways change as the lung volume changes. We have assumed hitherto that the flow in the airways (the mean flow in the turbulent case) is quasi-steady; it is important to examine the validity of this assumption.

The dimensionless parameter that indicates the fluid-mechanical importance of unsteadiness in laminar flow in a long straight tube of radius a is

$$\alpha = a(\omega/\nu)^{1/2}. \tag{5}$$

This represents the ratio between unsteady inertia forces and viscous forces in an oscillatory flow of angular frequency ω; $(\nu/\omega)^{1/2}$ is proportional to the thickness of the Stokes boundary layer formed on a rigid wall in oscillatory flow. Only if α is greater than about 1 is unsteadiness important. In Table 2 the values of α for several airways, with diameters taken from Weibel's model, are given for two frequencies; 0.25 Hz (15 breaths min^{-1}, corresponding to normal, fairly quiet breathing) and

Table 2   Values of α and ε

Airway	0.25 Hz, 10 L min^{-1}		3.0 Hz, 100 L min^{-1}	
	α	ε	α	ε
Trachea	2.9	0.57	10.0	0.68
Generation 1	2.0	0.21	6.9	0.25
Generation 2	1.3	0.08	4.5	0.10
Generation 3	0.9	0.03	3.1	0.04
Generation 4	0.7	0.06	2.4	0.07
Generation 5	0.6	0.06	2.0	0.07
Generation 10	0.2	0.12	0.7	0.14

3.0 Hz (180 breaths min^{-1}, corresponding to panting). At the lower frequency, all but the trachea and first two generations of bronchi have $\alpha < 1$; but during panting, $\alpha > 1$ in many airways.

However, the airways are not long straight tubes. In most airways, steady inspiratory flow is dominated by a boundary layer on the flow divider, of maximum thickness proportional to $(vl/\bar{u})^{1/2}$, where l is the length of the tube and $\bar{u}$ is the mean velocity in it. The flow should be quasi-steady as long as the Stokes layer is thicker than the (quasi-) steady boundary layer, i.e. if

$$\varepsilon = \omega l/\bar{u} \tag{6}$$

is less than 1. If we take the flow rate into the lung to be 0.16×10^{-3} m^3 sec^{-1} (10 litres min^{-1}) in quiet breathing and 1.6×10^{-3} m^3 sec^{-1} in panting, we can calculate $\bar{u}$ (and hence ε) in any airway from Weibel's data (Table 1). The values of ε obtained in this way are given in Table 2 and are considerably less than 1 for all airways. Thus, subject to experimental confirmation, inspiratory flow can be considered to be quasi-steady, at least as long as it is laminar. (This does not apply in the trachea, because the flow there is not like entry flow; nor is it laminar.)

The values of $\bar{u}$ used above are representative for most of the breathing cycle, but do not obtain throughout. In particular the velocity in every airway becomes zero each time the direction of flow reverses, so the flow cannot then be quasi-steady. However, as long as the time during which the velocity is close to zero is only a small fraction of one period, as in the airways, the overall effect will be small. A simple approximate method for estimating the importance of flow reversal in oscillatory boundary layers has recently been developed (Pedley 1976) but has not yet been applied to airways.

Very little information is available on turbulent flows with unsteady mean velocities. A rough assessment of the effect of unsteadiness can be made by calculating a turbulent counterpart, α^*, to the parameter α (Equation 5), using an eddy viscosity v^* in place of v. Pedley, Schroter & Sudlow (1977) estimated that $v^* > 0.015\bar{u}a$, so that

$$\alpha^* < (\omega a/0.015\bar{u})^{1/2}.$$

In the trachea, for the conditions assumed in calculating Table 2, α^* is at most 1.7 for quiet breathing and 1.9 for panting. In smaller airways the values are correspondingly smaller. Thus the effects of unsteadiness are not expected to be large even in the largest bronchi; unsteadiness may have some effect in the trachea, but not nearly as much as that predicted by Chow & Lai (1972), who assumed fully developed laminar flow.

Airway resistance is often measured by superimposing relatively high frequency oscillations on the normal breathing cycle and by measuring the pressure drop and flow rate at the "resonant" frequency of oscillation of the system, at which the elastic and inertial contributions to the pressure drop cancel out. This frequency is normally 4–7 Hz, and, as pointed out by Pedley, Schroter & Sudlow (1977), it is just about at this frequency that the fluid-mechanical effects of unsteadiness begin to be important. In patients whose "resonant" frequency is greater than this, significant distortion of the results might occur.

A few measurements have been made on unsteady laminar flow in model branched tubes. Jaffrin & Hennessey (1972) measured the variation of static pressure at several sites in a two-generation model when the flow rate was varied sinusoidally. The peak values of Re and α in the parent tube were taken to be 1560 and 2.7, respectively, in one case, and 1230 and 1.9 in another. There was a small phase shift between pressure drop and flow rate at the higher frequency, but no amplitude response. The measurements confirmed that transverse-pressure variations in a daughter tube are as great as the longitudinal variations in two generations. The pressure drop for expiratory flow was greater than that for inspiration. Brech & Bellhouse (1973) measured velocity profiles and shear stresses in pulsatile flow (with nonzero mean) through a single bifurcation, with parent tube Re = 750 and 1500, and α = 22. Even at this high value of α, quasi-steady flow was reported.

The above remarks concern flow in rigid tubes. The increase in airway lengths and diameters with lung volume is another way in which unsteadiness may be important. These changes certainly cause airway resistance to vary with lung volume, but this can be accounted for in a quasi-steady description as long as the air velocities induced by the dimension changes are small compared with the velocities already present. If the lung changes its volume by 12.5% in 2 seconds (as in normal, fairly quiet breathing), an airway will change its diameter by only 4%, inducing a velocity of less than 0.1 cm sec^{-1} if the diameter and length are 0.5 cm and 1.5 cm, respectively.

2.2 Expiratory Flow

RIGID BRANCHED TUBES   A disproportionately small amount of work has been done on expiratory flow patterns. It is as if everyone doing a model experiment on inspiration ran out of time or energy when it came to repeating the measurements for expiration. This is unfortunate because it means that not enough velocity profiles are available to estimate the dissipation, and therefore no model for predicting pressure drops, like the entry-flow model for inspiratory flow, can be assessed.

Steady, laminar expiratory flow in a single bifurcation has been studied by Schroter & Sudlow (1969), using both flow visualization and hot-wire velocity measurement. The flow was sucked through the junction by a pump downstream of the parent tube and entered the daughter tubes from a supposedly still reservoir. The velocity profiles on entry were presumed to be flat; a parent-tube Reynolds number of 700 was used. Flow visualization demonstrated the presence of two pairs of secondary vortices, generated by the curvature of the two fluid streams as they come together. The velocity profiles (Figure 8) confirm that there is a dip in the profile in the plane of the junction just downstream of the flow divider, but also show that it is rapidly transformed into a marked velocity peak. In fact the velocity near the walls is reduced, while the velocity in the center remains roughly constant: the profile in the perpendicular plane remains flat. The reason is presumably that the secondary motions cause the slower-moving fluid along the line of the flow divider to be swept out to the walls.

More recent measurements have been made on expiratory flow that has passed through several (up to 6) generations of branching (R. C. Schroter, personal communication). At the time of this writing, however, only qualitative results are

IN PLANE OF BIFURCATION

NORMAL TO PLANE

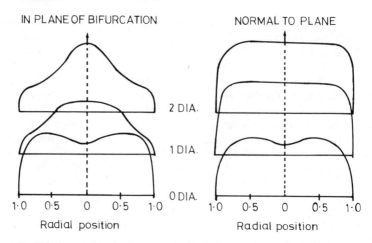

Figure 8   Velocity profiles in the parent tube during expiratory flow through a single bifurcation. Flat entry profile; Re = 700. Reprinted from Schroter & Sudlow (1969, p. 349), by courtesy of the North-Holland Publishing Co.

available. It was observed that the peak in the velocity profile in the plane of the junction is abolished after several generations, and the profile becomes virtually flat less than one diameter from the flow divider. This is perhaps not surprising, because the pattern of secondary motions is very complex, and its effect on the velocity profile may be expected to be similar to that of turbulence. The boundary layer that surrounds the flat core remains very thin, but its thickness has not yet been compared with that of the boundary layer in turbulent pipe flow at the same Reynolds number. The contraction of the cross-sectional area is certainly not enough by itself to keep the boundary layer thin. We may note that the thinness of this boundary layer means that flow unsteadiness will be even less important for expiration than for inspiration, according to the arguments of the last subsection. Note too that the experiments of West & Hugh-Jones (1959) and of Dekker (1961) in casts of the trachea and first few bronchi suggest that expiratory flow is less likely to become turbulent than inspiratory flow: the critical Reynolds number in the trachea was about 1500. No model experiments on turbulent expiratory flow have, as far as I know, been done.

No model for predicting pressure drop during expiration has yet been developed. However, the above observation of almost flat velocity profiles and complicated secondary motions suggests a model in which the pressure drop in each tube is approximately the same as for turbulent flow at the same Re. Since the tubes in the experiments are smooth and Re is not more than 10^5, we could as a first approximation use the well-known Blasius law, which in the notation of Equation (1) for a single tube is

$$\frac{D}{Q} = (\tfrac{1}{2}\rho \bar{u}^2)(l/d)(0.32 \, \text{Re}^{-1/4}).$$ (7)

Divided by the corresponding value for Poiseuille flow, this gives a value of Z (called Z_T) as follows:

$$Z_T = 0.005 \, \text{Re}^{3/4}. \tag{8}$$

The trouble with using (7) or (8) is that they are obtained for $\text{Re} > 3000$, and we wish to apply them for Re as low as 200. Nevertheless, it should give some indication of the overall drop in total head for a complete network, which is predicted to be proportional to the 1.75 power of the flow rate Q. In some preliminary experiments on expiratory flow through his models, Pacome (1975) found the total head loss to be proportional to $Q^{1.7}$, which is encouraging. We should also note that Z_T exceeds Z, as given by Equation (4) with $\gamma = 0.33$, for $\text{Re} > 1250$, but is less than Z for smaller Re. Both the experiments of Jaffrin & Hennessey (1972) and some preliminary measurements by M. F. Sudlow (personal communication) indicate the drop in total head (in models) to be greater for expiration than for inspiration at all flow rates, so clearly Equations (7) and (8) are too oversimplified.

Some theoretical work on expiratory flow in a bifurcation was done by Scherer (1972), who predicted an almost flat profile in the parent tube. However, his methods cannot account for the rapid change in the profile that occurs at the flow divider and are too crude to be very useful. The only other theory is by I. J. Sobey (personal communication), who has examined the flow just downstream of the flow divider by computing the development with longitudinal distance (x) of the axial profile $u = 1 + \frac{1}{4}(1 - z^2)(1 + 2y^2 - y^4)$ (representing a dip in the profile in one plane and not in the other) when acted on by the solenoidal secondary motions $v = -\frac{1}{2}y$, $w = \frac{1}{2}z$. He found that the dip was still present at $x = \frac{1}{2}$, but completely abolished by $x = 1$. The neglect of the tube walls of course means that this theory will have only very local validity.

COMPLIANT TUBES: FORCED EXPIRATION   During a forced expiration from TLC, the flow rate initially rises to a maximum at about 85% TLC, and then falls to zero as the lung volume is reduced toward its residual level (about 35% TLC). During this falling phase, the phenomenon of "flow limitation" takes place, in that an increase in expiratory effort does not result in an increase in flow rate for a wide range of lung volumes. The basic reasons for this are well understood, but the details have not been completely worked out.

During expiration the pressure in an airway (p) is smaller than that in the alveoli as a result of both viscous head loss (p_v) and kinetic-energy gains ($\frac{1}{2}\beta\rho\bar{u}^2$) (cf Equations 1 and 3). Now at any lung volume the pressure outside an intrathoracic airway is effectively the same as pleural pressure (p_{pl}), which is smaller than alveolar pressure by an amount equal to the static recoil pressure of the lung (p_{st}), which depends only on lung volume. When the flow rate is high enough for p to be smaller than p_{pl}, the airway will experience a compressive stress; above a critical value this can cause the airway to collapse to a much smaller cross-sectional area, increasing the resistance to flow. As the pleural pressure is increased, any tendency for the flow rate to increase further would be resisted by a tendency for collapse to occur more peripherally.

A similar phenomenon has been studied in model experiments (Figure 9) in which collapsible tubes are supported in a chamber whose pressure (p_0) can be adjusted independently of the upstream and downstream pressures (p_1 and p_2). In most of these experiments, the pressures are held constant and the collapse of the tube is treated as a quasi-steady phenomenon [except in some poorly understood cases, unlikely to be relevant to the lung, in which high frequency self-excited oscillations arise; see Conrad (1969)]. A simple analysis of the steady collapsed state uses Bernoulli's equation to relate p_1 and the upstream velocity u_1 to the velocity and pressure (u_2 and p_2) at the collapsed portion C; if the flow separates here, there is virtually no pressure recovery before the point where p_2 is measured (Lambert & Wilson 1972). Continuity is used to link the velocities and areas upstream and at C, and an independently measured (or calculated) relation is used to relate the area at C to the transmural pressure there. The flow rate corresponding to a given driving-pressure difference can then be calculated; Lambert and Wilson found good agreement between theory and their own model experiments.

To predict the pressure flow-rate volume relationships in the lung, however, is more complicated. One reason is that we still have very limited information on the effective compliance of airways when their cross-sectional area changes at a more-or-less fixed lung volume [Section 1.2 above; Hughes et al (1974) reported some measurements of this distensibility, while Lambert & Wilson (1973) modeled it theoretically]. Another reason is that the site at which collapse takes place must itself be one of the predictions of the model. Clément, van de Woestijne & Pardaens (1973 and earlier papers) have developed a good quasi-steady model of forced expiration. A bronchial pathway is treated as a sequence of short segments, each of which has a given relationship between transmural pressure ($p - p_{pl}$) and cross-sectional area A (the *elastic* curve in Figure 10): a segment is taken to be very stiff at high transmural pressures when it is circular, very compliant as it collapses, and very stiff again when A is small (the minimum area may be zero or finite). On account of the pressure drop along the pathway and the elastic recoil of the lung,

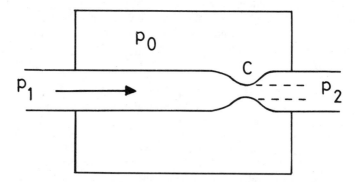

Figure 9   Experimental arrangement in which a collapsible tube is supported in a chamber where the pressure p_0 can be controlled independently of the upstream and downstream pressures p_1 and p_2. The dotted lines indicate a jet emerging from the collapsed portion C.

$(p - p_{pl})$ is also equal to $p_{st} - p_v - \frac{1}{2}\beta\rho\bar{u}^2$. This is plotted as the *fluid dynamic* curve in Figure 10. Its vertical asymptote represents the fact that $\beta\bar{u}^2$ becomes infinite as A tends to zero. At the other end the curve is not extended beyond the maximum value of A achieved by the segment. It is bounded above by the horizontal line $p - p_{pl} = p_{st}$, representing the situation at the beginning of expiration when the flow rate $\dot{V}$ is zero; this intersects the elastic curve at one position, X_0, representing a stable steady state. As p_{pl} and hence $\dot{V}$ is increased, the straight line becomes a curve, and the stable steady state (X) moves down the elastic curve. At certain p_{pl}, there are three points of interesection, two stable (X and Z) and one unstable (Y), but when p_{pl} (and $\dot{V}$) exceed a critical value, the two curves in Figure 10 once more

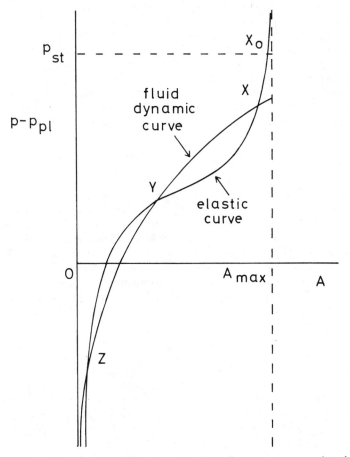

Figure 10   Schematic diagram of the *elastic* curve (distending pressure $p - p_{pl}$ plotted against cross-sectional area A) of a segment of airway. Also shown is the *fluid-dynamic* curve, where the same quantities, calculated from fluid-dynamic equations, are plotted. See text for further details.

have at most only one intersection, Z, representing the (stable) collapsed state. Clément et al proposed that the jump that must then occur from X to Z (or to zero area) represents collapse. As p_{pl} is increased above the critical value at which a single element first collapses, more segments collapse, as expected. However, the flow rate is then predicted to fluctuate sharply with p_{pl}, which is not observed.

What this theory ignores is the time-dependence of expiration and of airway collapse. In a subsequent paper (Clément, Pardaens & van de Woestijne 1974) these authors included terms representing the reactive pressure (i.e. local fluid inertia), wall inertia, and wall viscoelasticity in a model of a single airway of uniform properties; a numerical solution of the equations shows that the fluctuations are removed. In fact this theory makes it clear that the dynamic collapse of airways is analogous to the choking of supersonic flow in a nozzle, with the sound speed replaced by the speed of propagation of elastic waves in the tube wall. S. V. Dawson (personal communication) is analyzing the phenomenon from this point of view.

Finally, there is some radiographic evidence that airway collapse in normal subjects occurs in the trachea and primary bronchi (Macklem, Fraser & Bates 1963), which are inside the thorax but outside the lung tissue itself. This is consistent with the above arguments, because these airways experience a large transmural pressure without the stiffening influence of the surrounding tissue.

3   AIRWAY RESISTANCE

The resistance of an airway, group of airways, or the whole lung is defined as the pressure drop between its ends divided by the flow rate through it. In general it will vary with lung volume V as the airway dimensions change and with the overall flow rate $\dot{V}$ into (or out of) the lung, because of the nonlinear fluid mechanical effects already described. Thus resistance will vary throughout the breathing cycle, but nevertheless physiologists often speak of a single number, *the* airway resistance for quiet breathing about a particular lung volume, say FRC (in practice this is often taken to be the average slope of the pressure-drop/flow-rate curve between 0 and 0.5×10^{-3} m^3 sec^{-1}). In particular, even those whose intention is to test the predictions made on the basis of fluid dynamical experiments often ignore the fundamental differences between inspiratory and expiratory flow and assume that the resistance is the same during each phase. For this reason it is difficult to use many of the measurements as tests of the theory; what can be done is outlined below.

Overall airway resistance, from mouth to alveolus, is often partitioned into *upper* airway resistance, from mouth to trachea, and *lower* airway resistance. The latter is sometimes further divided into *central* and *peripheral* airway resistance, the division being taken in airways of about 3 mm diameter (however, the only experiments in which these quantities have been measured separately were with dogs, so their relevance to the human condition may not be great).

3.1   *Theoretical Predictions*

The most complete predictions can be made for pressure drop in the conducting airways during inspiration, as a result of the model experiments described above.

Pedley, Schroter & Sudlow (1970b) used Equation (4) with $\gamma = 0.33$, and Equation (3) with $\beta = 1.7$, to predict *both* the viscous pressure drop ($p_v = D/Q$) in every branch of Weibel's lung model *and* the variation of static pressure along it, and hence also the total lower-airway pressure drop. Fully developed turbulent flow (in a smooth pipe) was assumed in the trachea. Poiseuille flow was assumed in the smallest airways: in fact Z is predicted to fall below 1 at a value of Re equal to about 32, so Poiseuille flow ($Z = 1$) is taken for every tube in which Re falls below 32. During air breathing, this occurs in 1.3-mm airways when $\dot{V} = 0.5 \times 10^{-3}$ m^3 sec^{-1}, and in 0.9-mm airways when $\dot{V} = 2.0 \times 10^{-3}$ m^3 sec^{-1}; if a dense gas is breathed, this dividing line occurs in smaller airways. These predictions can be summed up in a single equation for the viscous pressure drop along any one generation of airways:

$$p_v = K\mu^{2-a}\rho^{a-1}\dot{V}^a, \tag{9}$$

where K is a known constant, with dimensions $L^{-(a+2)}$, which depends only on the lengths and diameters of the airways concerned, and the exponent a is equal to $\frac{7}{4}$ in the trachea, $\frac{3}{2}$ in the central bronchi, and 1 in the peripheral bronchi. Indeed, if the pressure drop in the *upper* airways is assumed to be dominated by a narrow constriction (in the nose or at the larynx), then it too can be represented by (9) with $a = 2$. Wood et al (1976) suggested that (9) be used everywhere, with a taken to decrease gradually but more or less continuously from 2 at the mouth to 1 in the peripheral bronchi. Predictions of inspiratory viscous pressure drop can be converted to those of static pressure drop by the subtraction of the kinetic-energy term, which is also of the form of (9) with $a = 2$.

The main predictions made by Pedley et al were as follows. (*a*) The viscous resistance of the first four generations is uniformly high, but that of smaller airways decreases progressively and rapidly, until it is negligible by the tenth generation. This is particularly true at higher flow rates: for example, at $\dot{V} = 0.17 \times 10^{-3}$ m^3 sec^{-1} (10 l min^{-1}), two-thirds of the lower-airway viscous pressure drop occurs in the first 6 generations, while over three-quarters occurs there at $\dot{V} = 1.7 \times 10^{-3}$ m^3 sec^{-1} (100 l min^{-1}). (*b*) At all but the lowest flow rates, the lower-airway resistance is much greater than would be predicted assuming Poiseuille flow everywhere, as shown in Figure 11, where it can be seen that the pressure drop at a flow rate of 1.0×10^{-3} m^3 sec^{-1} (1 l sec^{-1}) is about 60 N m^{-2}. The nonlinearity of the pressure–flow rate relationship is not in fact very marked over quite a wide range of physiological flow rates, largely because the kinetic-energy term is subtracted from the viscous pressure drop.

We should note that ever since the classic paper by Rohrer (1915), physiologists have tried to fit their experimental results on airway pressure drop (at a given lung volume) to a curve of the form

$$\Delta p = K_1\dot{V} + K_2\dot{V}^2.$$

The above considerations suggest that a more physically based curve would have the form:

$$\Delta p = K_1'\dot{V}^{3/2} + K_2\dot{V}^2, \tag{10}$$

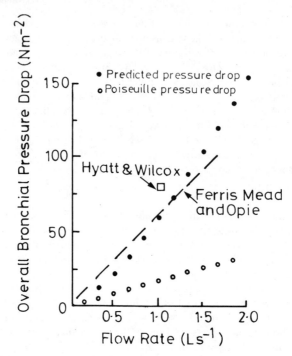

Figure 11 Predicted variation of the overall bronchial pressure drop with flow rate on inspiration at constant volume, compared with the Poiseuille viscous pressure drop and with two experimental values.

where K_2 would be negative for lower airways, but may be positive for the whole lung because of the constriction in the upper airways.

The main errors in these predictions are likely to arise from two factors. One is turbulence in the largest bronchi, which we are unable to account for properly: it was suggested above that it it might reduce the pressure drop, at least for tracheal $Re < 10^4$; however, Jaffrin & Kesic (1972), modeling the conducting airways as a rough pipe, indicated an enhancement of pressure drop for values of Re above about 4×10^3. The other source of error is the asymmetry of the bronchial tree. Two very crude ways of accounting for it have been put forward by Olson, Dart & Filley (1970) and by Pedley, Schroter & Sudlow (1977). Both methods indicate up to a 100% difference between the viscous pressure drops along different pathways, but neither properly accounts for the asymmetric fluid mechanics; for example Pedley et al applied Equation (4), still with $\gamma = 0.33$, to the pathways defined by Horsfield et al (1971). While the physics embodied in Equation (4) is presumably still applicable, the constant γ will inevitably depend on the geometry of each junction. The *average* pressure drop was still close to that predicted from a symmetric model.

It must be remembered that Weibel's data, and hence all of the above predictions, were developed for a lung volume of 75% TLC. Hence in comparing the results with experiment a correction for lung volume should be made. This can be done for each term of the form (9), if it is assumed that airway dimensions vary iso-tropically with lung volume. K has dimensions $L^{-(a+2)}$ and hence is proportional to $V^{-(a+2)/3}$.

No predictions comparable with the above have been made for expiratory flow. However, if the above speculations about its similarity to turbulent flow in a smooth pipe are correct, we would expect the expiratory lower-airway viscous pressure drop to be dominated by a term of the form (9) with $a = \frac{7}{4}$.

3.2   Comparison with Physiological Experiments

The absolute value of the lower-airway pressure drop at a flow rate of 1.0×10^{-3} m^3 sec^{-1}, as predicted above, is in reasonably good agreement with the few experiments in which it has been measured. The results of two such experiments (Hyatt & Wilcox 1963; Ferris, Mead & Opie 1964) are shown in Figure 11. Reason-able agreement was also obtained in atropinized subjects (who therefore had no bronchomotor tone) by Vincent et al (1970). These authors did not report non-linearity of the pressure–flow rate curve, but they were not looking for it. Overall airway resistance is well known to be flow-rate dependent (Rohrer 1915).

Experiments in which the effect of flow rate on lower-airway resistance has been examined have also involved the breathing of gas mixtures of different density and viscosity, in order to increase the range of Reynolds numbers examined. Jaeger & Matthys (1969) used four gas mixtures whose densities ranged between 0.45 kg m^{-3} (He and O_2) and 4.2 kg m^{-3} (SF$_6$ and O_2); viscosity was less variable, ranging between 2.9×10^{-5} kg m^{-1} sec^{-1} and 1.7×10^{-5} kg m^{-1} sec^{-1}, the densest gas being the least viscous. Two flow rates, of about 0.2×10^{-3} m^3 sec^{-1} and 3.0×10^{-3} m^3 sec^{-1}, were used for each gas. Pedley, Schroter & Sudlow (1970b) showed that these results agreed well with their predictions (Equation 9 with $a = \frac{3}{2}$), except for the densest gas mixture breathed at the higher flow rate, when the measured pressure drop was four times the predicted. However, in that case Re was 55,000 in the trachea and 3,000 in generation 10, and was too high for the theory to be applicable. The turbulent-flow theory of Jaffrin & Kesic (1972) agreed more closely.

Wood et al (1976) have also investigated the effect of gas density and flow rate on lower-airway resistance in three subjects. They made their measurements by generating 1.7 Hz oscillations in flow rate, about a zero mean, with an amplitude somewhat greater than 1.0×10^{-3} m^3 sec^{-1} and by averaging the inspiratory and expiratory pressure drops at two stages of the cycle corresponding to two different flow rates (0.5 and 1.0×10^{-3} m^3 sec^{-1}). This means that the kinetic-energy term cancels out, but any difference that may exist between inspiration and expiration is obscured. Thus their results are not strictly comparable with the predictions. Nevertheless, they found a definite increase of resistance with both ($\mu\rho$) and $\dot{V}$, although not as great as that predicted by (9) with $a = \frac{3}{2}$. These results were confirmed and extended by experiments on open-chested dogs, in which the pressure drop along airways of diameter greater and less than 4 mm could be measured separately (central and peripheral airways). These showed that although (9) with $a = \frac{3}{2}$ did not fit the total lower-airway pressure drop, it was quite a good fit for central airway pressure drop, especially at the highest gas densities, while peripheral airway pressure drop fitted the equation with a close to 1. This suggests that the entry-flow model is good for central airways, while a linear model is better for peripheral airways. The dividing line in Wood et al's experiments seemed to come at a Reynolds number of about 100 (not 32, as predicted). Note, however, that these experiments were performed in dogs, so that quantitative agreement with human predictions is not to be expected, although the basic physics is thought to be the same.

Wood et al's experiments are consistent with earlier findings of Macklem & Mead (1967), and confirm the prediction that the resistance of central airways exceeds that of peripheral airways, by a factor of about 2 on average.

Lower-airway resistance is predicted to vary with lung volume as V^{-1} in the (peripheral) airways for which $a = 1$ in Equation (9), and as $V^{-7/6}$ in the central airways for which $a = \frac{3}{2}$. In physiological experiments with a lot of scatter, it would be impossible to distinguish between these two relationships. Indeed the rough relation that conductance, G ($= 1$/resistance), is proportional to V is commonly assumed by physiologists: a more precise analysis of physiological data suggests

$$G = b(V - V_0)$$

where b and V_0 are constants. Blide, Kerr & Spicer (1964) found the slope b to be approximately 0.011 kg^{-1} m sec for lower-airway conductance, while Guyatt & Alpers (1968) found a value of about 0.0027 kg^{-1} m sec for total airway conductance. The predictions of Pedley, Schroter & Sudlow (1970b) suggest a value of 0.0032 kg^{-1} m sec for the lower airways, when $V = 5 \times 10^{-3}$ m^3 and $\dot{V} = 0.5 \times 10^{-3}$ m^3 sec^{-1}. Good agreement was found between the predicted variation of lower-airway resistance with lung volume and that measured by Vincent et al (1970) in atropinized subjects (with no bronchomotor tone), except near TLC where a slight increase in resistance was reported.

Finally, we turn to upper-airway resistance, which is about half of the total airway resistance (in mouth breathing) and has been measured by a number of workers. Spann & Hyatt (1971) found that this quantity fell as lung volume increased, with about half of the change being attributable to the change in cross section at the larynx; this was confirmed by Stanescu et al (1972). Both sets of workers also agreed that laryngeal resistance was smaller during panting than during quiet breathing. Jaeger & Matthys (1968) showed that upper-airway pressure drop increased non-linearly with flow rate, although not according to (9) with $a = 2$, as it would if the pressure drop were dominated by an orifice. Instead, for three gas mixtures spanning a 2.5-fold density range, they fitted fairly well to the equation with $a = 1.5$, and for an SF_6 mixture four times as dense as the densest of the others a value of $a = 1.9$ was more appropriate. For nose breathing, Swift & Proctor (1977) stated that, except at flow rates below 0.05 m^3 sec^{-1}, virtually all of the upper-airway resistance is concentrated at the constriction of the ostium internum, and that the resistance of the nasal passage accounts for about half of the total airway resistance. This is contrasted with the oral passage which accounts for about one-fifth during mouth breathing, the larynx accounting for at least another fifth.

4   MASS TRANSPORT

There is an enormous physiological literature in which the many factors affecting pulmonary gas exchange have been investigated. In this article it is appropriate to consider only those in which airway fluid dynamics is implicated, and we therefore concentrate exclusively on the convection-diffusion process that determines the concentrations of gases presented to the alveolar membrane. Model experiments are virtually nonexistent, and physiological experiments are difficult to interpret, so the aim will be to decide what constitutes an appropriate theoretical description of the process, consistent with physiological evidence.

Two situations are normally envisaged from the theoretical point of view. One is normal steady-state breathing of air, during which a little O_2 is absorbed and a little CO_2 is released each breath, while the N_2 concentration remains virtually uniform. The other is a "single-breath" experiment in which a new gas mixture is presented at the mouth at the beginning of or during inspiration; this is inspired and then fully expired again following a known period of breath holding. The concentration of something is measured at the mouth during expiration (such measurement may continue for several breaths, whether or not the new gas mixture is breathed continuously). The "new gas mixture" may be $100\% O_2$, and it might be N_2 whose mouth concentration is measured (the so-called "single breath N_2 wash-out test"). The gas mixture may contain O_2 at the same partial pressure as air, but the N_2 may be replaced by another inert gas (He, Ar, SF_6) of different density, viscosity, and diffusivity. Alternatively the gas mixture may consist of normal air with just a trace or a bolus of some foreign gas within it.

The concentration of N_2 during the first expiration after inspiration of $100\% O_2$ is initially very low, as the pure O_2 most recently inspired, which did not come into contact with the N_2-rich air already present, is expired again (Figure 12). Then there is a very sharp rise of N_2 concentration, representing the smeared out "interface" between fresh air and the air that was already present (and occurring roughly when the expired volume is equal to the volume of O_2 inspired), followed by a much flatter region, which is nevertheless usually sloping. This "alveolar plateau" is supposed to represent alveolar gas, and there has been great controversy about the cause of its slope.

One explanation is based on the proven regional inhomogeneities within the lung: if one region of the lung (normally the lower part if inspiration is from FRC; the upper part if from minimum, residual volume) is more compliant than another, then it will receive a greater proportion of the inspired gas [unless the inspiration is very rapid—see Pedley, Sudlow & Milic-Emili (1972)], and at the end of inspiration the N_2 concentration in it will be smaller. If, during expiration, this region is expired somewhat before the other region ("sequential emptying"), the N_2

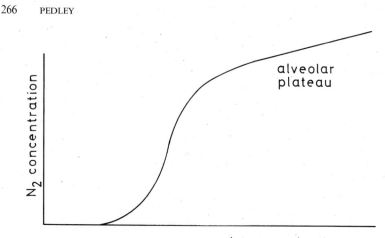

Figure 12 Schematic plot of N_2 concentration against volume expired after a single inspiration of pure O_2.

concentration in the first part of the alveolar expirate will be relatively low, gradually increasing as the other regions contribute more to the expirate. Thus the rising slope of the alveolar plateau is explained. If the regions of higher N_2 concentration are expired first, the curve will slope downwards. Glaister et al (1973) demonstrated convincingly that sequential emptying, and hence a sloping alveolar plateau, will occur if and only if different regions of the lung have different intrinsic elastic properties, i.e. if and only if a change from one given distending pressure to another results in different relative volume changes. They showed that, in general, the elastic properties of dog lungs are nonuniform; the evidence in man is less clear-cut.

The alternative mechanism would operate in the absence of regional inhomogeneity, when the slope of the alveolar plateau would reflect the concentration gradient in the terminal air spaces, which would have to be present for O_2 diffusion to take place. This mechanism is called diffusive, or "stratified," inhomogeneity. There are certainly some experiments that can be explained only in terms of diffusive inhomogeneity. For example, Cumming et al (1967) introduced a mixed bolus containing both Ne and SF_6 into the inspirate and measured the concentrations of these gases during the subsequent expiration. All lung regions would be expected to receive the same relative proportions of the two test gases, so that if the only inhomogeneity were regional, the ratio of Ne to SF_6 should remain constant on expiration. In fact the expired gas became progressively richer in Ne as time passed, indicating that Ne had diffused further into the alveolar spaces than SF_6, on account of its greater molecular diffusivity in air. Furthermore the magnitude of the effect is diminished after breath-holding.

Controversy centers on whether diffusive inhomogeneity is important physiologically: Farhi (1969) reviewed the evidence and concluded that it was. This is just as well, since virtually all theoretical models of mass transport in the lung have totally ignored regional inhomogeneity (because it is difficult to incorporate) and have consisted of more or less (often less) soundly based analyses of diffusion between the mouth and the alveoli, assuming that all pathways are the same.

In all the theories developed so far, the bronchial tree, at least as far as the respiratory bronchioles, has been modeled as a one-dimensional system, whose cross-sectional area $A(x)$ (where x is distance measured along a pathway) is at any x equal to the net cross-sectional area of all airways of the generation in which x lies, as given by Weibel (1963). $A(x)$ remains close to A_0, the tracheal cross-sectional area 2.5 cm^2, until about the end of the fourth generation (taken to be $x = 0$), when it starts to increase rapidly, to about 940 cm^2 in generation 19 when $x \approx 6$ cm. All models have hitherto ignored the fact that airway dimensions vary during respiration, and A is taken to be independent of time t. Thus the average fluid velocity in any airway, $\bar{u}(x, t)$, is equal to $\dot{V}(t)/A(x)$, where $\dot{V}(t)$ is the volume flow rate into the lung, which is positive during inspiration (gas compressibility is also neglected).

Furthermore, the variation in the concentration of any gas across the cross section of an airway is ignored, and the average concentration $C(x, t)$ is used as the dependent variable. In fact the transverse variations will be small, because of the vigorous mixing induced by turbulence in the upper airways and by secondary motions in the conducting airways. Assuming that longitudinal mixing in an airway is a diffusive process, described by means of an effective diffusivity $D(x, t)$, the convection-diffusion equation for C is then

$$\frac{\partial C}{\partial t} + \bar{u}\frac{\partial C}{\partial x} = \frac{1}{A}\frac{\partial}{\partial x}\left(AD\frac{\partial C}{\partial x}\right). \tag{11}$$

Many early models neglected all mixing in the airways ($D = 0$) and all convection in the respiratory units (which we have not yet incorporated into the model). More recent models (e.g. Baker, Uttman & Rhoades 1974; Paiva 1973) have used (9), but have everywhere replaced D by the molecular diffusivity D_{mol}. Only very belatedly has recognition been given to the fact that longitudinal mixing in a tube is much greater than can be described by the molecular diffusivity, because of the dispersing effect of a velocity profile, ultimately limited by *transverse* diffusion. This is the process described for fully developed laminar and turbulent flow by Taylor (1953, 1954) and introduced in a respiratory context by Wilson & Lin (1970) and Pedley (1970). To assume that such mixing can be described by a single diffusivity $D(x, t)$ is already an enormous simplification, but so far no one has found a way of improving on it in the airways.

Hitherto, two approaches have been used to evaluate $D(x, t)$ realistically in the airways. Pedley (1970) chose D to be equal to Taylor's (1954) prediction for fully developed turbulent flow, D_T, in the trachea and largest bronchi, and equal to D_{mol} in the smallest airways; in between, purely for analytical convenience, he chose $D/D_T \propto 1/A$. Butler (1974) used an equally convenient exponential form for A^2D in his analytical model. In a purely numerical model, however, Butler used values inferred from some experiments by Scherer et al (1975). These authors measured the dispersion of a bolus in a 5-generation network of branched tubes, at various laminar flow rates, and fitted the results by the formula

$$D = D_{mol}(1 + k\,Pe),\qquad\qquad(12)$$

where Pe is the Peclet number $(= \bar{u}\,d/D_{mol})$ in the parent tube and k is a constant equal to 1.08 on inspiration and 0.37 on expiration. In fact the values of D predicted by this approach (with Pe given its appropriate value in each generation) are slightly larger but of similar order of magnitude to those used by Pedley. They are much larger than D_{mol}, but much smaller than Taylor's (1953) laminar values.

Some authors have questioned whether in fact Taylor dispersion is important in the airways, because of their small volume and the consequent small time spent in them by fluid particles. However, Lacquet, van der Linden & Paiva (1975) found a considerable difference between predictions of the dispersion of SF_6 made using Scherer et al's values for D and those made using D_{mol}. Taylor dispersion is also the only plausible explanation for the results of certain physiological experiments on mass transport in different gas mixtures (Johnson & van Liew 1974; Kvale, Davis & Schroter 1975).

It is in their description of the respiratory regions, to which the above model of the conducting airways has to match, that the theoretical models differ significantly. The most complete has been given by Davidson & Fitz-Gerald (1974) and Davidson (1975), who computed the O_2 concentration everywhere during a single inspiration, followed by a single expiration, each at constant flow rate. A single pathway in the respiratory region was modeled as a circular cylinder, closed at one end, with annular partitions regularly spaced along it to represent the alveolar sacs (Figure 13). The length and radius of this unit vary with time during breathing, to accommodate the assumed changes in lung volume. Since only a single breath is modeled, during which only a very small amount of O_2 is taken up because the normal gradient of O_2 concentration $(\partial C/\partial n)$ is small, the boundary condition $\partial C/\partial n = 0$ was imposed on the wall. This is suitable in any case for physiologically inert gases. At the entrance to the unit, Davidson & Fitz-Gerald imposed the condition of matching with the results of Pedley's airway model. They solved the complete axisymmetric convection-diffusion problem in the unit by a finite-difference technique, using the velocity field previously computed (1972). Their results showed

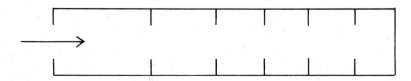

Figure 13   Partitioned cylindrical model of a respiratory unit; after Davidson & Fitz-Gerald (1974).

that convection in the respiratory units is important in improving O_2 uptake, at all flow rates above about 10^{-3} m^3 sec^{-1}.

Other models of the respiratory unit are less realistic in that they all ignore convection there, and all except those of Chang, Cheng & Farhi (1973) ignore the detailed geometry. Butler (1974) modeled this region merely as a well-mixed reservoir, but he did allow for a (small) flux of O_2 out of the system during each breath; this is required for steady-state breathing (which his model is intended to simulate) to be possible. He is the only author not to impose $\partial C/\partial n = 0$ at the alveolar membrane. He and Baker, Uttman & Rhoades (1974) are also the only ones to consider time-dependent (in fact sinusoidal) flow rates throughout respiration, instead of the uniform inspirations and expirations examined by others.

The models are completed by imposing a boundary condition at the mouth. On inspiration this must clearly be that C is equal to its value in the inspired gas. On expiration, however, this would impose an unnatural constraint, and Pedley (1970) suggested that $\partial C/\partial x = 0$ might be appropriate, implying that all the flux out of the mouth is convective not diffusive; this was used by Davidson (1975). Butler (1974) used a more complicated but roughly equivalent condition.

The authors of each of the different models have used them to simulate different physiological experiments, so it is not possible to compare their results directly. They all showed qualitative agreement with experiment, but none suggested a crucial test of his theory as opposed to another's. Nevertheless, some of the results are of interest and are well worth developing further. For example, Davidson (1975) showed that, according to his model, there is no slope to the alveolar plateau of expired O_2, suggesting that other factors (presumably regional inhomogeneity) cause it to appear in vivo. It would be extremely interesting to see if Davidson's model could reproduce the observations of Cumming et al (1967) in which regional inhomogeneities should have no effect. Lacquet, van der Linden & Paiva (1975) modeled the mixing of both H_2 and SF_6 in air and predicted considerable differences in their expired concentration curves. Butler (1974) obtained reasonable agreement with experiment on the variation of alveolar O_2 concentration with the frequency and amplitude of breathing. If the best features of all of these models were pooled into one, the result would, I believe, give a faithful simulation of a lung in which all pathways are identical. The incorporation of some regional inhomogeneity would then be necessary.

Mention should finally be made of some interesting experiments by Engel et al (1973a,b), further discussed by Fukuchi et al (1976), who directly measured N_2 concentrations in small (~ 2 mm) airways of artificially ventilated dogs. They demonstrated that longitudinal dispersion was five times greater when the dogs were alive than when they were not. The only physical difference between the breathing maneuvers in the two cases lay in the relatively high-frequency pulsations super-imposed on the ventilatory cycle by the beating heart, in the former case. This is consistent with the theoretical work of E. J. Watson and Chatwin (Chatwin 1975), which shows how longitudinal pulsations in a long straight pipe can greatly enhance mixing.

5 FUTURE STUDY

As far as flow in symmetric branched tubes is concerned, it will be necessary to rely on model experiments for several more years, until a reliable (and not too expensive) method has been found to calculate all of the details numerically. The theories of Smith, Sobey, etc., can be used to guide the numerical and experimental work and to help interpet the results, but I will be surprised if they can be developed far enough to yield, for example, the secondary velocity field as a function of branch angle and area ratio. Experimentally, one looks forward to a systematic investigation of the relationship between energy dissipation (i.e. pressure drop) and branch angle, area ratio, generation number, *and asymmetry* (as measured by the ratio of diameters and flow rates in the two daughter tubes). The hope, still untested, is that all these effects will be small. It is still necessary to resolve the difference between the results of Pedley, Schroter & Sudlow (1970a) and of Berger, Calvet & Jacquemin (1972) for laminar inspiratory flow, but it is more important to study (a) turbulent flow and (b) expiratory flow in detail, especially in the range of tracheal Reynolds numbers between 10^3 and 10^4. Observations on how far turbulence persists down a branched-tube network, at Reynolds numbers below the nominal critical value, would be most welcome and quite straightforward. Otherwise, I cannot foresee any useful and systematic way of studying how the flow in the upper airways affects that in the trachea (and hence in the bronchi), although it is clearly important.

Theoretical understanding of flow limitation during forced expiration will probably develop rapidly because of the parallel interest in similar phenomena in the circulation (see Caro, Pedley & Seed, 1974, for example). The most fruitful line of attack is that of Dawson (see above), but quantitative detail must await more complete data on the distensibility of airways whose volume changes independently of that of the surrounding lung tissue. It is perhaps in this area that mechanical theory is most likely to lead to new physiological tests for abnormality.

Mass-transport modelers should assiduously try to simulate all relevant physiological experiments, with the aim of devising a crucial test of the importance of the different features incorporated in their models. The next big step must be the marriage in one model of both diffusive and regional inhomogeneity, so that their relative effects can be systematically assessed. Here too is an area where a better understanding of the mechanics should lead to increasingly sensitive tests of abnormal lung function.

The big challenge confronting both respiratory physiologists and fluid dynamicists is achieving an understanding of the mechanics of the diseased lung. All the work reviewed in this article has concerned the normal lung; it is now time for a collaborative effort to determine the important anatomical features of the different diseases and their quantitative effect on the mechanical function of the lung.

ACKNOWLEDGMENTS

I wish to record my gratitude to R. C. Schroter and M. F. Sudlow, who have put up with me as a colleague for many years, and in collaboration with whom I have learned all I know about the lungs.

Literature Cited

Baker, L. G., Uttman, J. S., Rhoades, R. A. 1974. Simultaneous gas flow and diffusion in a symmetric airway system: a mathematical model. *Respir. Physiol.* 21:119–38

Berger, C., Calvet, P., Jacquemin, C. 1972. Structure d'écoulements de gaz dans des systèmes tubulaires bifurques. *Rep. DER en mesures. ONERA-CERT, Toulouse*

Blake, J. R. 1975. On the movement of mucus in the lung. *J. Biochem.* 8:179–90

Blide, R. W., Kerr, H. D., Spicer, W. S. 1964. Measurement of upper and lower airway resistance and conductance in man. *J. Appl. Physiol.* 19:1059–69

Brackenbury, J. H. 1972. Physical determinants of airflow pattern within the avian lung. *Respir. Physiol.* 15:384–97

Brain, J. D., Proctor, D. F., Reid, L., eds. 1977. *Respiratory Defense Mechanisms.* New York: Dekker

Brech, R., Bellhouse, B. J. 1973. Flow in branching vessels. *Cardiovasc. Res.* 7:593–600

Butler, J. P. 1974. *Oxygen transport in the human lung.* PhD thesis. Harvard Univ., Cambridge, Mass. 107 pp.

Caro, C. G., Pedley, T. J., Seed, W. A. 1974. Mechanics of the circulation. In *Cardiovascular Physiology,* ed. A. C. Guyton, Chap. 1. *MTP International Review of Science, Physiology,* Ser. 1, Vol. 1. London: Butterworth

Chang, H. K., Cheng, R. T., Farhi, L. E. 1973. A model study of gas diffusion in alveolar sacs. *Respir. Physiol.* 18:386–97

Chatwin, P. C. 1975. On the longitudinal dispersion of passive contaminant in oscillatory flows in tubes. *J. Fluid Mech.* 71:513–27

Chow, C.-Y., Lai, Y. C. 1972. Alternating flow in the trachea. *Respir. Physiol.* 16:22–32

Cinkotai, F. F. (with appendix by E. J. Watson). 1974. Fluid flow in a model alveolar sac. *J. Appl. Physiol.* 37:249–51

Clarke, S. W., Jones, J. G., Oliver, D. R. 1972. Factors affecting airflow through branched tubes. *Bull. Physio.-Pathol. Respir.* 8:409–28

Clément, J., Pardaens, J., van de Woestijne, K. P. 1974. Expiratory flow rates, driving pressures and time-dependent factors, simulation by means of a model. *Respir. Physiol.* 20:353–69

Clément, J., van de Woestijne, K. P., Pardaens, J. 1973. A general theory of respiratory mechanics applied to forced expiration. *Respir. Physiol.* 19:60–79

Comroe, J. H. 1974. *Physiology of Respiration,* 2nd ed. Chicago: Yearb. Med.

Conrad, W. A. 1969. Pressure-flow relationships in collapsible tubes. *IEEE Trans. Biomed. Eng.* BME-16:284–95

Cotes, J. E. 1965. *Lung Function, Assessment and Application in Medicine,* Chap. 14. Oxford: Blackwell

Cumming, G., Horsfield, K., Jones, J. G., Muir, D. C. F. 1967. The influence of gaseous diffusion on the alveolar plateau at different lung volumes. *Respir. Physiol.* 2:386–98

Davidson, M. R. 1975. Lung gas mixing during expiration following an inspiration of air. *Bull. Math. Biol.* 37:113–26

Davidson, M. R., Fitz-Gerald, J. M. 1972. Flow patterns in models of small airway units of the lung. *J. Fluid Mech.* 52:161–77

Davidson, M. R., Fitz-Gerald, J. M. 1974. Transport of O_2 along a model pathway through the respiratory region of the lung. *Bull. Math. Biol.* 36:275–303

Dekker, E. 1961. Transition between laminar and turbulent flow in the human trachea. *J. Appl. Physiol.* 16:1060–64

Dollfuss, R. E., Milic-Emili, J., Bates, D. V. 1967. Regional ventilation of the lung, studies with boluses of [133]Xenon. *Respir. Physiol.* 2:234–46

Douglass, R. W. 1973. *Flow in a human lung model at high Reynolds numbers.* MS thesis. Duke Univ., Durham, NC 135 pp.

Douglass, R. W., Munson, B. R. 1974. Viscous energy dissipation in a model of the human bronchial tree. *J. Biomech.* 7:551–57

Duncker, H.-R. 1974. Structure of the avian respiratory tract. *Respir. Physiol.* 22:1–19

Engel, L. A., Menkes, H., Wood, L. D. H., Utz, G., Joubert, J., Macklem, P. T. 1973a. Gas mixing during breathholding studied by intrapulmonary gas sampling. *J. Appl. Physiol.* 35:9–17

Engel, L. A., Wood, L. D. H., Utz, G., Macklem, P. T. 1973b. Gas mixing during inspiration. *J. Appl. Physiol.* 35:18–24

Farhi, L. E. 1969. Diffusive and convective movement of gas in the lung. In *Circulatory and Respiratory Mass Transport,* ed. G. Wolstenholme. London: Churchill

Ferris, B. G., Mead, J., Opie, L. H. 1964. Partitioning of respiratory flow resistance in man. *J. Appl. Physiol.* 19:653–58

Fishman, A. P., Hecht, H. H., eds. 1969. *The Pulmonary Circulation and Interstitial Space.* Chicago: Univ. Chicago Press

Fukuchi, Y., Roussos, C. S., Macklem, P. T., Engel, L. 1976. Convection, diffusion and cardiogenic mixing of inspired gas in the lung, an experimental approach. *Respir. Physiol.* 26:77–90

Fung, Y. C., Sobin, S. S. 1969. Theory of sheet flow in lung alveoli. *J. Appl. Physiol.* 26:472–88

Glaister, D. H., Schroter, R. C., Sudlow, M. F., Milic-Emili, J. 1973. Transpulmonary pressure gradient and ventilation distribution in excised lungs. *Respir. Physiol.* 17:365–85

Guyatt, A. R., Alpers, J. H. 1968. Factors affecting airways conductance: a study of 752 working men. *J. Appl. Physiol.* 24:310–16

Hoppin, F. G., Hildebrandt, J. 1977. Mechanical properties of lung parenchyma. See West 1977, Chap. 1

Horlock, J. H., Lakshminarayana, B. 1973. Secondary flows: theory, experiment, and application in turbomachinery aerodynamics. *Ann. Rev. Fluid Mech.* 5:247–80

Horn, L. W., Davis, S. H. 1975. Apparent surface tension hysteresis of a dynamical system. *J. Colloid Interface Sci.* 51:459–76

Horsfield, K., Cumming, G. 1967. Angles of branching and diameters at branches in the human bronchial tree. *Bull. Math. Biophys.* 29:245–59

Horsfield, K., Cumming, G. 1968a. morphology of the bronchial tree in man. *J. Appl. Physiol.* 24:373–83

Horsfield, K., Cumming, G. 1968b. Functional consequences of airways morphology. *J. Appl. Physiol.* 24:384–90

Horsfield, K., Dart, G., Olson, D. E., Filley, G. F., Cumming, G. 1971. Models of the human bronchial tree. *J. Appl. Physiol.* 31:207–17

Hughes, J. M. B., Hoppin, F. G., Mead, J. 1972. Effect of lung inflation on bronchial length and diameter in excised lungs. *J. Appl. Physiol.* 32:25–35

Hughes, J. M. B., Jones, H. A., Wilson, A. G., Grant, B. J. B., Pride, N. B. 1974. Stability of intrapulmonary bronchial dimensions during expiratory flow in excised lungs. *J. Appl. Physiol.* 37:684–94

Hyatt, R. E., Rodarte, J. R., Wilson, T. A. 1975. Effect of increased static lung recoil on bronchial dimensions of excised lungs. *J. Appl. Physiol.* 39:429–33

Hyatt, R. E., Wilcox, R. E. 1963. The pressure-flow relationship of the intrathoracic airways in man. *J. Clin. Invest.* 42:29–39

Jaeger, M. J., Matthys, H. 1968. The pattern of flow in the upper airways. *Respir. Physiol.* 6:113–27

Jaeger, M. J., Matthys, H. 1969. The pressure-flow characteristics of the human airways. In *Symposium on Airways Dynamics*, ed. A. Bouhuys, Springfield, Mass: Thomas

Jaffrin, M. Y., Hennessey, T. V. 1972. Pressure distribution in a model of the central airways for sinusoidal flow. *Bull. Physio-Pathol. Respir.* 8:375–90

Jaffrin, M. Y., Kesic, P. 1972. Airway resistance: a fluid mechanical approach. *MIT, Dep. Mech. Eng., Fluid Mechanics Lab., Publ. No. 72-12*

Johnson, L. R., van Liew, H. D. 1974. Use of arterial P_{O_2} to study convective and diffusive gas mixing in the lungs. *J. Appl. Physiol.* 36:91–97

Kvale, P. A., Davis, J., Schroter, R. C. 1975. Effect of gas density and ventilatory pattern on steady-state CO uptake by the lung. *Respir. Physiol.* 24:385–98

Lacquet, L. M., van der Linden, L. P., Paiva, M. 1975. Transport of H_2 and SF_6 in the lung. *Respir. Physiol.* 25:157–73

Lambert, R. K., Wilson, T. A. 1972. Flow limitation in a collapsible tube. *J. Appl. Physiol.* 33:150–53

Lambert, R. K., Wilson, T. A. 1973. A model for the elastic properties of the lung and their effect on expiratory flow. *J. Appl. Physiol.* 34:34–48

Lew, H. S., Fung, Y. C. 1969. On the low Reynolds number entry flow into a circular cylindrical tube. *J. Biomech.* 2:105–19

Lighthill, M. J. 1975. *Mathematical Biofluiddynamics.* Philadelphia: SIAM

Linderoth, L. S., Kuonen, E. A., Nuckols, M. L., Johnson, C. E. 1975. Heat and mass transfer in the human respiratory tract at hyperbaric pressures. *Duke Univ. Eng. Rep.*

Macklem, P. T. 1971. Airway obstruction and collateral ventilation. *Physiol. Rev.* 51:368–436

Macklem, P. T., Fraser, R. G., Bates, D. V. 1963. Bronchial pressure and dimensions in health and disease. *J. Appl. Physiol.* 18:699–706

Macklem, P. T., Mead, J. 1967. Resistance of central and peripheral airways measured by a retrograde catheter. *J. Appl. Physiol.* 22:395–401

Macklem, P. T., Permutt, S., eds. 1977. *The Lung in the Transition Between Health and Disease.* New York: Dekker

Olson, D. E. 1971. *Fluid mechanics relevant to respiration: flow within curved or elliptical tubes and bifurcating systems.* PhD thesis. Imperial Coll., London

Olson, D. E., Dart, G., Filley, G. F. 1970. Pressure drop and fluid flow regime of air inspired into the human lung. *J. Appl. Physiol.* 28:482–94

Olson, D. E., Iliff, L. D., Sudlow, M. F. 1972. Some aspects of the physics of flow in the central airways. *Bull. Physio-Pathol. Respir.* 8:391–408

Owen, P. R. 1969. Turbulent flow and particle deposition in the trachea. In *CIBA Symposium on Circulatory and Respiratory Mass Transport.* London: Churchill

Pacome, J.-J. 1975. *Structures d'écoulement et pertes de charges calculées dans le modèle d'arbre bronchique de Weibel.* Doctoral thesis. Paul Sabatier Univ., Toulouse 102 pp.

Paiva, M. 1973. Gas transport in the human lung. *J. Appl. Physiol.* 35:401–10

Pedley, T. J. 1970. A theory for gas mixing in a simple model of the lung. In *Fluid Dynamics of Blood Circulation and Respiratory Flow. AGARD Conf. Proc. No. 65*

Pedley, T. J. 1976. Viscous boundary layers in reversing flow. *J. Fluid Mech.* 74:59–79

Pedley, T. J., Schroter, R. C., Sudlow, M. F. 1970a. Energy losses and pressure drop in models of human airways. *Respir. Physiol.* 9:371–86

Pedley, T. J., Schroter, R. C., Sudlow, M. F. 1970b. The prediction of pressure drop and variation of resistance within the human bronchial airways. *Respir. Physiol.* 9:387–405

Pedley, T. J., Schroter, R. C., Sudlow, M. F. 1971. Flow and pressure drop in systems of repeatedly branching tubes. *J. Fluid Mech.* 46:365–83

Pedley, T. J., Schroter, R. C., Sudlow, M. F. 1977. Gas flow and mixing in the airways. See West 1976, Chap. 2

Pedley, T. J., Sudlow, M. F., Milic-Emili, J. 1972. A non-linear theory of the distribution of pulmonary ventilation. *Respir. Physiol.* 15:1–38

Proctor, D. F. 1964. In *Handbook of Physiology,* Chap. 8. Sect. 3, Vol. I: *Respiration.* New York: Am. Physiol. Soc.

Rohrer, F. 1915. Der Strömungswiderstand in den menschlichen Atemwegen. *Pflügers Arch. ges. Physiol.* 162:225–59

Rowe, M. 1970. Measurements and computations of flow in pipe bends. *J. Fluid Mech.* 43:771–83

Sackner, M. A., Rosen, M. J., Wanner, A. 1973. Estimation of tracheal mucus velocity by bronchofiberoscopy. *J. Appl. Physiol.* 34:495–96

Scherer, P. W. 1972. A model for high Reynolds number flow in a human bronchial bifurcation. *J. Biomech.* 5:223–29

Scherer, P. W., Shendalman, L. H., Greene, N. M., Bouhuys, A. 1975. Measurement of axial diffusivities in a model of the bronchial airways. *J. Appl. Physiol.* 38:719–23

Schreck, R. M., Mockros, L. F. 1970. Fluid dynamics in the upper pulmonary airways. In *AIAA Fluid and Plasma Dynamics Conf., 3rd, Los Angeles*

Schroter, R. C., Sudlow, M. F. 1969. Flow patterns in models of the human bronchial airways. *Respir. Physiol.* 7:341–55

Smith, F. T. 1976a. Steady motion through a branching tube. *Proc. R. Soc. London Ser. A.* In press

Smith, F. T. 1976b. On entry-flow effects in bifurcating, blocked or constricted tubes. *J. Fluid Mech.* In press

Smith, F. T. 1976c. Upstream interactions in channel flows. *J. Fluid Mech.* In press

Smith, F. T. 1976d,e. Flow through constricted or dilated pipes and channels, I, II. *Q. J. Mech. Appl. Math.* 29:343–64; 365–76

Smith, F. T. 1976f. Pipeflows distorted by non-symmetric indentation or branching. *Mathematika.* In press

Sobey, I. J. 1976. Inviscid secondary flow in a tube of slowly varying ellipticity. *J. Fluid Mech.* 73:621–39

Spann, R. W., Hyatt, R. E. 1971. Factors affecting upper airway resistance in conscious man. *J. Appl. Physiol.* 31:708–12

Stanescu, D. C., Pattijn, J., Clément, J., van de Woestijne, K. P. 1972. Glottis opening and airway resistance. *J. Appl. Physiol.* 32:460–66

Stehbens, W. E. 1959. Turbulence of blood flow. *Q. J. Exp. Physiol.* 44:110–15

Swift, D., Proctor, D. F. 1977. Access of air to the Respiratory Tract. See Brain et al 1976, Chap. 3

Taylor, G. I. 1953. Dispersion of soluble matter in solvent flowing slowly through a tube. *Proc. R. Soc. London Ser. A* 219:186–203

Taylor, G. I. 1954. The dispersion of matter in turbulent flow through a pipe. *Proc. R. Soc. London Ser. A* 223:446–68

Vincent, M. J., Knudson, R., Leith, D. E., Macklem, P. T., Mead, J. 1970. Factors influencing pulmonary resistance. *J. Appl. Physiol.* 29:236–43

Weibel, E. R. 1963. *Morphometry of the Human Lung.* Berlin: Springer

Weibel, E. R., Gil, J. 1977. See West 1977, Chap. 7

West, J. B. 1974. *Respiratory Physiology— The Essentials.* Oxford: Blackwell

West, J. B., ed. 1977. *Bio-engineering Aspects*

of *Lung Biology.* New York: Dekker

West, J. B., Hugh-Jones, P. 1959. Patterns of gas flow, in the upper bronchial tree. *J. Appl. Physiol.* 14: 753–59

West, J. B., Matthews, F. 1972. Stresses, strains, and surface pressures in the lung caused by its weight. *J. Appl. Physiol.* 32: 332–45

Widdicombe, J. G., ed. 1974. *Respiratory Physiology. MTP International Review of Science, Physiology,* Ser. 1, Vol. 2. London: Butterworth

Wilson, T. A., Lin, K.-H. 1970. Convection and diffusion in the airways and the design

of the bronchial tree. In *Airway Dynamics,* ed. A. Bouhuys, pp. 5–19. Springfield, Mass: Thomas

Wood, L. D. H., Engel, L., Griffin, P., Despas, P., Macklem, P. T. 1976. The effect of gas physical properties and flow on lower pulmonary resistance. *J. Appl. Physiol.* In press

Zeller, H., Talukder, N., Lorenz, J. 1970. Model studies of pulsating flow in arterial branches and wave propagation in blood vessels. In *Fluid Dynamics of Blood Circulation and Respiratory Flow. AGARD Conf. Proc. No. 65*

Ann. Rev. Fluid Mech. 1977. 9:275-96

FLOW AND TRANSPORT IN PLANTS

✷8104

M. J. Canny

Department of Botany, Monash University, Clayton, Victoria 3168, Australia

INTRODUCTION

This review is not a discussion in the language of fluid mechanics about esoteric properties of flow in plants, but an attempt by a botanist to explain, with as little technical botanical language as possible, how the life and construction of plants raise problems of flow, some of which are still unresolved. In order to understand these problems it will be necessary to become acquainted with some of the details of plant construction (morphology, anatomy), and some of the conditions of plant life (physiology). At several places in the explanation I have felt that the special knowledge of readers of this Review might illuminate matters that are still obscure to botanists. The symbol (†) indicates such calls for help. References are scarcely quoted since they are written in a technical jargon that would be barely understood. Rather the names are given of a few works that may serve as entry points into the literature for any reader who wishes to follow particular lines of thought, and these are grouped in a Bibliography with an indication of their scope.

Calculations by plant physiologists about the flows to be discussed are always done in terms of Poiseuille's equation. We are not sensitive to the limitations of this approach or aware of other relations that might be more appropriate in the fluid spaces of the plant (†). Technical botanical terms are italicized at their first appearance.

TREE LIFE AND ARCHITECTURE

Let the reader concentrate his attention in fact or in fancy on a sunlit summer tree, on its outward vesture of leaves, spread, with almost no mutual shading, in a field of radiant energy, waved by the wind in a dry air, and supported by a more dimly perceived architecture of anastomosing stalks, twigs, branches, and trunks, which disappear below ground as a simple pillar. The life of the tree is concentrated in these leaves and in them are generated the forces that produce the flow processes. What the leaves are doing is making sugar from the carbon dioxide (0.03%) in the air using a small fraction of the radiant energy in the sunlight. The carbon dioxide gets into the reaction centers (*chloroplasts*) of the leaf cells in three stages: first by gas-phase diffusion through holes in the leaf surface and the open-textured air spaces within the leaf; second by solution in water at the outside of the leaf cells, and third by liquid-phase diffusion from there to the chloroplasts (Figure 1). Simultaneously, water evaporating from the wet cell surfaces inside the leaf is diffusing in the opposite direction to the dry air outside. The leaves are able to control the diffusive exchange by opening and closing the holes (*stomata*). In dry conditions or in the dark, the holes will be shut and neither CO_2 uptake nor water loss takes place. These two processes in the cells of the leaf, the loss of water and the formation of sugar, lead to the two major transport processes of the tree: movement of water up the tree to replace what is lost by evaporation, and movement of sugar away from the leaves to be converted into the body materials and to form the extensions of the body by new leaves and branches, thicker trunks, and extending roots. The water movement is called *transpiration*; the sugar movement is called *translocation*. The water and sugar move in opposite directions in separate specialized transport tissues, the *xylem* and *phloem* tissues respectively. The two tissues are always found together, from the finest veins of the leaf through midribs, leaf stalk, twigs, branches, and trunk and on into the branching roots below ground, forming a continuous plumbing system that connects all parts of the tree body to the two processes going on in all the leaves.

In the twigs, branches, trunk, and main and minor roots, the central core woody

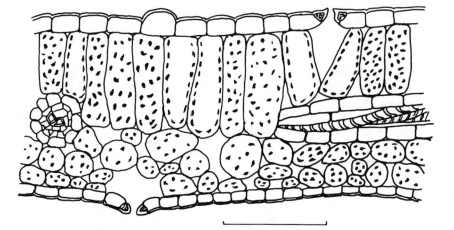

Figure 1   Section through a leaf such as a potato. The sugar-synthesizing organelles, the chloroplasts, are shown as black grains. Two of the stomatal pores are shown in the epidermis, connecting with air spaces between the leaf cells. By this path CO_2 diffuses in to the chloroplasts, and water vapor diffuses out from the wet cell walls. The stomata open and close, controlling these diffusions, especially in response to the water content of the leaf. The conducting veins are shown at left and right in transverse and longitudinal section. The tracheary elements are shown with banded thickenings. The phloem sieve elements in veins as small as this are very small and inconspicuous but are found below the tracheary elements. Scale 100 μm.

substance is xylem tissue—mostly old and no longer functional—the familiar "wood" of everyday experience. The phloem is in the inner layers of the bark. Where the bark peels away from the wood is located the *cambium,* a weak zone of immature and unspecialized cells that alone retain the capacity to divide. This sheathing cambium tissue is the universal source of new xylem and phloem tissues in all limbs of the plant body. It forms the xylem within its own circumference by the specialization of the daughter products of its cell divisions, producing an accreting cylinder of successive shells that are the familiar annual rings of wood. It forms the phloem outside its circumference by the specialization of its daughter cells on that side. However, these outside cells are short-lived: since the diameter of the cylinder is constantly increasing, they are soon crushed from within and flake off with the bark. The transpiration and translocation transports are carried out by narrow rings of freshly formed xylem and phloem within a few millimeters of the cambium. Where the branches of this plumbing system become very fine—in the veins of the leaves and at the tips of the fine roots—the cambium is no longer present, and a few cells of each tissue compose the network of vascularization that extends to serve all the nontransporting cells of the body. No cell of the leaf, for example, is more than, say, 100 μm away from a branch of this transport network, a distance over which diffusive equilibration in water takes only a few milliseconds.

THE TWO PHASES—APOPLAST AND SYMPLAST

All plant cells, the living protoplasmic parts that carry on the metabolic processes of life chemistry, surround themselves with a matrix of nonliving material, the *cell wall,* which forms a bag in which they always live (see Figure 2). This cell wall is made of a meshwork of open-textured, negatively charged cellulose microfibrils with a high water content. It may be strengthened and made less permeable in some mature cells by impregnation with a condensed phenolic polymer called lignin. It is lignin that gives its strength to wood. Where two cells are in contact, therefore, it is their walls that are joined. The protoplasts remain remote from each other, separated by the thickness of the two walls. Thin walls may be only 0.5 μm; thick walls, several μm thick. However, the protoplasts maintain connection with each other through the dead space of the two walls by means of fine threads of protoplasm called *plasmodesmata.* There are many hundreds of plasmodesmata per wall, and each is about 50 nm in diameter. Thus the whole body of the plant consists of two spatial continua: the continuum of wall material formed by the contacts of all the cell walls with parts of the walls of their neighbors; and the contained continuum of all the protoplasts connected by plasmodesmata (refer again to Figure 2). The first is known as the *apoplast* (or sometimes, *free space*); the second, as the *symplast.* The apoplast, which is nonliving and wet, permeates the thickness of the cell walls and is interspersed freely with air spaces where the cell wall contacts are dissolved; the symplast is living, metabolically vigorous, and everywhere contained within the apoplast. The interface between the two is the outer boundary of the protoplasm, a highly specialized membrane called the *cell membrane* or *plasmalemma,* across which all the important exchanges of substance must be made between the cells and their environment. Because it is the boundary that keeps on the inside all those things necessary for life, and on the outside all the things that are liable to upset it, it is a boundary of the utmost importance. It does not, however, figure largely in the present discussion in its role as an exchange surface, but rather as the fairly impermeable boundary between the two space continua that keeps the substances in the apoplast apart from those in the symplast. Water exchanges quite readily across the cell membrane, but many solutes are strictly controlled by it. Inside the protoplast of plant cells there is a further nonliving water space, called the *vacuole,* which occupies a very large part of the cell volume and reduces the protoplasm to a thin layer lining the wall. A solution of metabolic products whose amounts differ widely in different cells, the vacuole is separated from the protoplast by another membrane, different in many ways from the plasmalemma, known as the *tonoplast.* The solutes of the vacuole, separated from the apoplast by the two membranes and

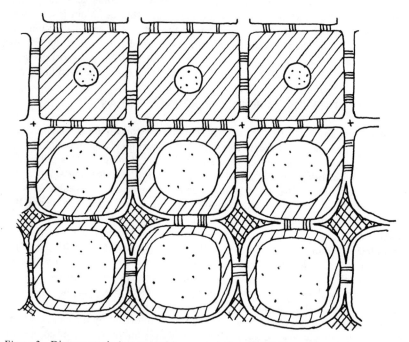

Figure 2   Diagrammatic (not to scale) representation of the tissue spaces in a plant. The cell walls (apoplast) are left white. The living protoplasm is shaded diagonally and connected by the grouped plasmodesmatal threads through the walls to form the symplasm continuum. Within each cell protoplast is a vacuole (dotted). A progression is indicated from young cells with small vacuoles at the top towards mature cells that are coming unstuck at their corners, forming another phase, the air-spaces (cross-hatched). Though the diagram suggests that the cells are isodiametric, many cells of the plant are hundreds of times longer than wide, and the reader may picture some of these as extending far at right angles to the page.

intervening protoplasm that together form a boundary impermeable to these solutes, constitute an osmotic system that has been much investigated by physiologists and physical chemists. Indeed Pfeffer used it as his measuring system in the investigations that led to the formulation of the ionic theory. Placed in water or some other environment of high water availability, the cell osmometer absorbs water and expands in volume against the confining bag of the cell wall, stretching it until the pressure exerted by the wall is equal and opposite to osmotic pressure of the cell contents. At this point the cell is in equilibrium with its water environment. Depending on its water content therefore, a plant cell's tendency to attract water varies from zero up to the maximum that is set by the osmotic potential of the vacuolar sap. The vacuoles of all cells are of course isolated from each other and form no continuum comparable to the apoplast and symplast. They are a third phase of space contained within the other two.

WATER TRANSPORT—TRANSPIRATION

Returning to our tree in the sun, its water loss will of course vary widely with many conditions, but an example may be instructive. A single maple tree 47 feet high, with 177,000 leaves (area 1/6 acre), was estimated to lose 58 gallons of water per hour on a summer afternoon. It will be noted that the evaporation of water from the wet cell walls within the leaf is a loss of water from the apoplast, and the transpiration stream of water to replace it from the soil up through the xylem tissue takes place within this phase continuum. The apoplast spaces of the leaf cell walls form a continuum with the specialized channels within the xylem leading all the way back to the fine branches of the roots. The xylem is a complex tissue of several cell types: most of them are living, but one set of them is dead, specialized out of life to form a part of the apoplast and constitute a low-resistance channel connecting the roots to the leaves. These cells are called *tracheary elements.* They alone of the cells of the xylem tissue are important to this treatment of transport.

Anatomy of the Xylem in Relation to Transpiration

Cambial cells, like all the other living cells, have an apoplastic wall containing the symplastic protoplasm. They are already elongated in the direction of the axis of the stem. Some of their progeny on the inside differentiate to become tracheary elements. The first stage of this change is an elaboration of the cell wall, both structurally and chemically: the protoplast lays down wall material in bands, rings, helices, or networks that greatly strengthen the wall to withstand the pressure difference of many atmospheres that will develop between outside and in. In the later stages, the impermeable lignin polymer is incorporated into parts of the wall, and other parts are eroded away. The result is a wide tubular cell with elaborately sculptured walls (Figure 3). The second important change is a refinement of the end walls where the cell is in contact with those next to it up and down the stem. If the reader's tree is a broad-leaved tree, on some cells these walls are perforated right through to give a continuous pipeline with a radius of about 150 μm and up to several meters in length, called a *vessel.* To convince ourselves of this, we can blow bubbles through an oak broomstick placed in a bucket of water. If the tree is a conifer, the conducting cells remain quite narrow and the end walls between successive tracheary elements are not perforated, but refined away to a thin membrane. These are called *tracheids* and have a radius of 20 μm and crosswalls every 1 mm. You cannot blow through even a short stick of pine. The third stage of changes involves the protoplast programming itself out of existence, destroying and dispersing the structures of organelles and membranes that have made it alive, and leaving the cell wall filled with water or a weak solution. By this last act, the whole interior space of the tracheary element is made part of the apoplast, joining its cell wall and those of other cells continuous with it in this phase, and its symplasm vanishes. The continuous water phase produced in this way in the tracheary elements pervades the whole xylem space of the tree from roots to leaf veins and constitutes the path of the transpiration stream. It will be apparent in a general way that the evaporative loss of water from the wet cell walls within the leaf can lead to a flow of water through the tracheary elements and a general transport of water from roots to leaves. This explanation of transpiration, first elaborated before the turn of the century by Dixon, is still called by the name he

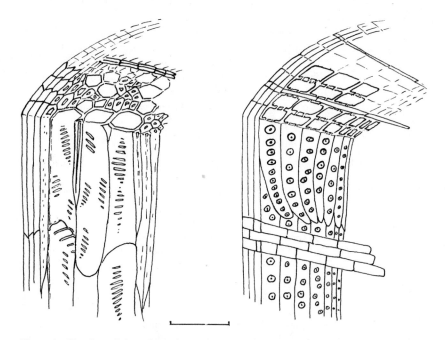

Figure 3   Sketches of the xylem of trees. The unspecialized arc of cambium is shown as the three-cell-layered periphery of the cylinder of xylem which it has formed. (Left) The xylem of a broad-leaved tree with vessels in a matrix of fibers. (Right) The xylem of a conifer composed mostly of tracheids. Lateral connections between both kinds of tracheary elements are by *pits* of characteristic shape. Scale 100 μm.

gave it, "The Cohesion Theory," and is generally accepted by plant physiologists today. Nevertheless, serious doubts about its validity have been raised from time to time, and many of its implications are still without experimental verification.

Implications of the Cohesion Theory

The forces impelling water about the plant body have been defined by plant physiologists in various ways, beginning with ideas of pressure arising from the study of cell osmotic pressures. For many years the reigning concept was of a negative pressure envisaged as the tendency for pure water at STP to move into the space in question. Its units were atmospheres, and it was called in Germany "Saugkraft," in England, "Suction Pressure,' and in the US, "Diffusion Pressure Deficit." By its relationship to vapor pressure, osmotic pressure, and the inward pressure of stretched cell walls, it was able to describe satisfactorily the forces and gradients. In recent years this definition and terminology have given way to the concept of "Water Potential" (ψ), defined as $\psi = (\mu_w - \mu_w^0)/\overline{V}_w$, where μ_w is the chemical potential of water in the system, μ_w^0 is that of pure free water at the same temperature, and $\overline{V}_w$ is the partial molal volume of water [see Slatyer (1967)]. It is seldom expressed in its proper units of work or energy, but more commonly in the old units of atmospheres, or (to be modern) bars. It is zero in pure water at STP, and is made more negative by the presence of solutes and the curvature of water-air interfaces in narrow spaces. Hydrostatic pressure makes it more positive, tension, more negative; and there is a small effect of temperature.

   In these terms, the tree forms a bridge between the high water potential in moist soil (near to zero) and the much lower water potential in partially saturated air (-300 atm at 80% relative humidity, RH). Within the leaf where the RH might be as high as 95%, the ψ value would still be -68 atm and causes water to evaporate from the wet cell walls into the air spaces. It is supposed that the water surface of the partially dried wall retreats into the narrow (10 nm) spaces within the microfibrillar structure of the wall. A concave water surface curved into such a cavity has a minimum ψ of about -300 atm (see Table 1). This low ψ is now connected to the whole continuum of the apoplast (and also of course to the two pervaded phases, the symplast and vacuoles, which will adjust their water contents to equilibrium with the local value in the apoplast), and water will move along this

Table 1   The energy of curved water surfaces related to vapor pressure and capillary rise

Radius of curvature (m)	Relative vapor pressure	Height of capillary rise (m)	Equivalent tension in water (bars)
10^{-3}	1.000	1.5×10^{-2}	0.0015
10^{-5}	0.999	1.5	0.15
10^{-7}	0.989	1.5×10^2	1.5×10
10^{-9}	0.3305	1.5×10^4	1.5×10^3

gradient of ψ according to the resistances offered by the various paths. The Cohesion Theory postulates that the path of lowest resistance is the continuum of water within the tracheary elements, and that most of the transpiration stream will travel in the narrow tubes. Where interconnected water-filled spaces are interrupted by lignified cell walls of lower water permeability, air spaces, etc, the water will flow around them in other paths of lower resistance, probably within permeable cell walls.

It will be seen that the theory requires the following:

1. There should be a substantial gradient of ψ from tree base to top, 0.1 atm per meter to balance gravity, plus a further gradient to cause flow through the tracheary elements and other paths. The size of this gradient will depend very much on the values assumed for the velocity, tracheary element radii, and proportion of functional tracheary elements in the wood. Recent estimates place it around 0.8–2 atm m^{-1}.

2. In a tall tree (100 m) a gradient of ψ at the lower value will generate a tension of 80 atm in the column of water in the tracheary elements. It is the crucial assumption that the water columns will not break under such a stress that led Dixon to coin the name "cohesion" for his theory. Once such a column has broken, it is not easy to see how it can be mended again.

3. There must be sufficient unbroken columns in the xylem to permit the flux of enough water to supply that lost by the leaves.

4. The xylem should contain water flowing at the speeds necessary to supply the flux to the leaves. Clearly this interacts with the last requirement.

5. The leaves should be found to attain ψ values of up to -80 atm at the tops of tall trees when transpiration is active. Of course if water loss is limited by the shutting of the stomata, the ψ values will be much more positive than this.

6. The above five positions will be related together through a value of "conductivity" σ for xylem, which presumably can be measured in isolated pieces of tissue from estimates of the speeds of flow produced by various pressure gradients. This value must be consistent with measured values in real trees of $d\psi/dl$, the amount Q of water transpired by the leaves, the cross-sectional area α of water-filled tracheary elements, the speed v of sap flow, and the gravitational component ρg, where $d\psi/dl = v/\sigma + \rho g$ (where $\rho g = 0.1$ atm m^{-1}) and $\sigma = (Q/\alpha)/(d\psi/dl)$ (see Heine 1971).

Uncertainties of the Cohesion Theory

The experimental facts are probably consistent with these six propositions, and most plant physiologists accept the Theory. However, when one looks in detail for unequivocal evidence of the facts and measured values to put in the equations, a good many uncertainties appear. These stem primarily from the difficulties of measuring the parameters of the equations and from controversy over the reliability of such measurements as there are.

1. THE GRADIENT OF TENSION IN THE XYLEM   Until recently there was no satisfactory way of measuring the postulated tension of many atmospheres in the xylem sap. Attempts to insert manometric probes into the xylem are upset by air from inter-

cellular spaces which is drawn into regions under tension (†). Tree circumference can be shown to contract and swell in response to changes of evaporative stress (even with the passing of a cloud over the sun) in a way that suggests strong tensions within, but this cannot be related to useful units. Since 1965 increasing use has been made of a pressure-bomb technique. Twigs or leaves are detached from the tree, releasing the tension on the water columns, and the water retreats from the cut surface into the remote parts of the xylem as the stressed tissues relax. The piece is sealed in a pressure vessel with the cut end protruding, and the pressure in the vessel increased until sap can be seen returning to the tracheary elements at the cut surface. The tissues are now supposed to be stressed to the same degree as when the part was on the tree, and the pressure in the vessel gives an estimate of the former tension in that part of the xylem. The method, which has been carefully investigated and extended in scope, has been found to be reliable (Tyree & Hammel 1972) and has produced measures of ψ at different heights on the tree that are of the size required (Hellkvist, Richards & Jarvis 1974), though there are some disturbing exceptions.

2. THE TENSILE STRENGTH OF WATER (†) Near my house is a public park tended daily by a gardener. I had paused there to talk as he watched his hoses and discuss the stringencies of the season. "You know," he said, "on a hot day, the water in those trees *boils.*" Ever since the origin of the Cohesion Theory this has been the problem. If the tension in the water is 100 atm, why indeed does it not boil (or cavitate, as my engineering friends say)? If it does, what becomes of the low-resistance pathway in the tracheary elements?

Though the theoretical strength of water is very large (>1000 atm), attempts to measure it experimentally have always yielded low values (0.5 to 50 atm). Plant physiologists fall back on the explanations that the water in tracheary elements is especially clean (free of nucleation centers) since it has been filtered through the cell walls, or that the walls are particularly wettable, having never been exposed to the impurities in the air, so that here water may attain nearly its theoretical strength. A version of this argument has been put forward by Oertli (1971), who says that the tracheary sap is indeed superheated and supersaturated with air, but does not cavitate because the activation energy required to form the first small bubble is so large.

Against this, Milburn has attached a sensitive sound amplifier to plants and recorded clicks which he believes are the sounds of the cavitations occurring (Milburn 1973). The clicks are more frequent when transpiration is high, they cease after a total number has been recorded, and the capacity to produce clicks once more can be restored by vacuum infiltration.

3. THE CONTENTS OF TRACHEARY ELEMENTS (†) In the low-resistance wood of a broad-leaved tree with vessels one may calculate that an annular zone of xylem not more than a centimeter or so wide just under the bark is sufficient to supply water at the rate it is transpired by the leaves, provided most of its tracheary elements contain water. The central core of old wood is probably unnecessary and unused for water transport, and the tracheary elements there are filled with air. Wood,

even fresh wood, floats. Attempts to show that a majority of tracheary elements in the outer zone of the wood are in fact filled with sap have not been conspicuously successful. The most determined of these, by Preston, involved placing a small reservoir of india ink against the exposed wood, puncturing vessels below the ink surface with a fine knife, and filming the very rapid entry of the ink into the vessel as the tension was released (Preston 1952). The steady movement of the ink into the moving column in a water-filled vessel could mostly be distinguished by this means from the injection of an air space under reduced pressure, which leads to a hyperbolic fall of velocity with time during the invasion. Preston found to his surprise that injections of the former kind were rare, and those of the latter, much the more common. Even in the xylem nearest to the cambium it seems that continuous water columns are few. Here is another point of weakness of the Cohesion Theory, another parameter of the relational equations that has an uncertain value. In the higher-resistance conifer wood the larger zone known to foresters as sapwood is probably necessary for and involved in the flow.

For many years it was held that another kind of experiment was evidence against the Cohesion Theory; this was known as the "double saw-cut" experiment. Two saw cuts made into a tree from opposite sides, with each extending more than half way, will break the continuity of all water columns in the lumina of tracheary elements, and it might be expected that the water supply to the leaves would be interrupted and the leaves would wilt. The leaves do not wilt. As long as precautions are taken to keep the weakened trunk from breaking, a tree mutilated like this may remain healthy for many months. It is now known that this is not contradictory to the Theory: the apoplastic pathway in the wet cell walls is still continuous around the cuts, and tracer molecules will flow through it in a zig-zag path and re-enter the tracheary elements above the cuts. All this operation does is introduce a section of higher-resistance channel into the normal pathway. As always, water follows the path of least resistance, which, at the level of the cuts, is in the cell walls.

Other means of detecting flowing water in the wood, by magnetic or thermal disturbances, are perhaps better discussed in connection with the velocity of flow.

It is generally agreed that once the watery contents of a tracheary element, particularly a vessel, are broken and the space interrupted with vapor or air that will expand greatly under the tension, then the reestablishment of a water phase in this space will be difficult, perhaps impossible even under changed conditions. Not only high tensions under evaporative stress are a danger here, but also cold severe enough to freeze the sap will bring air out of solution, which may remain when the sap thaws. There is a force, so far unmentioned as a driver of the transpiration stream, which may be important in repairing such damage. This force is known as *root pressure*. A number of plants at certain times show evidence of a positive rather than a negative pressure in their xylem sap: grape vines in spring bleed profusely from a cut stump; the sugar maple at the same season yields many gallons of a weak sugar solution; and many plants in damp greenhouses exude drops of water from their leaves (maize seedlings, nasturtiums, tomatoes). These positive pressures have been measured to reach several atmospheres, but are not thought to be generally interpreted as a means of supplying the leaves with water. The magnitude of the pressure is too small, and its duration ephemeral; when the leaf most needs the water, the pressure is replaced by a tension. However, these root pressures could be of the utmost value in refilling tracheary elements that have lost their water continuum under some previous stress, and may explain how the clicks observed by Milburn can indeed represent breaking columns that are in fact repairable.

4. THE VELOCITY OF THE TRANSPIRATION STREAM (†) For over a century the ascending sap stream in the xylem has been studied by standing cut plant shoots in solutions of dye, and following the progress of color up the tracheary elements. From such studies it is clear that solutions can travel there at many centimeters an hour. The more relevant question of how fast the stream moves in an uncut plant whose apoplastic continuum has not been disturbed is more difficult to answer. The common method is to bare the outer wood and heat it locally, and monitor the changing temperature above and below the heated patch. The analysis of the resulting temperature-time curves into contributions from conduction through the wet cell walls and convection in a flowing stream in an unknown proportion of the tracheary elements is fraught with the usual uncertainties (Heine & Farr 1973). Some support for the values comes from the movement of injected radio tracer. These estimates are the best we have, and must be used in the relational equations. Maximum velocities in coniferous tracheids are in the range 20–40 cm hr^{-1}, while in the vessels of broad-leaved trees, much higher speeds have been recorded, and 5 m hr^{-1} is quite common. The highest is 44 m hr^{-1} in an oak.

Attempts to measure the passing flow by the disturbance it causes in a magnetic field have not yet reached the level of a practical field method.

5. THE WATER POTENTIALS OF LEAVES From what has been said so far it might be expected that the water potentials of leaves at the top of the transpiring tree might be considerably more negative than those at the bottom, because they have to effect the transfer of water through a greater length of xylem and a larger gravitational component. This is not usually observed. The leaf water potential varies in the range -5 to -25 atm with the weather conditions (irradiance, RH, soil moisture, etc) rather than with position on the tree or size of the plant. This is seen not as a contradiction of the Theory, but as evidence that the growth and ramification of the xylem, as the plant body extends further from the soil, add more conducting tissues to the pathway and keep a constant ratio between the leaf area that is losing water and the area of conducting wood that is supplying it. The large negative values of leaf water potential that have been measured in many leaves are one of the surest foundations on which the general acceptance of the Cohesion Theory rests.

6. THE CONSISTENCY OF MEASURED VALUES IN THE RELATIONAL EQUATIONS It should be plain that the estimation of each of the necessary parameters is a business of considerable experimental difficulty and theoretical uncertainty, and that the values obtained are likely to vary over a wide range. Further, to measure them all for one tree under specified conditions is a marathon task. Exact confirmation is not to be expected. Broad general consistency has certainly been achieved, but is

not wonderfully reassuring when a tenfold uncertainty in one parameter may be compensated by another tenfold uncertainty in the other direction in a different one. One plant physiologist tried to resolve some of these uncertainties by building an artificial tree in the stairwell from glass capillaries sealed into plaster of paris and incorporating manometers at critical places in the path (Copeland 1902). He concluded that he could not understand the movement of water in this system either, and that the physics of the behavior of water and air bubbles in narrow tubes was quite unclear (†).

Values for the water conductivity of various woods under hydrostatic pressures have been measured under laboratory conditions for isolated pieces and provide values that are consistent with the estimates of velocities, fluxes, and gradients of water potential (Heine 1971). No field method or method that works for intact plants is known (†).

Conclusion on Water Transport

There is general agreement that the main features of the transpiration transport are understood. The water flows in nonliving apoplast down a gradient of water potential from the soil to the atmosphere along the path of least resistance, which is a combination of wet cell walls and water-filled tracheary elements, interspersed with air cavities both between cell walls, and (where the water continuum has been broken by stress) within the tracheary elements. The magnitude of the gradient is 0.5 to 2 atm m^{-1}, and produces flows of up to several meters per hour. The water continuum is strong enough to resist breakage for much of the time, but if it is broken the resistance to flow is increased. The broken column may in some conditions be restored by root pressure, or it may join the permanently air-filled spaces of the inner and inactive wood.

SUGAR TRANSPORT—TRANSLOCATION

The sugar made by the leaves in sunlight has to be distributed back to all the other parts of the tree, where it is turned into the units of tree structure: the cambium that divides to make more wood and bark, the buds that become new leaves and flowers, the root tips that explore the soil for more moisture. How this is done is still not known. There is no general agreement among those who study this transport on such basic matters as whether or not it is an advective process; what forces are applied to move the sugar or even whether they are of physical, chemical, or biochemical origin; what determines the direction or rate of the transport; or how this rate may be satisfactorily measured. This confusion has persisted for so long and become so much a part of the tradition of plant physiology that even if a correct explanation should appear it seems likely that its acceptance would spread extremely slowly. The impartial fluid mechanician who meets this problem in these pages must appreciate that the view presented is a personal one.

Each of the leaves on the tree has an output of sugar roughly equal to its own oven-dry weight every two or three days. This is collected by the phloem (see section on tree life and architecture) of the leaf veins and carried down the phloem of the leaf stalk to join the branch and trunk phloem that sheathes the whole structure. The rates of mass transfer that phloem can achieve have been measured many times, and whether one studies export from a leaf or import to a fruit or the passage along a trunk, the values are essentially the same. A square centimeter of phloem tissue is able to transport about 5 gm of dry weight through itself every hour. The chemical form of the material moving is in most plants predominantly the simple disaccharide sucrose, and the only internal parameter of the plant that has been shown to be related to the rate of movement is the concentration of sucrose in the phloem. The rate of mass transfer is proportional to the gradient of sucrose concentration in the phloem, and the constant of proportionality (analogous to a diffusion coefficient) can be as high as 0.07 cm^2 sec^{-1}. This implies that high rates of transfer are possibly only over quite short distances, and that over distances the size of tree trunks transfers would be very slow. There is surprisingly little investigation of this limitation.

Unlike the transpiration flow, translocation requires the continuity of living cells all along the path. Local killing or poisoning of the phloem blocks transport through the treated zone. In contrast, the transpiration stream is unaffected. An instructive demonstration of this is to kill a section of a leaf stalk with a match flame. If the dead region is supported mechanically by a wrapping of aluminium foil the leaf blade remains unwilted for a long time. Sugar can still be manufactured in the leaf but it is not exported, and accumulates, usually as starch.

The translocation is also especially sensitive to mechanical injury of the phloem, and cannot be shown to go on in isolated pieces of the tissue. When a leaf or twig is detached from a plant the capacity to translocate is lost in the phloem for a considerable distance from the cut. This dependence of translocation on living phloem is taken to mean that the movement goes on in the symplasm continuum; separate therefore from transpiration.

Though sugar transport is the major movement and the most studied, all kinds of organic substance are transported in the phloem, from amino acids and hormones to virus particles and foreign substances like weed killers. Any substance that can be got into the phloem without upsetting it will probably be transported by it away from the site of loading, down its gradient of concentration.

Anatomy of the Phloem in Relation to Translocation

The tissues made by specialization of cells from the cambium outside its circumference make up the sugar-transporting tissue, the phloem. Its cells are of two kinds, little-specialized, metabolically active *phloem parenchyma* cells whose walls remain thin, and the *sieve elements*. These cells are found only in the phloem, and since their discovery in 1837 have been regarded as especially concerned with sugar transport. They are quite unlike any other cells in the plant body in that (a) the cross wall separating two consecutive members of a file of sieve elements is a perforated plate known as a *sieve plate*; (b) they contain a solution of sugar (commonly sucrose) at a very high concentration (up to 30%); and (c) their internal structure of organelles is greatly degenerated (Figure 4). A file of sieve elements, each connected to the next through a sieve plate, is called a *sieve tube*. The characteristic features of sieve elements must be explained in more detail, since on them hinge all the arguments about the transporting capabilities of the tissue.

SIEVE PLATES   About half the area of a sieve plate is made up of the *sieve pores,* which are much larger than the normal connections through cell walls, the plasmodesmata (50 nm). In some plants, like pumpkins and ash trees, they can be enormous (5-µm radius); in most broad-leaved plants they have a radius of 0.8 µm; and in the conifers they can be very small (0.08–0.4-µm radius). Quite what it is that is connected from one sieve element to the next through the sieve pores is not certain (see below), but the widely accepted view at present is that there is a

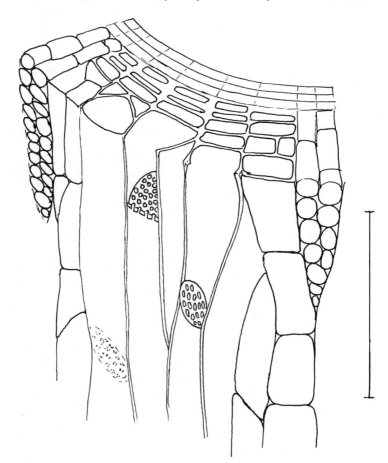

Figure 4   Sketch of the phloem of a tree outside the arc of cambium that has produced it. Three sieve tubes are embedded in a matrix of parenchyma cells. The cells that have just differentiated from the cambium are fibers, which are commonly interspersed in bands in the phloem. Scale 100 µm.

continuity of the sugar-rich sap together with microfilaments (5–10-nm radius) of a protein known as P-protein. The degree to which the pores are found to be filled with the P-protein filaments is much debated, and depends on how the cells have been prepared for microscopy.

Lining the pores, and in old sieve tubes extending to cover and block them, is a special carbohydrate polymer called *callose*. Many believe that it forms rapidly in response to damage to sieve tubes, and functions as a plugging agent to stop excessive sugar loss. When present it reduces the size of the pores still further. Most sieve plates are quite thick compared with other cell walls; a pore length of 5 μm is commonly assumed. There are about 20–50 plates per cm of sieve tube, and its diameter is in the range 10–25 μm.

(†) The presence of the sieve plates in the sieve tubes is one of the great puzzles of translocation. They are a universal feature; they are found in the early fossil land plants from the Devonian and yet their presence in a transport system, where it seems they must function as obstructions, is very difficult to explain. Only one of the several proposed mechanisms (see section on electro-osmotic mass flow), has been able to find a positive use for them.

THE SAP OF SIEVE ELEMENTS   What comes out of a punctured sieve tube is a solution of sucrose at a higher concentration than is known elsewhere in the plant. Many other minor organic constituents are present, and there is much potassium. The highest recorded concentrations of sugar in the sap are around 30% wt/vol, and it is seldom less than about 10%. Now a 30% solution of sucrose is about 1 M, and exerts an osmotic potential of about -30 atm. Since none of the water in the neighboring cells or apoplast is at such a low potential, the internal hydrostatic pressure generated by swelling against the elastic cell walls is very large. This pressure is released when a sieve element is cut or punctured and causes a surge of the contents that greatly disorganizes the structural components. From this fact stem both the difficulties of studying their internal structure and the sensitivity of the translocation function to cuts in the phloem. Clusters of particles on one side of a sieve plate or aggregates of P-protein plugging the pores are seen as evidence that the sieve tube has suffered damage of this kind. The simplest view of translocation (see sections on mass flow) is that this sap constitutes the moving stream in which sugar and everything else is carried along; more elaborate views see it as very closely related to the movement, but not itself the moving vehicle.

INTERNAL STRUCTURE OF SIEVE ELEMENTS   The sensitivity of both the structure and function of sieve elements to damage after the release of their high pressure was realized only slowly by the people who studied them. Early descriptions are mostly of states that would now be recognized as severely damaged. The observation of living and functioning sieve tubes with the light microscope is made very difficult by their sensitivity, since good images can be formed only of thin pieces of phloem, and the preparation of such pieces leads almost inevitably to damage. The more carefully the preparations are made the less there is to see in the way of cell organelles, particles, fibrils, etc. The functioning sieve element appears to be optically empty. Similar difficulties beset the preparation of sieve elements for examination in the electron microscope. Here the least-disturbed images show a great paucity of the normal organelles, no nucleus, no ribosomes, few and degenerate mitochondria, a few plastids, and little endoplasmic reticulum; and what organelles there are cling in a thin layer to the peripheral walls of the tube. The tonoplast membrane has long been thought to be absent, but better preparation methods suggest it may still be there. The central space of the cell contains wisps of P-protein. Presumably the sugar-rich sap fills this central space, bathing the P-protein, and connects through the sieve pores with the sap of the next element. Though reduced to a level of simplicity where their being alive can be questioned, they are indeed regarded as living because they do contain a great many active enzyme systems and maintain this level of structure for many weeks. Some of the synthetic functions may be taken on by the neighboring parenchyma cells, which are unusually rich in organelles, enzymes, and evidences of metabolic activity, and with which the sieve tubes connect sideways by plasmodesmata. The crucial questions raised by the possible mechanisms are the resistance to flow offered by the sieve pores (†), drag caused by dispersed P-protein (†), and the possibilities of active pumping or molecular acceleration offered by the vestigial protoplasmic structures.

The plasmalemma membrane still forms the outer boundary of the sieve element protoplasm, and by its semipermeability contains the sugar concentration and permits the formation of the high osmotic potential. It will be seen that the sieve tubes form part of the symplasm, consistent with the requirement for living phloem in the transport.

The Path of Translocation in the Phloem

It is generally assumed that the sieve tubes are the path of translocation, and all modern hypotheses start from this premise. The inferential evidence behind this assumption is plain from the exposition of the structure and contents of these cells. However, adequate proof that they are the sole or main path of translocation is lacking since they do not have a vigorous life separate from the accompanying parenchyma cells, nor can the functions of one set of cells be inactivated without jeopardy to the others. If we were sure that the transport is a flow phenomenon, then it would have to be in the sieve tubes since they are the only possible low-resistance path; but the major argument for the possibility of the flow is the presence of these tubes. If the tubes are not the path it cannot be flow; and if it is not flow the tubes may not be the principal path. Experimental attempts to localize the path by means of radioactive sugar are still ambiguous since one can detect only local accumulations of sugar; one cannot distinguish static from moving sugar, and the main accumulation is not necessarily in the direct path.

In order to compare mass transfers of dry weight one needs to assume a cross-sectional area of the path to place in the denominator. The values quoted above use the area of the whole phloem. If some part only of the tissue is the path (for example the sieve tubes) then the value should be increased in proportion. Now the proportion of sieve tubes in the phloem is quite variable. Percentages as low as 20 and as high as 70 are reported in different plants. Unfortunately this proportion is quite difficult and laborious to measure, and is not done as a routine in estimates of rate of translocation. With the double uncertainty of what this proportion should be, and whether it matters, I take the simplifying stand of using the whole phloem area as the denominator, including sieve tubes, phloem parenchyma cells, and their cell walls. There are no air spaces.

Proposed Mechanisms

MASS FLOW—PRESSURE DRIVEN   As already foreshadowed, this hypothesis proposes that the sieve tubes are pipes carrying a pressure-driven flow of sugar-rich sap. It explains the mass transfer values as the product of the speed of flow, the concentration of sugar, and the area of the tubes, translating 5 g hr^{-1} cm^{-2} in the phloem into, say, 7.5 g hr^{-1} cm^{-2} in the sieve tube component alone (66% of phloem area) and equating this with the flow of a 25% solution at 30 cm hr^{-1}, or a 10% solution at 75 cm hr^{-1}. The origin of the gradient of hydrostatic pressure to achieve this was originally conceived as osmotically generated by the high sugar contents of sieve elements in the leaves compared with lower sugar contents of sieve elements at the other end of the transport path. Many observations have cast doubt on the presence of differences in osmotic pressure of sufficient size, and many who believe in mass flow are still unprepared to be dogmatic about how the pressure is generated. Argument about this hypothesis centers on two questions: the first is what pressure gradient would be required and whether it exists; the second is whether experimental observations can be accommodated by a transport system which is of its nature one-directional.

Pressure gradients in sieve tubes   If the speed of flow is 100 cm hr^{-1}, the viscosity 1.5×10^{-2} poise, and the sieve-tube radius 12 μm, in the absence of sieve plates a pressure of 0.25 atm m^{-1} would be necessary to drive the flow. Adding 60 sieve plates per cm, 5 μm thick, half of whose area is pores of 2.5-μm radius (large), adds 0.32 atm m^{-1}; while for a radius of 0.8 μm (normal), an additional 3.1 atm m^{-1} is needed. In conifers with pore radii as small as 0.05 μm the figure rises to 800 atm m^{-1}. We physiologists assume, for want of better knowledge, that the same equations apply throughout this range of size (†).

All this happens if the pores are open. The presence of P-protein in the pores and lumen of the tubes clearly adds a further resistance whose size we only speculate about. The drag on viscous flow caused by a meshwork of fine filaments is thought to be "large," but precise estimates would be most welcome (†).

So in an average broad-leaved tree a pressure gradient of about 3 atm m^{-1} would be necessary to drive the mass flow. Grave doubts about its presence are felt by many, who also point out that there are certainly plants growing in which the pressures needed on this basis would be impossibly large. Many estimates of the changing concentration of sugar in the sieve-tube sap are available, and reveal gradients declining from leaves to base, but at a much lesser rate, typically around 0.2 atm m^{-1}. Attempts to measure hydrostatic pressure gradients in sieve tubes produced by some other means have shown only a general agreement with the osmotic values.

One-directional transport   The second controversial aspect of the mass-flow hypothesis is that of course in a flowing solution everything goes one way. However, it should be noted that in an organic body even as unspecialized as a plant's, materials are required to travel two ways. For example, a young leaf imports its substance, an old leaf exports its surplus, but even while so doing may require supplies of other nutrients from elsewhere. Of course the xylem transpiration stream is available as a reversed movement, but because it is in the apoplasm and the nonliving part of the body, it is not subject to fine control. In order to accommodate such requirements, protagonists of mass flow are driven to assume separate sieve tubes specialized for transport in opposite directions, and pressure gradients in opposite directions applied to each.

MASS FLOW—ELECTRO-OSMOTIC   An hypothesis of the mass-flow class that eliminates the first of these difficulties, the resistance of the sieve plates, places the driving force for the flow at the sieve plates. It suggests that each plate is negatively charged and that an emf is maintained across each, with all plates polarized in one direction in one sieve tube by the pumping of ions (possibly potassium) around a loop in the wall apoplast outside or through the next door cells. Then narrow pores are a positive advantage in that they generate quite large flows of solution towards the negative side of the plate. No increase of resistance occurs over long distances, as each plate propels the solution towards the next. This was a popular model for a while, but has now few adherents since it has been pointed out that such a system could not transport both anions and cations (which the plant appears to do), and that the circulating K^+ current would be too large for the cell membrane to accommodate.

MODELS BASED ON THE MOTION OF PROTOPLASM   A property of living protoplasm that has not so far been stressed is its power to generate motion, of which the swimming protozoon, the crawling amoeba, and the contracting muscle are familiar examples. Plant cells examined alive in the light microscope are quite commonly seen to have their protoplasm in motion, stirring around inside the cell wall, flowing to and fro in strands across the vacuole, carrying streams of visible organelles in channels that interlace like the traffic in some nightmare city. The forces that produce these motions (known as *protoplasmic streaming*) are believed to derive from the same two proteins, actin and myosin, that cause the contraction of muscle fibers. The motions, when corrected for the magnification of the microscope, are not very rapid, mostly around 5 cm hr^{-1}, with some as high as 30. Nevertheless, we do not really know the upper limit of the speed, since the motions are made visible by the particles carried in the stream, and the smaller the particles the faster they go. We cannot be sure that particles beyond the resolution of the light microscope ($<0.1\ \mu$m) are not moving much faster than the ones we can see. Many feel that this known capacity may form the basis of the translocation transport.

A major difficulty has been that, though the typical streaming motions can be plainly seen in the adjacent cells of the phloem, they are not present, or only doubtfully described in the sieve tubes. The more we learn about the structure of sieve tubes from either light or electron microscopy, the less likely it seems that the attenuated structures and activities there can generate motions that would need to be even more vigorous than the normal ones. For a while recently many felt that the P-protein filaments would prove to show the special properties demonstrable for actin, and manifest in the filaments of streaming cytoplasm and flowing slime moulds, but so far this hope has proved vain. The detailed cellular architecture of models of this type is not of interest in this account, but the general properties and limitations of such models need to be contrasted with those of the mass-flow class.

"Accelerated diffusion"   The relation of mass transfer to concentration gradient that was previously described and its analogy with diffusion was discovered in 1928 by a pair whose influence has dominated the field for fifty years, Mason & Maskell (1928a, 1928b). They were unwilling to crystallize this analogy in mechanical terms, held back by the doubts about protoplasmic streaming in the sieve tubes and the quantitative capacity of such mechanisms, and stressed only the revealed proportionality by using the name "accelerated diffusion." They had shown that the acceleration was between 10^4 and 10^5 compared with sugar in water. Many years ago I outlined a model of the streaming type which depended on the limited exchange of sugar between two sets of strands moving in opposite directions and the sieve-tube sap, which had all the necessary qualities for accelerated diffusion. It had even the right acceleration, and was able to account for the transfer of the necessary 5 g hr^{-1} cm$_{ph}^{-2}$, at least over short distances. I explain this not to defend this particular model, whose structural basis has been made doubtful by later microscopy, but to emphasize that models of this type can be shown to be quantitatively adequate.

Bi-directional movement   The important difference between such models and flow models is that they will, like diffusion, carry different substances at different rates in proportion to their gradients of concentration, and if the gradients are in opposite directions, then they carry substances simultaneously in opposite directions. As has been pointed out this is an important requirement in plant physiology, and a good deal of experimental evidence suggests strongly that it happens, and even perhaps that one sieve tube can carry two substances simultaneously in opposite directions. This is extremely difficult to prove unequivocally, but the experimental evidence has driven the mass-flow advocates to elaborate defensive explanations.

Speed of Translocation

The question of whether there is a speed of translocation in the same sense that there is a rate of mass transfer is extremely confused, both conceptually and experimentally. Models of the mass-flow type clearly have a definite speed component related to the maximum velocity of flow in the sieve tubes of different radius, and however unprecise this may be in experimental studies, it is directly related to the rate of transfer of mass. Models of the other type, however, like the process of molecular diffusion, have a speed component that is not necessarily related to mass transfer. For a fixed speed of protoplasmic streaming (or molecular motion) one may have large, small, zero, or negative transfers of mass, depending on the gradient of concentration.

Attempts to measure a speed component have shown that it is very variable. Almost the only practical measure is by the speed of spread of a radio tracer in the translocation system. Measured thus, the speed varies from centimeter to centimeter within the leaves and stems from as low as 2 cm hr^{-1} to as high as 200 or even 300. Most values fall in the range 20–100 cm hr^{-1}, and 100 is usually taken as the high average in speculative calculations. No studies have been made of the question raised in the last paragraph of whether high speeds are associated with high values of mass transfer. The tracer is not translocated as a step function, but attenuates by exchange and other processes into various shapes. These exchanges, within and between cells, and especially accumulations at the downstream end where labeled sugar is being made into other carbon compounds, confuse the interpretation, since a sizeable accumulation of static carbon easily obscures a fast but small component of mobile carbon.

Variation of Translocation with Distance

Both the pressure-driven mass-flow and the accelerated-diffusion types of model will be limited in their capacity for mass transfer over along distances. In both, the driving potential is a first differential with respect to distance, in the one case of pressure, and in the other of concentration. Only the electro-osmotic version is free from his limitation. Surprisingly little interest has been shown in this constraint, despite its importance for the life of a body the size of a tree. To put in some approximate figures, the high translocation rate of 5 g hr^{-1} cm$_{ph}^{-2}$ may be possible only over 30 cm or so; while for distances as great as 200 cm, the rate may be reduced to 0.2 g hr^{-1} cm$_{ph}^{-2}$. Since the relation is hyperbolic, very long distances are not greatly more limiting than moderate distances, and the slow rate is attained over both; it is only the very high rates that must be local. A study of the shapes of plants and of the positions of supplying leaves and consuming fruits and enlarging tubers suggests that the plant architecture may indeed be subject to a constraint of this kind. It is very difficult to find an arrangement of plant-body organs that suggests rapid transport over a distance more than about 30 cm. Rather, one finds chains of reservoirs of carbohydrate (starch) all along the path that can act as local sources of sugar over short distances as necessary. The tree especially is mostly composed of such reservoirs, and translocation in the reader's tree may not be at all in a stream moving at 100 cm h^{-1} from leaves to roots, but a slow filling and shufflling between massive reservoirs over quite short distances.

Conclusion on Sugar Transport

We have a system that we do not understand, and it is quite unlike other known transport systems, animate or inanimate (††).

SHORT-DISTANCE TRANSPORT

At the input and output of both transpiration and translocation systems there are many cell-to-cell transport processes whose properties are mainly conjectural, but which must involve flow processes over the scale of distances from 50 nm to 500 μm. The full exposition of these uncertainties would involve a further excursion into anatomical details, and there has been enough of that for a single review, but something may be based on what the reader already knows. If he will turn back to the diagram of a leaf (Figure 1) and the sketch of the tissue spaces (Figure 2) and picture the first amplified by the second, he will appreciate the possible pathways at this terminus of the long-distance systems. The water is arriving in the apoplastic spaces of the tracheary elements, and evaporating into air spaces from the apoplastic walls of the leaf cells. There is every reason to suppose that its movement between the two remains apoplastic in the wet cell walls, staying on the outside of the cell membranes, flowing round cells rather than through them, but of course equilibrating its water potential with them. Compared quantitatively with the opposite flux of sugar, the water flux is vastly greater. Of course both vary widely, but a rough average for each (in moles) shows a ratio in favor of the water of about 10,000 to 3. There seems good reason for keeping the fluxes separate, and the leaf micro-anatomy is such that this separation is probably achieved.

The sugar made in the chloroplasts of the leaf cells leaves the leaf within the symplasm of the phloem. There are three possible pathways between the chloroplasts and the phloem symplasm: to remain in the symplasm and travel within cells by diffusion helped by streaming from cell to cell via the plasmodesmata by either flow or diffusion or both; to cross the plasmalemma of the leaf cells into the apoplast and move there by flow or diffusion to the phloem and be pumped across a cell membrane into the symplast again; to travel mostly within cells, but at each junction of one cell with the next to cross the intervening space through the two cell membranes and the wall. In accord with the confusion in our understanding of long-distance sugar transport, you will find physiologists arguing for any one of these. I favor the first. The third involves sugar crossing many cell membranes, whose permeability to sugar is known to be exceedingly low; the second casts the sugar free of cellular control and makes it move in a space where 3000 times as much water is going the other way. The details of possible pressure flows and diffusion in tubes as small as the plasmodesmata raise questions outside the range of our training (†).

At the other terminus of the vascular system, the tips of roots, similar problems arise. Water enters from the soil and finds its way to the lumen of tracheary elements, and sugar arrives via the phloem and is unloaded to be processed by root cells and made into more root cells. The anatomy is quite different from that in the leaf, but space precludes a description of it. The patterns of movement are just as obscure. Though we concentrate our attention on these two termini, similar exchanges of water and sugar must occur into and out of the xylem and phloem all along the intervening path of twigs, branches, trunk, and roots for the maintenance of life there. Very little work has been done on these.

BIBLIOGRAPHY

For those interested in structure as a framework for the life processes generating flows, a textbook of general botany such as Holman & Robbins (1934) is useful; a more recent and pictorial work is O'Brien & McCully (1969).

Works on transpiration include the classics by Dixon (1914) and Maximov

(1929); the recent theoretical statements by Briggs (1967), Slatyer (1967), and Philip (1966); and a lucid balancing of the facts and problems by Preston (1952). Works on translocation range from the classic study revealing the diffusion analogy by Mason & Maskell (1928a, 1928b) to a very recent statement of a diversity of views including other versions of the models by Zimmerman & Milburn (1975); the opposing views are expressed by Crafts & Crisp (1971) (mass flow) and Canny (1973) (accelerated diffusion).

Literature Cited

Briggs, G. E. 1967. *Movement of Water in Plants.* Oxford, Engl: Blackwell. 142 pp.

Canny, M. J. 1973. *Phloem Translocation.* London: Cambridge Univ. Press. 301 pp.

Copeland, E. B. 1902. The rise of the transpiration stream: an historical and critical discussion. *Bot. Gaz. Chicago* 34:161–93

Crafts, A. S., Crisp, C. E. 1971. *Phloem Transport in Plants.* San Francisco: Freeman. 481 pp.

Dixon, H. H. 1914. *Transpiration and the Ascent of Sap in Plants.* London: Macmillan. 216 pp.

Heine, R. W. 1971. Hydraulic conductivity in trees. *J. Exp. Bot.* 22:503–11

Heine, R. W., Farr, D. J. 1973. Comparison of the heat-pulse and radio-isotope tracer methods for determining sap-flow velocity in stem segments of poplar. *J. Exp. Bot.* 24:649–54

Hellkvist, J., Richards, G. P., Jarvis, P. G. 1974. Vertical gradients of water potential and tissue water relations in Sitka spruce trees measured with the pressure chamber. *J. Appl. Ecol.* 11:637–67

Holman, R. M., Robbins, W. W. 1934. *Textbook of General Botany.* New York: Wiley. 626 pp. 3rd ed.

Mason, T. G., Maskell, E. J. 1928a. Studies on the transport of carbohydrates in the cotton plant. I. A study of diurnal variation in the carbohydrates of leaf, bark and wood, and of the effects of ringing. *Ann. Bot. London* 42:189–253

Mason, T. G., Maskell, E. J. 1928b. Studies

on the transport of carbohydrates in the cotton plant. II. The factors determining the rate and direction of movement of sugars. *Ann. Bot. London* 42:571–636

Maximov, N. A. 1929. *The Plant in Relation to Water,* transl. R. H. Yapp. London: Allen & Unwin. 451 pp.

Milburn, J. A. 1973. Cavitation in *Ricinus* by acoustic detection: induction in excised leaves by various factors. *Planta* 110:253–65

O'Brien, T. P., McCully, M. E. 1969. *Plant Structure and Development.* New York: Macmillan. 114 pp.

Oertli, J. J. 1971. The stability of water under tension in the xylem. *Z. Pflanzenphysiol.* 65:195–209

Philip, J. R. 1966. Plant water relations: some physical aspects. *Ann. Rev. Plant Physiol.* 17:245–68

Preston, R. D. 1952. Movement of water in higher plants. In *Deformation and Flow in Biological Systems,* ed. A. Frey-Wyssling, pp. 257–321. Amsterdam: North-Holland. 552 pp.

Slatyer, R. O. 1967. *Plant-Water Relationships.* London: Academic. 366 pp.

Tyree, M. T., Hammel, H. T. 1972. The measurement of turgor pressure and the water relations of plants by the pressure-bomb technique. *J. Exp. Bot.* 23:267–82

Zimmermann, M. H., Milburn, J. A., eds. 1975. *Transport in Plants.* I. *Phloem Transport.* Encycl. Plant Physiol. N.S. 1. Berlin: Springer. 535 pp.

Ann. Rev. Fluid Mech. 1977. 9:297-319

PARTICLE CAPTURE FROM ✺8105
LOW-SPEED LAMINAR FLOWS

Lloyd A. Spielman

Departments of Civil and Chemical Engineering, University of Delaware,
Newark, Delaware 19711

1 INTRODUCTION

The capture of small suspended particles from fluids in laminar flow is characterized
by the simultaneous action of forces of fluid-mechanical origin along with forces of
other origin that act between the particle and collector. These combined forces
govern the particle trajectories which in turn determine whether a particle will be
transported to and retained at the surface of a collector that is fixed in the flow.
There are important similarities between the capture of gas-borne particles and
liquid-borne particles, especially where the forces of fluid-mechanical origin are
concerned. Dissimilarities between the two arise primarily through the kinds and
magnitudes of those forces that are not of fluid-mechanical origin. To consider one
example, electrically charged aerosols (airborne particles) can have their trajectories
influenced significantly by electrostatic forces at comparatively large distances from
a collector, whereas electrostatic interactions in water are restricted to much shorter
separations because counterions in aqueous solution establish thin ion sheaths
(electrical double layers) on the solution side that locally neutralize surface charge.
These double layers accordingly confine the range of electric fields adjacent to
surfaces to rather short distances, typically on the order of a few hundred angstroms.
However, these short-range forces can greatly influence particle capture. Even
similarities in the fluid mechanics of gas- and liquid-borne particle capture should
not be overemphasized since the description of a gas as a viscous continuum breaks
down as certain characteristic dimensions, such as the separation between particle
and collector, approach the gas molecular mean free path (about 0.07 μm for air at
standard conditions). By comparison, a liquid behaves as a viscous continuum at
scales approaching the liquid molecular dimensions (a few angstroms for liquid
water). The importance of continuum breakdown on capture depends on the extent
over which other forces reach, but these differences between gas and liquid systems
are important in order to understand why certain approximations that may be
satisfactory for describing capture of airborne particles are not for waterborne
particles and conversely.

297

In the past, considerable attention has been given to describing airborne-particle capture in flow past simple collector geometries, especially cylinders. The principal motivation for this work has been to understand the mechanisms of aerosol capture in filtration by fibrous media. Among the first contributors to this field were Albrecht (1931), Kaufmann (1936), and Langmuir (1942), whose penetrating analyses still form much of the basis for modern aerosol filtration theories. These authors recognized that the sieving action of fibrous filters was of minor importance, since the particles being captured are characteristically smaller than the matrix pores; that Brownian diffusion to the collectors is important for capture of submicron particles; and that interception and particle impaction due to inertia can play important roles in the capture of larger, nondiffusing particles. The interception concept is based on a geometrical collision argument that seems to have been advanced independently by Stein (1940) and Langmuir (1942) to describe capture of waterborne and airborne particles respectively. This idea has played an important role in both early and modern particle-capture theories, yet clarification of its implicit approximations, especially those concerning fluid mechanics, has come about only recently. Inertial impaction can be important for gas-borne particles, but usually turn out to be unimportant for particles in liquids, because the particle-fluid density difference is smaller and the higher viscosity of liquids resists movement relative to the fluid.

This review is confined to particle capture from low-speed laminar flows with emphasis on particles that are very small compared with the collectors. Only certain aspects of aerosol deposition are covered. Aerosol deposition owing to electrical forces is not reviewed at all. This is not because the subject is unimportant, but because I preferred to place emphasis on those aspects of aerosol capture that have contributed to the understanding of waterborne-particle capture, which has undergone considerable advancement in recent years.

2 COLLECTOR FLOW FIELDS

2.1 *Geometric Considerations*

As has been noted above, much of existing progress made toward understanding particle capture by collectors has been motivated by the importance of fluid filtration. Here applications include purification of air and water (Davies 1973, Ives 1975) and industrial processing, as well as extracorporeal filtration of blood and other body fluids (Chang 1973, Foss, Messelt & Efskind 1963, Goldsmith & Kean 1966). Fibrous media that frequently are used to collect airborne particles are typically quite sparsely packed, usually greater than 90% voids, making it plausible to view them as assemblages of cylinders. Filtration of water supplies to remove particles is conventionally done using beds of unconsolidated granular media such as sand. The latter are frequently viewed as assemblages of interacting, but essentially separate spherical collectors, although the close proximity of individual grains and inevitable contact between them creates justified doubt as to the soundness of this approach. Nevertheless, the complexity of the actual situation has necessitated some form of simplification, and characterizing unconsolidated granular beds as assemblages of representative spheres presently seems to be the most widely accepted approach. Properly describing the flow fields near collectors in assemblages is an important issue, both because of practical applications and because most experimental data have been obtained under such conditions. Comparison of particle-capture theory with these experiments cannot be accomplished without resort to flow fields which represent reality rather crudely. Some experiments have been performed by various workers to measure particle capture from well-defined flows in order to eliminate the geometric uncertainties associated with assemblages. These include capture by isolated cylinders and the rotating disc, using both air and water.

2.2   Representative Collector Flow Fields in Filters

As a result of numerous experimental studies, it is well established that for low Reynolds number flow through fibrous or granular media, the pressure drop follows Darcy's law because it is proportional to the fluid viscosity, filter face velocity, and medium thickness. The coefficient of proportionality depends only on the medium geometry for no-slip Newtonian flow, but is independent of fluid density.

Early analyses (Langmuir 1942, Friedlander 1958) used the theoretical results of Lamb (1932) for low-speed flow past an isolated cylinder to describe flow near a fiber in a filter. Since the Stokes equations of creeping flow,

$$\nabla p = \mu \nabla^2 \mathbf{u},$$
$$\nabla \cdot \mathbf{u} = 0, \tag{1}$$

do not permit a solution for flow past an isolated cylinder, Lamb's solution necessarily incorporates fluid inertia. Use of Lamb's solution to predict the pressure drop therefore leads to an expression that contains the fluid density and conflicts with the experimental validity of Darcy's law. Langmuir (1942) recognized this and therefore used another method to describe fibrous-filter resistance, but inconsistently retained Lamb's velocity profile to predict particle capture within the medium. Over the range for which Darcy's law applies, it may be concluded that the Stokes equations of creeping flow govern the motion of fluid within fibrous media and that the neighboring fiber boundaries in the vicinity of a representative fiber must suppress fluid inertia near the latter so that it remains negligible when compared with viscous forces throughout the flow. Langmuir's analysis of particle capture, and that of many subsequent treatments, contain a dependence on the fiber Reynolds number that should not enter.

Stein (1940) first used Stokes' solution for creeping flow past an isolated sphere to describe flow near a representative grain in a granular filter. Because for an isolated sphere the creeping flow solution exists, fluid inertia does not enter and the objection based on the violation of Darcy's law that was raised against Lamb's solution above cannot be raised here. However, another objection to the use of isolated sphere or cylinder flow fields occurs because it implies that the drag force per sphere or per unit length of cylinder is independent of the medium packing density, which is not in accord with experiment. The preponderance of pressure-drop data for packed granular or fibrous media shows that the drag force per element increases as the packing density is increased (Happel & Brenner 1965).

To overcome both criticisms, several models based on the equations of creeping flow have been proposed to give better agreement with measured pressure drop and self-consistent expressions for particle capture. These include the cell models of Happel (1959) and Kuwabara (1959) as first employed by Fuchs & Stechkina (1963) and Stechkina & Fuchs (1966) to describe aerosol capture by fibrous media modeled as arrays of cylinders with axes perpendicular to the filter face velocity. Happel's model for arrays of cylinders associates a coaxial cylindrical cell with each representative solid-cylinder element. This permits a closed-form solution to the creeping-flow equations in the annular space. The radius of the cylindrical cell is determined by requiring the volume of fluid in the annular space to be in the same ratio to the solid-cylinder volume as the fluid-to-fiber volume ratio throughout the fibrous medium. The solid cylinder is taken to move within the cell with the superficial velocity U (which is equivalent to using a coordinate system that moves at the velocity U). Additional boundary conditions require no slip at the solid-cylinder surface and it is assumed that the outer cell boundary is a free surface, i.e., the shear stress tangential to the cell boundary vanishes there. Kuwabara's (1959) model is very similar except the free-surface assumption is replaced by that of vanishing vorticity at the outer boundary. Both models are equally adaptable to assemblages of spherical collectors with spherical cells (Happel & Brenner 1965). It is apparent that the cell models endeavor to account for the interference effect of neighboring collectors on the flow field near a representative collector in a very approximate way and the flow fields that result are expected to apply best near the collector surface. Fortunately, most mechanisms of particle capture are dominated by phenomena near the collector surface so these models prove useful. Using cylindrical coordinates with the origin at the center of a representative cylindrical collector and expanding in power series in the distance $(r - a_F)$ from the cylinder surface, the stream function takes the form

$$\psi = \frac{2A_F U}{a_F}(r - a_F)^2 \sin \theta, \qquad (2)$$

in which only the first nonzero term in the expansion has been retained. Here U is the superficial (face) velocity of the filter, a_F is the fiber radius, and A_F is a dimensionless coefficient that is an increasing function of the volume fraction α of fibers in the array. Happel's model gives $A_F = 1/[-\ln \alpha - 1 + \alpha^2/(1 + \alpha^2)]$ and Kuwabara's model gives $A_F = 1/[-\ln \alpha - 3/2 + 2\alpha - \alpha^2/2]$. The stream-function Equation (2), very near the collector, varies as $(r - a_F)^2$, which is a direct consequence of satisfying incompressible continuity with no slip at the collector surface. The effect of crowding by neighboring fibers is to compress the streamlines and increase the fluid speed adjacent to the central fiber because A_F increases with α.

Spielman & Goren (1968) rediscovered Brinkman's (1947a) model for flow through a swarm of spheres and applied it to cylinders. The essence of Brinkman's model is that, on the average, the fluid in proximity to a representative element within an array experiences a body damping force proportional to the velocity in addition to pressure and viscous forces, with the damping force accounting for the influence of neighboring elements on the flow. The Stokes Equations (1) are replaced by

$$\nabla p = \mu \nabla^2 \mathbf{u} - \mu k \mathbf{u},$$
$$\nabla \cdot \mathbf{u} = 0,$$

(3)

where p and $\mathbf{u}$ are viewed as the pressure and velocity averaged over all arrangements of the surrounding elements. The damping coefficient k is taken to be the Darcy resistance coefficient of the medium so that far from the central element where the average velocity becomes uniform and velocity gradients vanish on the average, the first of Equations (3) reduces to the differential form of Darcy's law. Very near the surface of the central object, the no-slip condition suppresses the damping term so Equations (3) reduce to Stokes Equations (1). Equations (3) are not theoretically rigorous, although Tam (1969) has deduced them for the case of spheres by assuming that the neighboring spheres act as homogeneously distributed point forces within the fluid. The coefficient k is not determined empirically but is obtained theoretically by requiring that the integrated drag on all the elements in a macroscopic control volume yield a Darcy resistance coefficient that is the same as that appearing in Equations (3). The latter are solved to obtain the local flow field and compute the drag on a representative element.

Brinkman's model breaks down as the packing density becomes too large, because then the neighboring elements are not distributed about the central element in approximately the same way they are distributed in the medium, i.e. uniformly. The model consequently overpredicts pressure drop for densely packed beds of spheres and its unmodified use for arrays of spheres is limited to sparse assemblages, which occur in hindered settling. Brinkman (1947b) extended the model to dense arrays of spheres by allowing a simple spatial variation of the damping coefficient near the central sphere, but at the expense of introducing a semiempirical fitting parameter to give agreement with pressure drop data. Fibrous media commonly contain 0.01 to 0.1 fraction fibers by volume. With this restriction, the model can be applied to arrays of fibers without the semi-empirical modification. The stream function for flow very near a cylindrical fiber within a uniform array with the fibers oriented perpendicular to the main flow is given by Spielman & Goren (1968) as Equation (2), where $A_F(\alpha)$ is given by the parametric equations $A_F = \beta K_1(\beta)/2K_0(\beta)$ and $\alpha = 1/[2+4K_1(\beta)/\beta K_0(\beta)]$. Although it has physical meaning the quantity β can simply be viewed as a dummy parameter, fixed values of which determine pairs of A_F and α. This parametric representation in terms of Bessel functions is computationally less convenient than the explicit algebraic forms resulting from the Happel and Kuwabara cell models; however, Brinkman's model is more straightforwardly extended to anisotropic arrays of cylinders (Spielman & Goren 1968) and to media with distributed fiber diameters. Spielman (1968) found that mixtures of cylinders having two different diameters are predicted by Brinkman's model to give lower resistance to flow than if the fibers are unmixed at the same porosity. Clarenburg & Werner (1965) performed experiments that confirm this aspect; however, the separate batches of fibers that were mixed in their experiments were themselves nonuniform in diameter and gas mean-free-path effects also entered so quantitative comparison was hampered.

At present, there is considerable disagreement among workers as to which, if any, of the different models outlined is best suited for use to predict particle capture by fibrous and granular filters. Davies (1973) advocates the Kuwabara model for fibrous media on the basis of flow model studies of Kirsch & Fuchs (1967a, 1967b). The latter carried out detailed experiments with glycerol and scaled-up cylinders in which they measured streamlines and resistance of sheets of parallel cylinders in various arrangements and found good agreement with Kuwabara's flow field. However, it is not clear that this agreement is the appropriate basis for choice because real fibrous media usually exhibit pressure drops that fall below those predicted by the Kuwabara, Happel, or Spielman & Goren flow fields, which overpredict the data in descending order. Best agreement with pressure drop recommends the flow field of Spielman & Goren, which overpredicts the least (Spielman & Goren 1968, 1972). Because the flow field is to be used to predict particle capture, agreement between theoretical predictions and capture experiments would seem to be the proper criterion for choice among flow fields. However, predicted differences in aerosol-particle capture by fibrous media, obtained using the different flow fields, are generally insignificant compared with either data scatter or systematic discrepancies between theory and experiment. That is, for the purpose of reproducing existing filtration experiments, all the flow fields are found to be about equally satisfactory or equally unsatisfactory. For this reason some workers have attempted to obtain theoretical formulas that express particle capture in terms of the coefficient A_F in Equation (2) when possible, without firm commitment to a particular flow field. Then comparisons can be made trying the different expressions for A_F.

For beds of granular collectors the situation is somewhat similar. Using spherical coordinates and retaining only the first nonzero term gives for the stream function near a representative sphere

$$\psi = \tfrac{3}{4} A_s U (r - a_s)^2 \sin^2 \theta. \tag{4}$$

Here U is again the face velocity and A_s is a dimensionless function of the bed voids, which varies from one flow model to another, as it does for the case of cylinders. For Happel's model, we have $A_s = 2(1-\gamma^5)/(2-3\gamma+3\gamma^5-2\gamma^6)$, with $\gamma = \alpha^{1/3}$. Among the representative sphere models for densely packed beds, Happel's model is algebraically convenient and it gives good agreement with pressure drop data (Happel & Brenner 1965, FitzPatrick 1972). Pfeffer (1964) and Pfeffer & Happel (1964) have used the model to predict heat and mass transfer in packed beds. However, unambiguous grounds for preferring it over other representative sphere models cannot be given.

Payatakes, Tien & Turian (1973) disapprove of representative sphere models for describing flows in densely packed granular beds. As an alternative, they present a constricted-tube model that requires numerical solution to obtain the velocity field. Payatakes (1973) has used this model to predict particle capture, with much poorer agreement with the experiments of FitzPatrick (1972) than obtained by the latter using Happel's model (FitzPatrick & Spielman 1973, Spielman & FitzPatrick 1974). This lack of agreement, the difficulties of performing calculations, and the lack of any clear improvement in representing real packed beds make this model less attractive than the representative sphere models.

2.3 Isolated Collector Flow Fields

Because of the ambiguity associated with flow fields in collector assemblages, some experimenters have studied particle capture from flow past simple collectors such as isolated cylinders (Gillespie 1955, Chang 1973) and deposition on a rotating disc (Marshall & Kitchener 1966, Hull & Kitchener 1969, Clint et al 1973, Toppan 1973, FitzPatrick & Harper 1976).

Low Reynolds number flow past an isolated cylinder has been used by Chang (1973) to study capture of waterborne particles. This flow field can be represented using Lamb's solution (Lamb 1932). To lowest order in the distance from the cylinder surface the Lamb stream function is conveniently given by Equation (2) with $A_F = 1/[2(2.0 - \ln \mathrm{Re})]$, where Re is the Reynolds number based on cylinder diameter. Creeping flow past an isolated sphere is given by Stokes' solution which, to lowest order, is given by Equation (4) with $A_s = 1$. It is evident that any analysis of particle capture by cylinders or spheres that depends on phenomena near enough to the collector surface that Equations (2) or (4) apply can be used to describe either isolated collectors or collector assemblages provided one uses an appropriate expression for A_F or A_s.

Cochran (1934) obtained the exact solution for laminar flow near a rotating disc. Expressed in cylindrical coordinates with origin at the disc center and retaining only the terms of lowest order in distance z from the surface gives the velocities very near the surface: $u_r = -0.510(\omega^{3/2}/v^{1/2})rz$, $u_\theta = -0.616(\omega^{3/2}/v^{1/2})rz$, $u_z = -0.510(\omega^{3/2}/v^{1/2})z^2$, where ω is the angular velocity and v is the kinematic viscosity.

3 PARTICLE CAPTURE BY ASSEMBLAGES

Theories of particle capture by assemblages focus on collection by a representative element within the array. Interpretation of filtration experiments requires relating capture by a representative element to that of the overall assemblage.

For a bed of uniform cylindrical fibers perpendicular to the mean flow, a suspended particle balance over a differential slice of bed having depth dx and containing many fibers gives for the differential change dn in suspended particle number concentration

$$-U\,dn = I\left(\frac{\alpha}{\pi a_F^2}\right)dx. \tag{5}$$

Here $\alpha/\pi a_F^2$ is the length of fiber per unit bed volume and I is the average rate of particle capture per unit length of fiber in the slice. The dimensionless fiber efficiency is conventionally defined as

$$\eta = I/2a_F Un. \tag{6}$$

Substituting Equation (6) into Equation (5) gives

$$-\frac{dn}{dx} = \left(\frac{2\alpha\eta}{\pi a_F}\right) n = \lambda n, \tag{7}$$

where $\lambda = 2\alpha\eta/\pi a_F = \alpha I/\pi a_F^2 U n$ is called the filter coefficient, first introduced to characterize deposition in filters by Iwasaki (1937); λ has units of reciprocal length. For all capture mechanisms where the suspended particles are independent of one another and capture results from individual encounters with the collectors, I is proportional to n, so both η and λ are independent of n. If the incoming number concentration is n_0 and the filter is uniform and unclogged, Equation (7) integrates to $n = n_0 \exp(-\lambda x)$. The quantity $1/\lambda$ is seen to be a characteristic depth of penetration of suspended particles. Experimentally λ is obtained by measuring n as a function of bed depth; then $\lambda = -d \ln n/dx$, and η or I may be obtained from Equations (6) and (7).

For a bed of collecting grains the situation is very similar. In this case, we let I be the rate of particles captured per grain. Then the representative grain collection efficiency is $\eta = I/\pi a_s^2 U n$, the filter coefficient is $\lambda = 3\alpha I/4\pi a_s^3 U n = 3\alpha\eta/4a_s$, and the relation $dn/dx = -\lambda n$ holds as for fibrous media.

4 CLASSICAL CAPTURE MECHANISMS

4.1 Brownian Diffusion

Submicron particles undergo Brownian motion, which promotes their deposition during flow past a collector. The Brownian diffusion coefficient of a particle with radius a_p is given by the Stokes-Einstein equation $D = kT/6\pi\mu a_p$ (Einstein 1926), where k is Bolzmann's constant and T is the absolute temperature. Langmuir (1942) first treated Brownian deposition on a cylinder by a very approximate method. Natanson (1957a) and Friedlander (1957) improved upon this independently by solving the boundary-layer form of the equation of steady-state convective diffusion,

$$\mathbf{u} \cdot \nabla c = D\nabla^2 c. \tag{8}$$

The analysis assumes that all impinging particles stick, which gives the condition $c = 0$ at the collector surface. Away from the cylinder surface we have $c \to n$, where n is the bulk-suspension concentration outside the concentration boundary layer. Brownian particle diffusivities are generally much smaller than molecular diffusivities, assuring large Peclet numbers so the approximation of a thin boundary layer may be used. This analysis treats the particles as if they are diffusing "points." The fiber capture efficiency that results from Natanson's solution (Natanson 1957a) of the thin boundary-layer equation and the definition, Equation (6), is

$$\eta_D = 3.64 A_F^{1/3}(D/2a_F U)^{2/3} = 3.64 A_F^{1/3} \, \text{Pe}^{-2/3}. \tag{9}$$

The value of the numerical coefficient obtained from the analysis of Friedlander (1957) is slightly different because an approximate integral boundary-layer method was used instead.

Levich (1962) solved the analogous problem of diffusion to an isolated sphere.

This analysis can be generalized straightforwardly using Equation (2) to adapt it to flow through assemblages (Pfeffer 1964). The spherical collector efficiency is then given by $\eta = 3.97 A_s^{1/3}(D/2a_s U)^{2/3}$. Levich (1962) also solved the problem of diffusion to a rotating disc with the result that the flux j which is uniform over the surface is given by $j = 0.621 D^{2/3} v^{-1/6} \omega^{1/2} n$. This result is restricted to thin diffusion layers compared to the momentum-layer thickness.

4.2 Interception

Capture by interception assumes that the center of a small nondiffusing spherical particle follows exactly an undisturbed fluid streamline near a larger collector until the particle and collector touch, whereupon the particle is retained by contact adhesion (Stein 1940, Langmuir 1942). Interception ignores increased hydrodynamic resistance between particle and collector which results from forced drainage of the viscous fluid from the narrowing gap during approach, but also ignores the finite reach of strong attraction arising through universal intermolecular forces. These opposing effects tend to cancel one another. Figure 1 depicts interception by a cylinder and the relationship between the grazing trajectory and the cylinder-capture efficiency given by Equation (6). Collection can be evaluated by tracing the limiting trajectory or by integrating the flux of particles into the front half of the collision envelope at $r = a_F + a_p$, assuming the particle centers move with the undisturbed fluid velocity. The latter procedure emphasizes that a reliable representation of the flow field near the collector is needed when a_p is small compared to a_F. The cylinder efficiency for capture by interception is

$$\eta_I = 2A_F(a_p/a_F)^2 = 2A_F R^2, \tag{10}$$

provided the interception number $R \ll 1$.

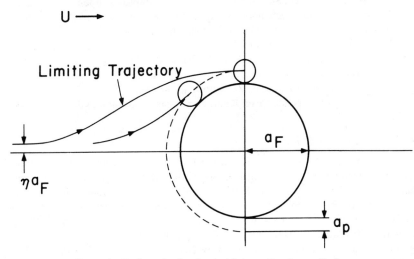

Figure 1   Trajectories for classical interception by a cylinder.

For the spherical collector the analogous calculation gives $\eta = (3A_s/2)(a_p/a_s)^2$ (see, e.g., Yao, Habibian & O'Melia 1971).

For interception by the rotating disc the particle flux to the surface is uniform and given by $j = 0.510(\omega^{3/2}/\nu^{1/2})a_p^2 n$ (see, e.g., Spielman & FitzPatrick 1973).

For small particles, interception predicts that the rate of collection by all collectors is proportional to a_p^2, which is a direct consequence of incompressible flow and the no-slip condition at the collector, because then the velocity normal to the collector varies as the square of the distance from the collector. For larger interception numbers, the calculation can be carried out but the resulting dependences are algebraically complex and require flow fields that apply at larger distances from the collector than Equation (2) (Davies 1973).

4.3   Simultaneous Diffusion and Interception

Friedlander (1958, 1967) and Stechkina & Fuchs (1966) realized that Brownian diffusion of "point" particles and interception can be viewed as limiting cases of Equation (8), but with the modified boundary condition, $c \to 0$ at the collision envelope rather than the collector surface. For the cylindrical collector this gives $c \to 0$ at $r = a_F + a_p$. Nondimensionalization and analysis of Equation (8) gives the interesting result

$$\eta R \, \text{Pe} = F(A_F R^3 \, \text{Pe}), \qquad (11)$$

which includes Equations (9) and (10) as special cases, since Equation (9) can be rewritten as $\eta R \, \text{Pe} = 3.64(A_F R^3 \, \text{Pe})^{1/3}$ and Equation (10) can be rewritten as $\eta R \, \text{Pe} = 2(A_F R^3 \, \text{Pe})$. Friedlander (1958) correlated the data of several experimenters by plotting the one-place function, Equation (11), although in that early work an unrealistic Reynolds number dependence appeared because Lamb's flow field for an isolated cylinder was used. Stechkina & Fuchs (1966) improved upon this by using the Happel and Kuwabara cell models (Section 2). Spielman & Goren (1968) adapted the same ideas to their flow field based on Brinkman's equation 3.

Using Kuwabara's flow field and Spielman and Goren's flow field, respectively, Pich (1966) and Chmielewski & Goren (1972) have extended the analysis of diffusion and interception and their simultaneous occurrence to incorporate gas mean-free-path effects through a slip boundary condition at the collector surface. Their results therefore include the dependence on Knudsen number in addition to those discussed above.

4.4   Inertial Impaction

Owing to inertia, sufficiently massive particles are unable to follow curvilinear fluid motion and tend to continue along a straight path as the carrier fluid curves around the collector. This gives rise to enhanced collection on the approach side. The importance of impaction for gas-borne particles was recognized quite early and given attention by numerous workers including Albrecht (1931), Sell (1931), Langmuir & Blodgett (1946), Mazin (1957), and Brun et al (1955), among others. Most of these early works are discussed by Herne (1960), Fuchs (1964), and Pich (1966).

In analyzing impaction it is usually assumed that the particle deviates from the fluid motion in accord with Stokes' law for a sphere in an unbounded fluid. This gives the equation of the particle trajectory,

$$\text{St}\,\frac{d\mathbf{u}_p}{dt} = \mathbf{u} - \mathbf{u}_p. \tag{12}$$

Here $\text{St} = mU/6\pi a_p\mu a$ is the Stokes number (impaction parameter), where m is the particle mass. In Equation (12), $\mathbf{u}_p$ and $\mathbf{u}$, the dimensionless particle and fluid velocities, are expressed as multiples of U, and t is the dimensionless time expressed as a multiple of (a/U) where a is the collector radius. Sometimes St includes the Cunningham slip factor, which accounts for slip-flow effects that enter for finite particle Knudsen numbers. Trajectories are obtained by integrating Equation (12) numerically, with the particle initially far on the approach size of the collector. In the earliest calculations, inviscid flow was assumed and the particles were treated as point masses. However, later calculations assume viscous flow and account for the finite size of the particle (interception effect). Harrop & Stenhouse (1969) and Stenhouse & Harrop (1970) did numerical calculations using the Happel and Kuwabara flow fields for cylindrical collectors, including the interception effect. Dawson (1969) carried out similar calculations using the flow field of Spielman & Goren (1968). In all such calculations, the stepwise integration requires complete forms of the flow fields, which apply at both large and small distances from the collector, rather than Equation (2), which is restricted to very near the collector. The numerical results are expressed as graphs or tables giving $\eta(\text{St}, R, \alpha)$. Davies (1973) gives a useful comparative tabulation of calculated results. As with diffusion and interception, the Kuwabara, Happel, and Spielman & Goren flow fields predict successively smaller capture efficiencies. For the Happel and Kuwabara flow fields, the cell boundary presents mathematical difficulties because the results are sensitive to initial conditions. Dawson (1969) avoided this problem by using the Spielman & Goren flow field, which is continuous to infinity where the streamlines become straight; however, the physical implications of this for inertial capture by fibrous media are not clear.

Impaction efficiency predictably increases with increasing particle size and fluid velocity. Impaction is important only for gas-borne particles. For liquid-borne particles the Stokes number is usually very small, primarily because the viscosity is much larger. An important consequence of capture by diffusion, interception, and/or impaction is the occurrence of a minimum in collector efficiency for increasing particle size. This is because the collision frequency increases with increasing particle size for interception and impaction, whereas Brownian diffusion decreases owing to the inverse dependence of diffusivity on particle size. Equations (9) and (10) show these reverse trends.

4.5 Gravitational Deposition

Gravitational deposition can be predicted by calculating the rate at which particles settle onto the collector, assuming that the particle settling velocity at the collision envelope is the same as that in an unbounded fluid. As for the other mechanisms discussed above, this ignores fluid drainage between the particle and collector and considers adhesion to act only upon contact, causing all collisions to result in sticking. With these assumptions, capture by combined interception and settling can be evaluated by vector addition of the fluid and settling velocities at the collision envelope. This implies that strict additivity of the two separate mechanisms gives their combined effect. The rate of settling onto the top face of a cylinder of unit length is $I = 2(a_F + a_p)u_s n$, where $u_s = 2\Delta\rho g a_p^2/\mu$ is the Stokes settling velocity. The definition, Equation (6), then gives the efficiency for settling codirectional and counterdirectional to the mean flow (Stechkina, Kirsch & Fuchs 1969),

codirectional:

$$\eta_G = \left(1 + \frac{a_p}{a_F}\right)\frac{u_s}{U},$$

counterdirectional:

$$\eta_G = -\left(1 + \frac{a_p}{a_F}\right)\frac{u_s}{U}. \tag{13}$$

For $(a_p/a_F) \ll 1$, the second term in parentheses can be ignored, giving $\eta_G \simeq u_s/U$. This last result also holds for settling of very small particles onto spherical collectors (Yao, Habibian & O'Melia 1971). Theory shows for settling counterdirectional to the flow that the negative contribution given by Equation (13) cannot exceed the positive contribution owing to interception; otherwise net collection is zero (FitzPatrick 1972). Yoshioka et al (1972) have treated combined capture by inertia and gravity for a cylindrical collector.

5   MODERNIZATION OF CAPTURE MECHANISMS FOR LIQUID SYSTEMS

5.1   Interception in Liquids

The classical analysis for capture by interception assumes that the center of a small particle follows exactly an undisturbed fluid streamline near a larger collector until contact occurs and adhesion retains the particle. However, a particle in close proximity to a collector must necessarily deviate from the undisturbed streamline. This is because the continuum description of fluid motion with no slip at both the collector and particle surfaces produces infinitesimally slow drainage of fluid from the gap between them as they approach under a finite force (Charles & Mason 1960). With the assumption that the fluid is a viscous continuum, one therefore reaches the anomalous conclusion that contact cannot occur. The anomaly disappears, however, owing to the action of attractive van der Waals forces that increase very rapidly as the particle approaches the collector and become strong enough to overcome the otherwise slow drainage. For liquids, the assumption of a viscous continuum with no slip should apply down to molecular dimensions, and the classical model for interception, which overlooks this, becomes inadequate to describe capture from liquids.

For gases the situation is somewhat different. Here the continuum description with no slip breaks down when the gap between the particle and collector becomes comparable to the mean free path of the gas molecules, which is about 0.07 μm for air at standard conditions. For gaps of this order and smaller the fluid resistance to closer approach is predictably lower than that for continuum flow with no slip (Goren 1973). It is believed that this reduced resistance promotes applicability of the classical theory to gas systems, where classical theory has achieved a measure of success, while it has not been successful for liquid systems (Spielman & Goren 1970, 1971; Goren 1973). Even for aerosols the classical theory leaves much room for improvement. Gallily & Mahrer (1973) discuss hydrodynamic interactions in connection with aerosols.

A first important step toward modernizing interception for liquid systems was taken by Natanson (1957b), who took into account the finite reach of intermolecular attraction by using Hamaker's (1937) expression for the unretarded London dispersion force acting between a sphere and a much larger collector, which is approximated as planar. The magnitude of the latter is given by

$$F_{Ad} = \frac{2Qa_p}{3(2a_p+h)^2h^2},$$

(14)

where h is the minimum separation between the particle and collector and Q is the Hamaker constant that depends on properties of the constituent molecules. The Hamaker constant ranges from about 10^{-14} to 10^{-12} ergs and can be estimated from material (especially optical) properties (Gregory 1969). However, Natanson's analysis (Natanson 1957b) ignores hydrodynamic resistance between the particle and collector since it was assumed that the particle deviates from the fluid in accord with Stokes' law for a sphere in an unbounded fluid.

Spielman & Goren (1970, 1971) incorporated the hydrodynamic interactions between the particle and collector rigorously by constructing the solution to Equations (1) that corresponds to the Stokes disturbance flow created by the particle in proximity to the collector, which is viewed as locally planar because the particle is assumed to be much smaller than the collector. In their analysis the boundary conditions are taken to be no slip at both the particle and collector surfaces with the undisturbed flow field, Equation (2), recovered outside the local disturbance. The net force and torque on the particle are taken to be zero. That is, both fluid and particle inertia are neglected, permitting the external force, Equation (14), and hydrodynamic forces to balance. The entrained particle is therefore propelled by both the external force and the moving fluid near the collector rather than artificially superimposed on the undisturbed fluid streamlines, so it translates and rotates according to its equations of motion. Neglect of fluid inertia in the analysis is justified by the smallness of the Reynolds number associated with the particle disturbance flow. Neglect of particle inertia is justified by its smallness compared with hydrodynamic forces acting on the particle, which results because the Stokes number is small for particles in liquids. The method of solution involves resolution of the local undisturbed flow into component flows that provide the far field boundary conditions for calculating the disturbance flows for a sphere near a plane using Equations (1). These are depicted for a cylindrical collector in Figure 2.

The net motion of the particle is thus obtained by superposition, which is permitted because Equations (1) are linear. The component Stokes flow problem depicted in Figure 2A corresponds to the motion of a sphere perpendicular to a plane in an otherwise stationary fluid. This was solved by Brenner (1961) who obtained the hydrodynamic interaction function $F_1(H) = F_1(h/a_p)$ shown in Figure 3. One can view F_1 as the ratio of the particle velocity under an applied force perpendicular to the collector to that which the particle would have in an unbounded fluid under the same force. The component flow problem shown in Figure 2B is that of a stationary sphere near a plane with the far field given by the symmetric resolute

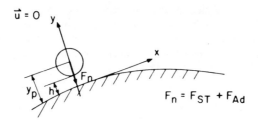

A. Particle moving under applied force.

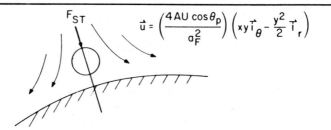

B. Stationary particle.

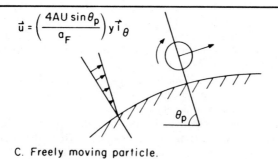

C. Freely moving particle.

Figure 2 Resolution of resultant particle motion into subsidiary Stokes flows near a much larger collector.

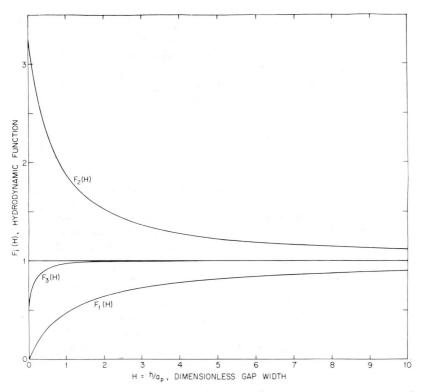

$F_2(H)$

$F_3(H)$

$F_1(H)$

H = $^h/a_p$, DIMENSIONLESS GAP WIDTH

Figure 3 Functions characterizing hydrodynamic interactions between a particle and much larger collector.

of the undisturbed flow shown in the figure. This problem was solved by Goren (1970) and Goren & O'Neill (1971), who calculated the hydrodynamic interaction function $F_2(H)$ shown in Figure 3. F_2 can be viewed as the ratio of the radial-directed force a stationary particle would experience near the collector to that it would experience if held stationary in an unbounded fluid moving uniformly with the radial velocity of the undisturbed flow [given by Equation (2)] evaluated at the point where the particle center is located. The component Stokes problem shown in Figure 2C corresponds to the free translation and rotation of a neutrally buoyant sphere near a plane with the far field flow having a uniform velocity gradient. This was solved by Goldman, Cox & Brenner (1967), who calculated the hydrodynamic interaction function $F_3(H)$ shown in Figure 3. The function F_3 can be viewed as the ratio of the velocity of the entrained particle parallel to the collector surface to the tangential velocity of the undisturbed fluid [given by Equation (2)] evaluated at the particle center position. The equation of the particle trajectories for motion around a cylinder with the external adhesion force, Equation (14), is (Spielman & Goren 1970, 1971; Spielman & FitzPatrick 1973)

$$-4(H+1)\frac{F_3(H)}{F_1(H)}\sin\theta_p\frac{dH}{d\theta_p} = 2(H+1)^2F_2(H)\cos\theta_p + \frac{N_{Ad_F}}{H^2(H+2)^2}. \qquad (15)$$

In Equation (15), $H = h/a_p$ is the dimensionless separation, θ_p is the angular position (Figure 4) and $N_{Ad_F} = Qa_F^2/9\pi\mu A_F Ua_p^4$ is the dimensionless adhesion number that measures the ratio of London attraction to hydrodynamic forces. Equation (15) gives rise to trajectories that qualitatively resemble those shown in Figure 4, although the scale is purposely exaggerated. Here the limiting trajectory, which divides capture trajectories from escape trajectories, no longer grazes as in Figure 1 for classical interception, but has a stagnation point at the rear of the collector. The efficiency is obtained by numerical integration of Equation (15) along the limiting trajectory (Spielman & FitzPatrick 1973). This gives $\eta/2A_F R^2$ as a function of N_{Ad_F} as shown in Figure 5 for neutrally buoyant particles. At large N_{Ad_F} the curve approaches the asymptote $\eta/2A_F R^2 = (3\pi N_{Ad_F}/4)^{1/3}$, which corresponds to Natanson's (1957b) result obtained by ignoring the hydrodynamic interactions. This occurs because the limiting trajectory gets displaced farther from the collector as N_{Ad_F} gets too large, diminishing the hydrodynamic interactions. The asymptote can therefore be obtained by setting $F_1 = F_2 = F_3 = 1$ in Equation (15), which can then be solved analytically (Spielman & Goren 1970).

It is interesting to compare Figure 5 with the corresponding result for classical interception, Equation (10). Because the ordinate of Figure 5 is $\eta/2A_F R^2$, Equation (10) would plot on Figure 5 as a horizontal line through unity. The modernized result is seen to be quite different, predicting a weaker dependence on particle size than $\eta \propto a_p^2$ given by Equation (10) and an inverse dependence on flow rate; classical interception predicts no dependence of η on flow rate. Chang (1973) did experiments using single fibers in water as collectors. These data demonstrate the importance of hydrodynamic resistance to approach and agree with the trend shown in Figure 5 much more closely than with classical interception, although complete agreement with the modernized theory was not found.

Smirnov & Deryagin (1967) and Deryagin & Smirnov (1971) independently attempted to incorporate hydrodynamic resistance to approach in particle capture. However, their analysis is not as rigorous as that leading to Equation (15) and

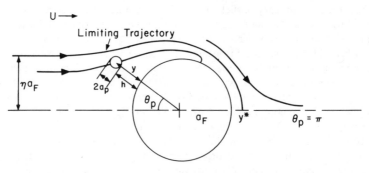

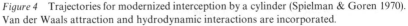

Figure 4   Trajectories for modernized interception by a cylinder (Spielman & Goren 1970). Van der Waals attraction and hydrodynamic interactions are incorporated.

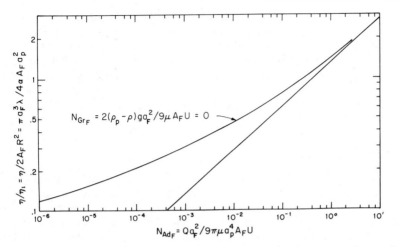

The y-axis is labeled $\eta/\eta_i = \eta/2A_F R^2 = \pi a_F^3 \lambda/4 a A_F a_p^2$ with values .1, .2, .5, 1, 2.

The x-axis is labeled $N_{Ad_F} = Q q_F^2/9\pi\mu a_p^4 A_F U$ with values 10^{-6}, 10^{-5}, 10^{-4}, 10^{-3}, 10^{-2}, 10^{-1}, 10^0, 10^1.

The curve is labeled $N_{Gr_F} = 2(\rho_p - \rho)g a_F^2/9\mu A_F U = 0$.

Figure 5 Normalized capture efficiency versus adhesion number for capture of neutrally buoyant particles by a cylinder (Spielman & FitzPatrick 1973). Hydrodynamic interactions and van der Waals attraction are incorporated.

makes incorrect assumptions for the hydrodynamic interactions that appear in Equation (15) and Figure 3. The predictions of their analysis therefore differ considerably from those of Spielman & Goren (1970, 1971) and Spielman & FitzPatrick (1973).

5.2 Gravitational Deposition in Liquids

Spielman & FitzPatrick (1973) also obtained numerical solutions to the analogs of Equation (15) that describe capture by a spherical collector and by the rotating disc, including gravitational forces on the particle in addition to the London-van der Waals force. The collection efficiencies that were calculated therefore include, besides the adhesion number, the dependence of a dimensionless gravitational parameter that has the formula shown equal to zero in Figure 5 for the cylindrical collector, since the latter applies only to neutrally buoyant particles. The experimental data of FitzPatrick & Spielman (1973) for waterborne particle capture by beds of glass spheres confirm the predictions of the modernized theory much better than the classical theory, although complete agreement was not found. FitzPatrick & Harper (1976) performed deposition experiments using the rotating disc, including particles large enough to test the modernized theory including hydrodynamic interactions for combined interception and gravity forces. Here too agreement between experiment and the modernized theory was generally found to be better than with the classical theory, but certain observations (such as nonuniform collection over the surface) are not explained by either theory.

5.3 Double Layer Repulsion

In addition to adhesive forces resulting from intermolecular attraction, surfaces in water commonly possess a surface charge that is neutralized locally by counterions that concentrate on the solution side to form an electrical double layer. When the double layers of approaching surfaces interact, the surfaces experience repulsion that can counteract intermolecular attraction and prevent adhesion (Verwey & Overbeek 1948). Spielman & Cukor (1973) incorporated London attraction and double-layer repulsion in the modernized trajectory equation for neutrally buoyant particle capture by the sphere and rotating disc collectors. The expression used for the repulsive double layer force was that derived by Hogg, Healy & Fuerstenau (1966) (see also Gregory 1975),

$$F_{DL} = \varepsilon \kappa a_p \zeta_1 \zeta_2 \frac{e^{-\kappa h}}{1 + e^{-\kappa h}}. \tag{16}$$

In Equation (16), ε is the dielectric constant of water, κ is the reciprocal Debye length, and ζ_1 and ζ_2 are the potentials marking the onset of the particle and collector Debye layers. Equation (16) is restricted to $\zeta_1 \simeq \zeta_2 < 50-60$ mV and $\tau = \kappa a_p > 10$. The potential energy of interaction that results when Equations (14) and (16) are combined appears as in Figure 6. The analysis for the spherical collector predicts that the efficiency should depend on two new groups, $N_\zeta = 3\varepsilon \zeta_1 \zeta_2 a_p/2Q$ and $\tau = \kappa a_p$, in addition to the adhesion number $N_{Ad_s} = Q a_s^2/9\pi \mu A_s U a_p^4$. The group N_ζ is a characteristic ratio of repulsive to attractive forces and τ is the ratio of the particle radius to the Debye length. Depending on the values of these three parameters, four possible modes of capture are predicted. Figure 7 shows the calculated regions of the parameters which correspond to each of the four modes. In the zone of no collection in Figure 7, particles are unable to surmount the repulsive barrier

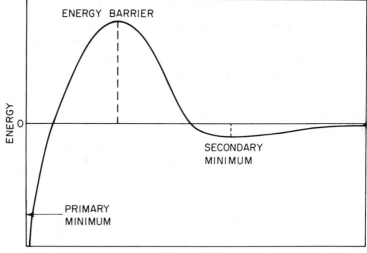

Figure 6   Interaction energy versus separation for combined van der Waals attraction and electrical double-layer repulsion.

of Figure 6 and all particles escape capture. In the zone of Figure 7 corresponding to capture by the primary minimum only, some approaching particles are propelled over the energy barrier of Figure 6 to be captured; all others escape. Capture neglecting double-layer repulsion is a special case of this. In the region of Figure 7 corresponding to capture by the secondary minimum only, no approaching particles are propelled over the barrier of Figure 6, but the secondary minimum of Figure 6 is sufficiently deep to prevent some particles from escaping. Finally, Figure 7 shows a tiny, thin region of combined collection for which particles that impinge near enough to the center of approach get propelled over the energy barrier of Figure 6, while those more slightly removed do not, but still are prevented from escape by

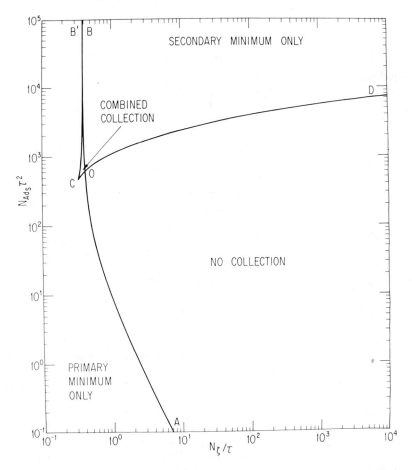

Figure 7 Locus of N_{Ad_s}, N_ζ, and τ distinguishing four regimes of capture of nondiffusing particles. Hydrodynamic interactions, van der Waals attraction, and double-layer repulsion are incorporated (Spielman & Cukor 1973).

the secondary minimum (Figure 6). The region of no collection obtained by Spielman & Cukor (1973) corresponds closely to the semi-empirical criteria of Spielman & FitzPatrick (1973) and FitzPatrick & Spielman (1973), which describe the conditions under which double-layer repulsion was found to be important in packed-bed collection experiments.

Payatakes (1973) performed calculations similar to those of Spielman and co-workers, including correction of the London dispersion force for electromagnetic retardation (e.g., Schenkel & Kitchener 1960) among other modifications. However, instead of using a representative sphere or cylinder as the collector, the constricted tube model of Payatakes, Tien & Turian (1973) was used. The calculations do not correlate FitzPatrick's (1972) data as well as calculations using Happel's flow field (Spielman & FitzPatrick 1973, 1974), but still provide useful insights because a wider range of external forces is considered.

5.4   Brownian Diffusion in Liquids

Spielman & Goren (1970, 1971) extended the modernized transport equation to include Brownian diffusion, but did not attempt to solve it. Prieve & Ruckenstein (1974) solved the transport equation for the spherical collector numerically, including London attraction, gravity, and Brownian diffusion, with hydrodynamic interactions fully incorporated. An important conclusion drawn by them is the adequacy of adding the classical result for Brownian capture (Section 4.1) to capture by London and gravity forces, ignoring diffusion; this was found to reproduce the fully coupled results with good accuracy. Yao, Habibian & O'Melia (1971) reached a similar conclusion, but their calculations for mechanisms, coupled and uncoupled, ignored hydrodynamic interactions between the particle and collector.

It is of considerable importance that experiments with waterborne particles in the submicron range where Brownian diffusion predominates and double-layer repulsion is absent are described very well by the classical theory discussed in Section 4.1. In particular, Yao, Habibian & O'Melia (1971) and Cookson (1970) found good agreement with the Pfeffer-Happel extension of Levich's theory for granular-bed filtration experiments. Also, Hull & Kitchener (1969) and Marshall & Kitchener (1966) found good agreement between rotating disc experiments with submicron particles and Levich's theory in the absence of repulsion.

Spielman & Friedlander (1974) and Ruckenstein & Prieve (1973) independently carried out analyses of deposition of submicron particles by Brownian diffusion in the presence of double-layer repulsion. When the interactions are confined sufficiently near the collector surface, the process is equivalent to ordinary convective diffusion in the bulk with an apparent first-order surface reaction at the collector. The surface reaction rate coefficient is expressed explicitly in terms of an integral involving the attractive-repulsive potential (Figure 6). Ruckenstein & Prieve (1973) incorporated spatial variation of the diffusivity owing to hydrodynamic interaction with the collector and approximated the integral to represent the reaction coefficient in the form of an Arrhenius rate expression. Numerical solutions are presented for collection by spheres and cylinders that go over to the results of Levich (1962) and Natanson (1957a) when double layer repulsion is absent (Section 4.1). Comparison of theory with the experiments of Hull & Kitchener (1969) shows large discrepancies. Ruckenstein & Prieve (1973) offer an explanation for the disagreement by pointing out the extreme sensitivity of predicted collection to small variations in surface chemical parameters.

6 SUMMARY

Because most experimental data on particle capture relate to collection by fibrous or granular media, several models for flow past the elements constituting assemblages have been developed. The earliest models idealized fibrous media as arrays of isolated cylinders and granular media as arrays of isolated spheres. More recent models realistically consider hydrodynamic interactions between the representative elements that make up the arrays, while some consider dense assemblages as bundles of constricted tubes. None of these models is without its questionable features, although the representative cylinder and sphere models show close similarities that indicate a degree of consensus. Existing experimental data do not agree so well with the predictions of any one flow model that it can be singled out as decisively superior to all others.

Recent experiments using well-defined flow fields show that modernization of classical capture mechanisms to include hydrodynamic interactions between particle and collector lead to improved agreement between theory and experiment for liquid systems. Although many aspects of airborne-particle capture experiments agree well with classical capture theories, overall agreement is not quantitative and considerable improvement might be obtained through acquisition of both highly reliable data and modernization of gas-borne-particle theory to incorporate hydro-dynamic interactions, including mean-free-path effects at small particle-collector separations.

Although waterborne-particle capture is better described by incorporating hydro-dynamic interactions than by ignoring them, there still remains significant discrepancy with experiment. In particular, when electrical double-layer repulsion is significant, agreement between experiment and theory is so poor as to cast doubt upon the widely accepted view that van der Waals attraction and electrical double-layer repulsion alone adequately characterize surface interactions. This view is presently undergoing evolution and carefully designed particle capture experiments are contributing significantly to progress in understanding surface interactions.

Literature Cited

Albrecht, F. 1931. *Phys. Z.* 32:48
Brenner, H. 1961. *Chem. Eng. Sci.* 16:242
Brinkman, H. C. 1947a. *Appl. Sci. Res.* A1: 27
Brinkman, H. C. 1947b. *Appl. Sci. Res.* A1: 81
Brun, R., Lewis, W., Perkins, P., Serafini, J. 1955. *Natl. Advis. Comm. Aeronaut. Rep.*, p. 1215
Chang, D. P. Y. 1973. *Particle collection* *from aqueous suspensions flowing over solid and permeable hollow fibers.* PhD thesis. Calif. Inst. Technol., Pasadena
Charles, G. E., Mason, S. G. 1960. *J. Colloid Sci.* 15:236
Chmielewski, R. D., Goren, S. L. 1972. *Environ. Sci. Technol.* 6:1101
Clarenburg, L. A., Werner, R. M. 1965. *Ind. Eng. Chem. Process Des. Dev.* 4(3):293
Clint, G. E., Clint, J. H., Corkill, J. M.,

Walker, T. 1973. *J. Colloid Interface Sci.* 44:121

Cochran, W. G. 1934. *Proc. Cambridge Philos. Soc.* 30:365

Cookson, J. T. 1970. *Environ. Sci. Technol.* 4:129

Davies, C. N. 1973. *Air Filtration.* London: Academic. 171 pp.

Dawson, S. V. 1969. *Theory of collection of airborne particles by fibrous filters.* ScD thesis (hygiene). Harvard Sch. Publ. Health, Cambridge, Mass.

Deryagin, B. V., Smirnov, L. P. 1971. In *Research in Surface Forces,* ed. B. V. Deryagin, 3:167. New York: Consultants Bureau

Einstein, A. 1926. *Investigations on the Theory of Brownian Movement.* New York: Dover. 119 pp.

FitzPatrick, J. A. 1972. *Mechanisms of particle capture in water filtration.* PhD thesis. Harvard Univ., Cambridge, Mass.

FitzPatrick, J. A., Harper, J. W. 1976. *Particle Deposition from Aqueous Suspension onto a Rotating Disc. Natl. Meet. AIChE, 81st, Kansas City, Mo.*

FitzPatrick, J. A., Spielman, L. A. 1973. *J. Colloid Interface Sci.* 43:350

Foss, O. P., Messelt, O. T., Efskind, L. 1963. *Surgery* 53:241

Friedlander, S. K. 1957. *AIChE J.* 3:43

Friedlander, S. K. 1958. *Ind. Eng. Chem.* 50:1161

Friedlander, S. K. 1967. *J. Colloid Interface Sci.* 23:157

Fuchs, N. A. 1964. *The Mechanics of Aerosols.* Oxford: Pergamon. 408 pp.

Fuchs, N. A., Stechkina, I. B. 1963. *Ann. Occup. Hyg.* 6:27

Gallily, I., Mahrer, Y. 1973. *Aerosol Sci.* 4:253

Gillespie, T. 1955. *J. Colloid Sci.* 10:266

Goldman, A. J., Cox, R. G., Brenner, H. 1967. *Chem. Eng. Sci.* 22:637, 653

Goldsmith, E. I., Kean, B. H. 1966. *Gastroenterology* 50:805

Goren, S. L. 1970. *J. Fluid Mech.* 41:619

Goren, S. L. 1973. *J. Colloid Interface Sci.* 44:356

Goren, S. L., O'Neill, M. E. 1971. *Chem. Eng. Sci.* 26(3):325

Gregory, J. 1969. *Adv. Colloid Interface Sci.* 2:396

Gregory, J. 1975. In *The Scientific Basis of Filtration,* ed. K. J. Ives, NATO Adv. Study Inst. Ser., Ser. E: Applied Sciences, Vol. 2. Leyden: Noordhoff

Hamaker, H. C. 1937. *Physica* 4:1058

Happel, J. 1959. *AIChE J.* 5:174

Happel, J., Brenner, H. 1965. *Low Reynolds Number Hydrodynamics.* Englewood

Cliffs, New Jersey: Prentice-Hall. 553 pp.

Harrop, J. A., Stenhouse, J. I. T. 1969. *Chem. Eng. Sci.* 24:1475

Herne, H. 1960. In *Aerodynamic Capture of Particles,* ed. E. G. Richardson. New York: Pergamon

Hogg, R., Healy, T. W., Fuerstenau, D. W. 1966. *Trans. Faraday Soc.* 62:1638

Hull, M., Kitchener, J. A. 1969. *Trans. Faraday Soc.* 65:3093

Ives, K. J., ed. 1975. *The Scientific Basis of Filtration,* NATO Adv. Study Inst. Ser., Ser. E: Applied Sciences, Vol. 2. Leyden: Noordhoff. 444 pp.

Iwasaki, T. 1937. *J. Am. Water Works Assoc.* 29:1591

Kaufmann, A. 1936. *Z. Verein Deutsches Ing.* 80:593

Kirsch, A. A., Fuchs, N. A. 1967a. *J. Phys. Soc. Jpn.* 22(5):1251

Kirsch, A. A., Fuchs, N. A. 1967b. *Ann. Occup. Hyg.* 10:23

Kuwabara, S. 1959. *J. Phys. Soc. Jpn.* 14(4):527

Lamb, H. 1932. *Hydrodynamics.* New York: Dover. 738 pp. 6th ed.

Langmuir, I. 1942. *Report on Smokes and Filters,* Sect. I. US Off. Sci. Res. Dev., 865:IV

Langmuir, I., Blodgett, K. B. 1946. *Army Air Forces Tech. Rep. 5418*

Levich, V. G. 1962. *Physicochemical Hydrodynamics.* Englewood Cliffs, New Jersey: Prentice Hall. 700 pp.

Marshall, J. K., Kitchener, J. 1966. *J. Colloid Interface Sci.* 22:342

Mazin, I. P. 1957. *The Physical Principles of Aircraft Icing.* Moscow: Gidrometeorizdat (In Russian)

Natanson, G. 1957a. *Dokl. Akad. Nauk SSSR* 112:100

Natanson, G. 1957b. *Dokl. Akad. Nauk SSSR* 112(4):696

Payatakes, A. C. 1973. *A new model for granular porous media. Application to filtration through packed beds.* PhD thesis. Syracuse Univ., New York

Payatakes, A. C., Tien, C., Turian, R. M. 1973. *AIChE J.* 19:58, 67, 1036

Pfeffer, R. 1964. *Ind. Eng. Chem.* 3:380

Pfeffer, R., Happel, J. 1964. *AIChE J.* 10:605

Pich, J. 1966. *Aerosol Science,* ed. C. N. Davies, Chap. IX. New York: Academic

Prieve, D. C., Ruckenstein, E. 1974. *AIChE J.* 20:1178

Ruckenstein, E., Prieve, D. C. 1973. *J. Chem. Soc. Faraday Trans. 2* 69:1522

Schenkel, J. A., Kitchener, J. A. 1960. *Trans. Faraday Soc.* 56:161

Sell, W. 1931. *VDI-Forschungsh.* 347:1

Smirnov, L. P., Deryagin, B. V. 1967. *Colloid J. USSR* 29(3):299

Spielman, L. A. 1968. Unpublished notes

Spielman, L. A., Cukor, P. M. 1973. *J. Colloid Interface Sci.* 43:51

Spielman, L. A., FitzPatrick, J. A. 1973. *J. Colloid Interface Sci.* 42:607

Spielman, L. A., FitzPatrick, J. A. 1974. *J. Colloid Interface Sci.* 49:328

Spielman, L. A., Friedlander, S. K. 1974. *J. Colloid Interface Sci.* 46:22

Spielman, L. A., Goren, S. L. 1968. *Environ. Sci. Technol.* 2(4):279

Spielman, L. A., Goren, S. L. 1970. *Environ. Sci. Technol.* 4:135

Spielman, L. A., Goren, S. L. 1971. *Environ. Sci. Technol.* 5:254

Spielman, L. A., Goren, S. L. 1972. *Ind. Eng. Chem. Fundam.* 11:73

Stechkina, I. B., Fuchs, N. A. 1966. *Ann. Occup. Hyg.* 9:59

Stechkina, I. B., Kirsch, A. A., Fuchs, N. A. 1969. *Ann. Occup. Hyg.* 12:1

Stein, P. C. 1940. *A study of the theory of rapid filtration of water through sand.* DSc thesis. Mass. Inst. Technol., Cambridge

Stenhouse, J. I. T., Harrop, J. A. 1970. *Chem. Eng. Sci.* 25:1113

Tam, C. K. W. 1969. *J. Fluid Mech.* 38:537

Toppan, W. C. 1973. *An experimental investigation of particle capture by the rotating disc (with application to water and wastewater treatment).* MS thesis. Clarkson Coll. Technol., Potsdam, New York

Verwey, J. W., Overbeek, J. Th. G. 1948. *Theory of the Stability of Lyophobic Colloids.* New York: Elsevier

Yao, K., Habibian, M. T., O'Melia, C. R. 1971. *Environ. Sci. Technol.* 5:1105

Yoshioka, N., Emi, H., Kanaoka, C., Yasunami, M. 1972. *Chem. Eng. Jpn.* 36:313

Ann. Rev. Fluid Mech. 1977. 9:321–37
Copyright © 1977 by Annual Reviews Inc. All rights reserved.

ELECTROKINETIC EFFECTS ×8106
WITH SMALL PARTICLES

D. A. Saville

Department of Chemical Engineering, Princeton University, Princeton, New Jersey 08540

INTRODUCTION

Electrokinetic phenomena is a generic term applied to effects associated with the movement of ionic solutions near charged interfaces. Although a respectable antiquity can be claimed for the subject, our understanding of the fluid mechanics begins with Smoluchowski's studies in the early part of the twentieth century. Much of the current interest stems from the use of electrokinetic measurements in biochemistry and biophysics. Although electrokinetic phenomena play important roles in many diverse natural and technological processes no attempt is made to deal with their ubiquitous nature here. Instead our understanding of two archetypal problems is reviewed. Other reviews may be consulted for insight into the diversity of the phenomena (see Abramson 1934; Abramson, Moyer & Gorin 1942; Overbeek 1950; Overbeek & Wiersema 1967; Dukhin & Deryagun 1974).

Another reason for current interest in these phenomena derives from their roles in the bulk behavior of suspensions. A focus of the work on suspensions is the use of microscale phenomena, e.g. flow around a single particle, to explain bulk behavior. Considerable attention has been paid to suspensions of uncharged particles but not to charged particles in ionic solutions. Here attention focuses on the description of the two prototypical problems that serve as the basis for an understanding of the bulk behavior. These problems are the motion of a small charged particle in response to an externally applied field and the behavior of a similar particle exposed to some sort of flow without any externally applied field. In both cases the essential complicating feature is the presence of diffuse space charge arising from the response of the ions in solution to the presence of the charged interface.

The basis for the description of the phenomena is reviewed first with emphasis on the ways in which electrokinetic effects alter the conservation equations. The current understanding of the two archetypal problems is reviewed in subsequent sections.

BALANCE LAWS

An understanding of the events near a small charged particle in an ionic solution requires, among other things, a description of the electric fields and forces in the fluid and on the particle. The description can be obtained from Maxwell's equations, simplified by the suppression of magnetic effects since the currents are invariably small. Furthermore, since the ions are free to move in response to electrical and Brownian forces, there must be an accounting for individual species.

Differential Equations

The motion of the liquid is described by solutions to the Navier-Stokes equation for an incompressible fluid with the extra body-force term, namely,

$$\rho \mathbf{v} \cdot \nabla \mathbf{v} = -\nabla p + \rho_e \mathbf{E} - \mu \nabla \times \nabla \times \mathbf{v}. \tag{1}$$

Here ρ_e stands for the bulk charge density and $\mathbf{E}$ for the electric-field strength. The electric field can be related to the charge density via an expression developed from Maxwell's equations:

$$\varepsilon \nabla^2 \phi = -\rho_e, \tag{2}$$

where ϕ is the potential and $\mathbf{E} = -\nabla \phi$; ε stands for the dielectric constant of the fluid.

Electrical attraction and repulsion combined with diffusion alter the distribution of ions in the vicinity of the charged interface so that a space charge exists. This charge is related to the local ion concentrations according to the expression

$$\rho_e = e \sum_1^N z^k n^k. \tag{3}$$

The detailed structure of the region very close to the interface is not fully understood. Evidently a thin region composed of adsorbed ions and molecules shields some of the surface charge, so that the potential at the outer edge of this region, often termed the zeta potential, differs from that at the surface of the particle. The structure of the stagnant region close to the surface as well as the region further removed, the diffuse double-layer, is discussed in general terms by Adamson (1967) and in more detail by Haydon (1964). Figure 1 depicts matters schematically. All the phenomena discussed here can be understood by supposing that there is a diffuse space charge adjacent to a thin charged layer immobilized on the surface of the particle, if one ignores, for the time being, the complicated structure of the surface layers.

For a completely ionized solute composed of N different species dissolved in a liquid, the charge per unit volume from ions of the kth type is $ez^k n^k$; e is the charge on a proton, z^k is the (signed) valance of the ion and n^k the concentration. For sodium chloride $z^1 = 1$, $z^2 = -1$. Each type of ion is carried along by the flow and moves, at the same time, in response to the gradients of the concentration and potential. Thus the flux of a given species is

$$n^k \mathbf{v} - \omega^k e z^k n^k \nabla \phi - \omega^k k T \nabla n^k.$$

In this expression $n^k \mathbf{v}$ accounts for transfer by bulk flow, $-\omega^k e z^k n^k \nabla \phi$ represents the flux due to the electric field, and $-\omega^k k T \nabla n^k$ is the diffusion flux (Booth 1950a). The definitions of the symbols are as follows: ω^k—the mobility of the kth species, i.e.

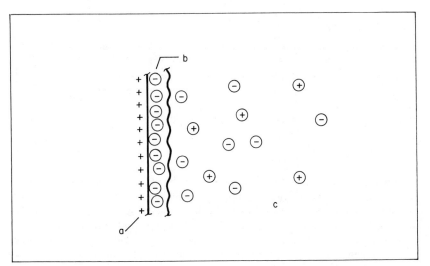

Figure 1   Schematic representation of the region near the particle-liquid interface showing (*a*) the charge layer near the particle surface, (*b*) an immobile layer adjacent to the surface, and (*c*) diffuse space charge.

the velocity per unit force; k—Boltzmann's constant; T—absolute temperature. Conservation of each species requires

$$\mathbf{v} \cdot \nabla n^k = \omega^k \nabla \cdot (ez^k n^k \nabla \phi + kT \nabla n^k). \tag{4}$$

Boundary Conditions

In the absence of electrochemical reactions the surface charge will remain constant, and the electrical current normal to the surface will be zero. According to Maxwell's equations, the potential gradient at the surface is proportional to the surface charge. For a spherical particle this means

$$-\varepsilon \frac{\partial \phi}{\partial r} = \frac{Q}{4\pi a^2}, \tag{5}$$

where Q is the total charge on the particle with radius a. Contributions due to the fields inside the surface are neglected on the grounds that the ratio of dielectric constants, $\bar{\varepsilon}\varepsilon^{-1}$, is small. For many polymeric particles in water the ratio is (ca.) 0.03, although in hydrocarbons the ratio can be near unity. The absence of current means that

$$\omega^k \left(ez^k n^k \frac{\partial \phi}{\partial r} + kT \frac{\partial n^k}{\partial r} \right) = 0. \tag{6}$$

The remaining electrical boundary condition on the surface is the continuity of the tangential components of $\nabla \phi$ (Sommerfeld 1964).

Reference Scales

Before turning to the individual phenomena, it is helpful to mention some of the scales involved. Our discussion deals with micron and submicron-sized particles, since for centimeter-sized particles the electrokinetic effects discussed here are negligibly small. The length scale characterizing the decay of the space charge from its maximum near the solid-liquid interface is termed the Debye length, κ^{-1}. From an analysis of the equilibrium double layer it is found that (Adamson 1967)

$$\kappa^{-1} = \left[e^2 \sum_1^N n_\infty^k (z^k)^2 / \varepsilon k T \right]^{-1/2} \tag{7}$$

The symbol n_∞^k stands for the concentration of the kth ionic species far from the surface where electrical neutrality prevails. The Debye length turns out to be roughly 0.3 μm for a 10^{-6} M solution of a 1-1 salt (e.g. NaCl) in water. Since the Debye length changes in inverse proportion to the square root of the concentration, an increase from 10^{-6} M to 10^{-4} M diminishes κ^{-1} by a factor of ten. For large particles, then, the double-layer thickness is but a small fraction of the particle radius, and for this reason electrokinetic effects vanish as $(a\kappa) \to \infty$.

Since the electrical potential energy of an ion in an electric field is proportional to $e\phi$, while the kinetic energy is proportional to kT, a natural scale for electrical potential is kTe^{-1}. Accordingly a stress scale is $\varepsilon(kT)^2(ea)^{-2}$ and an electrokinetic velocity scale is $\varepsilon(kT)^2(\rho v a)^{-1}e^{-2}$. In the presence of an external field, E_∞, the potential difference across a small particle, aE_∞, will generally be small compared with kTe^{-1}. Thus, where motion is caused by an external field the "Reynolds

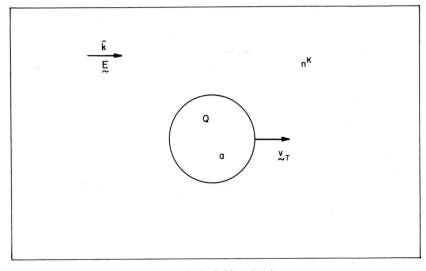

Figure 2   Definition sketch.

number" characterizing the magnitude of inertial and viscous effects, $\varepsilon(kT)^2\rho^{-1}(ev)^{-2}$, is usually small. An estimate of the relative importance of convection of charge compared with conduction and diffusion is $\varepsilon kT(\rho v\omega^0)^{-1}e^{-2}$, which is of the order of unity. In situations where an external flow is imposed in place of the electric field the conventional Reynolds and Peclet numbers appear.

The Phenomena

This review deals with rigid particles, mainly spheres. Fluid drops and bubbles are interesting subjects but micron-sized globules invariably behave as though their interfaces are inextensible. Some of the effects encountered with larger circulating droplets are discussed by, for example, Levich (1962).

What sort of phenomena can be expected? Consider first the effect of an externally imposed electric field. If the particle is free to move, the translation is called electrophoresis; and if the particle is stationary, the fluid motion is termed electro-osmosis. Two types of motion can be caused by the field. One arises from the action of the field on the particle's surface charge, another from the action of the electric field on the space charge in the liquid. Furthermore, both the field and the fluid motion tend to stretch out the double layer. This polarization of the charge cloud is offset by diffusion and migration of the ions in the induced electric field. The combined processes are called relaxation. When the external field is absent, deformation of the double layer results from the action of the externally imposed flow. This polarization produces additional drag forces and, in some cases, torques. It can also induce flow through its action on the space charge. Again, relaxation processes tend to counteract the polarization. The next sections review the current knowledge of the processes operative in these two situations.

MOTION DUE TO AN EXTERNAL ELECTRIC FIELD

An externally imposed field engenders motion of the particle, through its effect on the surface charge, and the fluid, through its effect on bulk charge in the liquid near the interface. The intensity of the motion will be proportional to the field strength E_∞. Measured using the scale kTe^{-1} for electric potential and a for the particle radius, this amounts to a dimensionless potential $aeE_\infty(kT)^{-1}$. If this dimensionless potential is small compared with the dimensionless surface potential, then, as a first approximation, a description of the phenomena can be constructed as a perturbation on the equilibrium double-layer situation. The analysis consists largely in finding solutions to the Navier-Stokes equation for an incompressible fluid

$$Nr\mathbf{v} \cdot \nabla\mathbf{v} = -\nabla p + (\nabla^2\phi)(\nabla\phi) - \nabla \times \nabla \times \mathbf{v} \qquad (8)$$

in which the electrical effects are embodied in $(\nabla^2\phi)\nabla\phi$. The electric field is established from the conservation equations for the several ionic species

$$Np\mathbf{v} \cdot \nabla n^k = \omega^k\nabla \cdot (z^k n^k\nabla\phi + \nabla n^k), \qquad k = 1,\dots,N \qquad (9)$$

along with the equation relating the field to the distribution of charge

$$\delta^2 \nabla^2 \phi = -\sum_1^N z^k n^k. \tag{10}$$

These equations are written in dimensionless form employing the particle radius as a length scale, $\varepsilon(kT)^2(\rho v a)^{-1} e^{-2}$ as a velocity scale, $\varepsilon(kT)^2(ea)^{-2}$ as a scale for stress, and kTe^{-1} as a scale for electric potential. The common scale for the various concentrations is n^0. Ion mobilities are normalized with respect to one of them, ω^0 (say) and the symbols Nr and Np stand for $\varepsilon(kT)^2 \rho^{-1}(ev)^{-2}$ and $\varepsilon kT(\rho v \omega^0)^{-1} e^{-2}$, with δ^2 representing $\varepsilon kT(n^0)^{-1}(ea)^{-2}$. The group Nr characterizes the magnitude of inertial forces as compared with viscous forces, and Np measures the effects of convection in distorting the ion distribution compared with the diffusive effects tending to restore equilibrium. The relative sizes of the electrical double layer and the particle are given by δ; $\delta \ll 1$ corresponds to a thin double layer as compared with the particle, and $\delta \gg 1$ to a thick one. In terms of a Debye thickness, κ^{-1}, $\delta = (a\kappa)^{-1}$.

The electrical boundary conditions at the surface of the particle are (a) continuity of the tangential components of $\nabla \phi$; (b) the presence of a uniform charge distribution,

$$-\frac{\partial \phi}{\partial r} = q; \tag{11}$$

and (c) the absence of any electric current (no electrochemical reactions)

$$z^k n^k \frac{\partial \phi}{\partial r} + \frac{\partial n^k}{\partial r} = 0. \tag{12}$$

In addition the usual conditions obtain regarding kinematics and dynamics, namely, rigid body motion and balances of forces and torques. Far from the surface the velocity must vanish while the ion concentrations approach uniform values, n_∞^k. At the same time the gradient of the potential tends towards the uniform value $-\beta \hat{k}$. Here q stands for $eQ(4\pi a \varepsilon kT)^{-1}$ and represents a dimensionless surface charge, while β is the dimensionless field strength, $aeE_\infty(kT)^{-1}$, and $\hat{k}$ is a unit vector in the direction of the external field.

The discussion of the various (approximate) solutions for this set of equations is facilitated by noting that for typical values of the various parameters[1] Nr and β are small. Since they are small, it is appropriate to disregard inertial effects and use perturbation methods to ascertain effects due to the field. Expansions for ϕ, n^k, p, and $\mathbf{v}$ in the forms

$$\Phi_0 + \beta \Phi + \cdots,$$

$$n_\infty^k + n_0^k + \beta n^k + \cdots,$$

---

[1] For $a = 0.1$ μm; $n^0 = 10^{-5}$ M; $E_\infty = 10$ v cm^{-1}, $\omega^0 kT = 10^{-5}$ cm^2 sec^{-1}, $\rho = 1$ g cm^{-3}, $\mu = 0.01$ g cm^{-1} sec^{-1}, $\varepsilon = 80$ ε_0, the values of Nr and β are 5×10^{-4} and 4×10^{-3}, respectively. Furthermore, $Np \doteq 0.5$ and $\delta^2 \doteq 2$. The value of q depends on the surface charge with a zeta potential of 25 mv corresponding to $q \doteq 1$.

$$p_0 + \beta p + \cdots,$$

and

$$\mathbf{0} + \beta \mathbf{v} + \cdots$$

are used with the O(1)-terms describing the equilibrium double layer.

The O(1)-Problem

Solutions for the equilibrium-double-layer problem for a spherical particle have been obtained in the form of a power series in q (Gronwall, La Mer & Sandved 1928) and by numerical methods (Loeb, Wiersema & Overbeek 1961). Asymptotic solutions are discussed by Dukhin & Deryagun (1974). Since this review is focused on electrokinetic phenomena, we regard the equilibrium situation to be known at the outset. Nevertheless it is worth noting that the analytical solution has the form

$$\Phi_0 = q\delta(1+\delta)^{-1} r^{-1} \exp\left[-(r-1)/\delta\right] + O(q^3) \tag{13}$$

for symmetrical electrolytes. In the present context, the surface or ζ-potential is given by the expressions

$$\zeta \equiv \Phi_0(1) = q\delta(1+\delta)^{-1} + O(q\delta^3), \qquad \delta \to 0 \tag{14}$$

and

$$\zeta \equiv \Phi_0(1) = q\delta(1+\delta)^{-1} + O(q^3/\delta^2), \qquad \delta \to \infty \tag{15}$$

if the surface charge, q, is fixed. Consequently the first term in (13), sometimes called the Debye-Hückel approximation, gives the correct limiting behavior as $\delta \to 0$ and $\delta \to \infty$ with q fixed. Furthermore, although calculations made using the Debye-Hückel approximation may be subject to considerable error for situations where δ is of the order of unity, the results will be correct in these limiting situations.

The O(β)-Problem

With the description of the equilibrium situation in hand we turn to a description of effects due to the external field. The equations to be solved are

$$0 = -\nabla p + (\nabla^2 \Phi_0)\nabla\Phi + (\nabla^2\Phi)\nabla\Phi_0 - \nabla \times \nabla \times \mathbf{v}, \tag{16}$$

$$Np\mathbf{v} \cdot \nabla n_0^k = \omega^k \nabla \cdot \{z^k(n_\infty^k + n_0^k)\nabla\Phi + z^k n^k \nabla\Phi_0 + \nabla n^k\}, \tag{17}$$

and

$$\delta^2 \nabla^2 \Phi = -\sum_1^N z^k n^k. \tag{18}$$

Complications arise from the coupled nature of the equations and frustrate the general use of analytical methods except in special circumstances. These circumstances offer considerable insight, however, and are reviewed in order. In each instance the particle being considered is a sphere.

Thick Double Layers Ignoring Charge Convection

Here interest centers on situations where $\delta \gg 1$ and $Np = 0$. Perturbations to the equilibrium ion distribution vanish, $n^k = 0$, and the equation for the potential is simply

$$\nabla^2 \Phi = 0. \tag{19}$$

Accordingly the potential field is

$$\Phi = -(r + \tfrac{1}{2}r^{-2})\cos\theta. \tag{20}$$

The electrical body force due to charge in the fluid is $O(\delta^{-2})$, and from this it follows that the velocity of the particle can be obtained by equating viscous drag to the electrical force on the surface charge. The translational velocity is

$$\mathbf{v}_T = \tfrac{2}{3}\beta q \hat{k}, \qquad \delta \to \infty, \tag{21}$$

a result first obtained by Hückel (1924). An alternate form is obtained by expressing the charge in terms of the surface potential (equivalent here to the so-called ζ-potential), namely, $q = \zeta$.

Thin Double Layers Ignoring Charge Convection

At the other extreme we have thin double layers, $\delta \ll 1$, corresponding to a large particle or a higher ionic strength. Two analytical approaches are possible, a perturbation scheme or the use of a more general solution due to Henry (1931), which is discussed shortly. In either case

$$\Phi = -(r + \tfrac{1}{2}r^{-2})\cos\theta. \tag{22}$$

Here, in contrast to the previous situation the action of the external field on the thin layer of charge $[O(a\delta)]$ does cause an electrokinetic flow near the particle. From a calculation of the drag one finds (Smoluchowski 1921)

$$\mathbf{v}_T = \beta q \delta \hat{k}, \qquad \delta \to 0. \tag{23}$$

Thus, as the double layer thins out (at constant charge), the translational velocity decreases. An equivalent relation follows from the relation between charge and potential for thin double layers, $\zeta = q\delta$, namely,

$$\mathbf{v}_T = \beta \zeta \hat{k}, \qquad \delta \to 0. \tag{24}$$

This expression can be misleading in that it appears to imply that $\mathbf{v}_T$ is $O(1)$ as $\delta \to 0$, which is not the case unless the charge is allowed to increase proportionally.

Double Layers of Arbitrary Thickness without Charge Convection

If alterations in the equilibrium charge distribution are completely ignored, then the potential is $-(r + \tfrac{1}{2}r^{-2})\cos\theta$ and the body force distribution $(\nabla^2 \Phi_0)\nabla\Phi$ allows an analytical solution of the Stokes' problem for the velocity and pressure fields (Henry 1931). Henry's is a general solution in the sense that once the equilibrium-double-layer potential is known, velocity and pressure fields follow upon evaluating several integrals. Numerical techniques are required to cover the whole range of surface charges and double-layer thicknesses. Nevertheless, suppression of alterations in the distribution of ions is strictly valid only when $q \ll 1$, for it amounts to ignoring alterations (polarization) due to the external field. The fact that the field must alter the charge distribution can be seen from setting $n^k = 0$ in (17) (with $Np = 0$) and noting that for consistency we must have

$$\frac{\partial n_0^k}{\partial r} \frac{\partial \Phi}{\partial r} = 0. \tag{25}$$

This is the case at the surface and as $r \to \infty$ but not in between. Thus the use of Henry's solution beyond the Debye-Hückel approximation is not strictly valid since polarization is omitted.

An analytical result can be obtained using the Debye-Hückel approximation (Henry 1931) and yields the formula

$$\mathbf{v}_T = \tfrac{2}{3}\beta q \delta (1+\delta)^{-1} f_1(\delta)\hat{k}. \tag{26}$$

Results calculated from this expression are shown in Figure 3. Henry's solution also reduces to those of Smoluchowski and Hückel in the appropriate limits. Henry's solution can also be used to describe the electrically induced flow near a stationary particle and shows that electrically induced streaming (electro-osmosis) disappears in the limits $\delta \to 0$ and $\delta \to \infty$ with q fixed. This follows from the fact that the bulk free charge vanishes and the body force on the fluid is then zero.

Double Layers of Arbitrary Thickness with Charge Convection

Two processes serve to deform the spherically symmetric double layer: polarization by the external field, as mentioned in the previous section, and convection. Since Np, which serves to characterize the order of magnitude of convection compared with diffusion, is $O(1)$ it is impractical to introduce convection via a perturbation scheme. Overbeek (1943) and Booth (1950a) termed their methods successive approximation techniques. The same results can be derived formally using expansions in powers of q, the surface charge. The $O(\beta)$-problem is thus further decomposed using expressions of the form

$$\Phi_0 \sim q\lambda_1 + q^2\lambda_2 + \cdots,$$

$$n_0^k \sim qm_1^k + q^2m_2^k + \cdots,$$

$$\Phi \sim \psi_0 + q\psi_1 + \cdots, \tag{27}$$

$$n^k \sim qc_1^k + \cdots,$$

$$\mathbf{v} \sim q\mathbf{v}_1 + \cdots,$$

and

$$p \sim qp_1 + \cdots,$$

The equilibrium distributions, Φ_0 and n_0^k are known from, say, the Gronwall-La Mer-Sandved expansion; ψ_0 represents the external field modified by the presence of

the sphere, and the other terms represent effects due to polarization and relaxation.
This procedure generates a set of expressions that are $O(1)$, $O(q)$, etc. The $O(q)$ problem is equivalent to that solved by Henry (1931) for the Debye-Hückel approximation. Continuing, we find alterations due to polarization of the equilibrium charge cloud by the external field and deformation by flow. Overbeek's analytical solution deals with symmetrical and unsymmetrical electrolytes while Booth's is restricted to symmetrical solutes.

Overbeek chose to express his results in terms of the surface potential, and for symmetrical systems of binary electrolytes the result is

$$\mathbf{v}_T = \tfrac{2}{3}\beta\zeta\{f_1(\delta) - (z\zeta)^2 f_3(\delta) - \tfrac{2}{3}Np\zeta^2 f_4(\delta)\}\hat{k} \qquad (28)$$

when the mobilities of the positive and negative ions are equal. $f_1(\delta)$ is simply Henry's solution for the Debye-Hückel approximation, while $f_3(\delta)$ is the correction for polarization and $f_4(\delta)$ represents the correction for relaxation; all are tabulated by Overbeek (1943, 1950). This expression can be recast in a form more appropriate for current purposes by using the relation between ζ and q for the equilibrium situation. Booth's result, on the other hand, is already in this form, namely,

$$\mathbf{v}_T = \tfrac{2}{3}\beta q\{X_1(\delta) + z^2 q^2 [X_3(\delta) + Y_3(\delta)] + 4q^2 Np Z_3(\delta)\}\hat{k} \qquad (29)$$

for a binary electrolyte with equal mobilities. Booth's theory for symmetrical electrolytes is somewhat more comprehensive than Overbeek's since it is not restricted to binary electrolytes and includes terms of order $O(q^4)$ (not shown here). $X_3(\delta)$, $Y_3(\delta)$, and $Z_3(\delta)$ are all negative, which shows that polarization (X_3 and Y_3)

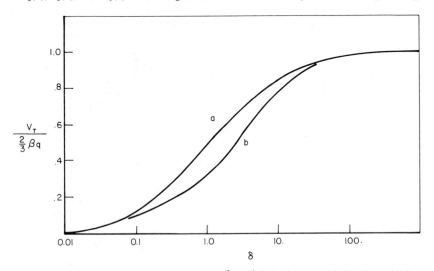

Figure 3 Scaled electrophoretic velocity, $v_T(\tfrac{2}{3}\beta q)^{-1}$, as a function of the dimensionless double-layer thickness, δ. The surface charge, q, is constant. (*a*) Henry's solution [Equation (26)]. (*b*) Booth's solution evaluated for a 1-1 electrolyte with $q = 6$ and $Np = 0.275$ [Equation (29)].

and relaxation (Z_3) all decrease the electrophoretic velocity throughout the range of double-layer thicknesses for symmetrical electrolytes. The three functions tend to zero for $\delta \to 0$ or $\delta \to \infty$. All this is shown graphically in Figure 3 along with Henry's result for comparison.

In spite of their considerable utility both theories are restricted in a strict sense to small surface charges or potentials. This led Wiersema, Loeb & Overbeek (1966) to use numerical methods on the $O(\beta)$-problem to avoid these restrictions. Their most extensive calculations are for 1-1 electrolytes, although results are given for other types.

Several general conclusions can be drawn. First, the two analytical theories generally overestimate the relaxation effect, although they are astonishingly accurate at surface charges (or ζ-potentials) larger than unity. Second, as the charge (or ζ-potential) increases at constant double-layer thickness, the electrophoretic velocity appears to reach a limiting value. Henry's theory predicts that the velocity ought to increase monotonically, while the analytical theories of Booth and Overbeek show a maximum beyond which the velocity decreases. This limiting value is a balance between an increased speed due to the larger charge and increased drag caused by polarization and relaxation. Another result is that the relaxation effect is increased by increasing the valance of the counter ions. Thus, with a positively charged particle a change in the valance of the counter ions in excess in the double layer from -1 to -3 can decrease the velocity by almost a factor of two for intermediate-sized double layers (e.g. $\delta = 0.2$).

Electrophoretic mobility is usually presented as a function of the double-layer thickness with the surface- or zeta-potential as a parameter. Some of the results calculated by Wiersema, Loeb & Overbeek (1966) are shown in Figure 4 to illustrate the difference between the behavior at constant charge (Figure 3) and constant potential. Analytical results are also shown to provide a comparison showing the accuracy of the approximate solution. As the double layer becomes thinner, relaxation effects reduce the mobility. At the same time, however, the surface charge increases and this tends to compensate since it results in a greater force on the particle. Finally the increased charge completely overshadows relaxation giving the Smoluchowski (1921) result as $(a\kappa) \to \infty$.

Experimental Studies

A strict confirmation of the theory as it now stands would require separate direct measurements of the surface charge (or ζ-potential) and the electrophoretic mobility. This does not appear possible at present, and so one of the tasks for the future is the development of experimental methods. Some earlier results are noted in the aforementioned reviews. Shaw (1969) also described results with model systems. Probably the best model systems are polymer lattices, and a recent study by Ottewill & Shaw (1972) involves such particles. In their extensive studies electrophoretic mobilities were measured over a wide range of double-layer thicknesses with two sorts of salt solutions. The main qualitative features agree with the theory but precise quantitative verification was not possible due to the difficulty in measuring the surface charge independently.

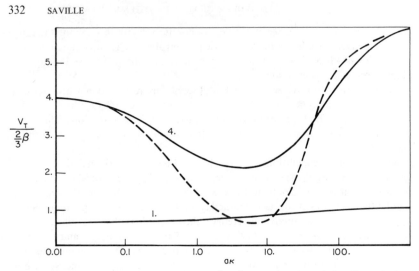

Figure 4   Scaled electrophoretic velocity, $v_T(\frac{2}{3}\beta)^{-1}$, as a function of the dimensionless double-layer thickness $(a\kappa)$. The surface potential is constant, and so the charge increases with $(a\kappa)$. (————) denotes numerical results by Wiersema, Loeb, and Overbeek for 1-1 electrolytes with $Np = 0.275$ and surface potentials. (- - - - -) denotes Overbeek's analytical solution for a 1-1 electrolyte with $\zeta = 4$ and $Np = 0.275$ [Equation (28)].

Studies of Other Situations

So far the emphasis has been on single isolated particles. Very little has been reported regarding systems with interactions between particles or a particle and a surface. Morrison & Stukel (1970) analyzed electrophoresis of a sphere perpendicular to a conducting plane for the case where the double layers are thin. The correction to Smoluchowski's relation (1921) turns out to be important when the separation is less than 3–4 radii. By way of comparison the correction to Stokes' law becomes important for separations less than (ca.) 10 radii (Happel & Brenner 1965).

Interactions between two equal-sized spheres with thin double layers can be analyzed using a straightforward adaptation of classical methods (Reed & Morrison 1976). Here there are electrical interactions due to the ways the two spheres alter the external field. Motion along and perpendicular to the line of centers is analyzed and again the effect on the mobility is small for separations exceeding 3–4 particle radii. They find the Smoluchowski relation for a single sphere to be exact for two spheres when the surface charge is the same on each sphere. According to Reed and Morrison this is due to the fact that the flow outside the thin double layers is irrotational.

For more concentrated systems Levine & Neale (1974a,b) employ a cell model to extend Henry's results. The analysis is, of necessity, approximate due to the usual problems associated with cell models, namely, specification of cell size, shape, and the selection of proper boundary conditions. Nevertheless, there is general agreement with experimental data by Long & Ross (1968).

The most systematic investigation of the electrophoresis of another particle shape is by Henry (1931), who studied cylinders as well as spheres. Here the problem is the indeterminant nature of the Stokes equations; therefore, Henry used Oseen's method. Some calculations relating to more complicated shapes, e.g. polyelectrolyte molecules are reported by Overbeek & Wiersema (1967).

ELECTROKINETIC EFFECTS ASSOCIATED WITH MOTION IN THE ABSENCE OF AN EXTERNAL FIELD

When the primary fluid motion is some sort of forced convection, uniform streaming or simple shear, for example, electrical effects arise solely from deformation of the double layer by the flow. Thus, ion conservation must be included from the outset. Two sorts of flow are discussed: simple shear and steady streaming. The results for simple shear can be used to estimate the electrokinetic contribution to the viscosity of a suspension. For steady streaming the alteration in the Stokes drag and the sedimentation potential can be calculated.

The equations are similar to those for electrophoresis except for the absence of an external field and the presence of a primary flow. The Navier-Stokes equations read

$$Re\mathbf{v} \cdot \nabla \mathbf{v} = -\nabla p + Ne(\nabla^2 \Phi)\nabla \Phi - \nabla \times \nabla \times \mathbf{v}. \tag{30}$$

The conservation equations for the ionic species and the electric field are the same as before except for the substitution of the Peclet number, Pe, for Np. The boundary conditions at the surface remain the same, while far from the surface the electric field vanishes and the velocity approaches the specified state, e.g. simple shear. The new symbols Re, Pe, and Ne stand for aUv^{-1}, $aU(\omega^0 kT)^{-1}$, and $\varepsilon(kT)^2(\rho v Ua)^{-1}e^{-2}$, respectively; U represents the speed of the external flow. The Reynolds and Peclet numbers are typically small for suspensions of small particles although for steady streaming past a fixed particle the Peclet number can be O(1) or larger even when $Re \ll 1$. Ne represents the ratio of electrical stress to viscous forces and can easily be of the order of magnitude of unity. Since the Reynolds number is small, inertial effects are safely ignored. Further simplifications are possible by introducing expansions for small q, namely,

$$q\Phi^{(1)} + q^2\Phi^{(2)} + \cdots,$$

$$n_\infty^k + qm_1^k + \cdots,$$

$$p^{(0)} + q^2 p^{(1)} + \cdots,$$

and

$$\mathbf{v}^{(0)} + q^2 \mathbf{v}^{(1)} + \cdots,$$

Here $p^{(0)}$ and $\mathbf{v}^{(0)}$ stand for the externally imposed flow field. For small Peclet numbers the variables are then expanded as

$$\Phi^{(1)} \sim \psi_0 + Pe\psi_1 + \cdots,$$

$$m_1^k \sim c_0^k + Pec_1^k + \cdots.$$

$$\mathbf{v}^{(1)} \sim Pe\mathbf{v}_1 + \cdots, \tag{31}$$

$$p^{(1)} \sim p_0 + Pep_1 + \cdots,$$

When these are substituted into the equations we find that the $O(1)$-terms constitute the description of the equilibrium double layer according to the Debye-Hückel approximation. The $O(Pe)$-terms lead to

$$0 = -\nabla p_1 + Ne[(\nabla^2\psi_0)\nabla\psi_1 + (\nabla^2\psi_1)\nabla\psi_0] - \nabla \times \nabla \times \mathbf{v}_1$$

$$\mathbf{v}^{(0)} \cdot \nabla c_0^k = \omega^k(z^k n_\infty^k \nabla^2\psi_1 + \nabla^2 c_1^k) \tag{32}$$

$$\delta^2\nabla^2\psi_1 = -\sum_1^N z^k c_1^k.$$

Solutions for Simple Shear and Suspension Viscosity

The $O(Pe)$-problem was solved, albeit in different form by Smoluchowski and Krasney-Ergen (see Booth 1950b) for thin double layers and their final results for viscosity differ by a numerical factor. Later Booth (1950b) solved the problem for an arbitrary double-layer thickness and calculated the viscosity of a dilute suspension of rigid spheres. For situations where the ionic mobilities are all the same his results can be expressed as

$$\frac{\mu^*}{\mu} = 1 + \tfrac{5}{2}[1 + q^2 NpZ(\delta)]c. \tag{33}$$

Here c denotes the volume fraction of spheres. The Z-function is given in analytical form and has the asymptotic behavior $1.5\delta^4$ for $\delta \to 0$, which shows that the electroviscous effect vanishes rapidly for thin double layers. In the other limit, $\delta \to \infty$, the expression shows a viscosity that increases in proportion to δ. Here, the restriction $c \ll \delta^{-3}$ is required to keep the double layers from overlapping and the effect appears too strong since the Einstein correction is lost as the double layer thickens.

Viscometric studies using polymer lattices offer the best opportunity for checking

Table 1 Primary electroviscous effect

δ	Experiment	Theory
0.11	0.3 ± 0.2	0.3
0.26	0.5 ± 0.2	0.3
0.62	0.0 ± 0.1	0.1
0.88	-2.0 ± 0.1	0.1
1.1	1.1 ± 0.3	0.7
1.1	1.0 ± 0.2	0.3
1.2	1.2 ± 0.4	0.5
1.2	0.8 ± 0.1	0.6

the theory since the surface charge can be measured by electrophoresis. A study by Stone-Masui & Watillon (1968) offers a good comparison between theory and experiment. Particle sizes ranging from 0.07 to 0.24 μm were used in a variety of ionic solutions. An important part of their study involved an attempt at reducing the double-layer thickness to very small values by increasing the amount of dissolved salt. This was done to check the Einstein correction, 2.5c. When this was tried, however, slow coagulation of the suspension took place to form aggregates and thwarted the test. At lower ionic strengths coagulation was apparently not a problem. The data in Table 1 have been abstracted from their results. Entries in the column labeled "experiment" were obtained by subtracting the Einstein correction from measured values of the slope of the effective viscosity-concentration relation. The column labeled "theory" was calculated by Stone-Masui and Watillon from Booth's theory. Considering the experimental uncertainty, the agreement seems generally fair.

Data by Stone-Masui and Watillon disclose that another effect, which is $O(c^2)$, may be very important in determining bulk rheology. The effect is due in part to overlapping of the double layers and since particle-particle interactions are involved the theory is necessarily complicated. Russel (1976) has developed the first part of a rigorous theory for the low shear limit.

Solutions for Uniform Streaming

Another flow field of interest corresponds to uniform streaming past a stationary particle. Streaming elongates the charge cloud in the direction of the flow so as to produce a potential that has a dipole form far from the sphere. This potential alters matters in two ways, by exerting a force directly on the particle and by producing an electrokinetic flow in the neighborhood of the sphere. Smoluchowski (1921) calculated both the potential and the induced drag for a thin double layer, but Booth (1954) gave the first theory for arbitrary double-layer thicknesses.

The tedious analytical solution of the $O(Pe)$-problem yields the induced potential, ψ_1, and the drag force. The form of the potential far from the sphere is

$$qPe\psi_1 \sim -\tfrac{1}{2}qPe\delta^2 S_3(\delta)r^{-2}\cos\theta. \tag{34}$$

$S_3(\delta)$ is given in analytical form and has the asymptotic forms 0.083δ for $\delta \to 0$ and $0.056 + O(\delta^{-1})$ for $\delta \to \infty$. The drag force is increased over the Stokes drag, 6π, by $-24\pi q^2 NpV_3(\delta)$; $V_3(\delta) \sim -0.25\delta^4$ for $\delta \to 0$ and $-0.028\delta^{-1}$ for $\delta \to \infty$. The alteration of the drag and the strength of the dipole are both small. Booth estimated, however, that the sedimentation potential generated by the slow settling of a suspension may reach 0.2 v cm^{-1}.

Booth's analysis is restricted to small Peclet numbers due to the type of perturbation scheme used. For larger Peclet numbers it turns out that a solution is possible using singular-perturbation methods for $\delta \ll 1$ as long as $Pe\delta^2$ stays $O(1)$ (Saville unpublished). The far-field form of the potential is

$$q\psi_1 \sim -\tfrac{3}{2}qPe\delta^2 r^{-2}\cos\theta, \tag{35}$$

and the additional drag is $4\pi q^2 Np\delta$. As expected, the electrokinetic effects here are much larger due to the strong flow.

CONCLUDING REMARKS

Extant theories appear to describe most aspects of the behavior of a single isolated particle. Although experimental studies designed to test these theories are hardly numerous, reasonable agreement exists in most instances. The most obvious limitations concern basic features of the model, in particular the physicochemical nature of the interfacial region.

Aside from the obvious need for experimental tests and numerical solutions to map out the full behavior, the most promising areas for future work concern the behavior of nonspherical particles and the systems with particle-particle or particle-wall interactions. Studies of other shapes provide insight into the behavior of macromolecules, cells and the like, subjects of intrinsic interest. Finally, the techniques recently developed in suspension mechanics will undoubtedly provide considerable insight into the electrokinetic behavior of suspensions.

ACKNOWLEDGMENTS

I am indebted to my colleagues W. B. Russel and W. R. Schowalter for numerous discussions and perceptive comments. Part of this review is based on work done on sabbatical leave in the Chemical Engineering Department at Stanford University, and I am pleased to express my appreciation to A. Acrivos.

Literature Cited

Abramson, H. A. 1934. *Electrokinetic Phenomena and Their Application to Medicine and Biology.* New York: Chemical Catalog Co. 331 pp.

Abramson, H. A., Moyer, L. S., Gorin, M. H. 1942. *Electrophoresis of Proteins.* New York: Reinhold. 341 pp.

Adamson, A. 1967. *Physical Chemistry of Surfaces.* New York: Interscience. 747 pp. 2nd ed.

Booth, F. 1950a. The cataphoresis of spherical, solid nonconducting particles in a symmetrical electrolyte. *Proc. R. Soc. London Ser. A* 203:514–33

Booth, F. 1950b. The electroviscous effect for suspensions of solid spherical particles. *Proc. R. Soc. London Ser. A* 203:533–51

Booth, F. 1954. Sedimentation potential and velocity of solid spherical particles. *J. Chem. Phys.* 22:1956–68

Dukhin, S. S., Deryagun, B. V. 1974. In *Surface and Colloid Science,* ed. E. Matijevic, 7:1–356. New York: Wiley. 356 pp.

Gronwall, T. H., La Mer, V. K., Sandved, R. 1928. Über den Einfluss der sogenanten höheren Glieder in der Debye-Hückelschen Theorie der Lösungen starker

Elektrolyte. *Phys. Z.* 29:358–93

Happel, J., Brenner, H. 1965. *Low Reynolds Number Hydrodynamics,* p. 331. Englewood Cliffs, NJ: Prentice-Hall. 553 pp.

Haydon, D. A. 1964. In *Progress in Surface Science,* ed. J. F. Danelli, K. G. A. Pankhurst, A. C. Riddiford, 1:94–158. New York: Academic. 414 pp.

Henry, D. C. 1931. The cataphoresis of suspended particles. *Proc. R. Soc. London Ser. A.* 133:106–29

Hückel, E. 1924. Die Kataphorese der Kugel. *Phys. Z.* 25:204–10

Levich, V. G. 1962. *Physicochemical Hydrodynamics,* pp. 472–531. Englewood Cliffs, NJ: Prentice-Hall. 700 pp.

Levine, S., Neale, G. H. 1974a. The prediction of electrokinetic phenomena within multiparticle systems. *J. Colloid Interface Sci.* 47:520–29

Levine, S., Neale, G. H. 1974b. Electrophoretic mobility of multiparticle systems. *J. Colloid Interface Sci.* 49:330–32

Loeb, A. L., Wiersema, P. H., Overbeek, J. Th. G. 1961. *The Electrical Double Layer Around a Spherical Colloid Particle.* Cambridge: MIT Press. 375 pp.

Long, R. P., Ross, S. 1968. The effects of

overlap of double-layers on electrophoretic mobility of polydisperse suspensions. *J. Colloid Interface Sci.* 26:434–45

Morrison, F. A., Stukel, J. A. 1970. Electrophoresis of an insulating sphere normal to a conducting plane. *J. Colloid Interface Sci.* 33:88–93

Ottewill, R. H., Shaw, J. N. 1972. Electrophoretic studies on polystyrene lattices. *J. Electroanal. Chem.* 37:133–42

Overbeek, J. Th. G. 1943. Theorie der Elektrophorese. *Kolloid-Beih.* 54:287–364

Overbeek, J. Th. G. 1950. In *Advances in Colloid Science,* ed. H. Mark, E. J. W. Verwey, III:97–135. New York: Interscience. 384 pp.

Overbeek, J. Th. G., Wiersema, P. H. 1967. In *Electrophoresis,* ed. M. Bier, II:1–52. New York: Academic. 553 pp.

Reed, L. D., Morrison, F. A. 1976. Hydrodynamic interactions in electrophoresis. *J. Colloid Interface Sci.* 54:117–33

Russel, W. B. 1976. Low shear limit of the secondary electroviscous effect. *J. Colloid Interface Sci.* 55:590–604

Shaw, D. J. 1969. *Electrophoresis.* New York: Academic. 144 pp.

Smoluchowski, M. 1921. In *Handbuch der Elektrizitat und des Magnetismus,* ed. L. Graetz, II:366–428. Leipzig: J. A. Barth. 772 pp.

Sommerfeld, A. 1964. *Electrodynamics,* p. 17. New York: Academic. 371 pp.

Stone-Masui, J., Watillon, A. 1968. Electroviscous effects in dispersions of monodisperse polystyrene lattices. *J. Colloid Interface Sci.* 28:187–202

Wiersema, P. H., Loeb, A. L., Overbeek, J. Th. G. 1966. Calculation of the electrophoretic mobility of a spherical colloid particle. *J. Colloid Interface Sci.* 22:79–99

Ann. Rev. Fluid Mech. 1977. 9:339–98

FLUID MECHANICS OF PROPULSION BY CILIA AND FLAGELLA

×8107

Christopher Brennen and Howard Winet

Division of Engineering and Applied Science, California Institute of Technology, Pasadena, California 91125

1 INTRODUCTION

1.1 Opening Remarks

Since the *Annual Review of Fluid Mechanics* first published a review on microorganism locomotion by Jahn & Votta (1972) considerable progress has been made in the understanding of both the biological and the fluid-mechanical processes involved not only in microorganism locomotion but also in other fluid systems utilizing cilia. Much of this knowledge and research, which has been built on the solid foundation of the pioneering work of Sir James Gray (1928, 1968) and Sir Geoffrey Taylor (1951, 1952a,b), has been reported extensively elsewhere, particularly by Gray (1928, 1968), Sleigh (1962), Lighthill (1975), and Wu, Brokaw & Brennen (1975). The subject is now sufficiently broad that it precludes any exhaustive treatment in these few pages. Rather, we restrict this review primarily to a summary of present understanding of the low-Reynolds-number flows associated with microorganism propulsion and the hydromechanics of ciliary systems. In this introductory section we wish to put such fluid-mechanical studies in biological perspective. Section 2 outlines the present status of low-Reynolds-number slender-body theory, and we discuss the application of this theory to biological systems in the final sections.

1.2 Ciliary and Flagellar Propulsion in Perspective

In the scheme of life the role of contractile elements is a major one. Some life functions are totally dependent on them and others are more efficient because of them. When considered as isolated structures, contractile elements are those that use up biochemical energy in doing mechanical work. But the artificiality of such isolation becomes evident when one considers the role of contractility in the two other kinds of work requiring biochemical energy, synthesis and concentration (Lehninger 1971). Consider, for example, the process by which bone is aided in its growth by motion-generated stresses (Black & Korostoff 1974). By such relationships contractility contributes significantly to biosynthesis. Consider also the recent demonstration that cilia lining the brain ventricles have a significant effect on transmural transport in the ependyma (Nelson 1975), and the link between contractility and concentration will be established. To be sure, concentration and biosynthesis can be said to generate some movements; the most familiar examples are the growth tropisms (geo- and photo-) and the opening and closing of stomates caused by turgor pressure changes, which is characteristic of the higher vascular plants. This review is not, however, concerned with noncontractile adaptations. In any case, it is a gross oversimplification to view contractile structures as isolated elements for they are an integral part of the life functions to which they contribute.

Contractile elements can be grouped into four classes for convenience: 1. prokaryotic flagella, 2. cytoplasmic filaments or microtubules, 3. eukaryotic cilia and flagella, and 4. smooth or striated muscle. Although only the first and third classes are the concern of this review, some perspective into the utility of each type of contractile element can be gained by an all-inclusive overview such as the one presented in Table 1. This table illustrates how natural selection has distributed the mechanisms of contractility among living things and their life functions. The life functions that directly include some aspect of propulsion are irritability, contractility, ingestion, digestion, circulation, reproduction, respiration, and excretion.

It may also be noted upon examination of the table that no mode of contraction has gained the exclusive right to serve a given life function. This diversity has come about not only because natural selection is opportunistic (i.e. whichever adaptation works at the "moment of truth" is the one selected for) but also because at least two classes of contractile elements—cytoplasmic microtubules and eukaryotic flagella—are interchangeable (e.g. the amoeboflagellate *Naegleria*).

In order for contractile elements to maintain required services for the life functions they attend and in order for them to assist in biosynthetic and concentration work, they must be provided with biochemical energy and structural replacements by biosynthesis (which includes chemical respiration) and concentration. Such interdependence is a natural consequence of the division of labor that is so characteristic of living systems and that must be kept in mind lest one be tempted to analyze in situ contraction as an isolated process. Inversely, this interdependence has the benefit of allowing one to learn some of the details about biosynthesis and concentration from their contributions to the contractile process.

In general the elucidation of propulsion by contractile elements is just as much an exercise in relating structure and function as is any other biological investigation. By contractile structure we mean the somewhat stable part of the contractile element—the components that actually move—while function refers to the motion, which is to say the performance of mechanical work. Such a division is, of course, artificial not only because changes in structure accompany all contractions (e.g. cytoplasmic streaming in amoebas) but also because the propulsive structure of the system—i.e. the part of the contractile system in contact with the fluid—and the physical structure of the fluid—density, homogeneity, viscosity, pressure, etc—interact to form a dynamic feedback relationship that is not always predictable from knowledge of structure alone.

Table 1   The occurrence of contractile elements in organism life functions[a]

Biological group	Locomotion	Ingestion	Digestion	Irritability	Circulation	Reproduction	Respiration	Excretion
Prokaryotes								
Eubacteria	flagellum	—	—	flagellum?	—	—	—	—
Spirochaetes	flagellum(?)	—	—	—	—	—	—	—
Blue-Green Algae	flagellum	—	—	flagellum?	—	—	—	—
Vascular Plants	—	—	—	—	cytoplasm	flagellum	—	cytoplasm
Protista								
Flagellata	flagellum, cytoplasm	flagellum	cytoplasm	flagellum?	cytoplasm	flagellum, cytoplasm, flagella	—	cytoplasm
Amoebida	cytoplasm, flagellum	cytoplasm	cytoplasm	cytoplasm?	cytoplasm	cytoplasm	—	cytoplasm
Ciliata	cilia, cytoplasm (spasmoneme)	cilia	cytoplasm	cytoplasm?	cytoplasm	cytoplasm, flagella, cilia	—	cytoplasm
Animals								
Porifera	flagella (larvae)	flagella, smooth muscle, cytoplasm	cytoplasm	—	flagella	cytoplasm, flagella	flagella	flagella
Cnidaria	striated muscle, smooth muscle, cilia (larvae)	striated muscle, cytoplasm	cilia	cilia (in ctenophores)	cilia	flagella, cilia	cilia	cilia
Platyhelminthia	smooth muscle, cilia	smooth muscle	smooth muscle, cytoplasm	cilia	—	smooth muscle, flagella, cytoplasm	—	cilia
Nematoda	smooth muscle	smooth muscle	smooth muscle	—	—	smooth muscle, flagella	smooth muscle	—
Annelida	smooth muscle, cilia (larvae)	smooth muscle, cilia	smooth muscle, striated muscle, cilia	cilia	cilia (where blood system absent), smooth muscle	flagella, smooth muscle	striated muscle (moves gills)	cilia, smooth muscle
Arthropoda	striated muscle, smooth muscle	smooth muscle, striated muscle	striated muscle	striated muscle (which moves sensory organs)	striated muscle	striated muscle, cilia (a few crustacea), flagella	—	cilia (few species), striated muscle
Mollusca	striated muscle, smooth muscle, cilia (larvae)	striated muscle, cilia	cytoplasm, cilia, smooth muscle	smooth muscle (chromophores)?	smooth muscle, striated muscle	cilia, flagella	smooth muscle, striated muscle, cilia (gills)	cilia, cytoplasm
Ectoprocta	cilia (larvae)	cilia	cilia	cilia (larvae)	—	flagella	—	cytoplasm
Echinodermata	striated muscle, cilia (larvae), smooth muscle	striated muscle, smooth muscle	cytoplasm, cilia, smooth muscle	—	cilia	flagella, cilia, cytoplasm, smooth muscle	cilia	cytoplasm
Chordata (vertebrates only)	striated muscle (of fibroblasts, white blood cells, etc), cytoplasm	striated muscle, cilia (in groups below reptiles), smooth muscle	smooth muscle, cilia (in groups below reptiles)	cilia (in membrane lining ventricles), smooth muscle (e.g. ciliary muscle)	smooth muscle, cardiac muscle, striated muscle	smooth muscle, flagella, cilia (both female and male)	smooth muscle, striated muscle	smooth muscle, cilia (in groups below birds)

[a] Compiled from Andrew (1959), Barber (1974), Bharier & Rittenberg (1971), Borradaile & Potts (1958), Bourne (1960), Gardner (1976), Prosser (1973), Rivera (1962), Smith et al (1971), and Weis-Fogh (1975).

1.3   An Overview of Structure and Function of Whiplike Contractile Elements

The contractile elements considered in this review are all slender oscillators that are responsible for propulsion of the organism in the fluid or propulsion of the fluid alone. They are called *cilia* or *flagella,* but the latter term is somewhat ambiguous because it is used for two evolutionarily unrelated structures: prokaryotic and eukaryotic flagella. Furthermore, cilia and eukaryotic flagella are closely related organelles having essentially the same structure (Section 5.1) for a given motion, and both utilize ATP as a primary energy source. The energy source for prokaryote motility is unknown (Larsen et al 1974), however, so there are few restrictions at present on the energy aspects of models for their motion. What can be said with some conviction is that this energy is devoted to helping the microbe move to a new environment, an ability that gives the motile prokaryote a distinct advantage over the nonmotile one.

2   FLUID MECHANICS OF SLENDER BODIES AT LOW REYNOLDS NUMBERS

2.1   Background

Since the oscillatory motions of cilia and flagella produce a mean translational motion it is important to define two Reynolds numbers, one for each kind of motion. The Reynolds number defined by the propulsive velocity, U, and the typical dimension of the organism L is UL/v where v is the kinematic viscosity of the organism's liquid environment; values range from 10^{-6} for many bacteria to about 10^{-2} for spermatozoa, and most of the organisms considered here lie within this range. Equally important is an oscillatory Reynolds number, Re, based on the radian frequency of beating of the organelle, ω, and the typical length of that organelle, l (Re $= \omega l^2/v$); typical values of this quantity are about 10^{-3}. Thus the fluid motions that result are dominated by viscous forces and the inertial forces usually play little part in the propulsive mechanisms. Of course there exist organisms in all ranges of Reynolds number, but the difficulties in the fluid-mechanical analyses when the Reynolds numbers approach unity are such that little quantitative work has as yet been done for natural swimming in this regime.

Before we can deal sensibly with the hydromechanics of cilia and flagella it is necessary to digress and discuss the fluid-mechanical basis for the analyses of the low-Reynolds-number flows past slender bodies. That we return to these basic principles is a reflection of the fact that the study of these biological systems has actually been one of the principal motivating factors for the development of slender-body theory at low Reynolds numbers (the other being studies of suspensions of elongated particles).

2.2   Fundamental Singularities

Analysis of the detailed hydrodynamics of low-Reynolds-number flows due to cilia and flagella has been greatly aided by the development of methods to construct the flow fields by means of distributions of fundamental singularities. For the purpose of describing these methods we must dwell briefly on the nature of the fundamental solutions to the equations of motion for an incompressible inertialess Newtonian fluid of viscosity μ. They consist of a continuity condition on the fluid velocity $\mathbf{u}$,

$$\nabla \cdot \mathbf{u} = 0, \tag{1}$$

and since there are no inertial forces, a condition of force equilibrium

$$\nabla p = \mu \nabla^2 \mathbf{u} \tag{2}$$

containing the fluid pressure, p. From this it follows that p is a harmonic function, and since $\nabla^4 \mathbf{u} = 0$, the velocity is a bi-harmonic function. The primary fundamental solution to these equations due to a single point force, $\mathbf{F}$, in an unbounded inertialess fluid was first obtained by Oseen (1927), developed further by Burgers (1938), and named a *stokeslet* by Hancock (1953). If one represents the strength and direction of the singular force at the origin of a coordinate system $\mathbf{x}$ by $8\pi\mu\boldsymbol{\alpha}$, where $\boldsymbol{\alpha}$ denotes the stokeslet strength and direction, the resulting fluid velocity and pressure are, respectively, (see, for example, Chwang & Wu 1975)

$$\mathbf{u}(\mathbf{x}; \boldsymbol{\alpha}) = \boldsymbol{\alpha}/r + (\boldsymbol{\alpha} \cdot \mathbf{x})\mathbf{x}/r^3,$$
$$p(\mathbf{x}; \boldsymbol{\alpha}) = 2\mu\boldsymbol{\alpha} \cdot \mathbf{x}/r^3, \tag{3}$$

where $r = |\mathbf{x}|$. It follows that a derivative of any order of this solution is also a solution to the basic equations. Thus one can construct higher-order singularities such as a Stokes doublet, Stokes quadrupole, etc. Batchelor (1970b) indicated how a Stokes doublet could be decomposed into an antisymmetric component representing the flow field due to a singular moment of strength $8\pi\mu\gamma$ and called a *couplet* [Chwang & Wu (1974) call this a *rotlet*] with velocity and pressure

$$\mathbf{u} = \gamma \times \mathbf{x}/r^3; \qquad p = 0 \tag{4}$$

and a symmetric component representing a pure straining or extensional motion of the fluid and termed a *stresslet*. Furthermore, a Laplacian of the stokeslet solution leads to a *potential doublet* of strength $\boldsymbol{\delta}$ for which

$$\mathbf{u} = -\boldsymbol{\delta}/r^3 + 3(\boldsymbol{\delta} \cdot \mathbf{x})\mathbf{x}/r^5; \qquad p = 0 \tag{5}$$

and which has zero vorticity. One sees that this has the same kinematic form as the conventional doublet in potential flow of an inviscid fluid but that its dynamic contribution to pressure is now zero because the inertia terms have been deleted.

Hancock (1953) seems to have been the first to attempt to use linear superposition of these singularities [permissible because the basic Equations (1) and (2) are linear] in order to construct the fluid mechanics of flagellated microorganisms; his classic work with Sir James Gray (Gray & Hancock 1955) remains a landmark in this respect for both biologists and fluid dynamicists. These works are the forerunners of slender-body theory and resistive-force theory as applied to microorganism locomotion; we return to these subjects shortly.

Chwang & Wu (1974, 1975) and Chwang (1975) have recently shown how solutions to many complex flows may be constructed by superposition of these fundamental singularities; indeed, for mathematically simple bodies such as spheroids in mathematically simple flow fields (uniform flow, shear flow, quadratic flow, extensional flow, etc) exact solutions are obtained. The simplest example is that of rectilinear translation (at velocity U) of a sphere of radius a, which requires at the center of the sphere only a stokeslet of strength $3aU/4$ in the forward direction and a potential doublet of strength $a^3U/4$ in the opposite direction in order to satisfy the no-slip boundary condition at the surface of the sphere. Indeed, one can visualize the stokeslet as simulating the drag on the body; this is the dominating effect in the far-field since a stokeslet, being the lowest-order singularity, decays least rapidly (like $1/r$). Furthermore, the potential doublet provides the finite geometry of the body in the near-field and its velocity contribution is necessary to satisfy the no-slip condition at the body surface (see Lighthill 1975, p. 48).

We reiterate Chwang & Wu's (1975) observation on the exact solution for the translation of a prolate spheroid of major axis, a, and minor axis, b. They observed that if the translational velocity is decomposed into components U_s and U_n parallel and perpendicular to the major axis and if one examined the force on an element of this spheroid contained between two planes perpendicular to the major axis and length ds apart, then this was composed of two components F_s and F_n in the same two directions where

$$F_s = -C_s U_s\, ds; \qquad F_n = -C_n U_n\, ds, \tag{6}$$

and C_s and C_n were simple constants dependent only on μ, a, and b and *independent* of the position of the element or the velocities U_s and U_n. This is a remarkable example of a case in which the resistive-force theory that we examine below holds exactly, irrespective of the slenderness of the body. For a slender prolate spheroid such that $b/a = \varepsilon \ll 1$, the resistive coefficients, C_s and C_n, become

$$C_s = \frac{2\pi\mu}{\ln(2a/b) - \frac{1}{2}} [1 + O(\varepsilon^2)] \tag{7}$$

$$C_n = \frac{4\pi\mu}{\ln(2a/b) + \frac{1}{2}} [1 + O(\varepsilon^2)] \tag{8}$$

2.3 Small Inertial Effects

It is wise to note at this point that the solutions above represent exact solutions only at *zero* Reynolds number. Introduction of the small contribution of inertia at low but finite Reynolds numbers necessitates re-examination of the far-field where the magnitude of the inertia terms becomes comparable with the viscous terms and leads, for example, to the well-known Stokes paradox for translation of an infinitely long cylinder. Linearization of these far-field inertia effects by means of Oseen's approximation still, however, permits the construction of flow fields by means of a modified set of fundamental solutions in which the *oseenlet* replaces the stokeslet. Developments of slender-body theory along these lines is only beginning [see a recent paper by Chwang & Wu (1976)]. Finally it should also be noted that, of course, there exists the fundamental solution of the entire Navier-Stokes equations for a singular force known as the *round laminar jet* and due to Slezkin (1934),

Landau (1944), and Squire (1951). The stokeslet is simply the limiting case of the round laminar jet for an inertialess fluid. However the nonlinearity of the Navier-Stokes equations does *not* permit superposition of these fundamental solutions.

2.4 Image Systems for Singularities

In the proximity of a boundary, whether it is a solid wall, a free surface, or a hypothetical boundary simulating a line of symmetry in the flow under consideration, it becomes advantageous to develop image systems for the fundamental singularities constructed so that the boundary conditions on that wall are automatically satisfied. In inviscid potential flow this is usually a simple matter since, for example, a solid-plane boundary requires only the identical singularity at the image point in order to satisfy the condition of zero normal velocity. At low Reynolds numbers one must also satisfy the no-slip condition, and the types of singularity required at the image point in order to accomplish this are not immediately obvious. Blake (1971c) obtained the image system for a stokeslet (at various orientations) in a stationary plane boundary, and more recently Blake & Chwang (1974) derived similar image systems for a couplet, a source, and a potential doublet. Some of these are indicated schematically in Figure 1. One of the important effects of the presence of the wall (or equivalently the image system) is that the nature of the far-field is altered. A stokeslet oriented parallel to a wall leads to a far-field, which is a stokes doublet

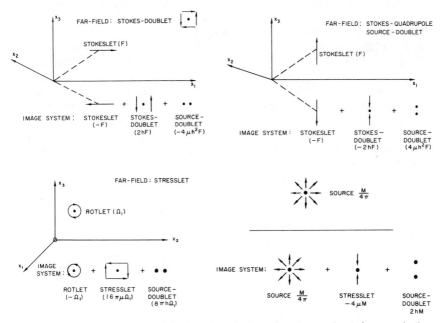

Figure 1 The image singularities in a no-slip boundary ($x_1 x_2$ plane) for a stokeslet tangential to the boundary, a stokeslet normal to the boundary, a rotlet whose axis is parallel to the boundary, and a source.

decaying like r^{-2} rather than the r^{-1} of a stokeslet in unbounded fluid. On the other hand the far-field of a stokeslet oriented perpendicular to a wall is even weaker and is like a stokes quadrupole or potential doublet. Blake & Sleigh (1974) believe that this has important consequences for the hydromechanics of cilia or for flagella near walls. The far-fields of the other singularities in the presence of a wall are similarly affected, the far-fields for both a rotlet and a source becoming stresslets (like r^{-2}); note that this differs from the far-field of a source near a wall in inviscid potential flow in the absence of a no-slip condition that is like a potential doublet (r^{-3}).

2.5 Slender-Body Theory

The objective of slender-body theory is to take advantage of the slenderness in order to achieve simplifications in obtaining approximate solutions for the flow around such bodies. The development of low-Reynolds-number slender-body theory evolved through the work of Burgers (1938), Broersma (1960), and Tuck (1964); recent work by Taylor (1969), Tillett (1970), Batchelor (1970a), Cox (1970), and Blake (1974b) has concentrated on construction of slender-body solutions by distributions of fundamental singularities along an axis of the body. [With the exception of Batchelor's (1970a) work on arbitrary cross-section, researchers have concentrated on bodies of circular cross-section.] A simple but elegant demonstration of low-Reynolds-number slender-body theory is given by Lighthill (1975, p. 49).

In choosing axes fixed relative to a particular section of the slender body under examination (Figure 2), one seeks the distribution of stokeslets, doublets, etc on the axis of the body that will satisfy the no-slip condition at points such as A on the surface of the slender body (local radius is a). The integrated induced velocity at such points must then be equated with the known or assumed translational velocity of the section under consideration. The result will, in general, be a system of complicated integral equations for the strength of the singularity distributions. The first simplification of slender-body theory results from the observation that the velocities induced at A by singularities outside a certain "near-field" will be dominated by the stokeslets in the "far-field" since their far-field effect (like r^{-1}) dominates that of the other singularities. Thus the primary distribution is one of stokeslets along the entire axis of the body. The boundary condition at the cross-section under consideration is satisfied by introducing a potential doublet (or if necessary other

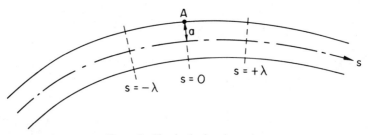

Figure 2   Slender-body schematic.

singularities) only within the near-field. In particular the integrated effect of singularities with a far-field decay faster than r^{-1} can be fairly accurately determined by terminating the integration at some distance $s = \pm \lambda$ from the section under consideration where $s_1, s_2 \gg \lambda \gg a$, s_1 and s_2 being the distances to the ends of the slender body. On the other hand, the integration for the velocity induced by the stokeslets cannot be truncated in this way and indeed yields a velocity with terms like $\ln (s_1 s_2/a^2)$. The reader is referred to Lighthill (1975, p. 49) for the forms of the integrated induced velocities. Note that this is another manifestation of Stokes' paradox for the translation of an infinitely long cylinder; when s_1 or s_2 tend to infinity, the boundary condition at the section under consideration cannot be satisfied. We must also note that such a construction is limited to sections sufficiently far from the ends of the slender body; Tillett (1970) has examined some of the problems associated with such "end effects."

The net result of these considerations is that one must seek the strength and direction of stokeslets distributed along the entire axis of the slender body plus the local distribution of higher-order singularities that satisfies the required boundary condition at every point on the slender-body surface. A useful approximate way of implementing this has been suggested by Lighthill in his John von Neumann lecture (June 1975 at Rensselaer Polytechnic Institute, Troy, New York) and by R. Johnson and T. Y. Wu (private communication). If the local radius of curvature of the body is large compared with a, then the combined effects of both the near- and far-field distributions may be replaced by a distribution of stokeslets alone in the far-field regions, $s > \delta$ and $s < -\delta$. For the components of the stokeslets normal to the axis, $\delta = a/2\sqrt{e}$, whereas for the components tangential to the axis, $\delta = a\sqrt{e}/2$. This observation considerably simplifies the algebra required in obtaining solutions for the motions of slender bodies of more complicated geometry.

The simplest solutions are those for the translation of straight slender cylinders as obtained by Tillett (1970) and Cox (1970). Defining force coefficients as the force per unit length of the body divided by the translational velocity, U, Cox (1970) improved on the original work of Burgers (1938) and Broersma (1960) to show that the force coefficient for a cylinder, with length $2l$ and maximum radius a, moving perpendicular to its axis was

$$C_n = \frac{4\pi\mu}{\ln (2l/a) + C_1} + O\left[\frac{\mu}{(\ln l/a)^3}\right], \tag{9}$$

while that for motion parallel with its axis, C_s, was

$$C_s = \frac{2\pi\mu}{\ln (2l/a) + C_2} + O\left[\frac{\mu}{(\ln l/a)^3}\right]. \tag{10}$$

The value of C_2 was $C_1 - 1$ and the value of C_1 depended on the axial variation of the radius of the cylinder. A uniform axial cylinder took a value $C_1 = \ln 2 - \frac{1}{2} = 0.193$, whereas a prolate spheroid yielded $C_1 = +\frac{1}{2}$. The latter agrees with the results of the exact solution for a spheroid, Equations (7) and (8); in this case the answers are more accurate than the error terms in (9) and (10) indicate.

2.6   Resistive Force Coefficients

The translation of any rigid slender body through a viscous fluid can be fairly readily analyzed by such methods, provided the radius of curvature is large compared with the body radius. In the present context it is useful to view the results by decomposing the velocities of each element relative to the fluid at infinity into normal and tangential components, U_n and U_s, and similarly decomposing the force on that local element into components involving normal and tangential force coefficients as defined in Equations (6). The resulting values of C_n and C_s always take the forms of Equations (9) and (10), but the coefficients C_1 and C_2 are dependent on the overall geometry of the body (through the integration of stokeslets along the entire axis). For example, a circular ring or torus moving in the direction of one of its major diameters takes values of $C_1 = 0.74$, $C_2 = 0.24$ (R. Johnson and T. Y. Wu, private communication).

This is the background for what has come to be known as resistive-force theory in which the force on any element of a slender body such as a cilium or flagellum is calculated from (a) motion of each elemental length of the organelle relative to the fluid at infinity and (b) force coefficients, C_n and C_s, which are determined from the geometry alone.

Hancock (1953) and Gray & Hancock (1955) made a major contribution to research on microorganism movement by applying slender-body theory to the analysis of a flagellum along which travelling waves were propagating (Figure 3). The motion of each individual element relative to the fluid at infinity is thus comprised of a combination of the oscillatory motions due to the passage of the wave and the steady translation of the flagellum through the fluid. The results of Hancock's analyses and the subsequent force coefficients derived by Gray & Hancock (1955) can be interpreted in a simple qualitative way by dividing the axial stokeslet distributions for such a flow into components due to each of these motions and by

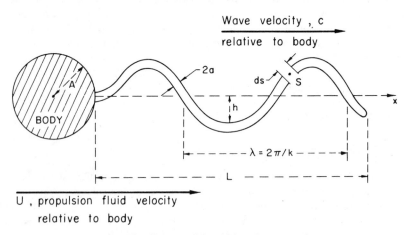

Figure 3   Flagellar propulsion with a planar waveform.

examining primarily the oscillatory motions for these generally involve larger velocities. It follows that the stokeslet distributions due to the oscillatory motions will be harmonic with distance s from the section at which the integrated velocity is being evaluated (see last section). Lighthill has pointed out that this will result in an integral that will converge more rapidly than those that gave terms like $\ln(s_1 s_2/a^2)$ in the last section; indeed, the resulting velocity will instead involve a term like $\ln(\lambda/a)$, so that the wavelength, λ, will be the effective body length rather than the overall flagellar length. The resulting force coefficients according to Gray & Hancock (1955) are

$$C_s = 2\pi\mu \Bigg/ \left[\ln\left(\frac{2\lambda}{a}\right) - \frac{1}{2} \right],$$ (11)

$$C_n = 2C_s.$$ (12)

Recently Lighthill (1975) has shown that the evaluation of the integrals harmonic in s leads to an "effective length" $l^* = 0.09\lambda$, so that

$$C_s = \frac{2\pi\mu}{[\ln(2l^*/a) - \frac{1}{2}]} = \frac{2\pi\mu}{[\ln(2\lambda/a) - 2.90]},$$ (13)

$$C_n = \frac{4\pi\mu}{[\ln(2l^*/a) + \frac{1}{2}]} = \frac{4\pi\mu}{[\ln(2\lambda/a) - 1.90]}.$$ (14)

These can, however, be regarded as only approximate; indeed, it is likely that more accurate coefficients, which are presently being sought, will also involve the total flagellar length, L. One indication of this is suggested above, since clearly the stokeslet components due to overall translation of the flagella will contribute terms like $\ln(2L/a)$ as in the case of the translation of rigid slender bodies.

We return later to the consequences of such analyses in the context of the fluid mechanics of biological slender bodies. But the section would not be complete without the addition of one other force coefficient, namely that due to rotation at angular velocity Ω of an element of a slender body about its own axis. Chwang & Wu (1974) have shown that the resulting moment M acting on the body about the axis is simply given by

$$M = C_M \Omega \, ds; \qquad C_M = 4\pi\mu a^2.$$ (15)

2.7   Wall Effects on Slender-Body Motions

The resistive coefficients on any body are clearly altered by the presence of a nearby boundary. Moreover, there are many situations in microorganism propulsion and in ciliary mechanics in which the slender bodies operate close to solid boundaries. Examples are (a) effects of the presence of the epithelial wall on ciliary dynamics (Blake & Sleigh 1974), (b) the motion of spermatozoa, in vivo, either in close proximity to a single wall or within narrow passages, and (c) the effects of a coverslip on studies of microorganism propulsion (Katz 1974 and Winet 1973). Concern about such wall effects has led to a significant number of recent papers on the influence of nearby boundaries on resistive coefficients for slender bodies.

When the distance of the center of the body from the wall, h, is large compared with its length, $2l$, the results of Brenner (1962) for the motion of bodies of arbitrary shape near walls are very useful. Brenner (1962) showed that the wall-affected resistive coefficient (drag/$2lU$) denoted by C^* was related to the coefficient, C, in the absence of the boundary by

$$\frac{C^*}{C} = \left[1 - Z\frac{C}{3\pi\mu}\frac{l}{h} + O\left(\frac{l}{h}\right)^3\right]^{-1}, \tag{16}$$

where Z was a function only of the geometry and the direction of particle motion. Examples of the numerical values of Z are (a) $Z = 9/16$ for motion parallel to a

Table 2 Wall effects on resistive coefficients for the translation of straight slender cylinders[a]

Orientation Parallel to the Wall

$l/h \ll 1$

$l/h \gg 1$

$$C_{s1} = 2\pi\mu \left/ \left[\ln\left(\frac{2l}{a}\right) - 0.807 - \frac{3l}{8h}\right]\right. \qquad C_{s1} = 2\pi\mu \left/ \left[\ln\left(\frac{2h}{a}\right)\right]\right.$$

$$C_{n2} = 4\pi\mu \left/ \left[\ln\left(\frac{2l}{a}\right) + 0.193 - \frac{3l}{4h}\right]\right. \qquad C_{n2} = 4\pi\mu \left/ \left[\ln\left(\frac{2h}{a}\right)\right]\right.$$

$$C_{n3} = 4\pi\mu \left/ \left[\ln\left(\frac{2l}{a}\right) + 0.193 - \frac{3l}{2h}\right]\right. \qquad C_{n3} = 4\pi\mu \left/ \left[\ln\left(\frac{2h}{a}\right) - 1\right]\right.$$

Orientation Normal to the Wall

$l/h \ll 1$

$l/h \to 1$

$$C_{n1} = C_{n2} = 4\pi\mu \left/ \left[\ln\left(\frac{2l}{a}\right) + 0.193 - \frac{3l}{4h}\right]\right. \qquad C_{n1} = C_{n2} \to 4\pi\mu \left/ \left[\ln\left(\frac{2l}{a}\right) - 0.75\right]\right.$$

$$C_{s3} = 2\pi\mu \left/ \left[\ln\left(\frac{2l}{a}\right) - 0.807 - \frac{3l}{4h}\right]\right. \qquad C_{s3} \to 2\pi\mu \left/ \left[\ln\left(\frac{2l}{a}\right) - 1.75\right]\right.$$

[a] Compiled from Brenner (1962), Katz & Blake (1975), Katz, Blake & Paveri-Fontana (1975), de Mestre (1973), and de Mestre & Russel (1975). Second subscript refers to direction of translation and force.

single solid plane wall, (b) $Z = 9/8$ for motion perpendicular to a single solid plane wall, (c) $Z = 1.004$ for motion parallel to and equidistant between two solid plane walls, and (d) $Z = 2.1044$ for motion along the axis of a cylindrical tube; other useful values are also given by Brenner (1962). First-order, or $O(l/h)$, corrections for wall effects on the resistive coefficients are therefore readily obtained by combination of the result (16) with the coefficients, such as (9) and (10) in the absence of walls. Some examples are listed in Table 2, with Cox's coefficients for slender cylindrical bodies (Cox 1970).

Another general result of particular importance for microorganism propulsion can be readily deduced from Brenner's result; we shall see that the value of the ratio $\gamma = C_s/C_n$ for a slender element of a cilium or flagellum is of considerable consequence to its propulsive capability. From (16) it is readily seen that the first-order wall effect on this ratio is given by

$$\gamma = C_s^*/C_n^* = \gamma_\infty \left[1 - Z \frac{(C_n^\infty - C_s^\infty)}{3\pi\mu} \frac{l}{h} \right], \tag{17}$$

where C_n^∞, C_s^∞ are resistive coefficients in the absence of the wall or walls and $\gamma_\infty = C_s^\infty/C_n^\infty$. Notice in particular that since $C_n > C_s$ and provided Z is positive, the effect of the nearby boundary always *decreases* γ. Note from Brenner's (1962) quoted values for Z that this quantity is invariably positive for solid boundaries.

When the slender body is closer to the wall so that l/h is no longer small, the geometry of the body becomes important and a more detailed analysis becomes necessary. Katz & Blake (1975) and Katz, Blake & Paveri-Fontana (1975) recently examined this situation by constructing the flow by a distribution of stokeslets along the axis of slender bodies and satisfying the no-slip condition at the wall or walls by adding the appropriate system of image singularities. The resulting integral equations are solved by the techniques developed by Tillett (1970) and Cox (1970). Solutions were obtained for slender cylinders parallel to a single-plane wall and between two plane walls when the distance, h, from the wall is much smaller than the length, $2l$ (but still much greater than the radius a). Their results are included in Table 2; it is significant to note that h now replaces l in the leading term for the coefficient. De Mestre (1973) and de Mestre & Russel (1975) have examined the wall effect for general values of l/h (both large and smaller) and orientations both parallel and perpendicular to the wall. Their results converge asymptotically to the simple results obtained by Brenner's relation at small l/h [provided some typographical errors in de Mestre & Russel (1975) are corrected] and to the results of Katz, Blake & Paveri-Fontana (1975) for parallel slender cylinders. The additional results for perpendicular slender cylinders as $l \to h$ are also incorporated into Table 2; it is reassuring that if one arbitrarily sets $l/h = 1$ in the expressions for $l/h \ll 1$, then the result differs only slightly from the more exact expressions for $l/h = 1$.

3 PROKARYOTIC CELL PROPULSIVE STRUCTURE AND FUNCTION

The flagella of bacteria are composed of a helical protein, flagellin. From one to eleven strands of flagellin coil together to form a single flagellum sheath (Gerber

1975) which has an amorphous core and a radius of $1.2–2.0 \times 10^{-6}$ cm. Both motile and fixed-and-stained flagella form a helix that has a pitch range of $1.5–2.5 \times 10^{-4}$ cm (Lowy & Spencer 1968). Each flagellum is attached to the cell at its base; the attachment site, called the "hook-basal body complex" (DePamphilis & Adler 1971), consists of four rings around the flagellar cylinder, each 2.25×10^{-6} cm in diameter as shown in Figure 4. The most important of these rings are apparently the S and M rings, which are located at the base of the hook.

The contractile mechanism for bacterial flagella has been a subject of recent controversy (Routledge 1975). Doetsch (1966) first proposed the rather startling hypothesis that the material of the flagellum rotates relative to the cell body, indeed that the hook rotates in the cell wall, thus providing a unique example in nature of continuous rotational deformation. Berg & Anderson (1973) and Berg (1974, 1975) have further examined the evidence for, and apparent quantitative features of, this bacterial motor system. The motor seems to consist of rotation of the S and M rings with the flagellum that they carry being driven by some mechanochemical process, presumably akin to the cross-bridge-stepping of heavy meromyosin on actin in striated muscle (Berg & Anderson 1973). However, some recent evidence (Larsen et al 1974) indicates that ATP is not the energy source for this process, so cross-bridge models may be premature. Nevertheless, the basic model of a bacterial flagellar motion appears to be gaining acceptance (Silverman & Simon 1974) at the expense of alternative hypotheses that the contraction consists of a helical wave

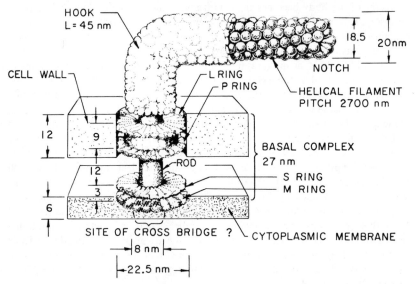

Figure 4   The hook-basal body complex at the junction of a prokaryotic cell and its flagellum. In the rotating-shaft models motion is presumed generated between the M and S portion of the hook and the cytoplasmic membrane. Possible sites of cross-bridges for a model analogous to the muscle sliding mechanism have been indicated. (Adapted from Routledge 1975.)

passing along the flagellum due to propagation of dislocations in the molecular structure of the outer sheath (Harris 1973, Calladine 1974). In terms of the external hydromechanics of the helical flagellum the two models differ only in the material motion of the surface of the flagellum. In the basal motor hypothesis the flagellum

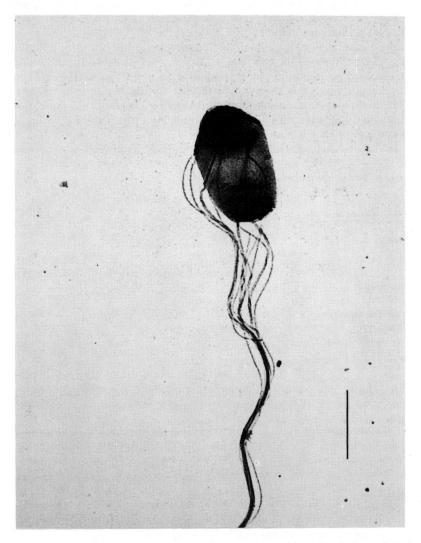

Figure 5   The flagellated bacterium *Salmonella abortus-equi* with its flagella bent aftward and associated in a flagellar bundle (Routledge 1975). This is a fixed specimen. The swimming organism would show less clearance between flagella. The scale bar is 1 μm. (We are indebted to Dr. L. M. Routledge for this photograph.)

is basically like a rigid corkscrew rotating relative to the head; in the wave-propagation model the material of the flagellum does not rotate relative to the material of the cell body, but the helix is formed by the helical conformation of the propagated wave. Unless one can observe the material rotation of the flagellum, the two motions appear identical and thus it is difficult to distinguish between them.

Many bacteria (e.g. *Escherichia coli, Salmonella*; see Table 3) have several flagella attached at points distributed over the surface of the cell (see Figure 5). When such bacteria are swimming, the separate flagella come together in a synchronous flagellar bundle, which propels the cell (Iino & Mitani 1966). In some strains, periods of concerted swimming are interrupted by brief periods of erratic wobbling ("twiddling"), which may be caused by the fact that the bundle has come apart and each flagellum is acting independently (Macnab & Koshland 1972, Adler 1976). Anderson (1975) has recently discussed the qualitative hydromechanical features of the formation of flagellar bundles.

The close association of rotating flagella in the bundle clearly implies the presence

Table 3 Prokaryote propulsion data

Prokaryote	Body Length (B) (μm) × width (μm)	Shape	Flagellar No. of flagella	Length, L (μm) or No. of waves × wavelength
Eubacteria				
Bacillus megaterium	3.0 × 1.5		~36	2.5 × 3.9
Bdellovibrio bacteriovorus	1.4 × 0.25		1	$L = 3.4$
Clostridium tetani	6.0 × 0.5		~15	4 × 1.8
Escherichia coli	3.0 × 1.5		6	2 × 2.7
Photobacterium phosphoreum	1.2		1	1.5 × 3.1
Pseudomonas aeruginosa	1.5 × 0.5		1	2 × 1.7
Salmonella typhosa	2.5 × 0.65		6	4.2 × 2.5
Sarcina ureae	2.0		1/cell	4 × 3.2
Serratia marcescens	1.0 × 0.5		>4	1.5 × 2.3
Spirillum serpens	3.0 × 1.0		>14	1.1 × 2.7
Spirillum volutans	13.5 × 1.5		>46, 200	1.1 × 6.5
Thiospirillum jenese	40.0 × 4.0		—	—
Vibrio cholera	3.0 × 0.45		1	1.0 × 2.4
Spirochaetes				
Cristispira balbianii	80.0 × 2.0		>100	3.5 × 6.0
Cristispira sp.	44.0 × 1.4		—	3.1 × 14.2
Leptospira icterohemorrhagiae	7.5 × 0.27		1	—
Spirochaeta litoralis	13.0 × 0.45		3	1.6 × 8.2
Spiroplasma citri[b]	6.0 × 0.16		?	4.1 × 0.97

[a] Flagellum was tethered.   [b] Observed in 0.25% agar.

of lubricating layers of fluid between the individual flagella and thus a significant fluid resistance internal to the bundle, especially in the basal motor model; to our knowledge the hydromechanics of this situation has not as yet been closely examined quantitatively, although Berg & Anderson (1973) discount it. Viewed from the exterior fluid the flagellar bundle could be considered as a single slender body whose mean surface rotates relative to the head if the basal motor model is assumed. Thus, whether the principal propulsive unit is a single flagellum or a bundle will have relatively minor effects on the external hydromechanics within the context of a particular contractile process. Finally, it is noteworthy that many bacteria exhibit an increased motility with small increases in viscosity of the surrounding medium and a subsequent decrease with larger increases in viscosity (Schneider & Doetsch 1974 and Shoesmith 1960).

Since the hydromechanics of bacterial flagella is best dealt with in conjunction with the hydromechanics of eukaryotic flagella, we delay the details until Section 4.6.

Table 3 *Continued*

Bundle			Organism		
Amplitude h (μm)	U/B	U/c	Body rotation, Ω (sec^{-1})	Conditions	References
0.94	6.7	—		19–25°C	*139, 183*
damped	100	—	600[a]		*173, 208*
0.24	—	—			*32, 139*
0.60	10	—	78[a]	19–25°C	*13, 139, 183*
0.40	—	—			*32, 139*
0.17	40	—		19–25°C	*139, 183*
0.17	10	—			*32, 62, 139*
0.38	10	—		19–25°C	*139, 183*
0.09	30	—		19–25°C	*139, 183*
0.55	6.7	[b]		19–25°C	*139, 183*
1.47	6.3	[b]	$\Omega/kc = 0.37$	20°C	*32, 62, 152, 158, 227, 233*
—	0.5	[b]		19–25°C	*183*
0.17	16.7	—			*32, 139, 158*
—	—	—			*31, 32, 120*
1.69	0.36	0.16	300	20°C	*31, 55*
—	2	—		19–25°C	*31, 32, 65, 158, 183*
0.84	0.46	0.08	300	20°C	*55, 104*
0.18	—	—		30°C	*67*

[a] Flagellum was tethered.   [b] Observed in 0.25% agar.

4   EUKARYOTIC CELL PROPULSIVE STRUCTURE AND FUNCTION

4.1   Structure of Cilia and Eukaryotic Flagella

Much more is known of the structure and function of cilia and eukaryotic flagella (we use the single word *cilia* for convenience) than is known for prokaryotic flagella. Since there are extensive books and review articles (Gray 1928; Sleigh 1962, 1971, 1974a; Holwill 1966a, 1974; Brokaw 1975; Brokaw & Gibbons 1975) on this subject, we attempt only the briefest overview aimed at the fluid mechanician. A typical cross-sectional view of a cilium or eukaryotic flagellum is shown in Figure 6. Within a membrane is the "axoneme," which consists of longitudinal fibrils or tubules (one of the structural elements of which is tubulin) arranged as a number of peripheral pairs plus a central pair. The number of outer pairs is often nine (hence the reference to a "9 + 2" pattern), although many other numbers and modifications of this basic pattern have been observed. "Arms" consisting of dynein project from the outer pairs of fibrils. The dynein and tubulin are believed to interact in a manner analogous to heavy meromyosin and actin in striated muscle, although the precise mechanical details of this interaction have yet to be clearly identified. It has, however, been well established that the energy source, namely ATP, is the same for both systems. The details of the sliding mechanism have not been fully determined as one can gather from the variety of models still being proposed (e.g. Brokaw 1975; Costello 1973a,b; Douglas 1975; Dryl 1975; Harris & Robison 1973; Satir 1974; Summers & Gibbons 1971). An important series of electron microscopy studies by Satir (1965, 1968) and Warner & Satir (1974) have demonstrated that the

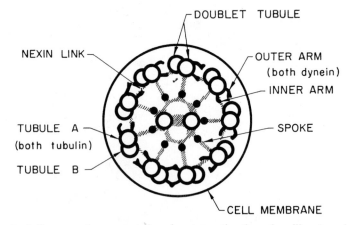

Figure 6   A diagrammatic representation of a cross-section through a cilium (or eukaryotic flagellum). Sliding is generally assumed to be generated longitudinally between the dynein arms and the B tubule across the gap spanned by the nexin link. The active role of the radial spokes in the contraction is not agreed upon. (Modified from Brokaw & Gibbons 1975.)

microtubules remain constant in length during bending and that the bending is associated with longitudinal switching of the "radial spokes"; Figure 7 (from Warner & Satir 1974) is a particularly good electron micrograph showing the rather faint,

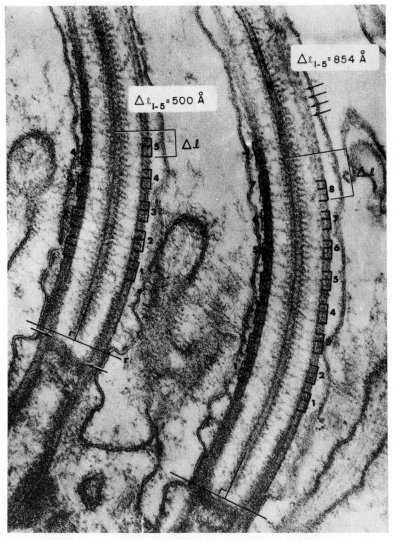

Figure 7   An electron micrograph of two bending cilia, which are longitudinally sectioned and viewed from the side (Warner & Satir 1974). Sliding displacement is indicated by Δl. The right-angle markers denote the upper part of the basal body. The radial spokes are the faint lines extending out from the core of each cilium. The cilia are about 0.18 μm wide. (We are indebted to Dr. F. D. Warner for this photograph.)

though visible, spokes on two bent cilia. These investigators have concluded that the radial spokes and their attachment to the central fibers are an important component in the generation of the sliding of peripheral subfibers past one another.

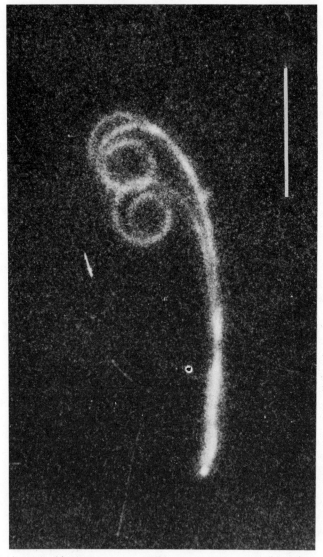

Figure 8 A sea urchin (*Tripneustes gratilla*) spermatozoon extruding microtubules by active sliding following treatment with ATP (Summers & Gibbons 1971). Scale bar is 10 μm. Note the two subfibers coiling as they are extruded. (We are indebted to Dr. K. E. Summers for this photograph.)

In an important series of physiological studies Summers & Gibbons (1971) have demonstrated the sliding phenomenon by inducing spermatozoa whose membranes have been partially digested to extrude subfibers by treating them with ATP; Figure 8 (from Summers & Gibbons 1971) shows this extrusion by sliding dramatically. These investigators propose that the total sliding force is generated between the dynein arms on one pair of peripheral fibrils and subfibril B of the adjacent pair. The discovery of motile spermatozoa lacking central fibrils (van Deurs 1974) appears to support the Summers & Gibbons form of the model. A more extensive account of the development of the "sliding filament" model since its proposal by Machin (1958) may be found in Brokaw & Gibbons' (1975) review.

It is evident that the actively generated bending moment in the contracting cilium is balanced by an internal resistance to motion (both elastic and viscous) and by the external viscous resistance. In this review we concentrate on the evaluation of the latter quantity, although it should be borne in mind that in the mechanics of cilia both elastic and viscous internal forces also appear to play significant roles and must be included in any attempt to extract knowledge of the basic activating force from knowledge of the motions of cilia and the fluid flow they create (see for example Brokaw 1970, 1971, 1972). The base of a cilium or eukaryotic flagellum is firmly imbedded in the cell membrane, and there is no question of relative motion between that base and the cell membrane as there was for prokaryotic flagella; propulsion is always achieved by propagation of waves along the cilia or flagella. The energy source for the motion, namely ATP, may either diffuse along the length of the flagellum or be diffused in from the surrounding fluid. Therefore, the principal unknown is the control mechanism. Much of the recent work has been directed toward identifying the control and feedback systems evidently associated with eukaryotic flagella and cilia (Sleigh 1966, 1969; Brokaw & Gibbons 1975).

4.2   Eukaryotic Flagellar Motions

In this section we concentrate on some of the characteristics of eukaryotic cell propulsion by single organelles, which we continue to call flagella at the risk of confusion with prokaryotic flagella; later we deal with propulsion by multiple organelles such as cilia.

The first fact to emphasize is the great variety of configurations of flagella and organisms (see Jahn & Votta 1972); here we can do no more than indicate some characteristic forms of flagellar motion and identify in particular those with different hydromechanical implications.

Many organisms, including spermatozoa, have long flagella along which they propagate either a planar wave (e.g. *Ceratium*) or a helical wave (e.g. *Trichomonas*) or some combination of the two (see Table 4); typically one finds about two wavelengths along the flagella as illustrated by the multiple exposure of sea urchin sperm in Figure 9 (Brokaw 1965) and the data of Table 4. Commonly the wave is propagated from the base to the tip, although the reverse has also been observed in the trypanosomes (Jahn & Votta 1972). Normally the direction of propulsion is opposite to the direction of wave propagation, although there exist counter examples, especially that of *Ochromonas* (Jahn, Landman & Fonseca 1964). This can

Table 4   Eukaryote propulsion by one to four flagella

Eukaryotes with 1–4 flagella	Length (B) (μm) × width (μm)	Shape	No. of flagella	Length, L (μm) or No. of waves × wavelength
"Exoflagellates"				
Ceratium fusus[a]	450 × 22		2	$L = 200$
Ceratium tripos[a]	225 × 332		2	2 × 125
Chilomonas paramecium	35 × 12		2	1.5 × ?
Chlamydomonas sp.	13		2	1 × 6.3
Codonosiga botrytis	15 × 5		1	$L = 30$
Dinophysis acuta[a]	65 × 55		2	$L = 65$
Dunaliella sp.				
Euglena gracilis	45 × 15		1	$L = 45$
Euglena viridis	52 × 17		1	1.5 × 35
Gonyaulax polyedra[a]	48 × 45		2	
Gyrodinium dorsum[a]	34.5 × 24.5		2	
Gyrodinium dorsum	34.5 × 24.5		1	
Menoidium cultellus	45 × 7		1	1.0 × 10
Monas stigmata	6		2	$L = 3, 15$
Monas stigmata	6		2	$L = 3, 15$
Ochromonas malhamensis	3		1	2.8 × 7
Peranema trichophorum	55 × 12		2	$L = 40$
Polytoma uvella	22 × 11		2	$L = 39$
Polytoma uvella	22 × 11		2	
Polytomella agilis	9.8 × 4.9		4	$L = 8.5$
Rhabdomonas spiralis	40 × 10		1	1.0 × 15
Strigomonas oncopelti	8.2 × 2.6		1	$L = 17$
Trypanosoma cruzi[b]	20 × 2		1	3 × 3.5
"Endoflagellates"				
Eimeria sp. merozoites	15 × 1.5		—	1.67 × 10
Plasmodium berghi sporozoites	10 × 2		—	$L = 8$

[a] Dinoflagellates with helical flagellum in peripheral groove. Note second *Gyrodinium* has no helical flagellum.
[b] Cell body propagates a wave, one wave of 11 μm length.

Table 4 *Continued*

| Flagellum | | Organism | | | | |
Amplitude h (μm)	Beat form	U/B	U/c	Ω/ω_a	Conditions	References
	helical+ planar	0.56			18–20°C	*133, 163*
	helical+ planar	1.11			18–20°C	*133, 163*
	helical	4.4			26°C, mastigonemes	*133, 137, 221*
		5.0				*116, 143*
				$\omega_a = 180$		*194, 198*
11	helical+ planar	sessile			18–20°C, mastigonemes	*163, 194*
		226/			20.5–21.5°C	*83, 116, 133*
	helical	3.6				
6	helical	1.5	0.19	0.08		*108, 116, 133*
	helical+ planar	5.2				*97, 133*
	helical+ planar	9.5		$\Omega = 9.4$		*99*
	planar	4.3		$\Omega = 8.2$		*99*
3	helical	4.3	0.47	0.06		*108*
	planar?	45		$\omega_a = 300$	in 3 mm deep chamber	*143*
	planar?	1.7		$\omega_a = 120$	between thin slides	*143*
1	planar	19.2		$\omega_a = 430$	18°C, mastigonemes	*107*
	3-D	0.36			mastigonemes	*133, 196*
	3-D	3.4			20–22°C	*40, 85*
	3-D	4.1		$\omega_a = 90$	20–22°C	*40, 85*
	breast stroke	8.4		0.09	20–22°C	*84, 85*
3.5	helical	3.0	0.32	0.056	mastigonemes	*108, 221*
2.4	planar	2.1	0.068	$\omega_a = 110$	22°C	*110*
0.5	planar	15.2	b		in blood, flexible body	*117*
5	planar	0.47			in bile	*30*
2.9	planar				in salivary gland fluid	*219*

[a] Dinoflagellates with helical flagellum in peripheral groove. Note second *Gyrodinium* has no helical flagellum.

[b] Cell body propagates a wave, one wave of 11 μm length.

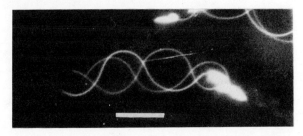

Figure 9 Multiple flash records of swimming tunicate (*Ciona intestinalis*) spermatozoa (Brokaw 1965). The flash rate is 50 Hz and the scale bar is 10 μm. (We are indebted to Dr. C. J. Brokaw for this photograph.)

be explained hydromechanically (Holwill & Sleigh 1967, Brennen 1976) because the flagellum of *Ochromonas* has attached to it a large number of rigid projections known as mastigonemes, which move through the fluid in response to the passage of the flagellar wave as indicated in Figure 10 (see Section 4.4).

We have listed some of the observed characteristics of the propulsion systems of eukaryotes with flagella in Tables 4 and 5 and depicted the general features of some spermatozoa in Figure 11 (for planar wave propagation $\omega_a = kc$ in these tables). Note again that few organisms are completely documented and even more rarely are all data for the same organism gathered under similar conditions.

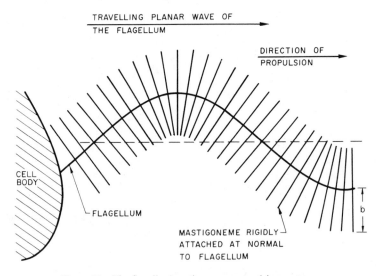

Figure 10 The flagellar/mastigoneme propulsion system.

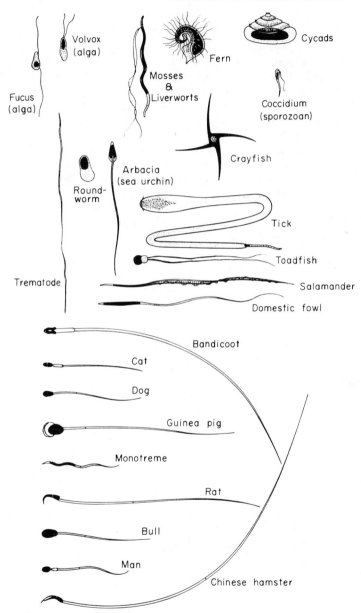

Figure 11   A sampling of the variety of spermatozoon body geometries. The mammalian spermatozoa—from bandicoot downward—are drawn at their relative sizes with the human spermatozoon 40 μm long. (Selected from Austin 1965.)

4.3   Hydromechanics of Flagella with Planar Waves

From a point of view of hydromechanical understanding it is simplest to begin by considering the propulsion of an organism by means of a single flagellum propagating planar waves from base to tip. A simple but idealized example might best serve as a starting point for the discussion. Suppose the spherical body of radius, A, of the idealized organism in Figure 3 is propelled by means of a flagellum with planar waves of wavelength $\lambda(k = 2\pi/\lambda)$, and wave amplitude, h, travelling at wave velocity, c relative to the body. We view an element of the flagellum, S, in a frame fixed in the body and assume the motion of S is purely normal to the direction x so that the motion of the element relative to the fluid at infinity (which has a velocity, U, corresponding to the velocity of propulsion) has components normal and tangential to the axis of the slender-body element S given by

$$q_n = U \sin \varphi - khc \cos \theta \cos \varphi, \tag{18}$$

$$q_s = U \cos \varphi + khc \cos \theta \sin \varphi, \tag{19}$$

Table 5   Spermatozoa propulsion

Spermatozoa of	Body		Flagellum
	Length (B) (μm) × width (μm)	Shape	Length, L (μm) or No. of waves × wavelength
Bos (bull)	10 × 5	⍴	1 × 60
Chaetopterus (annelid)	3.4 × 1.7	⍴	1.25 × 25.4
Ciona (tunicate)	4.1 × 2.4	⍴	1.25 × 32
Colobocentrotus (sea urchin)	8.2 × 2.9	⍴	1.5 × 30
Culex (mosquito)	17.1 × 1.4	⌒	3.3 × 15.7
Didelphis (oppossum)			
Mesocricetus (hamster)	13.8 × 3	⌒	$L = 236$
Homo (human)	5.8 × 3.1	◯	$L = 36$
Lygaeus (milkweed bug)	4.8 × 1.0	⌒	2.3 × 13.0
Lytechinus (sea urchin)	5.1 × 2.9	○	1.25 × 30
Mus (mouse)	33 × 5.5[a]	⍬	1.2 × 65
Ostrea (oyster)	2.6 × 2.8	○	$L = 47$
Ovis (ram)	10.6 × 6.2	○	$L = 59$
Psammechinus (sea urchin)	1.0	◿	1.25 × 24
Tenebrio [mealworm (beetle)]	6.2 × 1.7	⌒	4 × 11.7

[a] Midpiece included in "body."

where $\theta = k(x - ct)$ and $\tan \varphi = kh \cos \theta$. If we then assume known resistive coefficients C_n and C_s, the force on the element of length ds in the x direction is

$$C_n \, ds[U - (1 - \gamma)U \cos^2 \varphi - (1 - \gamma)khc \cos \theta \cos \varphi \sin \varphi] \tag{20}$$

at each instant in time where $\gamma = C_s/C_n$ is the ratio of the resistive coefficients. From an integration over one cycle in time it follows that each element is subject to a mean force in the positive x direction which can be integrated over the length L of the flagellum to yield a mean force on the flagellum equal to

$$C_n L[U - (1 - \gamma)c - (1 - \gamma)(U - c)(1 + k^2 h^2)^{-1/2}]. \tag{21}$$

If the organism were self-propelling, this would be equal to the drag $6\pi\mu U A$ on the head. Hence the propulsive velocity U is

$$\frac{U}{c} = \frac{(1 - \gamma)(1 - \beta)}{(1 - \beta + \gamma\beta + \delta)}, \tag{22}$$

where $\beta = (1 + k^2 h^2)^{-1/2}$ and $\delta = 6\pi\mu A/LC_n$. On the other hand, if the organism

Table 5 *Continued*

Flagellum				Organism		
Amplitude h (μm)	Beat form	U/B	U/c	Ω/ω_a	Conditions	References
11	3-D	9.7	0.075	$\omega_a = 135$	37°C, $\mu = 1$ cp	46, 170, 186
3.8	2-D	30.8	0.156	$\omega_a = 166$	16°C, $\mu = 1.4$ cp	41, 46
4.3	3-D & 2-D	40.2	0.147	$\omega_a = 220$	16°C, $\mu = 1.4$ cp	41, 46
2.8	2-D	27.8	0.237	$\omega_a = 200$	25°C, ATP reactivated	46, 81
6.4		0.36				165
	3-D				37°C, swim as pairs	164
	3-D			0.2	37°C	164
	3-D	8.6		0.5	37°C, U decreases 46% in cervical mucus	103, 164
2.1						165
4.6	2-D	31.0	0.185	$\omega_a = 180$	16°C, $\mu = 1.4$ cp	41, 46
15	3-D			0.5	37°C	164
4.7	2-D[b] 3-D	63.0		$\omega_a = 270$	23°C	70
7.4	2-D[b] 3-D	12.8		$\omega_a = 180$	35.5°C	70
4	2-D	190		$\omega_a = 220$		91, 93
4.2	2-D?	16.1	0.33	$\omega_a = 176$		6, 165

[b] Waves formed by flagellar beat are two-dimensional near a boundary (slide glass) and three-dimensional far from the boundary.

were restrained from moving, the thrust, T, developed by the flagellum in the positive x direction follows directly from (21) with $U = 0$ and is

$$\frac{T}{6\pi\mu Ac} = \frac{(1-\gamma)(1-\beta)}{\delta}. \tag{23}$$

Lighthill (1975, p. 55) shows that the results for more general waveforms do not differ from the above, provided one uses a more general definition for β as the mean value of the square of the tangential direction cosine of the waveform.

These results show the primary dependence of the performance of the flagellum on the wave velocity, c, the resistive-force-coefficient ratio $\gamma = C_s/C_n$, and the nondimensional wave amplitude, kh. The performance is clearly enhanced by decrease in both γ and β (the latter arising from increasing kh) not only in terms of uniform translational motion as given by U/c but also from the point of view of acceleration from rest and maneuverability, both of which could be characterized by T. Although there has been a tendency for the fluid-mechanical analyses to concentrate on the optimization of the propulsive system in terms of seeking that which would give maximum rectilinear propulsion per unit energy expenditure, it is not at all clear that this is necessarily the most important feature of the system for any particular microorganism. Indeed the ability to accelerate and maneuver could be an asset as important, if not more important, to the organism.

According to the relations (22) and (23), U/c and $T/6\pi\mu Ac$ increase monotonically with decreasing β or increasing nondimensional wave amplitude kh approaching asymptotic values of $(1-\gamma)/(1+\delta)$ and $(1-\gamma)/\delta$, respectively, for large kh. But the penalty paid for these enhanced propulsive effects is an increase in the energy required; the mechanical rate of work being done on the fluid can readily be obtained by integrating the increment of rate of work done per unit flagellar length $C_n(q_n)^2 + C_s(q_s)^2$ over one cycle of time and summing for the entire length of the flagellum. Lighthill (1975) has shown that this leads to a maximum efficiency of rectilinear propulsion by a general planar wave when

$$\beta = \gamma^{1/2}(1+\delta)/[\gamma+\delta+\gamma^{1/2}(1+\delta)], \tag{24}$$

for which $U/c = (1-\gamma^{1/2})/(1+\delta)$. Furthermore, this optimum value of β is rather insensitive to the values of either γ or δ and takes values for $\gamma = \frac{1}{2}$ of 0.586 for very small δ (i.e. an organism with a small cell body, A) and 0.471 for very large δ (i.e. an organism with a large cell body). In the case of a sinusoidal waveform, these values correspond to nondimensional wave amplitudes kh of 1.37 and 1.88, respectively, and it is of interest to observe that many organisms with planar flagellar waves appear to operate with wave amplitudes of this order. Similarly it is instructive to examine the maximum mean propulsive force in one direction that can be generated by a small element of a slender body whose position can oscillate sinusoidally in time within one plane and whose angle of inclination in that plane is also allowed to oscillate sinusoidally. One finds that the optimum propulsive force per unit energy expenditure occurs when the position oscillates normal to the direction of the required thrust, the mean inclination to this direction is zero, and the inclination oscillation is $\pi/2$ out of phase with the position oscillation. This corresponds precisely to the form of motion in a travelling wave, and one further finds that the optimum amplitude of the inclination oscillation, θ, is 63.9°, which for a sinusoidal travelling wave yields a value for kh of 2.

Gray & Hancock (1955) examined the propulsion for sea urchin spermatozoa (*P. milians*), which propagates a particularly sinusoidal waveform (see Figure 9), and observed an average propulsive velocity of 191.4 μm sec^{-1}, in excellent agreement with a value of 191 μm sec^{-1} computed by using the observed wave amplitude, length, and velocity and an expression similar to Equation (22) with $\gamma = 0.5$. Lighthill has since suggested, and Gray & Hancock were probably aware, that such agreement was in some sense fortuitous and misleading. First, the more sophisticated analysis of Hancock (1953) [see also Lighthill (1975) and Section 2.5] suggests that a more accurate value of γ is significantly higher (about 0.7), which in view of the factor $(1 - \gamma)$ in the expression (22) would cause significant disagreement. On the other hand, Gray & Hancock (1955) do mention that propulsion was occurring in close proximity to either the glass or the air surface; from Section 2.6 we have seen that the value of γ could be significantly reduced and propulsion enhanced by the proximity of a boundary and it would seem that the net result is a γ of order 0.5.

The last observation serves further to illustrate the difficulty of wall effects upon data obtained in the confined fluid of a microscope slide; it also further exemplifies the beneficial propulsive effect that can be obtained by a flagellated organism moving close to a solid boundary. The recent detailed analyses of this problem by Katz (1974) yielded further information on these wall effects for flagellated organisms. The results do not differ qualitatively from those expected on the basis of the result (17), although Katz has examined the waveforms on the flagellum that would lead to the maximum benefit in the presence of a boundary.

In concluding this section we must remark that while the simplicity of the resistive-force theory is a boon to biologists seeking approximate estimates, many potentially significant hydromechanical effects have been neglected in such an approach. First, there is the previously discussed uncertainty in the force coefficients, C_n and C_s, which in reality implies the necessity of abandoning such a simplistic approach in order to seek more accurate solutions. Secondly, the effect of the often large cell body on the flow experienced by the flagellum has been entirely neglected and is a problem clearly in need of attention. A more accurate analysis will require construction of the entire flow field due to both the cell body and the beating flagellum by means of fundamental singularities. Further evidence for the necessity of such an approach is provided by the observations of the flow field near a flagellum obtained by Lunec (1975). Lunec compared the actual flow near the flagellum of *Crithidia oncopleti* (as visualized by tracer particles) with a theoretical reconstruction based on a distribution of stokeslets along the flagellar axis, whose strength was obtained from Gray & Hancock's resistive-force coefficients. The resulting fluid velocities were in marked disagreement, and Lunec concluded that this could in part be due to the proximity of the cell body.

4.4   Hispid Flagella

Some eukaryotic organisms such as *Ochromonas* (Figure 10), which propagate planar waves, have rigid projections known as mastigonemes, which protrude from the flagellum. These mastigonemes move through the fluid as the waves pass along the flagellum, and their net effect is to propel the organism in the direction opposite to that which would occur in the absence of mastigonemes. Jahn, Landman & Fonseca (1964) suggested that a simple way of viewing the hydromechanical effect of the mastigonemes is that they increase C_s much more than C_n, resulting in values of γ greater than unity and thus propelling the organism in the direction opposite to that which occurs when $\gamma < 1$ [Equation (22)]. It is, however, a simple matter to apply resistive theory to the mastigonemes as Holwill & Sleigh (1967) and Brennen (1976) have done and to show that for rigid mastigonemes the result (22) is altered to

$$U/c = -(1-\beta)(\delta-\alpha)/(\delta+2\alpha+2\delta\alpha+\beta\delta-\beta\alpha), \qquad (25)$$

where $\alpha = 6\pi\mu A/C_n^m bn$, and n, b, and C_n^m are respectively the number, length, and normal resistive coefficient of the mastigonemes. Here γ has been assumed to be one half for both flagellum and mastigonemes. Clearly if the total length of all the mastigonemes together (bn) is greater than the flagellum length so that $\delta > \alpha$, then U/c is always negative and an organism with a hispid flagellum moves with its flagellum forward while propagating waves along the flagellum in the same direction. The result (25) yields a value of 60 μm sec^{-1} for *Ochromonas,* which is in good agreement with the observed values of 55–60 μm sec^{-1} (Holwill & Sleigh 1967); Brennen (1976) has also examined the case of flexible mastigonemes and concluded that while the mastigonemes of *Ochromonas* are probably thick enough to have sufficient rigidity for hydromechanical purposes, the smaller "hairs" on *Euglena* flagella are probably so flexible that they have little hydromechanical effect.

4.5   Helical Flagellar Propulsion

The propagation of a helical wave along any flagellum, as illustrated in Figure 12, gives rise to a net torque on the flagellum about the longitudinal axis; this causes the cell body to rotate (the *material* of the flagellum must rotate with the same angular velocity) so that an equal and opposite torque on the cell body is generated and the total torque on the organism is zero as it must be from mechanical first

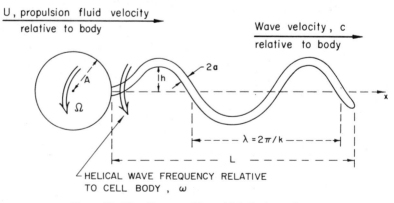

Figure 12   Flagellar propulsion with helical waveform.

principles. Although this point was fully appreciated by Gray (1953), it was left unresolved in some of the early resistive-theory analyses by Holwill & Burge (1963) and Holwill (1966b). Chwang & Wu (1971) (see also Schreiner 1971) first presented a complete solution in which both the condition of zero total longitudinal force and the condition of zero total torque were applied to obtain not only the ratio of the forward speed, U, to the helical wave velocity, c (relative to the cell body), but also the ratio of the angular velocity of spin of the cell body (equal to the *material* rotation of the flagellum), r, to the angular velocity of the helical wave propagation relative to cell body, ω (equal to kc where $k = 2\pi/\lambda$ and λ is the wavelength of the helical wave). These interconnected results are

$$\frac{c}{U} = \frac{1 + 2k^2h^2 + A^*}{k^2h^2} \left\{ 1 + \frac{2(1 + k^2h^2)^2 + (2 + k^2h^2)A^*}{(1 + 2k^2h^2 + A^*)B^*} \right\}, \tag{26}$$

$$-\frac{\omega}{\Omega} = 1 + \frac{(1 + 2k^2h^2 + A^*)B^*}{2(1 + k^2h^2)^2 + (2 + k^2h^2)A^*}, \tag{27}$$

where we have changed the sign of the second expression by defining values of ω and Ω to be positive in the same rotational sense in order to highlight the fact that, as a result of the torque balance, ω and Ω are naturally of opposite sign. In the above expressions h is the helical wave amplitude and

$$A^* = 3\mu A(1 + k^2h^2)^{1/2}/2\pi LC_s$$

$$B^* = 4\mu[\pi a^2 + A^3(1 + k^2h^2)^{1/2}/2\pi L]/h^2C_s$$

where A is the radius of the cell body (assumed spherical), L is the distance from the cell body to the end of the flagellum, a is the radius of the circular cross-section of the flagellum, and C_s is the tangential resistive coefficient. It was assumed that $\gamma = C_s/C_n$ was equal to $1/2$. These results show interesting asymptotic limits; with a vanishingly small head ($A \to 0$) the forward propulsion given by U/c will become small and the material tends to rotate with a velocity, Ω, almost equal and opposite to the angular wave velocity, ω. This particular limit has relevance to the propulsion of a spirochete, which, lacking a flagellum, propels itself by propagating a helical wave along its long thin body; apparently the torque arising from the helical wave is balanced by an opposite rotation of the surface of the body (Chwang, Winet & Wu 1974; Kaiser & Doetsch 1975; Wang & Jahn 1972). On the other hand, for a large cell body Ω tends to zero, but the propulsive velocity again becomes small due to the large drag on the cell body. Between these limits a maximum value of U/c occurs. For typical values of kh and kb of 1 and 0.1, respectively, this maximum occurs when the "head-to-tail" ratio, A/a, is between 10 and 20, which is apparently typical for many organisms.

As far as the helical flagellar propulsion of eukaryotic cells is concerned, there have been few comprehensive comparisons of the theory with observations; the obvious difficulty is that the material rotation, Ω, is extremely difficult to observe or measure. Some partial analysis for *Euglena* by Holwill (1966b) did, however, appear to yield propulsive velocities of the same order of magnitude as those observed, and the results of Chwang, Wu & Winet (1972) and Winet & Keller (1976) provide a detailed analysis of a more complex form of propulsion, namely that associated with the prokaryote *Spirillum*. In Tables 4 and 5 we have compiled some of the known data on eukaryote propulsion by flagella; invariably the helical wave propagation frequency quoted is the apparent wave frequency, ω_a, seen by the observer, and it should be noted that according to the present definitions for helical waves $\omega_a = \Omega + \omega$.

4.6   *Prokaryotic Flagellar Propulsion*

The above analysis applies to helical eukaryotic flagella and requires modification as far as prokaryotic flagellar propulsion is concerned; wave propagation along the flagellum should be replaced by relative material rotation between the cell body and the flagellum if the basal motor model is to be accepted. Although the results for this case do not appear explicitly in the literature, they should be readily obtained by the same methods used by Chwang & Wu (1971), and one can anticipate that the results will only differ from (26) and (27) because of the torque created by the different angular velocities of the material of the flagellum. The effect is probably small when the cell body is relatively large, and the torque due to the motion of the flagellar element through the fluid in an azimuthal direction is much larger than the torque due to flagellar material rotation.

Shimada, Yoshida & Asakura (1975) made a complete set of measurements for the bacteria *Salmonella* (many flagella forming a bundle) and *Pseudomonas* (single flagellum) and compared their observations with the expressions (26) and (27). The proper comparison might be with the expressions modified as suggested above; nevertheless, it is of interest to observe that while the agreement in the case of *Pseudomonas* appeared reasonable, the theory gave significantly lower values for U/c than those observed for the multiflagellated *Salmonella*. Although other explanations are possible, these results suggest that the effective γ for a flagellar bundle may be significantly less than $1/2$, a not unreasonable possibility.

5   THE HYDROMECHANICS OF CILIARY SYSTEMS

5.1   *Ciliary Motions*

In the previous section we discussed the hydromechanics of locomotion in organisms propelled by individual flagella or flagellar bundles. In some organisms with more than one flagellum the hydromechanical analysis could proceed along similar lines, provided the hydrodynamic interactions between the flagella are relatively weak. However, there are organisms with more than one flagellum that seemingly derive a beneficial propulsive effect by adjustment of the phase relationships of the beat patterns of their "flagella" [for example *Mixotricha* (Cleveland & Cleveland 1966) and *Volvox* (Hand & Haupt 1971)]. It appears that such beneficial interactions can also accrue to groups of individual organisms swimming close to one another; there are clearly analogous natural phenomena at high Reynolds numbers in the flight patterns of groups of birds and in the schooling of fish (Weihs 1975).

Cilia are essentially short flagella, which may beat back (recovery stroke) and forth (effective stroke) at different rates transcribing what is known as a "polarized" beat or which may oscillate in a manner indistinguishable from "eukaryotic flagella." They occur in large arrays, such as "ciliated epithelium," and produce fluid motion by collaborative action arising from a definite phase relationship between the beats of neighboring cilia. The presence of such a relationship is known as metachrony, which often results in a wave, known as a metachronal wave, travelling over the array. It may be that ciliary systems evolved from flagella, because of the beneficial hydrodynamic effects of the interactions of the cilia.

Cilia occur throughout the animal kingdom and indeed are extensively used not only for producing fluid motion but also for sensing motion. Examples of the former use are the cilia in the gills of nonmotile marine animals, used for ingestion of water (Aiello & Sleigh 1972, Sleigh & Aiello 1972), and the cilia lining the trachea and lungs that provide a cleaning mechanism by continuously propelling mucus up and out of the lungs. (Figure 13 is an electron micrograph of the cilia of frog lung mucosa.) Cilia also line the oviduct and contribute to the transport of the ova in that organ (Halbert, Tam & Blandau 1976; see also Dirksen & Satir 1972); the uterine wall is ciliated and the fact that spermatozoa appear to swim close to this wall may be because they derive a beneficial effect from the beating cilia. In addition, the cilia on the membrane lining the ventricles of the brain (the ependyma) have

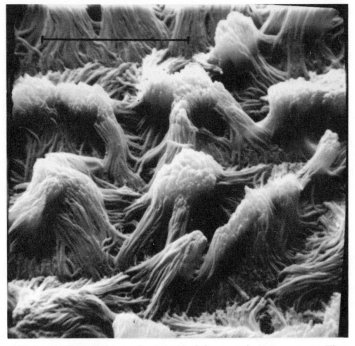

Figure 13   A scanning electron micrograph of frog respiratory mucosa. These fixed organelles display a typical metachronal pattern that reflects the beat-rhythm orientation. Scale bar is 10 μm. (We are indebted to Dr. P. P. C. Graziadei for this photograph.)

been shown to create sufficient local mixing to affect the thickness of the so-called unstirred layer, thereby enhancing the diffusion or transport of ions across the membrane (Nelson 1975). There has been little detailed hydrodynamic analysis of this system. Other functions in which motile cilia play a major role are cited in Table 1. So-called sensory cilia tend to be nonmotile and often display a 9+0 microtubular axoneme. They are found in sensory organs devoted to photoreception ("eyes"; the flagellum of some protozoa is also part of a photoreceptor system: e.g. Hand & Haupt 1971), chemoreception (olfactory organs: e.g. Reese 1965), and mechanoreceptors, which are adapted for detection of sound, touch, or orientation in a gravity field. In no case has it been demonstrated that the cilia are in fact "transducers" of stimulus energy into nerve transmission energy (Barber 1974). They all appear to "sensitize" the cells carrying them for the transduction process by either (a) distortion of the membrane so as to produce changes in the molecular organization of the excitable area, which cause changes in ion permeability leading to a depolarization, (b) deformation of adsorbed mucus so as to give rise to piezoelectric potential differences, or (c) initiation of electrical changes, which result from some "inherent mechanosensitivity" in the cilia (Barber 1974). Understanding of the operation of sensory cilia is beyond the limits of a strictly fluid-mechanical analysis, but the fluid mechanician needs to be aware of these limits to keep his analysis in perspective. The importance of ciliary systems extends even into ecological areas. For example, an individual California mussel *Mytilus californianus* can remove a significant amount of suspended mud and other matter from water, given an average filtration rate of 2.6 liters of water per hour through its cilia-lined gills (Fox & Coe 1943).

But perhaps the ciliary systems most readily observed are those that provide propulsion for eukaryotic cells (e.g. Figures 14 and 15). A sampling of the propulsive parameters of such cells is presented in Table 6. From a hydrodynamic point of view these systems are more readily understood because the fluid is usually Newtonian, or at least is readily adjusted to be Newtonian without placing the organism in an unusual environment. On the other hand, the fluid in mammalian ciliary systems is often highly non-Newtonian (e.g. the mucus in the lung). Although we concentrate here primarily on the locomotion of ciliated organisms, the future extension of the knowledge gained to the understanding of other ciliary systems in nature should be borne in mind.

Each individual cilium usually has a fairly regular beat pattern (see, for example, Sleigh 1960, 1962, 1968, 1972, 1973, 1974c; Parducz 1967), which often appears to be created by the propagation of a bend from the base to the tip of the cilia as illustrated by the beat pattern for *Opalina* in Figure 16. That phase of the beat in which each cilium is moving in a general direction so as to propel the organism is termed the *effective stroke*. Generally the cilium is straightened out during this phase and in the remainder of the beat known as the *recovery stroke* the cilium sneaks back to its starting point in a bent position so that a significant portion of each cilium is moving tangential to the fluid rather than normal to it as in the effective stroke. Such asymmetry immediately suggests that the cilia are taking advantage of the difference in the force coefficients for flow normal and tangential to a slender body. Furthermore, the motion is often three-dimensional with some recovery motion taking place out of the plane of the effective stroke, as is the case with *Paramecium* (Machemer 1972a,b; Tamm 1972). While precise information on the ciliary beat pattern represents necessary data prior to any hydrodynamic analysis, it is difficult to obtain from light microscopy studies. In this respect the beautiful electron microscopy studies originated by Tamm & Horridge (1970) and Tamm (1972) greatly add to our knowledge of ciliary motion (see Figures 14 and 15).

Often eukaryotic cilia ensembles exhibit metachrony: in one surface direction a cilium beats slightly out of phase with its neighbor so as to produce a metachronal wave (velocity, c) travelling over the surface (see Figure 14). To add to the complexity the direction of wave propagation may have almost any orientation relative to the direction of the effective stroke. Knight-Jones (1954) coined a series of terms to identify this relationship: when the metachronal wave propagation and the effective stroke are in the same direction, this is known as *symplectic* metachronism; if they are in opposite directions, it is termed *antiplectic*; and if the directions are normal to one another they are termed *diaplectic* (*dexioplectic* if the rotation from the metachronal wave direction to the effective stroke direction is 90° anticlockwise viewed from above and *laeoplectic* if 90° clockwise). Symplectic metachronism is illustrated by the electron micrograph of Figure 14 (Tamm & Horridge 1970) and the upper part of Figure 16 for *Opalina*; the lower part of Figure 16 represents an antiplectic approximation to the beat pattern of *Paramecium*, which does, however, contain a dexioplectic component as indicated in the electron micrograph of Figure 15 (Tamm 1972). On the other hand, there are ciliary systems in which metachrony is indiscernible. For example, Cheung & Jahn (1975) could not detect any organized metachrony in rabbit tracheal cilia, and Figure 17 from the frog's olfactory epithelium shows almost random cilia orientation (Graziadei 1971); on the other hand, another scanning electron micrograph from Graziadei (1971) of the cilia in the lung mucosa of the frog shows clear metachronism (Figure 13).

Many organisms have an avoidance response in which they reverse the direction of metachronal wave propagation and thus their direction of motion, a phenomenon known as ciliary reversal (Jahn 1975). This appears to be linked to their longitudinal electropotential gradient since it can be achieved by external imposition of an electrical potential. Other organisms such as *Opalina* appear able to vary continuously the direction of wave propagation and thus achieve greater maneuverability (Sleigh 1962). These responses in organisms with no identifiable and separate nervous system merely serve to highlight one of the great puzzles of ciliary systems, namely how these delicate phase relationships between the cilia are controlled. If one had to build a mechanical model to simulate such a system it would be extremely difficult, which probably accounts for the singular lack of mechanical model studies [the early work of Miller (1966) is the only work of this kind that we know of]. Nervous control of ciliated epithelium is one of the important problems to which these studies may be applied. In this context it may be noted that Murakami & Takahashi (1975) have shown that transient depolarization of the cell carrying the cilia by nervous activity is correlated with the "quick-arrest response" of the cilia.

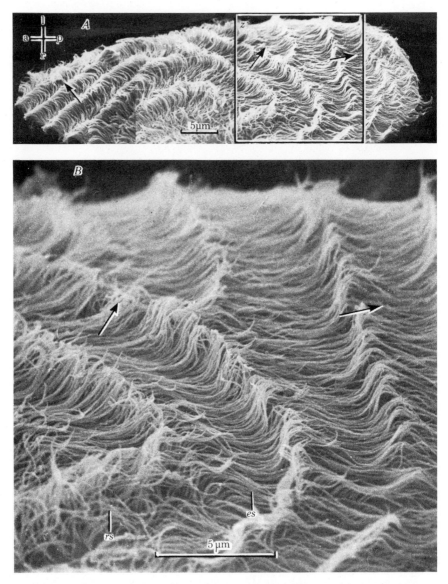

Figure 14   A scanning electron micrograph following rapid fixation of the ciliated protozoan *Opalina* (from Tamm & Horridge 1970). As in the preceding figure, the in vivo metachronal wave orientation is reflected in the pattern over the fixed specimen. Arrows indicate the directions of the metachronal wave. The key difference between the two specimens is that this figure is limited to all or part of a single cell. (We are indebted to Dr. S. L. Tamm for this photograph.)

Figure 15   A scanning electron micrograph following rapid fixation of the ciliated protozoan *Paramecium* (from Tamm 1972). The metachrony of this specimen is dexioplectic and/or antiplectic. A-P, anterior-posterior axis; D-V, dorsal-ventral sides. (We are indebted to Dr. S. L. Tamm for this photograph.)

In the following sections we confine ourselves to the hydromechanics of ciliary systems. We would, however, be remiss in not mentioning the important work in which attempts have been made to recover the internal motive force for ciliary beat patterns by working backwards from the known motion, the hydrodynamic forces, and the presumed elastic structure of the cilia (Holwill 1966a, Harris 1961, Sleigh & Holwill 1969, Rikmenspoel & Sleigh 1970, and Rikmenspoel 1975). Hopefully such analyses will make contact in the future with the electron-microscopy studies

Table 6 Eukaryote propulsion by cilia

Eukaryote	Body Length (B) (μm) × width (μm)	Shape	Cilia Cilia length l (μm)	Cilia No. cilia/μm² or distribution
Balantidium entozoon	106 × 55.6			
Coleps hirtus	66 × 30			
Colpidium sp.	79.1 × 38.6			
Condylostoma patens	371 × 102			
Didinium nasutum	126 × 87			2 circular rows
Euplotes patella[a]	202 × 124			
Frontonia sp.	378 × 213			
Halteria grandinella[a]	60 × 50			
Kerona polyporum[a]	107 × 64			6 rows cirri
Metopides sp.	115 × 33			
Nyototherus cordiformis	139 × 97.2			
Opalina ranarum	350 × 112		15	1.0
Ophryoglena sp.	250 × 92.8			
Opisthonecta henneg[a]	126 × 75			
Paramecium aurelia	125 × 31			
P. bursaria	126 × 57			
P. calkinsii	120 × 44			
P. caudatum	242 × 48		12	0.5
P. marinum	115 × 49			
P. multimicronucleatum	251 × 62			
Prorodon teres	175 × 160			
Spirostomum ambiguum[a]	1045 × 95		8.2	
S. polymorphus[a]	208 × 15.2		27.5	3.5 μm long. separation
Stylonichia sp.[a]	167 × 86			
Tetrahymena pyriformis	55.7 × 20		7	17–23 columns
Tillina magna	162 × 82			
Urocentrum turbo	90 × 60			2 circ. rows
Uroleptus piscis[a]	203 × 52			
U. rattulus[a]				
Uronema sp.	40 × 16			

[a] Has undulating membrane and/or membranelles

of the internal structure of cilia, particularly those of Warner & Satir (1974) and Warner (1974).

The fluid mechanics of ciliary systems is clearly quite complex and most of the detailed and quantitative analyses have been based on simplifying assumptions concerning the interaction of the cilia and the fluid. Most of these studies have concentrated on what we shall term local fluid/cilia interaction models; for these purposes most authors have considered an infinite flat surface upon which the cilia

Table 6 *Continued*

	Cilia				Organism	
Metachrony	Wavelength λ (μm)	Cilia beat freq. (Hz)	U/B	U/c	Condition	References
	11.6					149
			10.4			47
dexioplectic	10					149
			2.8		20°C	47, 149
dexioplectic			3.7			47, 149
			6.2			47
			4.3		21.5°C	47
			8.9			47
			4.3			47
dexioplectic	17.1		3.1			47, 149
dexioplectic	26.6					149
symplectic		4				47, 194
dexioplectic	11.4		16			149, 161
dexioplectic		10	9.5			119, 149
			16		21°C	47
			7.9		25°C	47, 48
			8.3			48
	12	29	10.9			48, 149, 199, 200
	10.8		8.1		19°C	47, 48
			11.3			48, 210
			6.1			47
antiplectic	8.5	30	0.78			47, 53
dexioplectic	13	33	4.6	1.26	20–22°C	47, 194, 203
	25.5	59	2.8		22°C	47, 146
dexioplectic	16.2	20	8.1		20°C	149, 162, 231
			12.3			47, 149
			7.8		28.5°C	47
			2.4		22°C	47
					21°C	47
						47

motions are spatially and temporally periodic so as to form metachronal waves. The manner in which such solutions should be applied to finite, ciliated organisms is not entirely obvious; we return to this later. For the present, we discuss the two principal kinds of local fluid/cilia interaction models, the so-called envelope and sublayer models.

5.2   The Envelope Model

The envelope model assumes that the cilia are sufficiently closely packed together, as in the case of *Opalina* in Figure 14, so that the fluid effectively experiences an oscillating material surface. This envelope is commonly assumed to be impenetrable and the motion of each "particle" on the surface is assumed to be roughly equivalent to the locus of the tip of an individual cilium. An analysis of the low-Reynolds-number flow due to such an oscillating sheet was made by Taylor (1951) as a rough two-dimensional model for flagellar propulsion; thus Taylor only pursued the solution in which the "particles" had motions normal to the plane of the surface. Subsequently in a short note Tuck (1968) delineated the nature of the solution in which the oscillatory motions were purely tangential to the surface. Since then, solutions of a more general kind with oscillatory particle motions or ciliary loci of more general form have been produced by Reynolds (1965), Blake (1971b), and Brennen (1974). Consider an arbitrary elliptical form for the ciliary tip locus

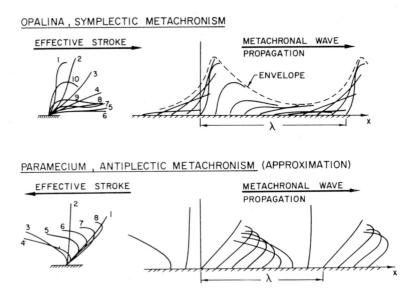

Figure 16   Approximate beat patterns for *Opalina* and *Paramecium* with the positions of an individual cilium at equal intervals in time on the left and the positions of an array of cilia at a given time on the right, showing the symplectic metachronism of *Opalina* and an antiplectic approximation to the metachronism of *Paramecium*.

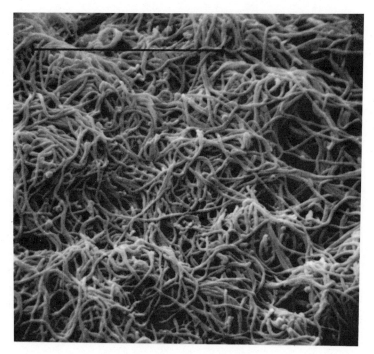

Figure 17   A scanning electron micrograph of the mucociliary epithelium of the olfactory surface of the frog. These cilia operate with no apparent metachrony, and the resulting random motion serves to stir and mix rather than propel the ambient fluid. Accordingly, they may affect transmembrane transport, an important consideration in any chemosensitivity analysis. The scale bar is 10 μm. (We are indebted to Dr. P. P. C. Graziadei for this photograph.)

(Figure 18) in which the tip performs a simple harmonic motion of amplitude h_s tangential to the surface and a simple harmonic motion of amplitude h_n normal to the surface. These deflections of frequency ω have a linear phase shift in the s direction along the surface so as to produce a metachronal wave of velocity, c, and wavelength, $\lambda(k = 2\pi/\lambda)$, travelling in the positive s direction. The mean position of the material surface (and thus of the ciliate) is fixed in this frame of reference. Far from the envelope these oscillatory motions produce a rectilinear translation of the fluid tangential to the sheet whose magnitude in the positive s direction we shall denote by U. If at any point, s, the motion tangential to the surface leads the motion normal to the surface by a phase angle $(\theta - \pi/2)$ and we define a parameter $K = (h_s^2 - h_n^2)/(h_s^2 + h_n^2)$, then the various cilia loci that are so described are indicated diagrammatically in Figure 18. Note that $K = -1$ corresponds to Taylor's (1951) solution, whereas $K = +1$ corresponds to Tuck's (1968) solution. Brennen (1974) has shown that, provided the amplitudes h_s and h_n are small compared with the wavelength λ and the Reynolds number $Re = \omega/k^2 v$ is small, the translation velocity, U, is given by

$$\frac{U}{c} = \tfrac{1}{2}k^2(h_s^2 + h_n^2)\left[\frac{(\beta+1)}{2\beta}(1 - K^2)^{1/2}\cos\theta - \frac{(\beta-1)}{2\beta} - K\right] \tag{28}$$

where

$$\beta = \{\tfrac{1}{2}[(1 + Re^2)^{1/2} + 1]\}^{1/2}.$$

Notice that this steady translation is quadratic in the nondimensional amplitudes kh_s, and kh_n; since these are assumed small, the velocity U is much smaller than the oscillatory velocities produced by motion of the envelope, which are first-order in kh_s and kh_n and which, incidentally, decay like e^{-kn} with normal distance, n, away

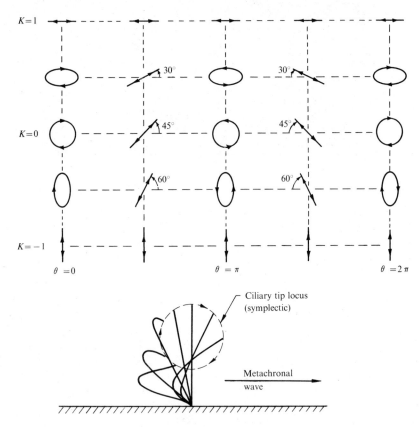

Figure 18 Variations of arbitrary elliptical ciliary tip loci with the parameters K and θ. The figure is correlated with the cell surface horizontal, the fluid above it and the metachronal waves travelling to the right. An example of a ciliary tip locus (symplectic) is indicated in the lower part of the figure.

from the envelope. The translation, U, arises from two different quadratic combinations of first-order oscillatory terms. The first and most important is in the quadratic term of the Taylor series expansion through which the velocity conditions on the envelope are satisfied. The second quadratic term is $O(Re)$ smaller and arises from the inertia terms in the Navier-Stokes equations. The latter disappears, therefore, when $Re \to 0$ and then

$$\frac{U}{c} \to \tfrac{1}{2}k^2(h_s^2 + h_n^2)[(1 - K^2)^{1/2} \cos\theta - K], \tag{29}$$

which is in agreement with Blake's (1971b) results. The variation of the propulsion velocity, U, with the (K, θ) parameters was investigated by Brennen and is shown in Figure 19 in which contours of constant $U/ck^2(h_s^2 + h_n^2)$ are plotted. Notice that this exhibits two optimum forms for the ciliary locus. When $K = -1/\sqrt{2}$ and $\theta = 0$ a maximum propulsive velocity of $ck(h_s^2 + h_n^2)/\sqrt{2}$ is achieved in the positive s direction, a situation that corresponds to symplectic metachronal propulsion. A simple Galilean transformation to bring the fluid at infinity to rest models a ciliate travelling in a direction opposite to the direction of wave propagation. On the other hand, maximum antiplectic propulsion of the same magnitude can equally well be achieved with a ciliary tip locus for which $K = +1/\sqrt{2}$, $\theta = \pi$ (see Figure 18). The energy expenditure per unit surface area, $\dot{E}$, for these motions is simply given by $\mu c^2 k^3(h_s^2 + h_n^2)$. It follows that the above optima are also the most efficient means of propulsion in terms of propulsive velocity per unit energy expenditure per unit area. Finally, we note that calculations based on the expression (28) for nonzero

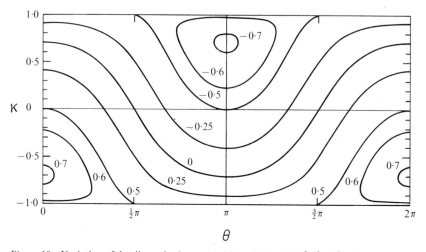

θ

Figure 19 Variation of the dimensionless propulsive velocity $U/ck^2(h_s^2 + h_n^2)$ with the ciliary tip locus parameters K and θ according to the envelope model result (29). Contours are shown for various labelled values of $U/ck^2(h_s^2 + h_n^2)$; the figure should be used in conjunction with Figure 18.

Reynolds number indicate that even when Re is of order unity, the contours are little changed from those of Figure 19 (Brennen 1974).

Envelope models of this kind have been applied to a wide variety of physiological situations. Blake (1971b) and Brennen (1974, 1975) considered their application to the locomotion of ciliated microorganisms. Katz (1972) and Shack & Lardner (1972) used the method to model the propulsion of fluid in the ciliated tubes of mammalian reproductive systems, both female and male (see also Blake 1973b). In this regard considerable attention has been given to the role of the cilia in the mammalian oviduct in propelling the ovum. Ross (1971) has also used an envelope model to study the propulsion of mucus by the ciliated epithelium of the trachea. Such analyses have much in common with peristaltic pumping (Jaffrin & Shapiro 1971) where the "envelope" is a real material surface; in this case, considerations of and conditions upon the extensibility of the envelope are often imposed.

Apart from other more general problems to be discussed below, one of the major difficulties in comparing results from envelope model analyses with observations is that most of these analyses are limited to amplitudes h_s and h_n that are small compared with the metachronal wavelength, λ; that is, kh_s and kh_n are small. On

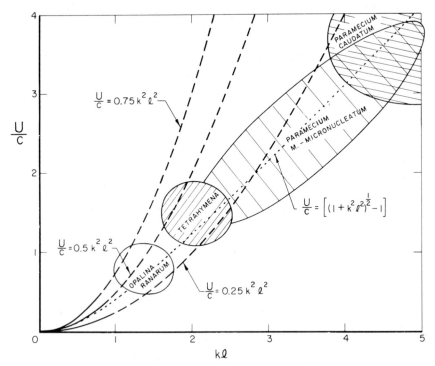

Figure 20 Comparison of the propulsive velocity of some ciliates (data given by areas denoting the scatter in the available information) with extrapolated predictions of envelope theories of the form $U/c \propto k^2 l^2$.

the other hand, as illustrated by the values in Table 6, most ciliary systems appear to operate with values of kh_s or kh_n that are of order one or greater. This point is exemplified by the data presented in Figure 20, where values of U/c are compared with kl, l being the cilia length; for purposes of comparison with the expression (29) and Figure 20 we may note that for many real ciliary beat patterns

$$l^2 \approx 4(h_s^2 + h_n^2).$$

Blake (1971b) has done some preliminary envelope analysis for larger amplitudes by evaluating higher-order terms in kh_s, kh_n. But further nonlinear analyses, which also incorporate higher-order harmonics in time in order to model the differences in the speeds of the "effective" and "recovery" strokes, are probably necessary before a conclusive evaluation of the utility of envelope models can be made.

5.3 Sublayer Models

A second distinct set of models has been proposed and developed for ciliary systems. These models concentrate on the interactions between individual cilia with the surrounding fluid and hence deal specifically with the flow among the cilia. For this reason they have been termed sublayer models. Blake (1972) first proposed such an analysis for the flow created by a regular array of cilia beating metachronously on an infinite plane wall.

By considering the relative motion between an element of a cilium and the surrounding fluid, sublayer models seek to establish the incremental force (or stokeslet strength) on each and every cilium element through use of resistive-force theory. The entire flow field, denoted by its velocity $\mathbf{q}(\mathbf{x})$ where $\mathbf{x}$ is a position vector, is then considered as having been created by this distribution of stokeslets and can be formally represented as an integral of the velocities induced at each point by each cilium element. Since these slender bodies operate close to a wall, the effect of the image system of singularities should also be included in order to satisfy a no-slip condition at that boundary. This has important consequences for ciliary propulsion because of the different forms of image systems for stokeslets oriented normal and tangential to the wall, as discussed in section 2.4. Since the far-field for normal stokeslets decays more rapidly (like r^{-3}) than that of tangential stokeslets (like r^{-2}), it follows that the tangential motion of the ciliary elements is of much greater consequence than the normal motion.

The fluid-mechanical problem is completely defined when the relative motion between an elemental length of cilium at a position $\mathbf{x}_0$ and its surrounding fluid is formulated in terms of a known cilium motion due to a specified beat pattern and the fluid velocity, $\mathbf{q}(\mathbf{x}_0)$. There is, however, a difficulty here because the model presupposes that after a Galilean transformation utilizing $\mathbf{q}(\mathbf{x}_0)$ the element is translating as though in fluid at rest far from the element. This implies that the methods developed up to the present time must be limited to situations in which the cilia are sufficiently widely spaced so that the local flow field around one cilium does not extend to the neighboring cilia. One is left with the impression that such difficulties are not entirely resolved in the existing literature. Indeed, Blake & Sleigh (1974) have pointed out that desirable improvements to the present sublayer models could be provided by studies of the translation of slender bodies in the presence of other similar bodies.

The above description suggests an iterative scheme that begins with a best guess for $q(x)$ and proceeds through evaluation of the stokeslets to the calculation of a "new" velocity field $q(x)$. Such iterative methods have been used by Keller, Wu & Brennen (1975). Alternatively the problem may be put in the form of an integral equation for $q(x)$ as originally demonstrated by Blake (1972).

In his sublayer analyses Blake (1972) included only a *steady* or time-averaged velocity of the fluid tangential to the wall, u, in his calculation of the stokeslet strength and obtained the steady velocity profile $u(x_3)$ (x_3 being the coordinate normal to the wall) that results from his solution of the integral equation. Keller, Wu & Brennen (1975) pointed out, however, that since oscillatory velocities of comparable and greater magnitude are created by the cilia motions these should be included in evaluation of the forces on individual cilium elements. They used a method somewhat different from that of Blake in which these forces are smoothed out to form a continuous body force field within the cilia layer and their solution is achieved by solving the Stokes equations in this layer with these body force terms included. The resulting iterative solution yields a velocity profile not only for the steady velocities but also for the oscillatory velocities both normal and tangential to the wall.

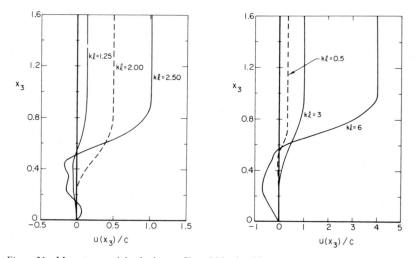

Figure 21   Mean tangential velocity profiles within the ciliary array (the normal coordinate, x_3, has been nondimensionalized by the cilia length l) for *Opalina* (*left*) and *Paramecium* (*right*) by using the planar beat forms of Figure 16. The dashed lines (Blake 1972) neglect the oscillatory velocity interactions of the cilia and were computed for cilia spaced at intervals d_1, d_2 on the cell surface, where $k^2 d_1 d_2$ was assumed to be 0.04 for *Opalina* and 0.00025 for *Paramecium*. The solid lines (Keller, Wu & Brennen 1975) include oscillatory velocities and are computed for values of $k^2 d_1 d_2$ of 0.082 and 1.3, values apparently more consistent with the observed cilia distributions.

The solutions are all obtained in a frame fixed in the organism so that $u(\infty)$ is the propulsive velocity of the ciliary system. Indeed a particular feature of all the present infinite-sheet sublayer models is that $u(x_3)$ remains constant beyond the maximum extent of the ciliary tips; we return to this point later. The majority of existing solutions are also for purely planar cilia beat patterns and for purely symplectic or antiplectic metachronism. As observed earlier, virtually all beat patterns contain to some greater or lesser degree displacements in all three coordinate directions and have metachronal patterns that deviate toward the diaplectic. Velocity profiles derived by Blake (1972) and Keller, Wu & Brennen (1975) for *Opalina* and *Paramecium* are shown in Figure 21; these are based on the planar beat patterns of Figure 16 and the assumption of symplectic and antiplectic metachronism for these two organisms, respectively. In both cases there are practical difficulties because the cilia of *Opalina* are too densely packed (see Figure 14) for one to have confidence in the sublayer model as presently constituted, and in the case of *Paramecium* because the metachronism is diaplectic and the beat three-dimensional (see above). The latter point was later rectified by Blake (1974a) who incorporated three-dimensionality in a sublayer model for *Paramecium*.

Apart from the individual imperfections of the two local fluid/cilia interaction models mentioned above, there are further difficulties in applying either to real physiological systems. We confine ourselves here to discussion of two of the principal difficulties to which some attention and thought has been given.

5.4   Ciliated Organisms

Apart from the infinite-sheet geometries discussed above, several attempts have been made to apply these models to finite organisms. In the first solution of this kind Blake (1971a), extending earlier work by Lighthill (1952), approximated small-amplitude travelling waves on a spherical body by combining two spherical-harmonic functions whose orders differed by one. This envelope model is rather restrictive in terms of the permitted variation of wave form and amplitude and it approximates travelling waves only near the maximum width of the body.

Subsequently Brennen (1974, 1975) has pointed out that the flow around most ciliated organisms for which the metachronal wavelength is small compared with overall dimensions will be comprised of two parts: (*a*) a relatively thin oscillatory boundary layer within which the oscillatory motions created by the cilia will decay rather rapidly with distance from the surface and (*b*) an outer steady Stokes flow around the organism. The problem is then to find some way of matching a local fluid/cilia interaction model within the boundary layer to the so-called complementary Stokes flow outside the boundary layer. For self-propelling organisms this complementary Stokes flow and the velocity of propulsion can only be obtained explicitly after application of the condition of zero total force on the organism; the velocity field far from the organism in this case probably cannot be like a stokeslet since the self-propelling organism exerts no net steady force on the fluid. It must be like a Stokes doublet or higher order. In this regard it is of interest to relate that some recent flow-visualization experiments with minute polystyrene tracer particles suggest that it is *not* like a Stokes doublet but more like that of a potential doublet (Keller & Wu 1976). Such a flow has less dissipation of energy than the Stokes doublet, which suggests that the ciliary system, at least in the organism observed, has been optimized to the extent of producing a complementary Stokes flow that does not contain a Stokes-doublet component. On the other hand, if the organism is pinned down, the complementary Stokes flow will be like that of a stokeslet in the far field and the thrust produced by the restrained organism can be computed. Brennen (1974, 1975) has applied such a matching technique to the propulsion of spherical and ellipsoidal ciliates using an envelope model for the local fluid/cilia interaction, and Blake (1973a) has also considered the effect of a finite cell body on results obtained for infinite-sheet models.

One particular feature of Brennen's results (1974, 1975) is that they allow evaluation of the thrust, T, that a restrained organism can produce. [With regard to this it is worth noting, as Taylor (1951) did, that there is zero net thrust in the infinite-sheet solutions.] Typical values from an envelope model are shown in Figure 22 for direct comparison with Figure 19. Note that the ciliary beat patterns for optimum thrust on restrained organisms differ from those for optimum rectilinear propulsive velocity. This raises questions, which will not otherwise be discussed here, of whether a large starting (or turning) thrust or an optimum steady propulsive velocity is of greater importance for individual species.

Recently some attempts have been made to measure the actual velocity profiles near the surface of ciliated microorganisms. Cheung & Winet (1975) report on some such measurements on *Spirostomum* using minute polystyrene tracer particles. One would hope that further quantitative data for other physiological situations will be obtained in order to allow detailed evaluation of the theoretical models. The *mean* velocity profiles of Cheung & Winet (1975) are shown in Figure 23. A particular

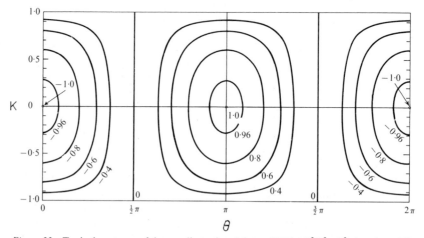

Figure 22 Typical contours of the nondimensional thrust, $3T/4\pi\mu k^2(h_s^2+h_n^2)Ac$, where T is thrust developed by a restrained ciliate of typical dimension $2A$, according to an envelope model for the fluid/cilia interaction (Brennen 1975). For comparison with Figures 18 and 19.

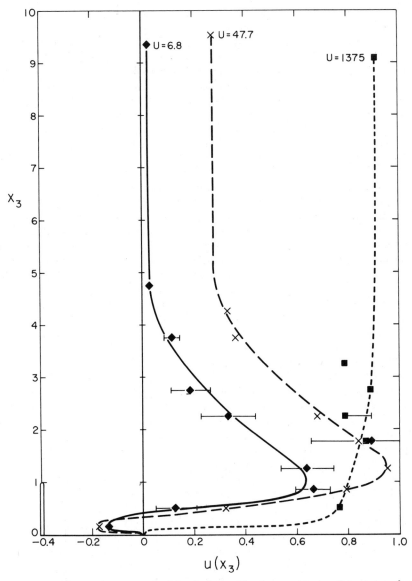

Figure 23 Measured mean tangential velocity profiles for a swimming *Spirostomum* for different swimming speeds U (in μm/sec) as indicated [from Cheung & Winet (1975), who observed polystyrene tracer particle motions]. The normal coordinate x_3 is given in cilia lengths and the horizontal velocity axis has been nondimensionalized by dividing by the fastest tracer particle value for each profile. Effective stroke toward the right.

feature of these results concerns the variation of the mean tangential velocity with distance, x_3, from the surface (here normalized with respect to cilia length) for the larger values of x_3. The corresponding envelope model for an infinite sheet would suggest that the velocity approaches its asymptotic value within a value of x_3 of about 3; the sublayer model for an infinite sheet has no variation beyond $x_3 = 1$. Thus it would seem that one is measuring the complementary Stokes flow at the higher values of x_3; hence the aforementioned need to understand the interaction between this predominantly steady flow and the more localized and unsteady interaction between the cilia and the fluid. In this context studies of the propulsion of ciliates in tubes can yield important information (Winet 1973).

5.5   Internal Flows and the Propulsion of Mucus

It appears that the primary difficulty regarding internal flows (and a number of external flows: Jahn et al 1965, Jahn & Hendrix 1969, Winet & Jones 1975) in organs with ciliated epithelia is that the fluid usually is non-Newtonian. This fluid virtually always consists of mucus or some other association of glycoproteins. Such systems are called *mucociliary systems*. When the concentration of these long-chain polymers in the colloidal system is greater than about 1% (wt vol^{-1}), the system invariably displays significant viscoelastic, shear-thinning, or thixotropic effects. At higher concentrations gel particles and eventually gel networks form. At all mucin concentrations up to and including the ones that produce gelation—the network-formation process—liquefaction (gel $\rightarrow$ sol transformation) will occur under applied stress (see Frey-Wyssling 1952; Eliezer 1974; Litt 1970; Hwang, Litt & Forsman 1969; and others).

The propulsion of mucus in mammalian trachea is one such situation that appears to be dominated by viscoelastic effects. The conventional view is that a blanket of highly viscoelastic mucus lining the airway is propelled by cilia that are surrounded by a much less viscous fluid (serous fluid). The cilia appear to move the blanket by contacting it only during the effective stroke (Cheung & Jahn 1975). Experiments by Sadé et al (1970) have indicated that the propulsion of mucus is quite sensitive to the form and state of the mucus, with optimum propulsion occurring when the concentration of glycoprotein is close to that of the sol $\rightarrow$ gel transformation. Ross & Corrsin (1974) constructed a theoretical two-fluid-layer model (inner serous-fluid layer and outer viscoelastic blanket of mucus) for mucus transport employing an envelope model for the cilia/serous fluid interaction. Their calculations indicate that propulsion is enhanced when the mucus is fairly rigid and when the ciliary tip loci have predominantly horizontal motions. Blake (1975) has also applied his sublayer method to model the interaction of the cilia and the serous fluid (see also Miller 1969 and Barton & Raynor 1967) and the subsequent propulsion of a solid slab representing the mucus blanket; he also gives a qualitative discussion of the effects of the elasticity of the slab.

However, the observation by Cheung & Jahn (1975) of more direct mechanical propulsion of the mucus blanket appears to require a reevaluation of the fluid mechanics of tracheal mucus propulsion. If the cilia penetrate the mucus only during the effective stroke, this provides a much more direct mechanism for mucus pro-

Table 7 Some characteristics of internal ciliary systems

			Cilia					Typical vol. flow (μm^3/sec), vol. flow/unit width, Q (μm^2/sec) or velocity, U (μm/sec)	References
	Length l (μm)	Beat freq. (Hz)	Concentration No. cilia/μm^2	Distribution	Metachrony	Wavelength (μm)	Velocity (μm/sec)		
Bos ductus efferentes cilia	10	20	—	—	—	—	200	33000	135
Canis tracheal cilia	—	—	—	—	—	—	—	$U = 600$	130
Cardium gill cilia	—	—	—	—	—	—	—	14×10^{10}	160
Chicken tracheal cilia	—	—	—	—	—	—	—	$U = 552$	130
Crassostrea gill cilia	—	—	—	—	—	—	—	$5.5\text{–}46 \times 10^{10}$	160
Dressena gill cilia	—	—	—	—	—	—	—	$0.14\text{–}5 \times 10^{10}$	160
Felis tracheal cilia	—	—	—	—	—	—	—	$U = 600$	51, 130
Homo sapiens tracheal cilia	6	—	17.5	200/cell	—	—	—	—	38, 135, 169
Monkey oviduct ampulla cilia	—	—	800	50% ciliated	—	—	—	—	11, 37
tracheal cilia	—	—	—	—	—	—	—	$U = 352$	130
Mya gill cilia	—	—	—	—	—	—	—	$1.6\text{–}36 \times 10^{10}$	160
Mytilus abfrontal gill cilia	62	20	—	$17\ \mu m^{-1}$	laeoplectic	6	100	$Q = 19 \times 10^4$	179, 194, 203, 235
lateral gill cilia	15	—	—	—	—	15	—	$Q = 5.8 \times 10^4$	86, 87, 202, 203
gill cilia and cirri	—	—	—	—	—	—	—	22×10^{10}	88, 194
Oris ductus efferentes	10	20	—	—	—	—	200	3.3×10^4	135
Pecten gill cilia and cirri	—	20	—	—	—	—	—	90×10^{10}	160
Rabbit oviduct ampulla cilia	5	20	5	—	—	—	—	$U = 110$	95, 132, 135, 157
tracheal cilia	6	20–25	—	—	—	—	—	$U = 250\text{–}410$	52
Frog pharyngeal cilia	6	12.5–20	3–11	$0.3\text{–}0.4\ \mu m^{-1}$ tufts	antiplectic	10–12	100–200	$U = 200\text{–}300$	82, 192, 228
brain ependyma cilia	20	21.4	30–40/cell	—	—	20	—	—	156
Rat ductus efferentes cilia	10	20	—	—	—	—	200	$6 \times 10^6\ gm^{-1}$ testis	135
tracheal cilia	5	22	8.4	—	—	—	—	$U = 250\text{–}350$	130, 202
Sabellaria gill cirrus (compound cilia, width 1 μm)	32–56	20	—	—	dexioplectic	—	—	—	194

pulsion. Furthermore, such a mechanism would not require any particular organization of the metachrony of the cilia, unlike most of the fluid-mechanical models. Indeed, some confirmation of this direct-contact mechanism is provided by the fact that propulsion seems to take place in the absence of organized metachronism (Cheung & Jahn 1975), which is often hard to observe in tracheal cilia. The observations suggest that the cilia attached to each ciliated cell (which are usually interspersed with secretion cells) beat in synchrony but that any relationship between the phase of the cilia on different cells is not readily apparent. It should be noted, however, that the tissue utilized for this study was observed in vitro (i.e. removed from its site in the organism and macerated to create a layer thin enough for observation on a thin slide preparation), and in vivo influences on ciliary motion such as the nervous effect described by Murakami & Takahashi (1975) could not have been taken into account. In any case a reexamination of the fluid mechanics of tracheal mucociliary propulsion should also show awareness that both the mucus blanket and the distribution of ciliated cells on the epithelium can be quite nonuniform.

In mucociliary systems where mucus is an incidental component of the propelled object, e.g. the mammalian oviduct in which ova and perhaps spermatozoa are the primary objects of ciliary activity, the role of mucus is not clear. It has long been assumed, for example, that mucus acts as a lubricant for the transport of ova down the isthmus of the oviduct. No quantitative test of this assumption appears to have been conducted in situ. A model system utilized recently for measuring the lubrication effect (Winet 1976) consists of a ciliated spheroid swimming down a mucin-filled tube. Observations of this system indicate not only a lubrication effect for small clearances but also an optimized drag reduction effect at larger clearances. Although we have concentrated here on tracheal mucus flows, we should mention in closing that there are many other internal flows in ciliated tubes such as the ductus efferentes and the oviduct for which some modification of an infinite-sheet model may suffice (e.g. Blake 1975). Some of the characteristics of internal ciliary systems have been collected in Table 7. However, neither sufficiently detailed observations nor complete quantitative information are presently available to allow confident examination of most of these ciliary systems.

6   CONCLUDING REMARKS

In closing we should emphasize again that, although we have concentrated in this review on the fluid mechanics of cilia and flagella, a complete understanding of the life functions of these biological systems requires much more than fluid mechanics. At the same time, fluid mechanics is an integral component of any quantitative analysis for the contraction processes and indicates where ciliation and flagellation give selective advantages to organisms not only in terms of their ability to propel fluids but also in terms of biosynthesis and concentration. Indeed, the ubiquity of these systems as indicated by Table 1 would itself be a worthy study by comparative physiologists.

We have thus attempted to give an overview of the fluid mechanics of these biological slender bodies. It should be clear that the subject is still in a developmental stage and much remains to be done. One expects to see further developments in the basic fluid mechanics of slender bodies at low Reynolds numbers, especially when these operate close to a large "cell" body. Furthermore, the present understanding of the fluid mechanics of ciliary systems is still rather limited and we have tried to indicate that there are many other ciliary systems that have not yet received attention from a fluid-mechanical point of view. But in conclusion we must stress that a multidisciplinary approach is very necessary for any assessment of the fluid-mechanical models and, consequently, for a fuller understanding of cilia and flagella and the life functions they attend.

ACKNOWLEDGMENTS

The authors gratefully acknowledge support from the National Science Foundation under Grants ENG 74-23008 A01 and AEN 72-03586 A01 and from the Office of Naval Research under Contract N 00014-76-C-0157 during the preparation of this paper. We should also like to thank Professor T. Y. Wu, Professor C. J. Brokaw, Dr. A. T. Chwang, and Mr. R. E. Johnson for their assistance as well as those who permitted us to use their photographs.

Literature Cited

1. Adler, J. 1976. The sensing of chemicals by bacteria. *Sci. Am.* 234:40–47
2. Aiello, E., Sleigh, M. A. 1972. The metachronal wave of lateral cilia of *Mytilus edulis. J. Cell Biol.* 54:493–506
3. Anderson, R. A. 1975. Formation of the bacterial flagellar bundle. See Ref. 234, pp. 45–56
4. Andrew, W. 1959. *Textbook of Comparative Histology.* New York: Oxford Univ. Press. 652 pp.
5. Austin, C. R. 1965. *Fertilization,* pp. 34–35. Englewood Cliffs, NJ: Prentice-Hall
6. Baccetti, B. et al 1975. Motility patterns in sperms with different tail structure. In *The Functional Anatomy of the Spermatozoon,* ed. B. A. Afzelius, pp. 141–50. Elmsford, NY: Pergamon
7. Barber, V. C. 1974. Cilia in sense organs. See Ref. 199, pp. 403–36
8. Barton, C., Raynor, S. 1967. Analytical investigation of cilia-induced mucous flow. *Bull. Math. Biophys.* 29:419–28
9. Batchelor, G. K. 1970a. Slender body theory for particles of arbitrary cross-section in Stokes flow. *J. Fluid Mech.* 44:419–40
10. Batchelor, G. K. 1970b. Stress system in a suspension of force-free particles. *J. Fluid Mech.* 41:545–70
11. Beck, L. T., Boots, L. R. 1974. The comparative anatomy, histology and morphology of the mammalian oviduct. In *The Oviduct and Its Functions,* ed. A. D. Johnson, C. E. Foley, pp. 1–52. New York: Academic
12. Berg, H. C. 1974. Dynamic properties of bacterial flagellar motors. *Nature* 249: 77–79
13. Berg, H. C. 1975. Bacterial movement. See Ref. 234, Vol. I, pp. 1–11
14. Berg, H. C., Anderson, R. A. 1973. Bacteria swim by rotating their flagellar filaments. *Nature* 245:380–82
15. Bharier, M. A., Rittenberg, S. C. 1971. Chemistry of axial filaments of *Treponema zuelzerae, J. Bacteriol.* 105:422–29
16. Black, J., Korostoff, E. 1974. Strain-related potentials in living bone. *Ann. NY Acad. Sci.* 238:95–120
17. Blake, J. R. 1971a. A spherical envelope approach to ciliary propulsion. *J. Fluid Mech.* 46:199–208
18. Blake, J. R. 1971b. Infinite models for ciliary propulsion. *J. Fluid Mech.* 49: 209–22
19. Blake, J. R. 1971c. A note on the image system for a Stokeslet in a no-slip boundary. *Proc. Cambridge Philos. Soc.* 70:303–10
20. Blake, J. R. 1972. A model for the micro-structure in ciliated organisms. *J. Fluid Mech.* 55:1–23
21. Blake, J. R. 1973a. A finite model for ciliated microorganisms. *J. Biomech.* 6:133–40

22. Blake, J. R. 1973b. Flow in tubules due to ciliary activity. *Bull. Math. Biol.* 35: 513–23

23. Blake, J. R. 1974a. Hydrodynamic calculations on the movements of cilia and flagella. Part I. *Paramecium. J. Theor. Biol.* 45: 183–203

24. Blake, J. R. 1974b. Singularities of viscous flow. Part II. Applications to slender body theory. *J. Eng. Math.* 8: 113–24

25. Blake, J. R. 1975. On the movement of mucus in the lung. *J. Biomech.* 8: 179–90

26. Blake, J. R., Chwang, A. T. 1974. Fundamental singularities of viscous flow. Part I. The image systems in the vicinity of a stationary no-slip boundary. *J. Eng. Math.* 8: 23–29

27. Blake, J. R., Sleigh, M. A. 1974. Mechanics of ciliary locomotion. *Biol. Rev.* 49: 85–125

28. Borradaile, L. A., Potts, F. A. 1958. *The Invertebrata.* Cambridge: Univ. Press. 795 pp. 3rd ed.

29. Bourne, G. J. 1960. *The Structure and Function of Muscle,* Vol. I, *Structure.* New York: Academic. 472 pp.

30. Bovee, E. C. 1965. Swimming movement of *Eimeria* sp. merozoites. *Progress in Protozoology, 2nd Int. Conf. Protozool.* Abstr. No. 166, p. 152

31. Bradfield, J. R. G., Cater, D. B. 1952. Electron-microscopic evidence on the structure of spirochaetes. *Nature* 169: 944

32. Breed, R. S., Murray, E. G. D., Smith, N. R. 1957. *Bergey's Manual of Determinative Bacteriology.* Baltimore: Williams & Wilkins. 1094 pp.

33. Brennen, C. 1974. An oscillating-boundary-layer theory for ciliary propulsion. *J. Fluid Mech.* 65: 799–824

34. Brennen, C. 1975. Hydromechanics of propulsion for ciliated microorganisms. See Ref. 234, pp. 235–52

35. Brennen, C. 1976. Locomotion of flagellates with mastigonemes. *J. Mechanochem. Cell Motil.* 3: 207–17

36. Brenner, H. 1962. Effect of finite boundaries on the Stokes resistance of an arbitrary particle. *J. Fluid Mech.* 12: 35–48

37. Brenner, R. M. 1969. Renewal of oviduct cilia during the menstrual cycle of the Rhesus monkey. *Fertil. Steril.* 20: 599–611

38. Brinkman, G. E. 1966. Discussion. *Am. Rev. Respir. Dis.,* 93, no. 3, pt. 2: 60

39. Broersma, S. 1960. Viscous force constant for a closed cylinder. *J. Chem.*

40. Brokaw, C. J. 1963. Movement of the flagella of *Polytoma uvella. J. Exp. Biol.* 40: 149–56

41. Brokaw, C. J. 1965. Non-sinusoidal bending waves of sperm flagella. *J. Exp. Biol.* 43: 155–69

42. Brokaw, C. J. 1970. Bending moments on free-swimming flagella. *J. Exp. Biol.* 53: 445–64

43. Brokaw, C. J. 1971. Bend propagation by a sliding filament model for flagella. *J. Exp. Biol.* 55: 289–304

44. Brokaw, C. J. 1972. Computer simulation of flagellar movement. *Biophys. J.* 12: 564–86

45. Brokaw, C. J. 1975. Spermatozoan motility: A biophysical survey. *Biol. J. Linn. Soc.* 7: Suppl. 1, pp. 423–39

46. Brokaw, C. J., Gibbons, I. R. 1975. Mechanisms of movement in flagella and cilia. See Ref. 234, Vol. I, pp. 89–132

47. Bullington, W. E. 1925. A study of spiral movement in the ciliate infusoria. *Arch. Protistenkd.* 50: 219–74

48. Bullington, W. E. 1930. A further study of spiraling in the ciliate Paramecium, with a note on morphology and taxonomy. *J. Exp. Zool.* 56: 423–51

49. Burgers, J. M. 1938. On the motion of small particles of elongated form suspended in a viscous fluid. *K. Ned. Akad. Wet. Verhand.* 16, no. 4: 113–84

50. Calladine, C. R. 1974. Bacteria can swim without rotating flagellar filaments. *Nature* 249: 385

51. Carson, S., Goldhamer, R., Carpenter, R. 1966. Mucus transport in the respiratory tract. *Am. Rev. Respir. Dis.,* 93, no. 3, pt. 2: 86–92

52. Cheung, A. T. W., Jahn, T. L. 1975. Determination of the movement pattern of the epithelial cilia of rabbit trachea and the clearance mechanism of the tracheal muco-ciliary clearance system. See Ref. 234, Vol. I, pp. 289–300

53. Cheung, A. T. W., Winet, H. 1975. Flow velocity profile over a ciliated surface. See Ref. 234, Vol. I, pp. 223–34

54. Chwang, A. T. 1975. Hydromechanics of low-Reynolds-number flow. Part 3. Motion of a spheroidal particle in quadratic flows. *J. Fluid Mech.* 72: 17–34

55. Chwang, A. T., Winet, H., Wu, T. Y. 1974. A theoretical mechanism for spirochetal locomotion. *J. Mechanochem. Cell Motil.* 3: 69–76

56. Chwang, A. T., Wu, T. Y. 1971. A note on the helical movement of microorganisms. *Proc. R. Soc. London Ser. B*

178:327–46
57. Chwang, A. T., Wu, T. Y. 1974. Hydromechanics of low-Reynolds-number flow. Part 1. Rotation of axisymmetric prolate bodies. J. Fluid Mech. 63:607–22
58. Chwang, A. T., Wu, T. Y. 1975. Hydromechanics of low-Reynolds-number flow. Part 2. Singularity method for Stokes flows. J. Fluid Mech. 67:787–815
59. Chwang, A. T., Wu, T. Y. 1976. Hydromechanics of low-Reynolds-number flow. Part 4. Translation of spheroids. J. Fluid Mech. 75:677–89
60. Chwang, A. T., Wu, T. Y., Winet, H. 1972. Locomotion of spirilla. Biophys. J. 12:1549–61
61. Cleveland, L. R., Cleveland, B. T. 1966. The locomotory waves of Koruga, Deltotrychonympha and Mixotricha. Arch. Protistenkd. 109:39–63
62. Coakley, C. J., Holwill, M. E. J. 1972. Propulsion of microorganisms by three-dimensional flagellar waves. J. Theor. Biol. 35:525–42
63. Costello, D. P. 1973a. A new theory on the mechanics of ciliary and flagellar motility. I. Supporting observations. Biol. Bull. 145:279–91
64. Costello, D. P. 1973b. A new theory on the mechanics of ciliary and flagellar motility. II. Theoretical considerations. Biol. Bull. 145:292–309
65. Cox, P. J., Twigg, G. I. 1974. Leptospiral motility. Nature 250:260–61
66. Cox, R. G. 1970. The motion of long slender bodies in a viscous fluid. Part 1. General theory. J. Fluid Mech. 44:791–810
67. Davis, R. E., Worley, J. F. 1973. Spiroplasma: Motile helical microorganism associated with corn stunt disease. Phytopathology 63:403–8
68. de Mestre, N. J. 1973. Low-Reynolds-number fall of slender cylinders near boundaries. J. Fluid Mech. 58:641–56
69. de Mestre, N. J., Russel, W. B. 1975. Low-Reynolds number-translation of a slender cylinder near a plane wall. J. Eng. Math. 9:81–91
70. Denehy, M. A. 1975. The propulsion of non-rotating ram and oyster spermatozoa. Biol. Reprod. 13:17–29
71. DePamphilis, M. L., Adler, J. 1971. Fine structure and isolation of the hook-basal body complex of flagella from Escherichia coli and Bacillus subtilis. J. Bacteriol. 105:384–95
72. Dirksen, E. R., Satir, P. 1972. Ciliary activity in the mouse oviduct as studied by transmission and scanning electron microscopy. Tissue Cell 4:389–404
73. Doetsch, R. N. 1966. Some speculations accounting for the movement of bacterial flagella. J. Theor. Biol. 11:411–17
74. Douglas, G. J. 1975. Sliding filaments in sperm flagella. J. Theor. Biol. 53:247–52
75. Dryl, S. 1975. Local microtubules interaction theory of ciliary and flagellar motion. Bull. Acad. Polon. Sci. Ser. Sci. Biol. 23:339–46
76. Eliezer, N. 1974. Viscoelastic properties of mucus. Biorheology 11:61–68
77. Fox, D. L., Coe, W. R. 1943. Biology of the California sea-mussel (Mytilus californianus). Biol. Bull. 83:205–49
78. Frey-Wyssling, A. 1952. Deformation and Flow in Biological Systems. New York: Wiley-Interscience
79. Gardner, C. R. 1976. The neuronal control of locomotion in the earthworm. Biol. Rev. 51:25–52
80. Gerber, B. R. 1975. Towards a molecular mechanism for the movement of bacterial flagella. See Ref. 234, Vol. I, pp. 69–87
81. Gibbons, B. H., Gibbons, I. R. 1972. Flagellar movement and adenosine triphosphate activity in sea urchin sperm extracted with Triton X-100. J. Cell Biol. 54:75–97
82. Gilboa, A., Silberberg, A. 1975. In situ rheological characterization of epithelial mucus. Biorheology 13:59–65
83. Gittleson, S. M., Hotchkiss, S. K., Valencia, F. G. 1974. Locomotion in the marine dinoflagellate Amphidinium carterae (Hulburt). Trans. Am. Microsc. Soc. 93:101–5
84. Gittleson, S. M., Jahn, T. L. 1968. Flagellar activity of Polytomella agilis. Trans. Am. Microsc. Soc. 87:464–71
85. Gittleson, S. M., Noble, R. M. 1973. Locomotion in Polytomella agilis and Polytoma uvella. Trans. Am. Microsc. Soc. 92:122–28
86. Gosselin, R. E. 1958. Influence of viscosity on metachronal rhythm of cilia. Fed. Proc. 17:372
87. Gosselin, R. E. 1966. Physiologic regulators of ciliary motion. Am. Rev. Respir. Dis., 93, no. 3, pt. 2:41–59
88. Gosselin, R. E., O'Hara, G. 1961. An unsuspected source of error in studies of particle transport by lamellibranch gill cilia. J. Cell Comp. Physiol. 58:1
89. Gray, J. 1928. Ciliary Movement. Cambridge: Univ. Press. 162 pp.
90. Gray, J. 1953. Undulatory propulsion. Q. J. Microsc. Sci. 94:551–78

91. Gray, J. 1955. The movement of sea-urchin spermatozoa. *J. Exp. Biol.* 32: 775–801

92. Gray, J. 1968. *Animal Locomotion.* London: Weidenfeld & Nicolson. 479 pp.

93. Gray, J., Hancock, G. J. 1955. The propulsion of sea-urchin spermatozoa. *J. Exp. Biol.* 32:802–14

94. Graziadei, P. P. C. 1971. Olfactory mucosa of vertebrates. In *Handbook of Sensory Physiology, Vol. IV, pt. 1. Olfaction,* ed. L. M. Beidler, pp. 27–58. Berlin: Springer

95. Halbert, S. A., Tam, P. Y., Blandau, R. J. 1976. Egg transport in the rabbit oviduct: The role of cilia and muscle. *Science* 191:1052–53

96. Hancock, G. J. 1953. The self-propulsion of microscopic organisms through liquids. *Proc. R. Soc. London Ser. A* 217:96–121

97. Hand, W. G., Collard, P. A., Davenport, D. 1965. The effects of temperature and salinity change on swimming rate in the dinoflagellates. *Biol. Bull.* 128:90

98. Hand, W. G., Haupt, W. 1971. Flagellar activity of the colony members of *Volvox aureus* Ehrbg. during light stimulation. *J. Protozool.* 18:361–64

99. Hand, W. G., Schmidt, J. A. 1975. Phototactic orientation by the marine dinoflagellate *Gyrodinium dorsum* Kofoid. II. Flagellar activity and overall response mechanism. *J. Protozool.* 22: 494–98

100. Harris, J. E. 1961. The mechanics of ciliary movement. In *The Cell and the Organism,* ed. J. A. Ramsay, V. B. Wigglesworth, pp. 22–36. Cambridge: Univ. Press

101. Harris, W. F. 1973. Bacterial flagella, do they rotate or do they propagate waves of bending? *Protoplasma* 77: 477–79

102. Harris, W. F., Robison, W. G. Jr. 1973. Dislocations in microtubular bundles with spermatoxon of the coccid insect *Neosteingelia texana* and evidence for slip. *Nature* 246:513–15

103. Harvey, C. 1960. The speed of human spermatozoa and the effect on it of various diluents with some preliminary observations on clinical material. *J. Reprod. Fertil.* 1:84–95

104. Hespell, R. B., Canale-Parola, E. 1970. *Spirochaeta litoralis* sp. n. a strictly anaerobic marine spirochaete. *Arch. Mikrobiol.* 74:1–18

105. Holwill, M. E. J. 1966a. Physical aspects of flagellar movement. *Physiol. Rev.* 46: 696–785

106. Holwill, M. E. J. 1966b. The motion of *Euglena viridis*: The role of the flagella *J. Exp. Biol.* 44:578–88

107. Holwill, M. E. J. 1974. Hydrodynamic aspects of ciliary and flagellar movement. See Ref. 199, pp. 143–76

108. Holwill, M. E. J. 1975. The role of body oscillation in the propulsion of microorganisms. See Ref. 234, Vol. I, pp. 133–41

109. Holwill, M. E. J., Burge, R. E. 1963. A hydrodynamic study of the motility of flagellated bacteria. *Arch. Biochem. Biophys.* 101:249–60

110. Holwill, M. E. J., Silvester, N. R. 1965. The thermal dependence of flagellar activity in *Strigomonas oncopelti. J. Exp. Biol.* 42:537–44

111. Holwill, M. E. J., Sleigh, M. A. 1967. Propulsion by hispid flagella. *J. Exp. Biol.* 47:267–76

112. Hwang, S., Litt, M., Forsman, W. 1969. Rheological properties of mucus. *Rheol. Acta* 8:438–48

113. Iino, T., Mitani, M. 1966. A mutant of *Salmonella* possessing straight flagella. *J. Gen. Microbiol.* 49:81–88

114. Jaffrin, M. Y., Shapiro, A. H. 1971. Peristaltic pumping. *Ann. Rev. Fluid Mech.* 3:13–36

115. Jahn, T. L. 1975. New problems in propulsion of micro-organisms. See Ref. 234, Vol. I, pp. 325–38

116. Jahn, T. L., Bovee, E. C. 1967. Motile behavior of protozoa. In *Research in Protozoology,* ed. T. T. Chen, pp. 41–200. New York: Pergamon

117. Jahn, T. L., Bovee, E. C. 1968. Locomotion of blood protists. In *Infectious Blood Diseases of Man and Animals,* ed. D. Weinman, M. Ristic, pp. 393–436. New York: Academic

118. Jahn, T. L., et al 1965. Secretory activity of the oral apparatus of ciliates: Trails of adherent particles left by *Paramecium multimicronucleatum* and *Tetrahymena pyriformis. Ann. NY Acad. Sci.* 118: 912–20

119. Jahn, T. L., Hendrix, E. M. 1969. Locomotion of the Telotrich ciliate *Opisthonecta henneguyi. Rev. Soc. Mex. Hist. Nat.* 30:103–31

120. Jahn, T. L., Landman, M. D. 1965. Locomotion of spirochetes. *Trans. Am. Microsc. Soc.* 84:395–406

121. Jahn, T. L., Landman, M. D., Fonseca, J. R. 1964. The mechanism of locomotion in flagellates. II. Function of the mastigonemes of *Ochromonas. J. Protozool.* 11:291–96

122. Jahn, T. L., Votta, J. J. 1972. Loco-

motion of protozoa. *Ann. Rev. Fluid Mech.* 4:93–116
123. Kaiser, G. E., Doetsch, R. N. 1975. Enhanced translational motion of *Leptospira* in viscous environments. *Nature* 255:656–57
124. Katz, D. F. 1972. On the biophysics of in vivo sperm transport. 'PhD thesis. Univ. Calif., Berkeley. 219 pp.
125. Katz, D. F. 1974. On the propulsion of micro-organisms near solid boundaries. *J. Fluid Mech.* 64:33–49
126. Katz, D. F., Blake, J. R. 1975. Flagellar motions near walls. See Ref. 234, Vol. I, pp. 173–84
127. Katz, D. F., Blake, J. R., Paveri-Fontana, S. L. 1975. On the movement of slender bodies near plane boundaries at low Reynolds number. *J. Fluid Mech.* 72:529–40
128. Keller, S. R., Wu, T. Y. 1976. A porous prolate spheroidal model for ciliated microorganisms. *J. Fluid Mech.* In press
129. Keller, S. R., Wu, T. Y., Brennen, C. 1975. A traction layer model for ciliary propulsion. See Ref. 234, Vol. I, pp. 253–72
130. Kensler, C. J., Battista, S. P. 1965. Chemical and physical factors affecting mammalian ciliary activity. *Am. Rev. Respir. Dis.*, 93, no. 3, pt. 2:93–102
131. Knight-Jones, E. W. 1954. Relations between metachronism and the direction of ciliary beat in metazoa. *Q. J. Microsc. Sci.* 95(4):503–21
132. Koester, H. 1970. Ovum transport. In *Mammalian Reproduction*, ed. H. Gibian, E. J. Plotz, pp. 189–228. New York: Springer
133. Kudo, R. R. 1954. *Protozoology*. Springfield, Ill.: Thomas. 966 pp.
134. Landau, L. D. 1944. A new exact solution of the Navier-Stokes equations. *C. R. (Dokl.) Acad. Sci. URSS* 43:286–88
135. Lardner, T. J., Shack, W. J. 1972. Cilia transport. *Bull. Math. Biophys.* 34:325–35
136. Larsen, S. H., Adler, J., Gargus, J. J., Hogg, R. W. 1974. Chemomechanical coupling without ATP: The source of energy for motility and chemotaxis in bacteria. *Proc. Natl. Acad. Sci. USA* 71:1239–43
137. Lee, J. W. 1954. The effect of pH on forward swimming in *Euglena* and *Chilomonas. Physiol. Zool.* 27:272–75
138. Lehninger, A. L. 1971. *Bioenergetics*. Menlo Park, Calif.: Benjamin. 245 pp.
139. Leifson, E. 1960. *Atlas of Bacterial Flagellation.* New York: Academic. 171 pp.
140. Lighthill, M. J. 1952. On the squirming motion of nearly spherical deformable bodies through liquids at very small Reynolds numbers. *Commun. Pure Appl. Math.* 5:109–18
141. Lighthill, M. J. 1975. *Mathematical Biofluiddynamics*. Philadelphia: SIAM. 281 pp.
142. Litt, M. 1970. Flow·behavior of mucus. *Ann. Otol. Rhinol. Laryngol.* 80:330–35
143. Lowndes, A. G. 1944. The swimming of *Monas stigmatica* Pringsheim and *Peranema trichophorum* (Ehrbg.) Stein and *Volvox* sp. Additional experiments on the working of a flagellum. *Proc. Zool. Soc. London* 114A:325–38
144. Lowy, J., Spencer, M. 1968. Structure and function of bacterial flagella. *Symp. Soc. Exp. Biol.* 22:215–36
145. Lunec, J. 1975. Fluid flow induced by smooth flagella. See Ref. 234, Vol. I, pp. 143–60
146. Machemer, H. 1969. Filmbildanalysen 4 verschiedener Schlagmuster der Marginalcirren von *Stylonychia. Z. vergl. Physiol.* 62:183–96
147. Machemer, H. 1972a. Ciliary activity and the origin of metachrony in *Paramecium*; effects of increased viscosity. *J. Exp. Biol.* 57:239–60
148. Machemer, H. 1972b. Temperature influences on ciliary beat and metachronal coordination in *Paramecium. J. Mechanochem. Cell Motil.* 1:97–108
149. Machemer, H. 1974. Ciliary activity and metachronism in protozoa. See Ref. 199, pp. 199–287
150. Machin, K. E. 1958. Wave propagation along flagella. *J. Exp. Biol.* 35:796–806
151. Macnab, R. M., Koshland, D. E., Jr. 1972. The gradient-sensing mechanism in bacterial chemotaxis. *Proc. Natl. Acad. Sci. USA* 69:2509–12
152. Metzner, P. 1920. Die Bewegung und Reizbeantwortung der bipolar begeisselten Spirillen. *Jahrb. Wiss. Bot.* 59:325–412
153. Miller, C. E. 1966. Flow induced by mechanical analogues of mucociliary systems. *Ann. NY Acad. Sci.* 130:880–90
154. Miller, C. E. 1969. Streamlines, streak lines, and particle path lines associated with a mechanically-induced flow homomorphic with the mammalian mucociliary system. *Biorheology* 6:127–35
155. Murakami, A., Takahashi, J. 1975. Correlation of electrical and mechani-

cal response in nervous control of cilia. *Nature* 257:48–49

156. Nelson, D. J. 1975. *The distribution, activity and function of the cilia of the brain.* PhD thesis. Univ. Calif., Los Angeles. 196 pp.

157. Nilsson, O., Reinius, S. 1969. Light and electron microscopic structure of the oviduct. In *The Mammalian Oviduct,* ed. E. S. E. Hafez, R. J. Blandau, pp. 57–83. Chicago: Univ. Chicago Press

158. Ogiute, K. 1936. Untersuchungen über die Geschwindigkeit der Eigenbewegung von Bakterien. *Jpn. J. Exp. Med.* 14:19–28

159. Oseen, C. W. 1927. *Neuere Methoden und Ergebnisse in der Hydrodynamik.* Leipzig: Akad.-Verlag. 337 pp.

160. Owen, G. 1974. Feeding and digestion in the bivalva. *Adv. Comp. Physiol. Biochem.* 5:1–36

161. Parducz, B. 1964. Swimming and its ciliary mechanism in *Ophryoglena* sp. *Acta Protozool.* 2:367–74 (3 plates)

162. Parducz, B. 1967. Ciliary movement and coordination in ciliates. *Int. Rev. Cytol.* 21:91–128

163. Peters, N. 1929. Über Orts- und Geisselbewegung bei marinen Dinoflagellaten. *Arch. Protistenkd.* 67:291–321

164. Phillips, D. M. 1972. Comparative analysis of mammalian sperm motility. *J. Cell Biol.* 53:561–73

165. Phillips, D. M. 1974. Structural variants in invertebrate sperm flagella and their relationship to motility. See Ref. 199, pp. 379–402

166. Prosser, C. L. 1973. *Comparative Animal Physiology.* Vol. II. Philadelphia: Saunders. 554 pp.

167. Reese, T. S. 1965. Olfactory cilia in the frog. *J. Cell Biol.* 25:209–30

168. Reynolds, A. J. 1965. The swimming of minute organisms. *J. Fluid Mech.* 23:241–60

169. Rhodin, J. A. G. 1965. Ultrastructure and function of the human tracheal mucosa. *Am. Rev. Respir. Dis.* 93(3): Pt. 2, pp. 1–15

170. Rikmenspoel, R. 1962. Biophysical approaches to the measurement of sperm motility. In *Spermatozoan Motility,* ed. D. W. Bishop, pp. 31–54. Washington: AAAS

171. Rikmenspoel, R. 1975. Contraction model for cilia. See Ref. 234, Vol. I, pp. 273–88

172. Rikmenspoel, R., Sleigh, M. A. 1970. Bending moments and elastic constant in cilia. *J. Theor. Biol.* 28:81–100

173. Rittenberg, S. 1974. Personal communication

174. Rivera, J. A. 1962. *Cilia, Ciliated Epithelium and Ciliary Activity.* New York: Pergamon. 167 pp.

175. Ross, S. M. 1971. *A wavy wall analytic model of muco-ciliary pumping.* PhD thesis. Johns Hopkins Univ., Baltimore, Md. 305 pp.

176. Ross, S. M., Corrsin, S. 1974. Results of an analytical model of mucociliary pumping. *J. Appl. Physiol.* 37:333–40

177. Routledge, L. M. 1975. Bacterial flagella: Structure and function. In *Comparative Physiology—Functional Aspects of Structural Materials,* ed. L. Bolis, H. P. Maddrell, K. Schmidt-Nielsen, pp. 61–73. Amsterdam: North-Holland

178. Sadé, J., Eliezer, N., Silberberg, A., Nevo, A. C. 1970. The role of mucus in transport by cilia. *Am. Rev. Respir. Dis.* 102:48–52

179. Satir, P. 1963. Studies on cilia: I. The fixation of the metachronal wave. *J. Cell Biol.* 18:345–66

180. Satir, P. 1965. Studies on cilia: II. Examination of the distal region of the ciliary shaft and the role of the filaments in motility. *J. Cell Biol.* 39:77–94

181. Satir, P. 1968. Studies on cilia: III. Further studies on the cilium tip and a "sliding filament" model of ciliary motility. *J. Cell Biol.* 39:77–94

182. Satir, P. 1974. The present status of the sliding micro-tubule model of ciliary motion. See Ref. 199, pp. 131–42

183. Schneider, W. R., Doetsch, R. N. 1974. Effect of viscosity on bacterial motility. *J. Bact.* 117:696–701

184. Schreiner, K. E. 1971. The helix as a propeller of microorganisms. *J. Biomech.* 4:73–83

185. Shack, W. J., Lardner, T. J. 1972. Cilia transport. *Bull. Math. Biophys.* 34:325–35

186. Shahar, A., et al 1975. Effect of Δ^9-tetrahydrocannabinol (THC) on the kinetic morphology of spermatozoa. In *The Functional Anatomy of the Spermatozoon,* ed. B. A. Afzelius, pp. 189–94. New York: Pergamon

187. Shimada, K., Yoshida, T., Asakura, S. 1975. Cinematographic analysis of the movement of flagellated bacteria. See Ref. 234, Vol. I, pp. 31–43

188. Shoesmith, J. G. 1960. The measurement of bacterial motility. *J. Gen. Microbiol.* 22:528–35

189. Silverman, M., Simon, M. 1974. Flagel-

lar rotation and the mechanism of bacterial motility. *Nature* 249:73–74

190. Sleigh, M. A. 1960. The form of beat in cilia of *Stentor* and *Opalina. J. Exp. Biol.* 37:1–10 (1 plate)

191. Sleigh, M. A. 1962. *The Biology of Cilia and Flagella.* New York: Macmillan. 242 pp.

192. Sleigh, M. A. 1965. Ciliary coordination in Protozoa. *Progress in Protozoology, 2nd Int. Conf. Protozoology,* Abstr. No. 112, pp. 110–11

193. Sleigh, M. A. 1966. Some aspects of the comparative physiology of cilia. *Am. Rev. Respir. Dis.* 93:Suppl., pp. 16–31

194. Sleigh, M. A. 1968. Patterns of ciliary beating. In *Aspects of Cell Motility, Symp. Soc. Exp. Biol.* 22:131–50

195. Sleigh, M. A. 1969. Coordination of the rhythm of beat in some ciliary systems. *Int. Rev. Cytol.* 25:31–54

196. Sleigh, M. A. 1971. Cilia. *Endeavour* 30:11–17

197. Sleigh, M. A. 1972. *Features of ciliary movement of the ctenophores Beroe, Pleurobrachia and Cestus.* In *Essays in Hydrobiology,* ed. R. B. Clark, R. S. Wootton, pp. 119–36. Exeter, Engl.: Univ. Exeter

198. Sleigh, M. A. 1973. *The Biology of Protozoa.* London: Arnold. 315 pp.

199. Sleigh, M. A. ed. 1974a. *Cilia and Flagella.* New York: Academic. 500 pp.

200. Sleigh, M. A. 1974b. Metachronism of cilia of metazoa. See Ref. 199, pp. 287–304

201. Sleigh, M. A. 1974c. Patterns of movement of cilia and flagella. See Ref. 199, pp. 79–92

202. Sleigh, M. A. 1976. Fluid propulsion by cilia and physiology of ciliary systems. In *Perspectives in Experimental Biology. Vol. 1, Zoology,* ed. P. S. Davies, pp. 125–34. New York: Pergamon

203. Sleigh, M. A., Aiello, E. 1972. The movement of water by cilia. *Acta Protozool.* 9:265–77

204. Sleigh, M. A., Holwill, M. E. J. 1969. Energetics of ciliary movement in *Sabellaria* and *Mytilus. J. Exp. Biol.* 50:733–44

205. Slezkin, N. A. 1934. On a case of integrability of the complete differential equations of a viscous fluid. *Moskov. Gos. Univ. Uch. Zap.* 2:89–90

206. Smith, J. E., Carthy, J. D., Chapman, G., Clark, R. B., Nichols, D. 1971. *The Invertebrate Panorama.* London: Weidenfeld & Nicolson. 406 pp.

207. Squire, H. B. 1951. The round laminar jet. *Q. J. Mech. Appl. Math.* 4:321–29

208. Stolp, H. 1968. *Bdellovibrio bacteriovorus*—ein räuberischer Bakterienparasit. *Naturwissenschaften* 55:57–63

209. Summers, K. E., Gibbons, I. R. 1971. Adenosine triphosphate-induced sliding of tubules in trypsin-treated flagella of sea-urchin sperm. *Proc. Natl. Acad. Sci. USA* 68:3092–96

210. Tamm, S. L. 1972. Ciliary motion in *Paramecium.* A scanning electron microscope study. *J. Cell Biol.* 55:250–55

211. Tamm, S. L., Horridge, G. A. 1970. The relation between the orientation of the central fibrils and the direction of beat in cilia of *Opalina. Proc. R. Soc. London Ser. B* 175:219–33

212. Taylor, G. I. 1951. Analysis of swimming of microscopic organisms. *Proc. R. Soc. London Ser. A* 209:447–61

213. Taylor, G. I. 1952a. Analysis of long and narrow animals. *Proc. R. Soc. London Ser. A* 214:158–83

214. Taylor, G. I. 1952b. The action of waving cylindrical tails in propelling microscopic organisms. *Proc. R. Soc. London Ser. A* 211:225–39

215. Taylor, G. I. 1969. Motion of axisymmetric bodies in viscous fluids. In *Problems of Hydrodynamics and Continuum Mechanics,* pp. 718–24. Philadelphia: SIAM

216. Tillett, J. P. K. 1970. Axial and transverse Stokes flow past slender axisymmetric bodies. *J. Fluid Mech.* 44:401–17

217. Tuck, E. O. 1964. Some methods for flows past slender bodies. *J. Fluid Mech.* 18:619–35

218. Tuck, E. O. 1968. A note on a swimming problem. *J. Fluid Mech.* 31:305–8

219. Vanderberg, J. P. 1974. Studies on the motility of *Plasmodium* sporozoites. *J. Protozool.* 21:527–37

220. van Deurs, B. 1974. Spermatology of some Pycnogonida (Arthropoda), with special reference to a microtubulenuclear envelope complex. *Acta Zool.* 55:151–62

221. Votta, J. J., Jahn, T. L., Griffith, D. L., Fonseca, J. R. 1971. Nature of the flagellar beat in *Trachelomonas volvocina, Rhabdomonas spiralis, Menoidium cultellus* and *Chilomonas paramecium. Trans. Am. Microsc. Soc.* 90:404–12

222. Wang, C. Jahn, T. L. 1972. A theory for the locomotion of spirochetes. *J. Theor. Biol.* 36:53–60

223. Warner, F. D. 1974. The fine structure

of the ciliary and flagellar axoneme.
See Ref. 199, pp. 11–38
224. Warner, F. D., Satir, P. 1974. The
structural basis of ciliary bend formation. Radial spoke positional changes
accompanying microtubule sliding. J.
Cell Biol. 63:35–63
225. Weis-Fogh, T. 1975. Principles of contraction in the spasmoneme of vorticel-
lids. A new contractile system. In Comparative Physiology — Functional
Aspects of Structural Materials, ed.
L. Bolis, H. P. Maddrell, K. Schmidt-
Nielsen, pp. 83–98. Amsterdam: North
Holland
226. Weihs, D. 1975. Some hydrodynamic
aspects of fish schooling. See Ref. 234,
Vol. II, pp. 703–18
227. Williams, M. A., Chapman, G. B. 1961.
Electron microscopy of flagellation in
species of Spirillum. J. Bact. 81:195–
203
228. Wilson, G. B., et al 1975. Studies on
ciliary beating of frog pharyngeal epithelium in vitro. II. Relationships
between beat form, metachronal coordination, fluid flow and particle
transport. See Ref. 234, Vol. I, pp.

301–16
229. Winet, H. 1973. Wall drag on free-
moving ciliated microorganisms. J.
Exp. Biol. 59:753–66
230. Winet, H. 1976. Ciliary propulsion of
objects in tubes: Wall drag on swimming Tetrahymena (Ciliata) in the
presence of mucin and other long-
chain polymers. J. Exp. Biol. 64:283–
302
231. Winet, H., Jahn, T. L. 1974. Geotaxis
in protozoa I. A propulsion-gravity
model for Tetrahymena (Ciliata). J.
Theor. Biol. 46:449–65
232. Winet, H., Jones, A. R. 1975. Muco-
cysts in spirostomum (ciliata: hetero-
tricha). J. Protozool. 22:293–96
233. Winet, H., Keller, S. R. 1976. Spirillum
swimming. Theory and observations of
propulsion by the flagellar bundle. J.
Exp. Biol. In press
234. Wu, T. Y., Brokaw, C. J., Brennen, C.,
eds. 1975. Swimming and Flying in
Nature, Vols. 1, 2. New York: Plenum.
1005 pp.
235. Yoneda, M. 1962. Force exerted by a
single cilium of Mytilus edulis II. Free
motion. J. Exp. Biol. 39:307–17

Ann. Rev. Fluid Mech. 1977. 9:399–419

OPTIMUM WIND-ENERGY CONVERSION SYSTEMS

×8108

Ulrich Hütter

Institut für Flugzeugbau, Universität Stuttgart, and Forschungsinstitut
Windenergietechnik FWE der INGEST an der Universität Stuttgart,
7000 Stuttgart 80, West Germany

ADVANTAGES OF HIGH SPECIFIC SPEED

The history of energy-converting machines is the story of consistently increasing specific speed. This is also true for wind-energy converters if we neglect the step backwards between 1868 and approximately 1910, when comparatively small low-speed multiblade steel windmills were introduced, as a temporary help mainly for farming in semiarid areas.

It was several centuries ago, during the first severe confrontation between Occident and Orient after the decline of the Roman Empire, that the European knights, aside from candied fruits and decimal figures, discerned the importance of those strange windmills, most probably with sail-wing rotors. At all events there does exist a document of 1105, six years after the end of the first crusade, which granted to the Benedictine monastery of Savigny a windmill privilegium (Bilau 1933 and Golding 1955).

From then on up to the end of the eighteenth century a consistently empirical development led finally to those well-known Dutch windmills with four-bladed rotors of from 18 up to 26 m diameter, optimum tangential tip-speed ratios $1.8 \leq \lambda_{\omega,\mathrm{RTR}} \leq 2.4$, and maximum speed ratios when idling at zero output of up to 3.2 times the wind velocity v_{FFL} in free flow. We define

$$\lambda_{\omega,\mathrm{RTR}} = r_{\mathrm{TIP}} \cdot \omega_{\mathrm{RTR}}/v_{\mathrm{FFL}}, \tag{1}$$

where r_{TIP} is the blade-tip radius, ω_{RTR} the rotor's angular velocity, and $\lambda_{\omega,\mathrm{RTR}}$ the "speed ratio" or "specific speed" of the rotor, which is related to a definite radius, indicated by a corresponding subscript, and perhaps to a position in relation to the rotor's disc of rotation by a second subscript. Only rotors of the size and specific speed $\lambda_{\omega,\mathrm{RTR}}$ of the above-mentioned "Dutch windmills" yielded even in moderate breezes sufficient RPM and power to drive the huge millstones adequately.

Better understanding of aerodynamics since the early 1920s has directed the main emphasis to improvements in aerodynamics, sometimes overestimating the new possibilities that appeared. Yet in the meantime wind-energy converters (WECs) have been more and more considered as heterogeneous and multiparametric but necessarily integral systems, whose economy depends on minimum total investment at the utmost optimum efficiency of any contributing element. A WEC's economy is moreover dependent on the precision, reliability, and redundancy of its control, its ease of transportation to, and assembly at, remote places of installation, and its lifetime with respect to fatigue as well as attrition. So not only aerodynamically possible high efficiencies, but also the feasibility of constructing and controlling with ease very large rotors. attain, in comparison with the past, extremely high specific speeds at satisfactory efficiencies.

A number of WEC systems that have been actually operating for months or years, feeding electricity to public networks, involved the system groups shown below. The investment for each group is related to the total investment as follows:

Rotor blades and hub	$27\% \pm 6\%$
Main shaft, bearings, transmission gear, and generator	$30\% \pm 4\%$
Turntable, wind direction and general control	$22\% \pm 4\%$
Tower, machine, and ground foundations	$21\% \pm 3\%$

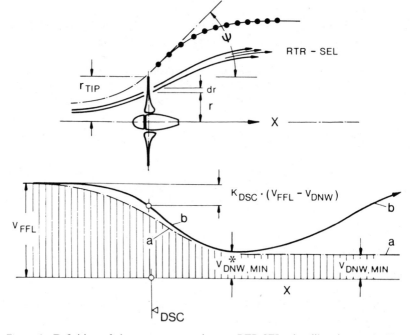

Figure 1   Definition of the rotor-stream element RTR-SEL, the dilatation angle Ψ at $x = 0$, the velocities v_{FFL} and v_{DNW}, and the rotor-disc velocity coefficient k_{DSC}. (*a*) Velocity v_x with low axial load, $\Psi \simeq 0$, $k_{DSC} = 0.5$, no decay of vortex layers (Betz & Prandtl 1919). (*b*) v_x, $0 < \Psi < 60°$, $k_{DSC} \lessgtr 0.5$, decay of vortex layers. v_{DNW}^* if vortex decay is taken into consideration.

The relative investment for the first two groups, and especially for the second one, is a function of the magnitude of the rated power of the total system and of the lowest value of the RPM given by the rated wind speed and the rated specific speed $\lambda_{\omega,\mathrm{RTR}}$ at the rotor shaft. The advantage of applying the largest feasible machinery units is well known, for these have higher overall efficiencies (Meier 1975, among others).

The size of the machinery that converts the rotor's output into electricity, heat, stored water, etc. is determined by the RPM and mechanical power P_{RSH} at the rotor shaft. That power is a function of the wind velocity v_{FFL}, the air density ρ, the rotor's effective tip radius r_{TIP}, and the rotor's power coefficient $c_{P,\mathrm{RTR}}$. This last is in turn a function of $\lambda_{\omega,\mathrm{RTR}}$, of $\zeta_{(r)}$ where $r_{\mathrm{HUB}} \leq r \leq r_{\mathrm{TIP}}$, of k_{DSC}, of the number z of rotor blades, and of the lift-to-drag ratios at the respective radii, and is given by

$$c_{P,\mathrm{RTR}} \equiv P_{\mathrm{RSH}} \cdot 2 \cdot (\rho \cdot v_{\mathrm{FFL}}^3 \, r_{\mathrm{TIP}}^2 \, \pi)^{-1} \tag{2}$$

$$= 2\eta_z \cdot \int_{r_{\mathrm{HUB}}/r_{\mathrm{TIP}}}^{r/r_{\mathrm{TIP}}=1} r/r_{\mathrm{TIP}} \cdot c_{P(r)} \, d(r/r_{\mathrm{TIP}}), \tag{3}$$

where necessary definitions are given in Equations (4)–(10). We define

$$\eta_z \approx (1 - 1.36 \cdot z^{-1} \cdot \lambda_{\mathrm{DSC}}^{-1})^2, \tag{4}$$

$$c_{P(r)} = 2\lambda_{(\omega,r)}^2 (R_{(r)} - 1) \cdot [1 + k_{\mathrm{DSC}}(\zeta_{(r)} - 1)] \cdot \eta_{\mathrm{LDR}(r)}, \qquad \text{with } 0 \leq \zeta_r \leq 1 \tag{5}$$

and

$$\eta_{\mathrm{LDR}(r)} = [k_{\mathrm{DSC}}^{-1} - 1 + [1 - (1 - \zeta_{(r)}^2)(1 + \mathrm{LDR}_{(r)}^{-1} \cdot \lambda_{\mathrm{DSC}(r)}^{-1})]^{0.5}]$$
$$\cdot (k_{\mathrm{DSC}}^{-1} - 1 + \zeta_{(r)})^{-1}(1 - \mathrm{LDR}_{(r)}^{-1} \cdot \lambda_{\mathrm{DSC}(r)}). \tag{6}$$

The root $R_{(r)}$, an abbreviation, is defined by

$$R_{(r)} \equiv [1 + (1 - \zeta_{(r)}^2)\lambda_{(\omega,r)}^{-2}]^{0.5}, \qquad \text{with } 0 \leq \zeta_r \leq 1 \tag{7}$$

and $\mathrm{LDR} = c_{L,\mathrm{AFL}}/c_{D,\mathrm{AFL}}$ is the lift-to-drag ratio of the rotor blade airfoil section at the radius (r). We also define

$$\lambda_{\mathrm{DSC}(r)} = \lambda_{(\omega,r)} \cdot (k_{\mathrm{DSC}}^{-1} + R_{(r)} - 1) \cdot (k_{\mathrm{DSC}}^{-1} + \zeta_{(r)} - 1)^{-1}, \tag{8}$$

where $\lambda_{\mathrm{DSC}(r)}$ is the specific speed or speed ratio with respect to velocities relative to the rotor blades in the disc of rotation of the rotor. Important parameters in Equations (5), (6), (7), and (8) are the retardation ratio ζ, where

$$\zeta \equiv v_{\mathrm{DNW}}/v_{\mathrm{FFL}}, \tag{9}$$

and the rotor-disc velocity coefficient

$$k_{\mathrm{DSC}} = [(v_{\mathrm{DSC}}/v_{\mathrm{FFL}}) - 1][(v_{\mathrm{DNW}}/v_{\mathrm{FFL}}) - 1]^{-1}, \tag{10}$$

where v_{DNW} is the mean axial velocity component of the air that has passed the rotor disc at a distance "far" downwind, and v_{DSC} is the axial component of the mean velocity at the rotor disc. Instead of the mean values, the values in a rotor stream element passing through the rotor between the radii r and $r+dr$ can also be employed (see Figure 1).

A rotor's actual $c_{P,RTR}$ data, plotted for a number of given general pitches versus the speed ratio $\lambda_{\omega,RTR}$ and called the rotor characteristics, define its qualities. For only a limited number of such characteristic families do there exist experimental results, obtained by wind-tunnel tests or measurements on actual operating WEC systems (Prandtl & Betz 1932; Darrieus 1933; Champly 1933; Hütter 1942; Fateev 1948; Sabinin 1953; South & Rangi 1972; Hütter 1964; Rosen, Deabler & Hall 1975; Thomas & Sholes 1975). See Figure 2. However, means of calculating the characteristics numerically are now developed to such reliability that the influence of parameter variations can be effectively studied.

Assuming a hypothetical ideal horizontal-axis WEC rotor with an infinite number of blades and an overall lift-to-drag ratio $= \infty$ (i.e. frictionless airfoils), we get a generally valid optimum diagram for all rotors. This diagram on the one hand can be used as a reference for all measured as well as calculated rotor characteristics; on the other hand it can be distorted in such a way that a bonus or penalty may be assigned, with a bonus for less expensive converting systems at high-input RPM, and with an appropriate penalty for more expensive equipment if only low-input RPM is available (see Figure 2).

The result of this valuation is that, up to specific speeds where other limitations turn out to be essential, rotors with higher optimum rated speeds are doubtless by far the best solution. The above-mentioned advantages are evident, especially

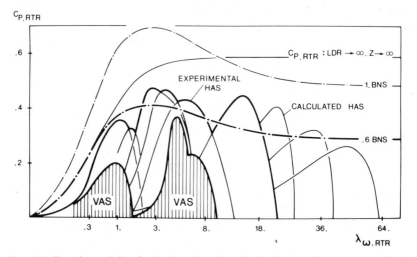

Figure 2 Experimental data for horizontal- and vertical-axis $c_{P,RTR}$, $\lambda_{\omega,RTR}$ characteristics. The abscissa is distorted by the one-third power to enable also a presentation of calculated data from rotors with airfoils having $LDR_{OPT} = 150$, $\lambda_{\omega,RTR,OPT} > 20$. The bonus/penalty (BNS) reference is given by distortion of the $c_{P,RTR,IDEAL}$ line and its proportionals. For vivid expression emphasis is given to the bonus-reference line of 60%. VAS means vertical-axis systems, HAS horizontal-axis systems. Envelopes for optimum reported data for VAS as well as for HAS are marked especially (Fateev 1948, Hütter 1964, Perli 1938, Templin 1974, Weber 1975).

compared with systems that are essentially limited to very small optimum $\lambda_{\omega,RTR}$ values, like the Savonius rotor. Experimental results from Savonius rotors as well as the first reported data from wind-tunnel tests of Darrieus rotor systems seem so far not to be at all competitive, not even with experimental results reported in the early thirties.

LIMITATIONS FOR SPECIFIC SPEED

The relatively low specific energy density of those winds that yield the steadiest source of energy requires for each WEC unit what the Dutch windmill builders provided: the largest feasible area swept by the moving rotor blades, an area that is taken perpendicular to the actual airstream. The maximum circumferential speed of the rotor, $r_{TIP} \cdot \omega_{RTR}$, that is permissible with respect to lateral buckling of blades as well as to hurricane wind loads at rest, flutter limits, Mach-number-effects, and so on, is limited by the fact that compulsory correlations of the rotor blade's slenderness exist. This is represented by a blade-shape function, or local "density" value of a blade:

$$\theta_{(r)} \equiv c_{L,AFL(r)} \cdot f_{AFL(r)} \cdot z/(2\pi r). \tag{11}$$

Let us consider again the rotor stream element (RSE) already mentioned, which involves the hypothetical airspace included between two stream areas consisting of streamlines that extend indefinitely in the direction of the rotor axis. These stream areas are, by definition, considered to be axially symmetric about the rotor axis. Applying blade-element theory as well as momentum considerations to this RSE, and defining

$$\theta_{(1,r)} \equiv \theta_{(c_{L,AFL}=1,r)}, \tag{12}$$

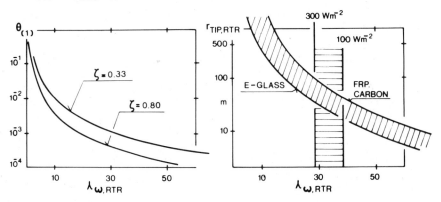

Figure 3 Comparison of blade-shape function $\theta_{(1,r)}$, plotted against $\lambda_{(\omega,r)}$ with operation limited by transverse buckling (Timoshenko 1936) under own weight and by Mach number. Parameters for the rotors considered are: the WEC's specific rated power Π: P_{Rated} $(r_{TIP}^2 \cdot \pi)^{-1}$, where r_{TIP} is a rotor-blade tip radius; and the material specific stiffnesses, E/ρ and G/ρ, of the rotor's structural materials.

we obtain:

$$\theta_{(1,r)} = 2(R_{(r)} - 1)\lambda_{(\omega,r)} \cdot [1 + k_{DSC}(\zeta_{(r)} - 1)]^{-1}(\lambda^2_{DSC} + 1)^{-1/2}. \tag{13}$$

[For definitions see Equations (1), (7), (8), (9), and (10).]

In Equations (7) and (8), if we assume $\lambda_{(\omega,r)} \gg (1 - \zeta^2_{(r)})$, we get as rough approximations $\lambda_{DSC(r)} \simeq 1.5 \cdot \lambda_{(\omega,r)}$ and $\theta_{(1,r)} \simeq 1.33 \cdot (1 - \zeta_{(r)}) \cdot \lambda^{-2}_{(\omega,r)}$, thus obtaining an impression of how strongly a rotor blade's local density is dependent upon $\lambda_{(\omega,r)}$ (Figure 3).

The use of optimum airfoils having thickness ratios $\leq 18\%$ at radii $> 0.5\ r_{TIP}$ and the choice of large values of the specific speed $\lambda_{\omega,RTR}$, desirable for economic reasons, yield a tendency to exceed the elastic and/or dynamic stability limits for the individual rotor blades. Lateral buckling (Timoshenko, 1936) under the rotor's own weight at rest with the rotor axis horizontal, and for a given structurally efficient value of the square root $(E/\rho)^{1/2}$ of the ratio of Young's modulus to density ratio, limits the maximal feasible rotortip radii with respect to the outlay speed ratio $\lambda_{\omega,RTR}$ for optimum c_P (see Figure 3, right diagram).

The flutter parameter or reduced torsional frequency $\Omega^*_{RTR} = \frac{1}{2}\omega_{TORSION} \cdot f^*/w_{DSC} \cdot (1 - Ma^2)^{-1/2}$ as a measure of the tendency of rotor blades to exceed stall flutter boundaries is, for a given value of the square root $(G/\rho)^{1/2}$ of the ratio of shear modulus to density, strongly reduced with increasing $\lambda_{\omega,RTR}$, since the reference airfoil chord f^*_{AFL} is proportional to $\theta_{(1,r*)}$. In addition, the blade-section resultant velocity w_{DSC} is also proportional to $\lambda_{(\omega,r)}$. Thus, if one assumes speed ratios $\lambda_{\omega,RTR}$ at or beyond 25, and takes into account that v_{FFL} is proportional to (rated power)$^{1/3}$, Mach numbers are easily reached that cause significant reductions of lift-to-drag ratios of airfoils (Figure 3, Figure 2). If one considers such limitations carefully, it seems that the application of speed ratios $\lambda_{\omega,RTR}$ beyond 30 might cause a number of serious difficulties. This is to be expected even with rotor blades constructed from optimal carbon composites. Equation (11) indicates that increasing the number of blades reduces these limits to even smaller values of $\lambda_{\omega,RTR}$.

ROTOR CHARACTERISTICS AT STALL STATE

In the early thirties Prandtl & Betz (1932) and Darrieus (1933) made wind-tunnel tests of high-speed horizontal-axis rotors with blade numbers $2 \leq z \leq 6$. The results, represented by plotting $c_{P,RTR}$ and the rotor's drag coefficient $c_{D,RTR}$ against the tip-speed ratio $\lambda_{\omega,RTR}$ [see Equation (1)], showed two typical discrepancies when compared with low-speed rotors (Figure 4). For $\lambda_{\omega,RTR} < \lambda_{\omega,RTR}$ at $c_{P,RTR,OPT}$, there are considerable losses as a consequence of increasing the angle

$$\Phi_{DSC} = \arctan(\lambda^{-1}_{DSC}) \tag{14}$$

[see also Equation (8)] at low $\lambda_{\omega,RTR}$ values, thus causing stall and sharply rising rotor-blade drag. For $\lambda_{\omega,RTR} > \lambda_{\omega,RTR}$ at $c_{P,RTR,OPT}$ there are significant losses, resulting from the appearance of a partial vortex ring at the rotor.

Since neither phenomenon has any influence on the optimum values of a rotor's $c_{P,RTR}$, nobody worried too much about them in the twenties and thirties. Problems of the total system, namely optimum reliable and efficient control, fatigue life of rotor blades, endurance and long-time lubrication of bearings, lightning damage, adaptation and maintenance problems, and so on were standing too strongly in the foreground. However, the stalling of fixed-pitch high-speed rotors at low $\lambda_{\omega,\mathrm{RTR}}$ values did in fact cause problems, especially starting difficulties. Consequently wind-tunnel tests with variable pitch were performed in the late thirties and early forties by several persons already mentioned with respect to Figure 2.

General closed solutions for analysing the stalling behavior of high-speed rotors had no chance of being developed then and make no sense now, since numerical rotor-element calculations can easily be performed by computers. Thus the approach via the rotor-element theory, which had been initiated for propellers by Drzewiecki (1892) with a feedback from Rankine's momentum theory, provides realistic solutions if adequate iterations are used. This approach has proved to be effective.

The post-stall behavior of airfoils used for rotor blades is not only nonlinear, and generally given empirically, but is also decisively influenced by the local Reynolds number, which varies along the radius of the rotor:

$$\mathrm{Re}_{(r)} = v_{\mathrm{FFL}(r)} \cdot [1 + k_{\mathrm{DSC}}(\zeta_{(r)} - 1)] \cdot (1 + \lambda^2_{\mathrm{DSC}(r)})^{1/2} \cdot 2\pi r\, \theta_{(1,r)}(z \cdot c_{\mathrm{L,AFL}(r)} \cdot v)^{-1}, \qquad (15)$$

where v is the kinematic viscosity.

The author (Hütter 1942, 1961) has been able to obtain solutions using graphs of general validity for the relevant parameters,

$$(\lambda_{(\omega,r)},\ \zeta_{(r)},\ k_{\mathrm{DSC}},\ \theta_{(1,r)},\ \Phi_{\mathrm{DSC}(r)},\ \alpha_{\mathrm{AFL}(r)}\ \text{and}\ c_{\mathrm{L,AFL}(r)}),$$

that have been determined so as to be consistent with the given geometric conditions as well as with Euler's and d'Alembert's principles (Figure 5). A basic assumption for these graphs considered is that the induction from the rotor's general

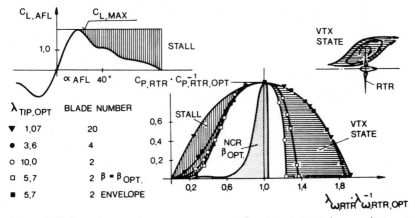

Figure 4 Stall at $\lambda_{\omega,\mathrm{RTR}} < \lambda_{\omega,\mathrm{RTR}}$ at $c_{\mathrm{P,RTR,OPT}}$ and vortex state at $\lambda_{\omega,\mathrm{RTR}} > \lambda_{\omega,\mathrm{RTR}}$ at $c_{\mathrm{P,RTR,OPT}}$ cause limitation. Evidence of this is given by comparison with envelopes obtained by variation of general pitch and with a low-speed rotor characteristic (Prandtl & Betz 1932, Champly 1933, Weber 1975). Abscissa: $\lambda_{\omega,\mathrm{RTR}}/\lambda_{\omega,\mathrm{RTR,OPT}}$; ordinate: $c_{\mathrm{P,RTR}}/c_{\mathrm{P,RTR,OPT}}$.

vortex system is regarded as given, or as defined for the elements of a contemplated rotor-stream element (RSE) by the values of $k_{DSC(r)}$ and $\zeta_{(r)}$. This enables one to separate for the present the vortex-sheet analysis from the other considerations. An adequate iteration during the analysis is evidently necessary for refinement, as the findings with regard to the disc velocity factor k_{DSC} also will explain. Equations (3), (4), (5), (6), (7), and (8) yield final data for plotting the rotor characteristics under the assumptions given.

With regard to Equation (6) one should mention a typical discrepancy between turbine and propeller analyses. Friction, expressed by the airfoil drag coefficient $c_{D,AFL}$, has no doubt the effect of reducing the power-effective tangential force at the blade by a certain friction factor defined by $1 - \lambda_{DSC(r)}/LDR_{(r)}$ [see also Equation (6)], where LDR is the lift-to-drag ratio of the considered airfoil at the angle of attack α_{AFL}, given by the data. But unlike the case of a propeller, axial force on a WEC's turbine rotor does not affect its efficiency directly. However, an indirect effect exists, as the axial component of the airfoil drag reduces slightly the mass flow through the annular area of the RSE.

Thus we obtain for a rotor element at a given radius the local value of the power coefficient given by Equations (5) and (6), which involves the condition that

$$(1 - \zeta^{*2})/(1 - \zeta^2) = 1 + LDR_{(r)}^{-1} \lambda_{DSC(r)}^{-1}. \tag{16}$$

Here ζ^* is the value of the retardation coefficient due to the drag of the rotors' airfoils.

THE ROTOR'S DISC-VELOCITY COEFFICIENT

Betz & Prandtl (1919) indicated that in the disc of blade rotation of a very lightly loaded propeller the axial component of velocity could be taken with sufficient accuracy as the mean value of the undisturbed wind velocity v_{FFL} and of v_{DNW}, the minimum value of the axial velocity "far" downwind. Thus, with the definition of ζ from Equation (9) we have $s_{DSC} = v_{FFL} \cdot (1 + \zeta)/2$. This has later been widely accepted for the description of the behavior of propellers at near-optimum efficiency. Betz (1926) also used this assumption as a premise when deriving a fundamental equation for windmills that was based upon the Rankine-Froude momentum theory. He defined an ideal "Leistungsbeiwert," the "power coefficient" that indicates the maximum possible extraction of energy from a given airstream and a given dimension of the rotor, and stated that this coefficient cannot exceed the value $c_P = 0.592592$ at $\zeta = 1/3$ for an ideal system with infinite blade number, no rotation of the airflow, and no friction losses at the blades, for which

$$c_{P,RTR} = (1 - \zeta^2)(1 + \zeta)/2. \tag{17}$$

This unfortunately involves almost a contradiction, since for $\zeta = 0$ the power factor comes out to be $c_P = 0.5$. This evidently is not very probable as it means dissipation into an infinite space without remarkable reaction to the incoming airstream.

It is also not satisfying that this contradiction loses any practical consequences for rotor outlay design when one considers a more general description of flow through a friction-free rotor with infinite blade number and also takes into account the rotation in the slipstream (Hütter, 1942). The reason for this discrepancy, not discerned earlier for the application to WECs, is that the assumption of a low axial load, or in other words, negligible change of velocity in the airstream in the vicinity of the rotor, is not in accord with reality for WEC rotors operating close to optimum conditions. The existence of small values of ζ means wide dilatations of the stream tube that touches the rotor-blade tips (see Figure 1) and results in significantly changed induced velocities.

A consideration, using moderately idealized contours of the vortex sheets downstream that are also adequately shaped to satisfy the continuity condition at any value of the retardation ratio for $0 \leqq \zeta \leqq 1$, results in more realistic values for v_{DSC} and v_{DNW} when integrating the axial components of the induced velocities with regard to the rotor disc as well as to a parallel area far downwind. From that the disc-velocity factor k_{DSC} [Equation (10)] proves to be a function of ζ, as Figure 6 shows. As a result of viscosity effects k_{DSC} does not jump discontinuously from zero to one.

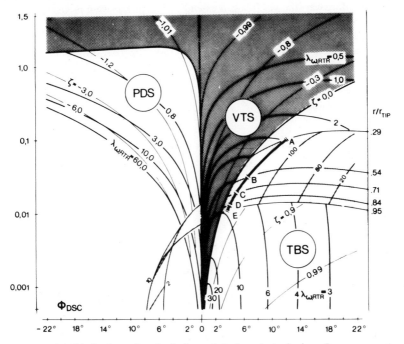

Figure 5   The "blade-shape function" $\theta_{(1,r)}$ plotted against abscissa $\Phi_{DSC} = $ arc cotan $\lambda_{DSC(r)}$ with $\zeta_{(r)}$ and $\lambda_{(\omega,r)}$ as parameters. Shaded is the range of "vortex state," VTS. Left of the zero Φ_{DSC} coordinate ($\zeta_{(r)} = -1$) is the range of "powered descent" PDS. The $c_{L,AFL}$ over α_{AFL} diagrams for a sample rotor blade for five equivalent positions A, B, C, D, E from blade root to tip are plotted in relative positions in the diagram for an optimum operation state $\lambda_{\omega,RTR} = 10$.

Experiments of Blenk & Trienes (1956) with fences for protection of soil against wind erosion, and of Pelser (1975) with models of multirotor plants, using fine screens to simulate the operating rotor discs, indicate that for systems with large regions of viscous flow the disc-velocity factor k_{DSC} satisfies $0.5 \leqq k_{DSC} \leqq 1.0$ for the whole range of ζ considered. One consequence of this result is that the Betz optimum power factor is not a definite upper limit. It also can be verified, by introducing the functional correlation of $k_{DSC} = k_{DSC}(\zeta)$ into Equation (5), that the velocity retardation ratio ζ for $c_{P,OPT}$, i.e. for maximum harnessing of the available energy, is shifted to smaller values than have so far been assumed to be best.

Resulting overall improvements of the theoretical optimum of from 4% up to 13% seem to be attainable. However, that is not the only reason it makes sense to choose the smallest feasible values of ζ for the design of WEC rotors. Lower values of ζ at optimum $c_{P,RTR}$ will make possible a tighter grouping of WECs in fields with favorable wind conditions. This is because turbulent exchange via the downwind vortex system will enable recovery after a phase of retardation within shorter distances in the airstream of a sufficient proportion of the original wind velocity. Moreover, the vortices will extend into higher elevations than the rotor-blade tips, and thus will transport energy from there into the stream towards the next WECs within an array.

By using the equations of Joukowski and Oseen (1911), Albring (1968) discusses

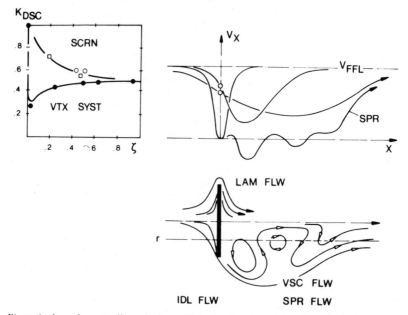

Figure 6 k_{DSC}, the rotor disc-velocity coefficient as function of ζ_{RTR} (mean value). Calculated data. Comparison with experimental data, v_x of screens (Blenk & Trienes 1956, Pelser 1975). Flow with negligible viscosity and early state laminar flow around a circular disc show $k_{DSC} = 1.0$, with flow separation due to von Kármán vortices giving even bigger values.

the latter phenomenon in detail with respect to the interpretation of an experimentally refined evaluation of kinematic viscosity in the rolling up of vortices. In order to define the initial strength and the subsequent decay of rolled-up near-tip vortex sheets, we find that the intensity of the downwind vortices that induce disturbance velocities is

$$(u \cdot r)_{(n)}/(u_0 r_0)_{max} \cdot [1 - \exp(-p_{DCY} \lambda_{\omega,RTR} \cdot z/n)], \tag{18}$$

where $n = x/(2r_{TIP})$, x being the distance behind the rotor disc and p_{DCY} the decay coefficient, to be defined from experiments analogous to those that have been performed recently by Cook (1973), Ciffone & Orloff (1975), and Mokry (1975). In the light of these findings, a result (Hütter 1942) of wind-tunnel measurements of the axial velocity component up to three rotor radii downstream of a three-blade horizontal-axis rotor model permitted a new interpretation. Significant irregularities, initially assumed to be just crude scatter, appeared to be a consequence of vortex-sheet roll-up (see Figure 7).

VORTEX-STATE PROBLEMS

As mentioned in the section on rotor characteristics at stall state, $\theta_{(1,r)}$ [see Equations (11), (12), and (13)] is a well-defined function of $\Phi_{DSC(r)}$, the angle between the rotor disc and the vector w_{DSC} of relative frequency with respect to the rotor blades, which is given by

$$w_{DSC} = v_{FFL(r)}[1 + k_{DSC}(\zeta_{(r)} - 1)] \cdot (1 + \lambda_{DSC(r)}^2)^{1/2}. \tag{19}$$

Explicit parameters of $\theta_{(1,r)}$ and $\Phi_{DSC(r)}$ are $\zeta_{(r)}$ and $\lambda_{(\omega,r)}$. An implicit parameter is k_{DSC}, as it is a function of $\zeta_{RTR,mean}$. Since $\theta_{(1,r)}$ is well defined, this is doubtlessly true for the turbine state of operation $0 \leqq \zeta_{(r)} \leqq 1$ and is quite reasonable for that

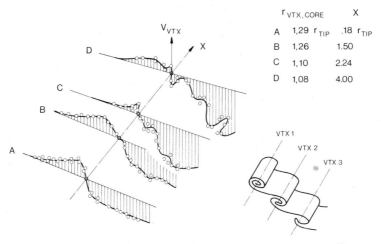

	$r_{VTX,CORE}$	X
A	1,29 r_{TIP}	.18 r_{TIP}
B	1,26	1.50
C	1,10	2.24
D	1,08	4.00

Figure 7 Tip vortices: maximum tangential velocity decay and vortex roll-up pattern downwind of a three-blade $\lambda_{\omega,RTR} = 3.6$ at $c_{P,OPT}$ rotor model. Wind-tunnel test (Hütter 1942).

state, when fully developed, which is similar to that of a "powered descent" (PDS) of helicopter rotors. This state is reached if locally at a rotor blade the values of the retardation ratio are negative: $-\infty \leqq \zeta_{(r)} \leqq -1$. For WECs this is a very hypothetical state, for the rotor is consuming enormous power if it is rotating under such conditions at $\lambda_{(\omega,r)}$ values similar to those reached at the optimum operation as a WEC. Thus operation at a PDS state is extremely improbable. However, if cases of a loaded rotor are considered, a PDS might have some importance in so far as extreme conditions, for example, malfunction of pitch control when connected with a strong public network, have to be taken into consideration for safety reasons.

The vortex state can easily occur in the interval $-1 \leqq \zeta_{(r)} \leqq 0$. If once begun, it exists over a certain extension from the rotor-tip radius r_{TIP} down to a zero-flow radius, where the return flow ($v_x < 0$) turns into the normal airstream direction ($v_x > 0$) of the turbine state. The radius where the flow direction changes we call the turning-point radius r_{TPT} (Figure 8). It is this turnback of the flow that initiates the closed-ring vortex, which is a torus with its axis close to the wing-tips. The vortex torus receives energy constantly from that part of a WEC rotor ($r_{TIP} \geqq r_{VTX\,RING} \geqq r_{TPT}$) that is seized by it. This energy is, compared with average $c_{P(r)}$ values, comparatively small. Expressing it in terms of the power factor, we obtain

$$\Delta c_{P,RTR,VTX} = p_{VDS}(c_{L,AFL}\,\theta_{(1,r)})^2[1-k_{DSC}(\zeta_{(r)}-1)]^3(1+\lambda_{DSC(r)})^{3/2}z^{-1}. \tag{20}$$

The terms are referred to $r = (r_{TPT}+2r_{TIP})/3$ as an approximation. Here p_{VDS} is a proportionality factor. If we use the rough approximations mentioned when explain-

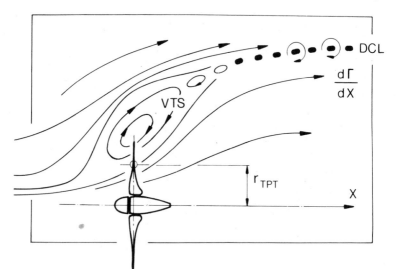

Figure 8   The radially, axially trailing vortices generated at the rotor-blade tips by a vortex-state vortex and forming a discontinuity layer (DCL) between free flow at velocity $\approx v_{FFL}$ and the flow with velocity v_{DNW} in the wake of the rotor. Definition of the turning point at r_{TPT}.

ing Equation (13), we understand that $c_{P,\text{VTX}}$ is roughly $\simeq c_{L,\text{AFL}}^2/(\lambda_{(\omega,r)} \cdot z)$. The power coefficient for a rotor with an existing vortex state is:

$$c_{P,\text{RTR}} = 2 \int_{r_{\text{HUB}}/r_{\text{TIP}}}^{r_{\text{TPT}}/r_{\text{TIP}}} (r/r_{\text{TIP}}) c_{P(r)} d(r/r_{\text{TIP}}) - \Delta c_{P,\text{RTR,VTX}}. \tag{21}$$

As the vortex ring is basically unstable with respect to its position, it is constantly renewed by the rotor tip and fades away into the discontinuous vortex sheet that separates the free flow from the wake of the rotor's inner part (Figure 8).

If we consider the whole range of ζ_{RTR} that is within reach if output from as well as input to the rotor shaft is admitted, we obtain at a given general pitch positive and negative values of $c_{P,\text{RTR}}$ as a function of ζ_{RTR} (Figure 9).

THE NEAR-CRITICAL OPERATING ROTOR (NCR)

It can be seen from the synoptic presentation of general momentum conditions and blade-airfoil behavior in Figure 7 that rotor elements drop into stall if they are in a state with $\zeta_{(r)}$ close to $+1$ and that even a very small change of $\Phi_{\text{DSC}(r)}$

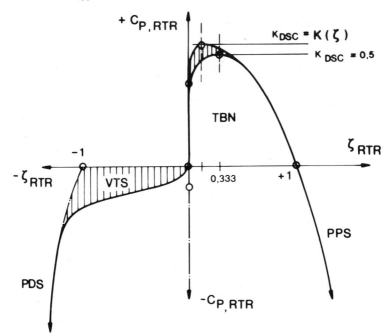

Figure 9 Schematic presentation of $c_{P,\text{RTR}}$ plotted against ζ_{RTR} for $-1.3 < \zeta_{\text{RTR}} < +1.3$ at a given $\lambda_{\omega,\text{RTR}}$. For $\zeta_{\text{RTR}} > +1.0$ a real propeller state (PPS) is entered, for $\zeta_{\text{RTR}} < -1.0$ a powered-descent state (PDS) takes place. For $-1 < \zeta < 0$, the Lock (1928) and Glauert (1935) vortex state (VTS) is encountered. At VTS only a relatively low negative power is required. It causes small $-c_{P,\text{RTR}}$ values. For $\zeta_{\text{RTR}} \simeq 0$, ζ_{RTR}, understood as mean value, includes in the defining integral positive and negative data of v_x.

brings a rotor element from optimal normal operation as a turbine into a power collapse as a consequence of a rapidly expanding vortex state.

There is no doubt that the high-specific-speed, high $c_{P(r)}$ rotor configurations are those that yield maximum energy at minimum investment for given wind conditions. Moreover, certain excellent lift-to-drag-ratio airfoils have a sharp stall boundary. Thus rotors operating under near-critical conditions are the most promising ones.

Figure 10 together with Figures 2 and 4 explains how fine adjustment of general pitch permits rotors that would otherwise lose efficiency to operate at the peak of a $c_{P,RTR}$ envelope for all $\lambda_{\omega,RTR}$ available. This fine adjustment makes it possible to maintain operation of even high-speed NCRs at optimum speed and tight grouping with respect to output.

Starting such rotors, even in weak winds, is no problem if the pitch is adequately adjusted during run-up. A rotor's torque coefficient $c_{T,RTR} = c_{P,RTR} \cdot \lambda^{-1}_{(\omega,RTR)}$), evidently a function of β_{RTR} as well as $\lambda_{\omega,RTR}$, reaches at $\lambda_{\omega,RTR} = 0$ maximal values at those pitch angles where at the outer part of the rotor blades $c_{L,AFL,max}$ is reached (Figure 11).

REFINEMENTS, SIMPLIFICATIONS

Refinements of the shape of rotor blades do not burden the production cost at all significantly if adequate technology is applied. Modern composite technology

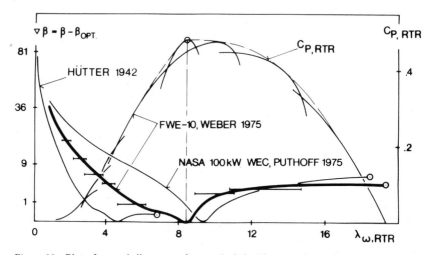

Figure 10 Plot of several diagrams of general pitch $\Delta\beta$ versus $\lambda_{\omega,RTR}$ from experimental data. Evidence is given: (*a*) that following the envelope of maximum available $c_{P,RTR}$ data a continuous fine pitch change is to be recommended. (*b*) that near the real optimum a slight change in pitch might result in a remarkable gain in $c_{P,RTR}$. The $\Delta\beta$, $\lambda_{\omega,RTR}$ near $\lambda_{\omega,RTR}$ at $c_{P,RTR,OPT}$ has a marked "valley" (Hütter 1942, Weber 1975, Puthoff & Sirocky 1975).

(Hütter 1960, 1963, 1972) permits production of even extremely slender blades of any desired shape, including taper and twist, and airfoils with any required accuracy and surface quality, defined by rigorous demands with respect to roughness as well as waviness.

The problems of feasibility, fatigue life, and load induction have reached the state of general applicability to hardware. Blade root joints, designed on the basis

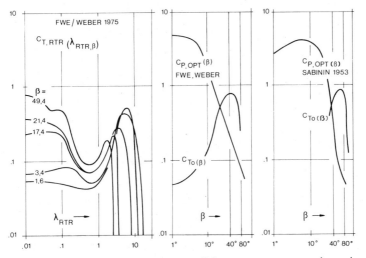

Figure 11 Calculation of a rotor's torque coefficients $c_{T,RTR}$ near zero $\lambda_{\omega,RTR}$ is quite reliable, for $c_{L,AFL}$ is sufficiently accurately defined at α_{AFL} close to 90°. Experimental data from Sabinin (1953), calculated data from Weber (1975).

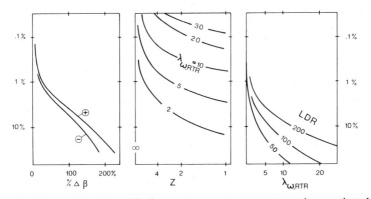

Figure 12 Comparison of losses if refinement parameters are not at optimum values. Left: losses due to mistakes concerning optimum twist $\Delta\beta$. Positive and negative values of $\Delta\beta$ up to no twist at all. Middle: losses due to a finite blade number z. Reciprocal scale for blade numbers. Right: losses of $\lambda_{\omega,RTR}$ at $c_{P,OPT}$ due to indicated values of lift-to-drag ratio LDR versus $\lambda_{\omega,RTR}$.

of endless loops, flat, toroidal, or arranged in flanges, have proved to be reliable devices, unaffected by long-time creep (Hütter 1962, 1963, 1964). Composite rotor blades have withstood climatic impact for more than 10 years without deterioration of the blade surface, without evidence of lightning traces, and without leading-edge rain or hail erosion worth mentioning. Crude blade-shape design to reduce the price of rotor blades seems not to be necessary at all.

However, comparing so-called simplified solutions of blade structure with respect to selected parameters such as the number of rotorblades z or optimum twist $\Delta\beta = \Delta\beta_{(r)}$, it turns out that the number of blades has the smallest influence on overall efficiency if $\lambda_{\omega,RTR}$ is beyond 8 (Figure 12). As far as optimum twist is

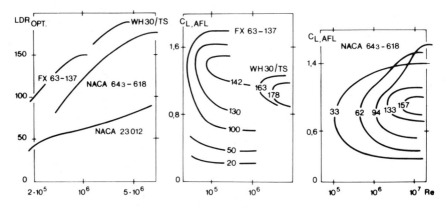

Figure 13   Influence of Reynolds number on optimum lift-to-drag ratio of several airfoils (Hütter 1954, Abbott & von Doenhoff 1959, Althaus 1972).

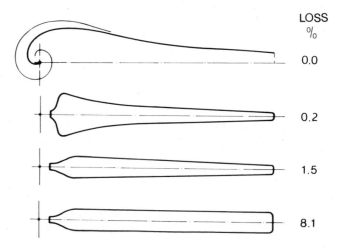

Figure 14   Loss due to nonoptimum blade shape in percent of $c_{P,RTR,OPT}$.

concerned, even a deviation of $\pm 30\%$ from the calculated optimum twist reduces the optimum values of the power factor $c_{P,\text{RTR}}$ by less than 1%. By contrast, the influence of the lift-to-drag ratio LDR, optimum and average, over the whole range of $c_{P,\text{RTR}}$ within the operation envelope is considerable especially at $\lambda_{\omega,\text{RTR}}$ values beyond 6 to 8.

As the LDR is a function of the chosen airfoil's lift coefficient and Reynolds number Re_{AFL}, the unfavorable influence on the overall efficiency of increasing the blade number can be considerable if a rotor operates at high values of $\lambda_{\omega,\text{RTR}}$. Mostly that influence is far more significant than the modest losses due to blade number from vortex-system considerations (Figure 13).

A blade contour that is optimal with respect to aerodynamic theories is generally not appropriate with regard to hub design and the necessary geometric clearences for pitch change. To avoid fatigue failures a steady and smooth change of shape is required, especially close to the blade root. However, no such shape adaptions

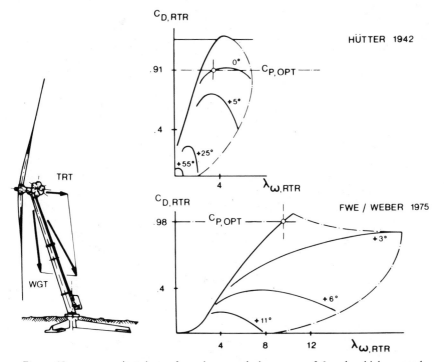

Figure 15 $c_{D,\text{RTR}}$ against $\lambda_{\omega,\text{RTR}}$ for a low speed, $\lambda_{\omega,\text{RTR,OPT}} = 3.6$ and a higher speed, $\lambda_{\omega,\text{RTR,OPT}} \simeq 10$ (Hütter 1942, Weber 1975; see also Rohrbach & Worobel 1975). $c_{D,\text{RTR}}$ for $c_{P,\text{RTR,OPT}}$ remains at approximately the same values as thrust $\times v_{\text{DSC}} =$ power. Here $v_{\text{DSC}} = v_{\text{FFL}} \cdot [1 + k_{\text{DSC}} \cdot (\zeta_{(r)} - 1)]$. Partial unloading of tower system by choice of optimal configuration: approximately same bending moment at no-power, no-thrust state $|-M_{\text{BDG}}|$ as at full power, maximum thrust $|+M_{\text{BDG}}|$. Condition $|-M| \simeq |+M|$. Hütter (1975). Thrust = TRT, Weight = WGT.

cause severe losses, except for really crude simplifications such as constant-chord blades (Figure 14).

No possibilities at all exist for influencing significantly the thrust-to-power relation with the aim of reducing that portion of the investment that is based on the amount for the turntable, machine bed plate, tower and ground foundation, that is, $(21+7)\% \pm 5\%$ of the total investment. The thrust or drag coefficient of a rotor element at radius (r) is:

$$c_{D,RTR(r)} = 2(R_{(r)} - 1)\lambda_{(\omega,r)}(\lambda_{DSC(r)} + LDR_{(r)}^{-1})c_{L,AFL(r)}. \tag{22}$$

However, advantages with respect to the necessary turntable, tower, and foundation investment are available via a general engineering concept: the weight of the rotor systems and the tower is used to compensate the bending moment of the tower (Figure 15).

OPTIMUM RATED POWER

Since the inherent energy of an airstream passing a certain area is proportional to the cube of v_{FFL}, the temptation is strong to choose the rated power of a WEC that is much too large. This is especially true for prototypes, for public opinion at first instinctively evaluates the importance of an energy-converting machine by its rated power, instead of considering the potential price and quality of energy. But the price of energy is dependent also on the relation between rate of demand and rate of availability.

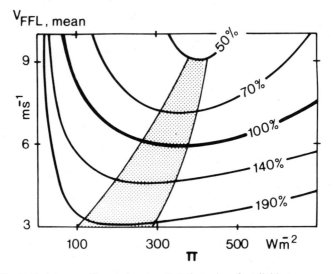

Figure 16   Optimizing specific rated power. Relative price of available energy as function of long-time average wind velocity and specific rated power Π. (See caption for Figure 3.) At given wind conditions a narrow optimum band of Π_{OPT} is defined. Hütter (1974).

HOURS / YEAR

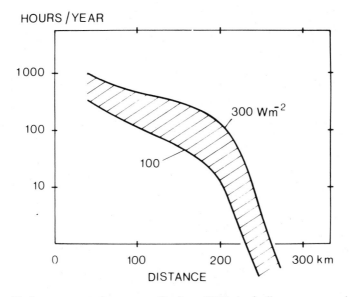

Figure 17 Improvement of energy quality from WECs by feeding energy supply from several units sufficiently far apart from each other into the same powerline. Defining the total of zero output hours per year as measure of energy quality. Π as parameter indicates that low specific rated power results in higher energy quality. Molly (1975).

High values of the specific rated power Π (= converter power capacity divided by the area swept by the rotor blades) mean at moderate average wind velocities at a certain place an unsteady supply of energy. Lower Π values mean steadier supply, or in other words, higher quality and thus potential price of the energy on the market. As a basis, there do exist minima of relative energy price values for Π at given average wind velocity, regardless of what the quality of the energy is expected to be (Figure 16).

With respect to the energy-quality problem, one of the most promising ways to smooth the energy supply is to link together on a public electricity supply system a number of WECs that have a certain average distance from each other (Figure 17). To obtain a measure for an average distance of a number of individual WECs scattered irregularly over a certain territory, the mean distance from the "center of gravity" of the group considered is chosen.

The energy quality is characterized by the total number of hours in one year during which the energy supply goes to zero. An evaluation based on a relatively small region in northwest Germany is shown in Figure 17. Realistic wind-velocity output data, obtained during long-time operation of a 100 kW WEC with $\Pi = 110$ Wm^{-2} (Hütter 1964) and from hourly means of the wind velocity from 28 meteorological stations for one year (the same year for all stations) have been used for the calculation (Molly 1975). The results seem to be most promising.

Literature Cited

Abbott, I. H., von Doenhoff, A. E. 1959. *Theory of Wing Sections.* New York: Dover. 693 pp.

Albring, W. 1968. Wirbelsysteme und Turbulenz. *Monatsber. Dtsch. Akad. Wiss. Berlin.* 10:934–45

Althaus, D. 1972. *Stuttgarter Profilkatalog I.* Inst. Aerodynam. Gasdynam., Univ. Stuttgart

Betz, A. 1926. Windenergie und ihre Ausnutzung durch Windmühlen. Göttingen: Vandenhoek & Ruprecht. 64 pp.

Betz, A., Prandtl, L. 1919. Schraubenpropeller mit geringstem Energieverlust. *Nacht. Ges. Wiss. Göttingen Math Phys. Kl.* 1919, pp. 193–217.

Bilau, K. 1933. *Windmühlenbau einst und jetzt.* Leipzig: Die Mühle

Blenk, H., Trienes, H. 1956. Strömungstechnische Beiträge zum Windschutz. *Grundlagen der Landtechnik* 8. Düsseldorf: VDI

Champly, R. 1933. *Les Moteurs à Vent.* Paris: Dunod. 270 pp.

Ciffone, D. L., Orloff, K. L. 1975. Far field wake vortex characteristics of wings. *J. Aircr.* 12:464–70

Cook, C. V. 1973, The structure of the rotor blade tip vortex. *AGARD-CP-111* pp. 3-1 to 3-14

Darrieus, M. 1933. Théorie élémentaire du moulin à vent. In *Les Moteurs à Vent,* R. Champly, pp. 93–118. Paris: Dunod

Drzewiecki, S. 1892. Méthode pour la détermination des éléments mécaniques des propulseurs hélicoïdaux. *Bull. Assoc. Tech. Marit.* 3:11–31

Fateev, E. M. 1948. *Vetrodvigateli i Vetroustanovki.* Moscow: Ogiz Sel'khozgiz. 543 pp.

Glauert, H. 1935. Airscrew theory. In *Aerodynamic Theory IV,* ed. W. F. Durand, pp. 170–82. Berlin: Springer

Golding, E. W. 1955. *The Generation of Electricity by Wind Power.* London: Spon. 318 pp.

Hütter, U. 1942. *Beitrag zur Schaffung von Gestaltungsgrundlagen für Windkraftwerke.* Dissertation Tech. Hochschule Wien

Hütter, U. 1954. Neue Wege im Segelflugzeugbau. *Z. Flugwiss.* 2:17–24

Hütter, U. 1960. Tragende Flugzeugteile aus glasfaserverstärkten Kunststoffen. *Luftfahrttechnik* 6, no. 2:34–44

Hütter, U. 1961. The aerodynamic layout of wing blades of wind-turbines with high tip-speed ratio. *UNO-Conference on New Sources of Energy, E/Conf. 35/W/31, New York*

Hütter, U. 1962. The transfer of force in highly stressed monocoque bodies of glass-reinforced plastics. *Trans. J. Plastics Inst.* 30:16–27

Hütter, U. 1963. Vermessung räumlicher Spannungszustände durch eingebettete Dehnmesstreifen in Glas/Harz-Konstruktionen. *Kunststoffe* 53:831–38

Hütter, U. 1964. Betriebserfahrungen mit der 100 kW-Windkraftanlage der Studiengesellschaft Windkraft e.V. Stuttgart. *Brennst.-Wärme-Kraft* 16:333–40

Hütter, U. 1972. Present and future possibilities of high strength and stiffness-to-weight ratio composites in primary structures *AGARD-Conf. Impact Compos. Mater. Aerosp. Veh. Propul. Syst. Preprint No. 112*

Hütter, U. 1974. *Optimum Design Concept for Windelectric Converters.* Presented at workshop on advanced wind-energy systems, Stockholm

Hütter, U. 1975. *Vom Wert der Windenergie.* Forschungsinstitut Windenergietechnik, Univ. Stuttgart

Lock, C. N. H. Photographs of streamers illustrating the flow around an airscrew in the vortex ring state. *Aeronaut. Res. Comm. Rep. Memo. No. 1167.* 13 pp.

Meier, R. C. 1975. Concept selection, optimization, and preliminary design of large wind-generators. *Second Workshop on Wind Energy Conversion Systems,* pp. 46–48. Mitre Corp.

Mokry, M. 1975. Calculation of vortex sheet roll up in a rectangular wind tunnel. *J. Aircr.* 12:750–56

Molly, J. P. 1975. *Nutzung der Windenergie in der B. R. Deutschland.* Forschungsinstitut Windenergietechnik, Univ. Stuttgart

Oseen, C. W. 1911. Über Wirbelbewegung in einer reibenden Flüssigkeit. *Ark. Mat. Astron. Fys.* 7 no. 14:1–13

Pelser, J. 1975. Wind energy research in the Netherlands. See Meier 1975 pp. 188–95

Perli, S. B. 1938. *Vetronasosnie i Vetroélektricheskie agregati*

Prandtl, L., Betz, A. 1932. *Ergebnisse der Aerodynamischen Versuchsanstalt zu Göttingen. IV. Lieferung,* pp. 118–23. München, Berlin: Oldenbourg

Puthoff, R. L., Sirocky, P. 1975. *Status Report of 100 kW Experimental Wind Turbine Generator Project. NASA Tech. Memo. TMX-71758*

Rohrbach, C., Worobel, R. 1975. *Performance Characteristics of Aerodynamically Optimum Turbines for Wind Energy Generators. Am. Helicopter Soc. Preprint*

No. S-996

Rosen, T., Deabler, H. E., Hall, D. G., 1975. *Economic Viability of Large Wind Generator Rotors.* Intersociety Energy Convers. Eng. Conf., 10th, Univ. Delaware, Hamilton Standard

Sabinin, G. Kh. 1953. Teoriya bystrokhodnogo stabilizatornogo *Vetryaka TsAGI*, pp. 19–20

South, P., Rangi, R. S. 1972. *A Windtunnel Investigation of a 14 ft. Diameter Vertical Axis Windmill. Natl. Res. Coun. Can., Labor Tech. Rep. LTR-LA-105*

Templin, R. J. 1974. *Aerodynamic Performance Theory for the NRC Vertical Axis Wind Turbine. Natl. Res. Coun. Can., Labor Tech. Rep. LTR-LA-160*

Thomas, R. L., Sholes, J. E. 1975. *Preliminary Results of the Large Experimental Wind Turbine Phase of the National Wind Energy Program. NASA Tech. Memo. TMX-71796*

Timoshenko, S. 1936. *Theory of Elastic Stability.* New York: McGraw-Hill. 518 pp.

Weber, W. 1975. Optimale Auslegung rotierender Flügel für horizontale Windenergiekonverter. *Z. Flugwiss.* 23:443–47

Ann. Rev. Fluid Mech. 1977. 9:421–45
Copyright © 1977 by Annual Reviews Inc. All rights reserved.

FINITE-ELEMENT METHODS   ✖8109
IN FLUID MECHANICS

Shan-fu Shen
Sibley School of Mechanical and Aerospace Engineering, Cornell University,
Ithaca, New York 14853

INTRODUCTION

The large strides made by the computer industry have continually whetted the appetite of engineers to solve practical problems of higher and higher complexity by numerical simulation. While the utopia of completely dispensing with the expensive full-scale experimental verification may be debatable, it should be no longer controversial to acknowledge the rapidly expanding domain of problems for which answers satisfactory to design engineers can be obtained by numerical studies alone. For both fluid and solid mechanics, important practical applications often involve complex geometrical configurations and nonhomogeneous material properties. Circumstances, however, have led to a natural contrapuntal development of the solution technique. There is the finite-difference method, time-honored and now superbly honed to deal with many fluid problems, and there is the finite-element method, with a late start but precipitously brought to an omnipresence in solid-mechanics literature. Not by default alone, however, has the finite-element method established its supremacy in the solid area. It should have the potential to carry over its advantages at least to certain classes of fluid problems.

Adaptations of the finite-element method for fluid problems, perhaps not surprisingly, have come mainly from the solid-mechanics community. Beginning with the crude solution of the potential flow of a uniform stream over a circular cylinder by Martin (1968), numerous papers have appeared in diverse applications. From a formal viewpoint, the casting of any set of differential equations of a properly posed problem into an algebraic system, according to a finite-element method, can always be made; the difficulty is primarily one of bookkeeping. But such is also the case if a finite-difference method is used. The difference lies in the flexibility of the finite-element method in permitting a great deal of innovation by the user. The ultimate form of the algebraic system may be thus quite individual. At any rate, reducing the differential equation to an algebraic system is only the first step, and it does not follow that the desired solution can be easily extracted. Most of all, singular behavior abounds in fluid flows, and flow instability and turbulence find little parallel in solid mechanics. It needs no emphasis that an

421

awareness of the rich content of fluid phenomena is a prerequisite in order to take full advantage and avoid the pitfalls of the finite-element formulation.

At present, the prevalent attitude of fluid dynamicists toward the finite-element method is one of both fascination and suspicion. The fascination is aroused by its apparent and universal success in a closely related sister field. The suspicion comes largely from a lack of acquaintance. In this article, a review of the finite-element method is made in broad strokes, and selected applications are briefly described for illustration. We hope to show how its advantages can be realized from a deeper insight into the special difficulties of a given problem. For details of the method, readers should consult such books as those by e.g. Zienkiewicz (1971), Gallagher (1975), Strang & Fix (1973), and Øden (1972), and the long mathematical article by Babuška & Aziz (1972). An overall impression of the types of fluid problems that have been attacked by the method may be gained by browsing through Gallagher et al (1975). The objective here is mainly to present to the curious fluid dynamicists the finite-element method as interpreted by one from among their own ranks.

BASIC FEATURES OF THE FINITE-ELEMENT METHOD

To fix ideas, it is useful to restrict our attention first to the simple case of the Laplace equation in a simply connected domain B with Dirichlet conditions specified along its smooth boundary ∂B. Let the unknown ϕ be governed by

$$\left.\begin{array}{ll} \nabla^2 \phi = 0 & \text{in } B \\ \phi = f, & \text{given on } \partial B \end{array}\right\}. \tag{1}$$

As a field problem, the solution ϕ must be found at each of the infinite number of interior points of B. Numerically it is possible to construct only an approximation $\hat{\phi}$ in terms of a finite number of parameters. Thus in the finite-difference method a generally regular grid work is laid over the domain B, and the parameters are chosen to be the set of $\hat{\phi}_i$ at the regularly spaced grid points.

The most important feature of the finite-element method is to have the freedom of using irregularly spaced grid points. It starts by dividing B into N nonoverlapping but contiguous "finite elements" $B_1, B_2, B_3, \ldots, B_N$ (see Figure 1): $B = \Sigma_{\alpha=1}^N B_\alpha$. Then the solution ϕ is sought in terms of the solutions that hold in the subdomains:

$$\left.\begin{array}{ll} \nabla^2 \phi^{(\alpha)} = 0 & \text{in } B_\alpha \\ \phi^{(\alpha)} = f^{(\alpha)}, & \text{given on } \partial B_\alpha \end{array}\right\}. \tag{2}$$

If the boundary between B_α and B_β is denoted as $\partial B_{\alpha-\beta}$, then $f^{(\alpha)}$ is actually known only along a part of ∂B_α that coincides with the given boundary ∂B. However, for a continuous solution there must hold:

$$f^{(\alpha)} = f^{(\beta)} \quad \text{along} \quad \partial B_{\alpha-\beta}. \tag{3}$$

At this point, the finite-difference method may be regarded conceptually as a special case in the manner in which the subdivision of B is achieved.

With irregularly shaped subdivisions and no size limitation, the finite-element method has great flexibility. The price is that the construction of an adequate approximation in each finite element is no longer feasible through replacing the differential operator by a simple difference operator. On the other hand, the method of attack can be custom-tailored to accommodate the perception of the user. In complex problems a selective emphasis to cope with local special difficulties can be built in. There is definitely more room for ingenuity in the preparation of the program. For a given computer capacity, a larger category of problems can be handled, or better accuracy can be achieved.

The determination of an approximate solution in each finite element is typically done through integral constraints. For the Laplace equation, and whenever possible, an extremum principle guaranteeing monotonic convergence is clearly the preferred route, and rigorous mathematical studies on the rate of convergence as a function of the mesh size can be made. It is well known that the solution of Equation (1) should, among all admissible functions satisfying the boundary condition, minimize the "energy functional" $\Pi(\phi)$, defined by

$$\Pi(\phi) = \int_B \tfrac{1}{2}(\nabla\phi)^2 \, dV, \tag{4}$$

where dV denotes an element of volume in B. With the subdivision of B, the volume integral can be evaluated as a sum

$$\Pi(\phi) = \sum_{\alpha=1}^{N} \frac{1}{2} \int_{B_\alpha} (\nabla\phi^{(\alpha)})^2 \, dV, \tag{5}$$

where in each element B_α one treats $\phi = \phi^{(\alpha)}$ as the solution of a Dirichlet problem with unknown boundary data along $\partial B_{\alpha-\beta}$ [see Equation (3)].

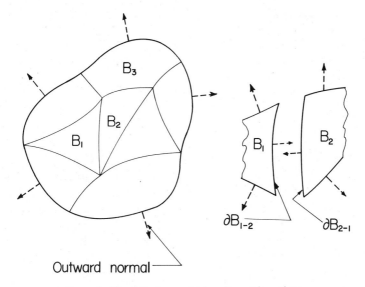

Outward normal

Figure 1   Discretization and inter-element boundary.

The actual construction of an approximation $\hat{\phi}$ with Equation (4) as the functional to be minimized is, of course, none other than the Rayleigh-Ritz procedure. Because of Equation (5), however, it suffices to construct $\hat{\phi}^{(\alpha)}$ individually, and the local approximation has to be close to the true solution only within each element. The use of piecewise continuous functions in the Rayleigh-Ritz procedure leads to a sparsely populated matrix in the resulting algebraic system, similar qualitatively to the one following from the finite-difference method. This feature is, of course, crucial for convenient practical computation.

For example, let the finite elements be laid out by a regular grid work, say, in triangular patterns for a two-dimensional problem as in Figure 2, and let a linear interpolation of $\hat{\phi}$ be used *in each triangle* in terms of the nodal values $\hat{\phi}_i$ at the "node" labeled "i." Equation (3) is evidently satisfied and the functional can be approximately calculated according to Equation (5). The nodal value $\hat{\phi}_0$ will make its contribution only in the triangular elements that share the node "0." Minimization of Π with respect to $\hat{\phi}_0$ needs to be carried out in four elements for pattern a, six for b, and eight for c. However, the net result for all three turns out to be the same equation

$$\phi_0 = (\phi_1 + \phi_2 + \phi_3 + \phi_4)/4,$$

the well-known 5-point central-difference formula. Thus the resultant coefficient

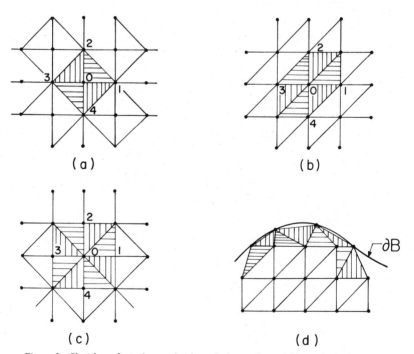

(a)   (b)

(c)   (d)

Figure 2   Sketches of regular mesh triangulation and special boundary elements.

matrix for the nodal unknowns is almost identical to that for the corresponding finite-difference equation. The deviation will appear, due to only those nodes adjacent to the boundary ∂B, to bring in the specified boundary values f. In contrast to the finite-difference method, no difficulty will arise for an arbitrary boundary contour, as long as the contour is approximated by a succession of linear segments. A layer of not-so-regular triangular elements can be introduced, such as the shaded ones in Figure 2d, and the contribution to the functional can be evaluated in a straightforward manner.

From this simple example, the central ingredients of the finite-element method can be seen: (a) discretization of the domain, (b) integral statement of the problem, and (c) local approximation in each element. These will now be discussed separately. Whether and how the resulting large matrix equation can be efficiently solved are important considerations for all numerical programs, but are omitted in this review.

Discretization of the Domain

Although often overlooked in the practice of the finite-element method with its myriad of details, the foremost advantage of the method is nevertheless its total flexibility in the discretization of the domain of interest. In parts of the domain, sometimes the solution can be better described or more easily constructed by other available methods. The discretization could and should exploit such properties. A finite-element method may proceed with a few "super-elements" of large size, in each of which the solution has been determined by a different approach. We then arrive at the final set of algebraic equations by properly patching the elements together.

The method of handling the conditions at infinity for external flows over a body, for example, is typically to prescribe the uniform flow, say, along a convenient outer contour at a large (but finite) distance from the body. It amounts to a discretization of the domain into two "super-elements," an inner one and an outer one with the chosen contour as the inter-element boundary. In the outer element that extends to infinity, the solution is assumed to be represented by the uniform flow with sufficient accuracy. An obvious improvement is to use, in the outer element, in place of the uniform flow the asymptotic form of the analytic solution that includes unknown parameters reflecting the presence of the body. Conceptually, the procedure is analogous to the matched asymptotic expansions. Of course, there is no small parameter; the inner solution is still to be numerically constructed, and the two solutions are to be *patched along* the inter-element boundary and not *matched* according to their behavior in approaching the inter-element boundary.

When local singularities are present in the domain of interest, it is likewise preferable to enclose each of them in a small but again finite-sized special element, provided the nature of the singularity can be analytically resolved. There is no basic difference from the above since the point at infinity may be regarded as a singularity too. Such techniques relieve the numerical computation from the odious task of accurately reproducing the singular behavior and are commonly incorporated in the more sophisticated finite-difference schemes.

As mentioned previously, a finite-difference scheme may be equivalent to certain finite-element methods that use a regular mesh layout. A considerable amount of effort by finite-element workers has been devoted to the development of elements of various shapes and degrees of complexity, to which we return later.

For the obvious reason of simplicity, the most popular basic element shapes in the plane are normally triangular and quadrilateral. Their counterparts in three-dimensions are of course the tetrahedron and the parallelopiped. Except in theoretical studies of convergence with respect to the size parameter, the mesh layout is seldom regular but guided by the user's intuition. One attempts to locate more elements where the solution is expected to change rapidly. There is also the general guideline that very slender elements are to be avoided. However, the more valid criterion should really be that the geometry of the element ought to reflect the local length scales, which need not be uniform in all directions. The optimal mesh pattern for a given number of elements of chosen shape is an intriguing question, but the answer appears to be elusive. It might be approached through an updating process as the true nature of the solution is revealed in successive trials.

Integral Statement of the Problem

VARIATIONAL PRINCIPLES   The equivalence of the original differential equations and its boundary conditions with an extremum principle, as in the above example, cannot be expected in all problems. For a linear-operator equation, defined in a domain B with appropriate boundary conditions on boundary ∂B, write

$$L\phi = F, \tag{6}$$

where L is the linear operator, ϕ the unknown, and F an assigned function. We allow only real-valued functions that are sufficiently smooth and square-integrable in B, so that the "inner product" of two admissible functions u and v, denoted by $\langle u, v \rangle$, can be defined:

$$\langle u, v \rangle \equiv \int_B uv \, dV. \tag{7}$$

Then the operator L^*, adjoint to L, is defined to have the property

$$\langle u, Lv \rangle = \langle v, L^*u \rangle, \tag{8}$$

and its form can be determined through integration by parts. An adjoint problem to Equation (6) is to find ψ obeying

$$L^*\psi = F^* \quad \text{in } B, \tag{9}$$

again with appropriate boundary conditions on ∂B and F^* denoting a suitably chosen forcing term. When L is self-adjoint, i.e. $L = L^*$, it is well known that the functional leading to a variational principle is

$$\Pi(\phi) = \langle \phi, L\phi \rangle - 2\langle \phi, F \rangle. \tag{10}$$

For an arbitrary small variation $\delta\phi$ of the solution ϕ, chosen from the space of admissible functions, it follows that

$$\delta\Pi = \langle\delta\phi, L\phi - F\rangle = 0. \tag{11}$$

In addition, if L is positive-definite so that for any admissible function u, $\langle u, Lu\rangle > 0$, then the second variation is positive, $\delta^2\Pi > 0$, and the solution ϕ of Equation (6) minimizes Π.

However, if L is not self-adjoint, $L^* \neq L$, the generalization of Equation (10) is to consider a pair of functions ϕ and ψ and the functional

$$\Pi(\phi, \psi) = \langle\psi, L\phi\rangle + \langle\phi, L^*\psi\rangle - 2\langle F, \psi\rangle - 2\langle F^*, \phi\rangle, \tag{12}$$

for which, analogous to Equation (11),

$$\delta\Pi = 2\langle\delta\psi, L\phi - F\rangle + 2\langle\delta\phi, L^*\psi - F^*\rangle. \tag{13}$$

Now if ϕ and ψ satisfy Equations (6) and (9), respectively, we have again a variational principle, though usually not an extremum.

Among nonlinear operators, those for which a variational principle, not necessarily yielding an extremum, can be constructed are referred to as "potential operators." The necessary and sufficient condition for an operator to be a potential operator is basically that its Gateaux differential should exhibit a certain symmetry property, corresponding to the integrability condition when a function of several variables is defined through its first-order partial derivatives. Exhaustive treatment can be found in Vainberg (1964).

THE LEAST-SQUARES METHOD   If we specialize in Equation (12) by taking

$$\psi = L\phi, \qquad F^* = L^*F,$$

it follows that

$$\Pi(\phi) = 2\langle L\phi - f, L\phi - f\rangle + \text{const.} \tag{14}$$

From Equation (14),

$$\delta\Pi(\phi) = 2\langle\delta L\phi, L\phi - F\rangle$$

and $\tag{15}$

$$\delta^2\Pi(\phi) = \langle\delta L\phi, \delta L\phi\rangle \geqq 0.$$

Thus the solutions of Equation (6) will minimize Π as defined by Equation (14), provided the variation is taken as $\delta L\phi$. This extremum principle is none other than the "least-squares method" in the construction of an approximate solution $\hat{\phi}$, which is to be chosen among the admissible functions that also satisfy the appropriate boundary conditions. However, the "residual" R defined by

$$R \equiv L\hat{\phi} - F \tag{16}$$

must now be square-integrable, thus requiring the admissible $\hat{\phi}$ to have higher-order differentiability. By rewriting Equation (15), it follows that

$$\delta\Pi = 2\langle L\delta\phi, L\phi - F\rangle = 2\langle L^*(L\phi - F), \delta\phi\rangle.$$

We thus see that the Euler-Lagrange equation for this variational principle is actually

$$L^*L\phi = L^*F. \tag{17}$$

It is not surprising that the sought solution has to be constructed from a basis of smoother functions.

Formulation in the least-squares sense has the advantage, on the other hand, in its ready extension to nonlinear problems. Since we may interpret the functional to be minimized as the "cost" due to $\hat\phi \neq \phi$, it can also be amended to include errors other than the pointwise residual in B. For example, when the boundary conditions are not met rigorously by $\hat\phi$ as required, a "penalty function" in the form of a boundary integral may be added to the cost functional. Such a formulation is described in Strang & Fix (1973, p. 132).

THE WEIGHTED-RESIDUAL METHOD   In a differential equation, sometimes a certain derivative of the solution may not exist locally. A broader interpretation of the differential equation is made possible by casting it into the "weak form": Equation (6), for example, is rewritten as

$$\langle v, L\phi \rangle = \langle v, F \rangle, \tag{18}$$

where v is any member of a set of test functions, required to be sufficiently smooth and capable of representing the desired solution in B through linear superposition. The solution ϕ of Equation (18) is the "weak solution." Note that v need not be differentiable, so long as the inner products are meaningful in Equation (18). If v is not differentiable, the "weak solution" of Equation (18) coincides with the classical one, and all derivatives in the expression $L\phi$ must exist. On the other hand, if v is infinitely differentiable and vanishes outside B, repeated integration by parts turns Equation (18) into

$$\langle \phi, L^*v \rangle = \langle F, v \rangle. \tag{19}$$

The "weak solution" ϕ of Equation (19) then need only be continuous.

Equation (18) obviously can be rewritten as

$$\langle v, L\phi - F \rangle = 0.$$

The weighted-residual method consists in determining an approximate solution $\hat\phi$, such that

$$\langle \hat{v}, L\hat\phi - F \rangle = \langle \hat{v}, R \rangle = 0 \tag{20}$$

where $\hat{v}$, the "weight function," is to be chosen from a subset of the admissible test functions v. The formal extension to nonlinear-operator equations is immediate. The classical Galerkin method further removes the choice of the test functions by identifying v with the admissible functions for $\hat\phi$. Equation (20) becomes

$$\langle \delta\hat\phi, R \rangle = 0, \tag{21}$$

where $\delta\hat\phi$ is an arbitrary variation of the approximate solution $\hat\phi$. Thus the overall residual in the domain B is set equal to zero on a weighted-average basis. If

$\hat{\phi}$ is defined by N undetermined parameters $\gamma_1, \gamma_2, \ldots, \gamma_N$, then $\delta\hat{\phi}$ is the set $\partial\hat{\phi}/\partial\gamma_i$, $i = 1, 2, \ldots, N$. Equation (21) then provides N integral constraints on $\hat{\phi}$. Like the least-squares method, the advantage of Equation (21) is that it can always be written regardless of the nature of the operator. However, Equation (21) is equivalent to an implementation of Equation (11) only if the variational principle in fact exists.

Ambiguity in the choice of the weight function arises when, for example, ϕ is a vector and Equation (6) is a set of simultaneous equations. As long as the operator L is linear, this ambiguity can be somewhat logically resolved by again appealing to a variational principle similar to Equation (12) via the adjoint function ψ, which of course is now also a vector. From this viewpoint, Equation (21) should preferably be rewritten as

$$\langle \delta\hat{\psi}, R \rangle = 0. \tag{22}$$

The adjoint ψ often may be interpretable as the solution of a "reverse flow." At any rate, by examining the adjoint equation, the components of $\hat{\psi}$ should have a correspondence with the components of $\hat{\phi}$. If the same admissible functions are permitted for the corresponding components, the weight functions can be logically selected. Within the spirit of the weighted-residual method, no further attention needs to be paid to the adjoint problem for ψ. But the simplicity is gained at a price. By giving up the refinements necessary for a variational principle, the weighted-residual method searches for the solution with a yardstick that tolerates first-order errors.

For nonlinear operators that are not potential, a guideline for the choice of the weight function can also be obtained. Suppose the vector equation is written as

$$A\phi - F = 0, \tag{23}$$

A being the nonlinear operator and F given. Let us introduce the Lagrange multiplier (vector) λ and write

$$\Pi(\lambda, \phi) = \int_B \lambda(A\phi - F)\, dV. \tag{24}$$

Then, performing the variations we have

$$\delta\Pi = \int_B \delta\lambda(A\phi - F)\, dV, \tag{25}$$

provided the Lagrange multiplier is made to satisfy

$$\int_B \lambda\delta(A\phi)\, dV = 0. \tag{26}$$

After integration by parts and suitable choice of the boundary conditions of λ along ∂B, Equation (26) can be reduced to

$$\int_B \delta\phi A^*(\phi, \lambda)\, dV = 0,$$

where $A^*(\phi, \lambda)$ is in general a nonlinear operator on both ϕ and λ. It follows that we may regard λ as the adjoint of ϕ, satisfying

$$A^*(\phi, \lambda) = 0 \qquad (27)$$

together with the suitably chosen boundary conditions. When A is linear, the result is of course equivalent to Equations (12) and (13). Again, we need only identify the corresponding components of the vectors ϕ and λ for the subsequent purpose of selecting the weight function.

The approximate solution $\hat{\phi}$ of Equation (23) now yields a residual R,

$$R \equiv A\hat{\phi} - F.$$

The counterpart of Equation (22) can be written as

$$\langle \delta\lambda, R \rangle = 0,$$

where $\delta\lambda$ are chosen from the basis of the approximation $\hat{\lambda}$ for the Lagrange multiplier λ as defined by Equation (27).

OTHER METHODS   When difficulties arise from the formulation of variational principles, a wide variety of alternatives are found in the literature. The most versatile technique is probably to employ Lagrange multipliers, well known in the calculus of variations. Our derivations following Equation (23) serve as one example. As another example, instead of satisfying the boundary conditions over an irregular contour, e.g. for Equation (1),

$$\phi = f \qquad \text{on } \partial B,$$

which is the "essential boundary condition" for trial functions to be used with the functional Equation (4), an alternative way is to amend the energy functional and use instead

$$\Pi(\phi, \lambda) = \int_B \tfrac{1}{2}(\nabla\phi)^2 \, dV + \int_{\partial B} \lambda(\phi - f) \, dS. \qquad (28)$$

In other words, the required boundary condition is considered as a constraint only. The technique is also useful at the element level; see the "hybrid model" of Pian & Tong (1969).

The penalty-function method is closely parallel, except in controlling the violation of a constraint through the assessment of a penalty. It is, of course, the direct application of a standard procedure in constrained optimization. For the same example of the Laplace equation that led to Equation (28), the new functional may be

$$\Pi(\phi) = \int_B \tfrac{1}{2}(\nabla\phi)^2 \, dV + \Lambda \int_{\partial B} (\phi - f)^2 \, dS$$

where Λ is an arbitrarily chosen large number. The exact solution is approached in the limit when Λ tends to infinity. In actual numerical calculations, an optimal choice should be consistent with the other errors of approximation, from discretization to quadrature. See Babuška & Aziz (1972, pp. 244–52).

Other intuitive and ad hoc procedures could lead to "restricted variational principles." The "restriction" is invariably to assign certain terms as provided by the exact though unknown solution, therefore not subject to subsequent variations. As an illustration, consider again potential flow but in a compressible fluid, so that Equation (1) becomes

$$\left.\begin{array}{ll} \nabla \cdot \rho \nabla \phi = 0 & \text{in } B \\[4pt] \phi = g, & \text{given on } \partial B \end{array}\right\} \tag{29}$$

where ρ is the density, depending on $(\nabla \phi)^2$. If ρ is assumed to be given by the exact solution, the counterpart of Equation (4) is simply

$$\Pi(\phi) = \int_B \tfrac{1}{2}\bar{\rho}(\nabla \phi)^2 \, dV, \tag{30}$$

where the notation $\bar{\rho}$ indicates that it is held fixed when ϕ is given a variation. In actual computation, however, an evaluation of $\bar{\rho}$ is necessary, and if it is evaluated in terms of the approximation $\hat{\phi}$, the error should be generally $O(\delta\phi)$. Such restricted variational principles provide no more than a recipe for a weighted-residual formulation.

For time-dependent problems it is advantageous to recognize the time and space variables and to distinguish between initial and boundary-value data. Regarding the solution as an evolution in time, at each instant we have again a boundary-value problem, and often a variational principle, or at least a weighted-residual formulation, can be set up with time appearing as a parameter. A finite-element method may be devised to handle the boundary-value problem, leading to a system of ordinary differential equations in time with prescribed initial conditions. This procedure is known as the "semi-discrete" finite-element method, clearly a counterpart of the method of integral relations discussed in Belotserkovskii & Chushkin (1965). The forward marching in time may sometimes be carried out in finite-element fashion, but the practical advantage is in general unlikely to be significant. For linear problems, the time dependence of course can be eliminated formally by appealing to the Fourier or Laplace transform; see e.g. Gurtin (1969). The resulting variational principle for the boundary-value problem in the transform plane carries the transform variable, instead of time itself, as a parameter. Numerical computation in the transform plane must then be done for the entire range of the transform variable, with sufficient resolution for the eventual inversion. Leaving the strategy of inversion aside, we see that the approach is still "semi-discrete" in nature.

For differential equations other than elliptic there will be a timelike space coordinate even in steady problems. The momentum-integral method in boundary-layer theory is a primitive weighted-residual formulation in the spacelike coordinates normal to the downstream (timelike) direction. Actually, a semi-discrete formulation is by no means restricted to cases where a timelike direction necessarily exists. The method of integral relations has in fact been frequently applied to transonic flows governed by an equation of the mixed type.

Local Approximation in Each Element

Although the finite-element method has the flexibility of allowing an arbitrary subdivision of the domain of interest, for large-scale computation it is still imperative to have available an arsenal of standard building blocks, which can be assembled in repetitive fashion for programming convenience. Each of these building blocks corresponds to a difference formula in the finite-difference method. The practice of the finite-difference method prefers to stay with the same difference formula. Similar reasoning demands that the same building blocks should be used as much as possible in a finite-element method.

GEOMETRY OF THE ELEMENT   It is self-evident that for two-dimensional problems triangles and quadrilaterals should be the primary candidates as the basic elements because of their simple geometrical definition. Curved boundaries are not well approximated by rectilinear triangles, thus other elements with curvilinear sides are introduced. A segment of the boundary curve can always be approximated by a polynomial. A particularly clever way to simplify the subsequent quadrature within the element leads to the so-called "isoparametric elements"; see Ergatoudis et al (1968).

Besides the shape of the elements, a choice must also be made as to the size. A size parameter "h" can be defined by, e.g., the diameter of the smallest circle enclosing the element. It plays the role of grid spacing in the finite-difference scheme whenever error analysis of the finite-element method is attempted. Thus, if all elements are of the same size, the situation is analogous to uniform grid spacing in the finite-difference method, and an error of $O(h^2)$, say, will also have the same meaning. However, all practical computations by the finite-element method try to achieve nearly uniform *local* accuracy by manipulating the size distribution, without altering the basic scheme for the construction of an approximation within each element. Any a priori knowledge of the expected solution may be exploited in the mesh layout, and a large number of small elements are usually placed in

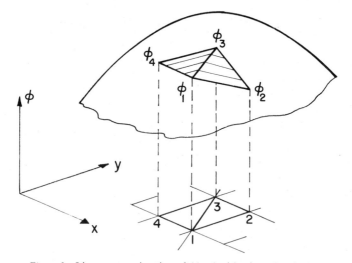

Figure 3   Linear approximation of $\phi(x, y)$ with triangular elements.

the region of rapid change. Formal guidelines for an optimal mesh layout appear to be still an unresolved issue.

These considerations enter into the mesh layout that subdivides the given domain B, and the result usually looks like a quilt of elements of the same shape but of different size. A number of reference points, called the nodes, are assigned along the inter-element boundaries, sometimes also within an element. In a "nodal" finite-element method, the solution is finally to be represented by information at the nodes. The discretization, in other words, replaces the original continuous spatial co-ordinates by a countable set of nodal points, irregularly distributed throughout the region of interest.

THE SHAPE FUNCTION   The shape function refers to the approximate representation of the unknown in a given element, usually by means of a polynomial. In the nodal finite-element method, the polynomial is constructed by interpolating the specified unknowns at the nodal points, dependent on the number of nodes and their location—therefore the "shape" of the element. Consider the example of the velocity potential $\phi(x, y)$ in the two-dimensional case, depicted in Figure 3 as the dome-shaped surface over the domain B. Triangulation of B and linear interpolation of ϕ from values assigned to the vertices as nodes is like erecting columns to reach the dome, and then laying flat triangular slabs from the anchor points as an approximation. This particular choice of shape and interpolation is referred to as the linear triangular element. Other popular simple elements are the 6-node quadratic triangle and the 4-node bilinear rectangle, with obvious indications of the number of nodes, type of interpolation, and shape of the element (see Figure 4).

Within the element α, it is clear that the approximation $\hat{\phi}$ is sought as

$$\hat{\phi}^{(\alpha)} = \sum_{i=1}^{k} \phi_i^{(\alpha)} N_i^{(\alpha)}(x, y), \tag{31}$$

where k is the number of nodes, $\phi_i^{(\alpha)}$ is the nodal unknown at the ith node, and $N_i^{(\alpha)}$ is called *the shape function* for $\phi_i^{(\alpha)}$. Since the $\phi_i^{(\alpha)}$'s are not yet required to satisfy any other constraint, and therefore are independent from each other, the shape

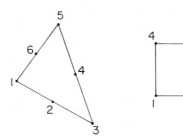

3-node linear triangle   6-node quadratic triangle   4-node bilinear rectangle

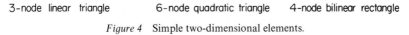

Figure 4   Simple two-dimensional elements.

functions must behave like an influence function:

$$N_i^{(\alpha)} \begin{cases} = 1 & \text{at the } i\text{th node of element } \alpha, \\ = 0 & \text{at other node points.} \end{cases}$$

With polynomial approximation, Equation (31) is but a recasting of a local Taylor-series expansion truncated to be determinate in terms of the nodal unknowns $\phi_i^{(\alpha)}$. If a completely arbitrary function is to be approximated as a Taylor series, a uniform truncation error of $O(h^4)$, say, requires that all the derivatives up to the third order should be determined from the data. Hence a complete cubic polynomial and ten nodal unknowns must be introduced. Practical elements are nevertheless often constructed without using complete polynomials; the 4-mode bilinear rectangle is an example. Another rule of thumb is that very slender elements are to be avoided. The basic reason for this is not hard to understand. As in numerical differentiation, the error in the derivatives increases if the evaluation makes use of not-very-accurate data at points close to each other. However, when the function is known to vary much more slowly in one direction, we clearly do not need many control points in the same direction. The effective size h for the truncation error is now associated with the element lengths in the other directions, and the slenderness rule should be interpreted accordingly.

THE ELEMENT AND GLOBAL MATRICES   Having selected an element shape and the shape functions, we must next construct a local solution for each element, by whatever means are at our disposal. In an integral formulation, the procedure is well illustrated in the specific example of the Laplace equation, say, for the velocity potential $\phi(x, y)$. For the element α, let us start from Equation (11),

$$\delta\Pi(\phi^{(\alpha)}) = \int_{B_\alpha} \delta\phi_i^{(\alpha)} \nabla^2 \phi^{(\alpha)} \, dV = 0.$$

By integration by parts, it is turned into

$$\tfrac{1}{2}\delta \int_{B_\alpha} (\nabla\phi^{(\alpha)})^2 \, dV - \int_{\partial B_\alpha} \delta\phi^{(\alpha)} \frac{\partial\phi^{(\alpha)}}{\partial n} \, dS = 0, \tag{32}$$

which says that either $\phi^{(\alpha)}$ or $\partial\phi^{(\alpha)}/\partial n$ of the *exact* solution must be known along the boundary. We shall account for the boundary integral later. Then, by substituting the assumed form of $\hat\phi^{(\alpha)}$, Equation (31), into Equation (32), and taking the variation with respect to the nodal unknowns $\phi_i^{(\alpha)}$, the result is a system of linear algebraic equations that can be written as

$$[K^{(\alpha)}][\phi_i^{(\alpha)}] + \int_{\partial B_\alpha} \cdots = 0, \tag{33}$$

where $[K^{(\alpha)}]$, a positive-definite symmetric square matrix, is called the element "stiffness matrix" from its origin in solid mechanics, and $[\phi_i^{(\alpha)}]$ is the column vector of the nodal unknown. The term $\int_{\partial B_\alpha} \cdots$ is simply to denote the contributions of the boundary integral not yet evaluated. Note that $[K^{(\alpha)}]$ depends only on the element and its shape functions, providing the coefficients for a special type of finite-difference formula.

It remains to sum up all the elements:

$$\sum_{\alpha=1}^{N} \left\{ [K^{(\alpha)}][\phi_i^{(\alpha)}] + \int_{\partial B_\alpha} \cdots \right\} = 0. \tag{34}$$

Note that along each inter-element boundary between element α and element β, the boundary integral appears twice in Equation (34), but with opposite normal direction (see Figure 1). Combining the two we have the typical term $\int_{\partial B_{\alpha-\beta}} (\delta\phi^{(\alpha)} - \delta\phi^{(\beta)})(\partial\phi/\partial n)\, dS$. Since the exact $\partial\phi/\partial n$ is not known, these typical terms can be made to vanish if $(\delta\phi^{(\alpha)} - \delta\phi^{(\beta)})$ is zero along the boundary $\partial B_{\alpha-\beta}$. From this comes the central requirement in the construction of the shape function: *There must be sufficient continuity in the piecewise approximation $\hat{\phi}$ so that all interior boundary integrals cancel.* In the present example, only continuity of the value of $\hat{\phi}$ across $\partial B_{\alpha-\beta}$ is needed, or $\hat{\phi}$ must have "C^0-continuity," and even the linear triangle element will do. For the biharmonic equation, the same consideration leads to the requirement that $\hat{\phi}$ must have C^1-continuity, both in the value and the normal derivative along any inter-element boundary. Elements possessing the required continuity are called "conforming elements."

With conforming elements, Equation (34) finally can be "assembled" into

$$[K][\phi_i] = [P], \tag{35}$$

where $[K]$ is the "global stiffness matrix," $[\phi_i]$ the column vector of all the nodal unknowns, and $[P]$ is the "load vector" arising from the specified boundary conditions on ∂B. The global stiffness matrix $[K]$ is banded, because of the piecewise continuity of $\hat{\phi}$. The evaluation of $[P]$ is straightforward in principle, but the roles of the essential and natural boundary conditions must be strictly respected.

We mention in passing that as an example of a conforming element with C^1-continuity, the complete quintic over a triangle has 21 degrees of freedom (nodal unknowns). For convenience, however, nonconforming elements are often used without correction. In spite of the deficiency, the result may converge to the correct solution if they can pass a "patch test" originally devised by Irons (see Irons & Razzaque, 1972, also Strang & Fix, 1973, pp. 309–12).

APPLICATION TO FLUID FLOWS

If the finite-element method is viewed as a technique to generate algebraic formulas for the unknowns at irregularly spaced but nearby node points, it is clear that the resulting system must basically resemble that obtained by the conventional finite-difference method. Given a problem, it may be anticipated that the difficulties besetting one method would generally have to be overcome in the other. The vast backlog of experience with application of the finite-difference method to fluid flows thus is of tremendous importance to the practice of the finite-element method. In fact, many difficulties common to both have been resolved by techniques that are at least conceptually equivalent.

A number of fluid-flow problems find their counterpart in solid mechanics by analogy. These may be solved by mimicking the procedures in solid-mechanics literature. There are nevertheless also aspects that distinguish fluid flows from standard solid-mechanics problems. In the following, our emphasis is on how some of the special features in fluid flows have been treated in the finite-element method. The illustrative examples are mostly familiar problems, such that their proper statements can be largely left as understood.

Incompressible Potential Flows

Governed by the Laplace equation and a well-known extremum variational principle, these problems with Dirichlet or Neumann boundary conditions are straightforward for the finite-element method. For external flows efforts have been made to cut down the size of the domain by employing the asymptotic behavior of the singularities that represent the body, as is often also done in finite-difference methods. This device amounts to replacing the outer region beyond the domain of computation by a special element, in which the solution has been parametrized by the unknown singularity strengths. The use of geometrical transformations to map an infinite domain to a finite one has no intrinsic advantage, as the proper asymptotic behavior still needs to be accounted for. (It may nevertheless facilitate the automatic generation of the mesh points, which is of significant practical convenience.) An alternative scheme is the concept of "picture frame" due to McDonald & Wexler (1972). They propose to relate the nodal variables along two contours, one within the other (Figure 5), by the classical formula

$$\phi\big|_{\partial B} = \frac{1}{4\pi} \int_{\partial B_1} \left(\phi \frac{\partial G}{\partial n} - G \frac{\partial \phi}{\partial n} \right) dS,$$

where G is Green's function in free space (the source). The separation between the two contours is to avoid the singular behavior of G, at a cost of increasing the bandwidth because of the coupling. It does have the potential of greatly reducing

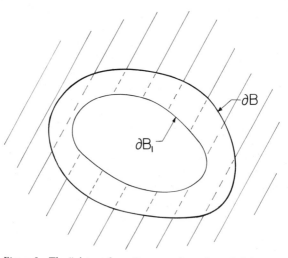

Figure 5   The "picture-frame" concept for unbounded domain.

the size of the domain of computation. Of course, the smallest number of nodal unknowns is attained by going to the limit of collapsing ∂B and ∂B_1 on to the body contour itself—and the integral equation in terms of a surface distribution emerges as the result.

A concave corner on the boundary also produces a local singular behavior, and the usual polynomial approximation is of poor accuracy. The error has been analyzed theoretically by Kellogg (1972). In practice either much smaller elements should be placed around the corner, or a special element should be constructed that has the proper analytic behavior suggested by a local analysis. The implementation is similar to that for the representation of the singular point at infinity mentioned above. However, unlike the two-dimensional case, the singular behavior at a corner in three dimensions is apparently unknown in full generality.

In applying the variational principle, the region considered should be simply connected. A barrier is thus necessary for external flows over a two-dimensional body, and several may be needed when multiple bodies are involved. The case of a lifting wing should have a barrier that excludes the sheet of shed vortices from the domain of the potential flow. The proper conditions to be specified along opposite sides of the barrier are provided by the statement of the problem, e.g. the jump of the velocity potential due to circulation for a lifting airfoil. In handling multiple-body problems, the flexibility of the finite-element method makes it sometimes feasible to prepare subprograms for finite "patches" that each enclose a single body, and to have them assembled subsequently. The patches now serve as special elements, or "super-elements" because of their size, that contain a body instead of a singularity in the interior. An infinite cascade of airfoils is an obvious example requiring such a technique, as has been successfully demonstrated by Habashi (1975). In his extensive study of solving airfoil problems by the finite-element method, the single airfoil is first mapped to a near circle by the Joukowski transformation. The working plane in polar coordinates (r, θ) is then mapped into a rectangular strip using (r, θ) as cartesian coordinates, and a near optimal triangulation in the working plane is laid out by a simple automatic mesh-generation routine. Some of his results are in Shen (1975).

More difficult to handle numerically is the class of problems where free boundaries are present whose locations must be determined together with the solution. The direct attack is usually through iteration, successively correcting an initial guess of the location. The jet issuing from an orifice, for instance, has been solved in this way by Chan & Larock (1973); the difficult case of a jet impinging on a curved surface, by Sarpkaya & Hiriat (1975). The mesh layout near the boundary must of course be adjusted successively to follow the free surface. The capability of the finite-element method to cope with arbitrary boundary shapes is of significant advantage in these circumstances.

Subsonic Compressible Potential Flows

The quasi-linear differential equation governing compressible potential flows not only is of practical interest in aerodynamics but also appears in lubrication theory with thermal effects. The variational principles of Bateman have been known for some time; see e.g. Rasmussen (1971). In terms of the velocity potential ϕ, we have the velocity $\mathbf{v} = \nabla\phi$, and the solution satisfies

$$\delta\Pi_1(\phi) = \delta \int_B p \, dV - \int_{\partial B} \rho \frac{\partial\phi}{\partial n} \delta\phi \, dS = 0, \tag{36}$$

where p is the pressure, to be understood as the well-known transcendental expression in terms of $\nabla\phi$ $[p = p(q), q \equiv |\mathbf{v}| = |\nabla\phi|]$. For two-dimensional cases, the stream function ψ may be used instead of ϕ; it satisfies

$$\nabla \cdot \rho^{-1}\nabla\psi = 0 \tag{37}$$

with the density ρ now expressed in terms of ψ $[\rho = \rho(q), q \equiv |\mathbf{v}| = \rho^{-1}|\nabla\psi|]$. The variational principle for the ψ-formulation is

$$\delta\Pi_2(\psi) = \delta \int_B (p + \rho q^2) \, dV - \int_{\partial B} \rho^{-1} \frac{\partial\psi}{\partial n} \delta\psi \, dS = 0, \tag{38}$$

where, again, $p + \rho q^2$ is to be understood as the well-known transcendental expression in terms of $\nabla\psi$. At the exact solution, Π_1 has a minimum and Π_2 has a maximum.

These variational principles were first applied to the test case of a circular cylinder in steady motion by Wang (1948), who used radial and trigonometric global trial functions, prior to the advent of the electronic computer. It is clear that for external problems, a technique is first needed to render the functional finite for the infinite domain. The numerical treatment is difficult due to the transcendental nature of the integrand. After suitable approximations, a large system of nonlinear algebraic equations is finally derived. If we resume the attack, the functional may be evaluated by a finite-difference or a finite-element procedure. Eventually it boils down to solving the nonlinear algebraic system via an iterative scheme. There are however no known iterative schemes that are infallible regardless of the nonlinearity. From this viewpoint, we should preferably start such considerations without losing sight of the particular differential equation itself.

Consider Equation (36), and let $\phi^{(n)}$ be the nth iterate of ϕ, etc. The following schemes are obvious alternatives:

$$\nabla^2\phi^{(n+1)} = \nabla \ln \rho^{(n)} \cdot \nabla\phi^{(n)} \tag{39}$$

and

$$\nabla \cdot \rho^{(n)}\nabla\phi^{(n+1)} = 0. \tag{40}$$

If the initial approximation is to take $\rho^{(0)} = \text{const}$, then Equation (39) corresponds to the Rayleigh-Jansen method of expansion in M_∞^2, M_∞ being the Mach number in the free stream, and Equation (40) identifies itself with the electrolytic-tank analogy of Taylor & Sharman (1928). Equation (39) has the virtue of simplicity in practical calculations, because of the persistence of the Laplace operator. It is popular as a finite-difference method and has been employed by most investigators for a variety of problems with the finite-element method (see Carey, 1975, Shen, 1975). The variational formulation of Equation (40) and computations for the circular cylinder using finite differences are represented by Gelder (1971); a finite-element version has been studied by Habashi (1975). In the case of the circular cylinder, the radius of convergence of the M_∞^2-expansion is shown by Hoffman (1974) to be only slightly greater than the critical Mach number. Consequently Equation (39) must become progressively worse as the critical Mach number is approached. Equation (40) will also break down, as already reported by Taylor & Sharman (1928). Generally speaking, however, the convergence rate of Equation (40) is much better than that of Equation (39), no doubt due to the more realistic variable density as opposed to the representation of its effect through a spatial source distribution.

The failure to converge owes its origin to the insistence of Equations (39) and (40) that the operator remain elliptic. Therefore the method is strictly applicable only to totally subsonic flows. A better approximation to the operator by treating the flow in each finite element locally by a Prandtl-Glauert approximation has been proposed by Shen & Habashi (1976). Accordingly, we have *for each element*

$$(1 - M_\alpha^2)\frac{\partial^2 \phi}{\partial x_\alpha^2} + \frac{\partial^2 \phi}{\partial y_\alpha^2} + \frac{\partial^2 \phi}{\partial z_\alpha^2} = 0, \tag{41}$$

where M_α is the local average Mach number in element α, and $(x_\alpha, y_\alpha, z_\alpha)$ are local cartesian coordinates with x_α oriented along the local average velocity vector. The unknown local velocity and the coordinates must be successively updated by iteration, and care must be taken to respect the inter-element continuity. In tests on a circular cylinder in subsonic flow, Habashi (1975) found that the same finite-element discretization would converge twice as fast, in computing time, based upon the "local linearization" of Equation (41) as compared with Equation (40). But more importantly, the change of type of the local operator for $M_\alpha > 1$ is clearly exhibited, paving the way for further refinement to overcome transonic difficulties. In spirit, the finite-element approach with local coordinates according to Equation (41) is akin to the "rotating difference" advocated by Jameson (1974) in the finite-difference method for transonic flows.

The need to introduce locally oriented coordinates can be relieved if prior information on the streamline direction is known. It is observed that the streamline pattern usually does not greatly alter because of the compressibility effects. For two-dimensional flows, the early finite-difference computation for a nozzle by Emmons (1946) was carried out by first adopting the "potential coordinates" (ξ, η), which are, respectively, the incompressible velocity potential and stream function for the given problem. Equations (29) and (37) are unchanged except that (ξ, η) replace (x, y), respectively, in the operator. Considered as a geometrical transformation, the use of potential coordinates has the advantage of rectifying the curved boundary of the body into a segment on the ξ-axis, although for finite bodies like an airfoil, the leading and trailing edges will appear as singular points. Taking account of the small inclination of the compressible streamlines in the (ξ, η)-plane, Shen & Chen (1976) proposed a "streamline" perturbation technique, which turns Equation (29) into an equation of the type

$$(1 - M^2)\phi_{\xi\xi} + \phi_{\eta\eta} + \cdots = 0, \tag{42}$$

where M is the local Mach number, and the three dots represent additional terms of lower-order derivatives and also the information of the body shape through its incompressible velocity field. A universal finite-element discretization thus can be used for different body shapes. The iteration is now to be carried out on the local Mach number only. Chen (1976) has developed highly efficient finite-element programs for both Equation (42) and an analogous equation in terms of the stream function ψ to handle nonlifting symmetrical bodies. A parallel development in the finite-difference literature is reported by Caughey (1974), in his computations for the transonic airfoil.

Supersonic and Transonic Flows

Totally supersonic flows are governed by differential equations of the hyperbolic type, requiring initial data in the timelike direction. The solution furthermore is determined only within the zone of influence delineated by the characteristics along which signals propagate. In finite-element formulations, this fact has to be recognized at the element level, and seems obviously difficult to accommodate, especially when the local characteristics depend on the unknown solution. To our knowledge, a finite element is yet to be developed that yields logically the counterpart of the 5-point backward-difference formula in finite-difference schemes, which is unconditionally stable for computation in the supersonic zone. What can be done at present appears to be limited to the semi-discrete type, in which one carries out the finite-element approximation only in the spacelike direction, essentially like the method of integral relations. An example is the calculation for the Emmons nozzle in supercritical flow by Chen (1976), briefly reported in Shen (1975). Here a semi-discrete program for the supersonic region after the throat is patched to another subsonic program upstream via a transonic solution in the throat neighborhood. The transonic solution actually used is the local series representation due to Oswatitsch & Rothstein (1949).

In general, the transonic flow is more difficult for numerical work, especially when a supersonic bubble appears locally and is surrounded by subsonic flow. The governing equation becomes of mixed type, with unknown boundaries for the sonic condition and possibly the shock surface. Here we need to look to the proven techniques in the finite-difference method. Since Murman & Cole (1971), the further advances can be gleaned from Jameson (1976). The line relaxation normal to the streamline direction plus subsequent sweeping is not basically different from semi-discretization by the finite-element method. But type-dependent difference schemes are employed, and there is shock smearing via a pseudo-viscosity. These artifices in the finite-element literature made their first appearance in Chan et al (1975). In an effort to stabilize the numerical scheme in the supersonic region over an airfoil, they constructed a finite-element program by the least-squares principle and imitated the backward-difference scheme by simply nullifying the coupling with the downstream-side nodal unknowns wherever the element is supersonic. The resulting discrete system was claimed to have dissipative properties, and a smeared shock was shown to appear. The pressure distribution compared well with experiment in the case of an NACA 6% symmetrical airfoil. The same method has also been applied to an oscillating airfoil by Chan & Brashears (1975), but the pressure distributions show large deviations from experimental values. The reason is unclear; they claim the accuracy of the experimental data is uncertain. Other approaches of both the shock-capturing and shock-fitting types are reported by several authors in the *Second International Symposium on Finite Element Methods in Flow Problems,* S. Margherita Ligure (Italy), June 14–18, 1976. The fruition may come in the near future.

Another possible outlet, yet to be developed but well grounded in theory, is to transform the governing equation into a symmetric positive-definite system, which was defined and analyzed by Friedrichs (1958). The main feature of such systems is that there is no distinction according to type, so that the difference approximations derived from them are equally applicable in hyperbolic and in elliptic regions. Studies of the finite-element method along this line can be found in Lesaint (1973) and Aziz & Leventhal (1976).

It may be of interest to note that, for Equation (29), all the finite-element methods devised for subsonic flows can in practice work slightly beyond the critical Mach number and predict the presence of a small supersonic bubble. The theoretical objection manifests itself only if the Mach number is further raised and the supersonic condition prevails in a significant number of elements. A plausible explanation may be that the local violation of a well-posed problem leads to only local numerical instability, therefore creating no havoc without a sufficient number of bad elements. Mesh refinement in such situations does not necessarily produce better accuracy!

Viscous Flows

For brevity, our discussions are restricted to incompressible Newtonian fluids. We begin with creeping flow without inertia effects, physically governed by equilibrium. Different forms of the differential equation, however, may be used. The most compact description is the two-dimensional case, in terms of the stream function ψ, satisfying the biharmonic equation

$$\nabla^4 \psi = 0 \quad \text{in } B \tag{43}$$

with the boundary conditions either on the velocity or the stresses along the boundary ∂B. That Equation (43) is the Euler equation of the following extremum variational principle is well known:

$$\delta\Pi(\psi) = \delta \int_B \tfrac{1}{2}(\nabla^2\psi)^2 \, dV - \int_{\partial B} \left(\delta\psi \, \frac{\partial}{\partial n}\nabla^2\psi + \delta\frac{\partial\psi}{\partial n}\nabla^2\psi \right) dS = 0. \tag{44}$$

Thus the approximate solution $\hat{\psi}$ must be piecewise *twice* differentiable; any prescription of ψ and $\partial\psi/\partial n$ along ∂B must be respected by $\hat{\psi}$, while correct $(\partial/\partial n)\nabla^2\psi$ and $\nabla^2\psi$ along ∂B are needed only if the contour integral is to be evaluated. The finite-element formulation is straightforward, except for the technical difficulty of constructing simple conforming elements with continuous inter-element normal derivatives. This difficulty can be bypassed if Equation (43) is split into two simultaneous equations for ψ and the vorticity ω:

$$\nabla^2 \psi = \omega, \qquad \nabla^2 \omega = 0, \tag{45}$$

for which the variational principal may be written as

$$\delta \Pi(\psi, \omega) = \delta \int_B (\tfrac{1}{2}\omega^2 + \nabla\omega \cdot \nabla\psi)\, dV - \int_{\partial B} \left(\omega\delta \frac{\partial\psi}{\partial n} + \psi\delta \frac{\partial\omega}{\partial n} \right) dS = 0. \tag{46}$$

The approximate solutions $\hat{\psi}$ and $\hat{\omega}$ are required to be piecewise differentiable only, so that even linear elements are acceptable. The number of nodal unknowns is doubled, but boundary conditions on the vorticity are not needed as long as along ∂B the essential condition $\partial\psi/\partial n$ and the natural condition ψ are specified for the problem. Still a third form may be used if we turn to the original momentum and continuity equations. In terms of the velocity $\mathbf{v}$ and the pressure p, i.e. the primitive variables, the variational principle is equivalent to the virtual (rate of) work principle:

$$\delta \Pi(\mathbf{v}, p) = \delta \int_B (\Phi - p\nabla \cdot \mathbf{v})\, dV + \int_{\partial B} (p\delta v_n - \tau_n \cdot \delta\mathbf{v})\, dS = 0, \tag{47}$$

where Φ is the viscous dissipation and τ_n the vector viscous stress, acting on the surface dS, at which the outward normal is, of course, the unit vector $\mathbf{n}$. Interelement continuity is now required on the velocity only. This formulation yields the pressure and velocity at nodal points directly, but is less convenient if the streamline pattern is desired. It is equally valid for three-dimensional flows, however. It should also be noted that when using Equation (47), the continuity requirement on $\mathbf{v}$ is more severe than that on p. Intuitively, the elements should reflect this fact by using, for instance, quadratic interpolation for $\mathbf{v}$ but linear for p. (Such is obviously implied if finite-differencing is applied to the original differential equations.) A numerical study of the effects in the case of a circular cylinder is reported by Hood & Taylor (1974).

As already mentioned in the even simpler case of the Laplace equation, attention should be paid to the singularity at corners of the boundary and the question of infinite domain for an unbounded flow. In the two-dimensional case, the singularity at the corner is more complex but can still be worked out (see e.g. Weinbaum, 1966). A similar singularity also occurs with discontinuous boundary conditions, e.g. at the exit of a jet as analyzed by Richardson (1970). Regarding infinity, it is recalled that the creeping-flow approximation itself is not always valid there and an Oseen correction may be logically necessary. However, most calculations (e.g. for a circular cylinder) appear to have ignored the correction. By assigning the condition at infinity to prevail at a finite distance, an error is introduced but apparently the theoretical difficulty is happily covered up.

The full Navier-Stokes equation is the ultimate and has been challenged by a sizeable number of investigators via the finite-element method. See e.g. Oden & Wellford (1972), Tong (1974), Baker (1973), Taylor & Hood (1973). Whenever reference is made to the "local potential" or a pseudo-variational principle, the common underlying concept can be identified as treating the acceleration terms as a *given* forcing function added to the creeping-flow equations. All three versions,

Equations (43–45), have their counterparts. Because of the rapid decay of vorticity away from the body, the version using ψ and ω is generally favored as the best compromise for convenience. As discussed previously, the procedure is really a weighted-residual method with the weight function chosen to be the same as that for the reduced creeping-flow problem. For steady flows, after the nonlinear algebraic system is derived for the nodal unknowns, an iterative scheme has to be applied to the acceleration terms. Indeed, parallel to the discussions above for the compressibility effects, the added term may be iterated as either

$$\left[\frac{D}{Dt}\mathbf{v}\right]^{(n)} \quad \text{or} \quad \left[\frac{D}{Dt}\right]^{(n)}\mathbf{v}^{(n+1)},$$

corresponding to the "Poisson iteration" [Equation (39)] and the "Taylor iteration" [Equation (40)], respectively. The latter updates the contribution due to convection, much as updating the density in Equation (40), and not surprisingly leads to faster convergence in test cases. Since the procedure preserves the creeping-flow operator, in essence the iteration has the nature of an expansion in the Reynolds number. It cannot be expected to work for higher-than-moderate Reynolds numbers, and eventually the parabolic boundary-layer type of behavior must enter into competition. If hydrodynamic instability can arise, it will further limit the range of validity. Furthermore, the boundary condition at infinity over an unbounded region, particularly in the downstream direction, requires more than casual attention. The vorticity is convected downstream, and the wake extends to infinity downstream in a steady flow.

As is well known, iteration is rather analogous to an evolution process, and an effective way to obtain the steady-state solution is often by solving the unsteady problem. The weighted-residual method in the last paragraph is then carried out in the space coordinates, the time variable occurring only as a parameter. The finite-element approximation that follows must also be in the space coordinates— therefore "semi-discrete" for the unsteady case. The resulting ordinary differential equations for the nodal unknowns are to be solved by marching in time.

If we switch over to the large-Reynolds-number limit, the steady laminar boundary-layer equation is also parabolic in type, the downstream direction being timelike. A general semi-discrete procedure for the unsteady Navier-Stokes equations thus can handle the boundary layer as a degenerate problem with one fewer spatial coordinate. The COMOC program described by Baker (1974) indeed has such a built-in feature, plus compressibility effects, etc., so long as there is but one timelike coordinate in the entire system of equations. It appears to be highly powerful in providing practical answers to a wide variety of complicated flows.

CONCLUDING REMARKS

We have presented above the salient features of the finite-element method and reviewed the current status of its application to typical fluid-flow problems. The inevitable question is how does it measure against the finite-difference method.

Perhaps there can never be a fair comparison, for much depends on the nature of the problem as well as the skill of the user. Theoretical analysis of the finite-element method, as in Babuška & Aziz (1972), always refers to the size parameter h. Regardless of the results, a uniform choice of h discriminates against the finite-element method by restraining its freedom in discretization. If, on the other hand, the finite-element method is regarded as a scheme to provide difference formulas for a near *optimal* grid-point distribution, its supremacy follows by definition. However, the hand now is stacked against the finite-difference method, since the necessary efforts to construct such a good finite-element program have to be included in the balance.

Research in the finite-element method must naturally center around how to ensure a "good" program for a given class of problems and how to reduce the effort in preparing the program. For fluid flows, it is the nonlinear terms in the equations of motion that give rise to the complicated phenomena not usually present in solid mechanics. These very same terms are responsible when troubles are encountered in the numerical solution of the final discretized version. If the formulation is properly done, the troubles should be signaling the inadequacy of the approximations in describing the actual phenomenon. When a shock wave occurs, for instance, the method can hardly be expected to work without allowing a possible surface of discontinuity in the representation.

The kinship with the finite-difference method would also suggest that whatever flow problems one method can solve, can also be solved by the other. The limit of either method is the resolution of fine structure that can be built into the discretized approximation. From this viewpoint, the finite-element method may have little advantage if only one length scale suffices to characterize the problem. Conversely, the flexibility of the finite-element method enables it to cope with the presence of widely different length scales much more effectively. The success reported in many test cases where the finite-element method achieves the same accuracy with much fewer unknowns is, we believe, largely due to this property. It has thus the potential of solving a more difficult problem with a given computer. Numerous areas remain to be conquered in fluid mechanics through computation. There can be little doubt of the increasing role to be played by the finite-element method in the hands of a competent fluid dynamicist.

ACKNOWLEDGMENTS

This work has been supported in part by the US Office of Naval Research under contract N00014-75-C-0375.

Literature Cited

Aziz, A. K., Leventhal, S. H. 1976. *Numerical Solution of Partial Differential Equations—III*, ed. B. Hubbard, pp. 55–88. New York: Academic

Babuška, I., Aziz, A. K. 1972. *The Mathematical Foundations of the Finite Element Method with Applications to Partial*

Differential Equations, ed. A. K. Aziz, pp. 3–345. New York: Academic

Baker, A. J. 1973. *Int. J. Num. Meth. Eng.* 6:89–101

Baker, A. J. 1974. *NASA Contract. Rep. CR-2391*, Washington, DC

Belotserkovskii, O. M., Chushkin, P. I. 1965.

Basic Developments in Fluid Dynamics, ed.
M. Holt, Vol. 1. pp. 1–126. New York:
Academic
Carey, G. F. 1975. *Finite Elements in Fluids,*
ed. R. H. Gallagher, J. T. Oden, C. Taylor,
O. C. Zienkiewicz, Vol. 2, pp. 159–77.
London: Wiley
Caughey, D. A. 1974. *AIAA Paper 74-541*
Chan, S. T. K., Brashears, M. R. 1975.
AIAA Paper 75-875
Chan, S. T. K., Brashears, M. R., Young,
V. Y. C. 1975. *AIAA Paper 75-79*
Chan, S. T. K., Larock, B. E. 1973. *ASCE
J. Hydraul. Div.* 99:(HY1) 81–97
Chen, H. C. 1976. PhD thesis. Cornell Univ.,
New York
Emmons, H. 1946. *NACA Tech. Note TN-
1003,* Washington, DC
Ergatoudis, I., Irons, B., Zienkiewicz, O. C.
1968. *Int. J. Solids Struct.* 4:31–42
Friedrichs, K. O. 1958. *Commun. Pure Appl.
Math.* 11:333–418
Gallagher, R. H. 1975. *Finite Element
Analysis, Fundamentals.* Englewood Cliffs,
NJ: Prentice-Hall
Gallagher, R. H., Oden, J. T., Taylor, C.,
Zienkiewicz, O. C., eds. 1975. *Finite
Elements in Fluids,* Vols. 1 & 2. London:
Wiley
Gelder, D. 1971. *Int. J. Num. Meth. Eng.*
3:35–43
Gurtin, M. 1969. *Arch. Ration. Mech. Anal.*
16:34–50
Habashi, W. G. 1975. PhD thesis. Cornell
Univ., New York
Hoffman, G. H. 1974. *J. Mech.* 13:433–47
Hood, P., Taylor, C. 1974. *Finite Element
Methods in Flow Problems,* ed. J. T. Oden,
O. C. Zienkiewicz, R. H. Gallagher, C.
Taylor, pp. 121–32. Huntsville, Ala.: UAH
Press
Irons, B., Razzaque, A. 1972. *The Mathe-
matical Foundations of the Finite Element
Methods with Applications to Partial
Differential Equations,* ed. A. K. Aziz, pp.
557–87. New York: Academic
Jameson, A. 1974. *Commun. Pure Appl. Math.*
27:283–309
Jameson, A. 1976. *Numerical Solution of
Partial Differential Equations—III,* ed. B.
Hubbard, pp. 275–320. New York:
Academic
Kellogg, R. B. 1972. *Numerical Solution of
Partial Differential Equations—II,* ed. B.
Hubbard, pp. 351–400. New York:

Academic
Lesaint, P. 1973. *Num. Math.* 21:244–55
Martin, H. C. 1968. *Proc. Conf. Matrix
Methods in Structural Mechanics, 2nd,
AFFDL-TR-68-150 517-535,* Wright-
Patterson Air Force Base, Dayton, Ohio
McDonald, B. H., Wexler, A. 1972. *IEEE
Trans. Microwave Theory Tech.* MTT-20:
841–47
Murman, E. M., Cole, J. 1971. *AIAA J.*
9:121–41
Oden, J. T. 1972. *Finite Elements of Nonlinear
Continua.* New York: McGraw-Hill
Oden, J. T. Wellford, L. C. Jr. 1972. *AIAA J.*
10:1590–99
Oswatitsch, K., Rothstein, W. 1949. *NACA
Tech. Memo TM-1215,* Washington, DC
Pian, T. H. H., Tong, P. 1969. *Int. J. Num.
Meth. Eng.* 1:3–28
Rasmussen, H. 1971. *R. Aircr. Estab. Tech.
Rep. TR-71-234*
Richardson, S. 1970. *Proc. Camb. Philos. Soc.*
67:477–89
Sarpkaya, T., Hiriat, G. 1975. *Finite Elements
in Fluids,* ed. R. H. Gallagher, J. T. Oden,
C. Taylor, O. C. Zienkiewicz, 1:265–279.
London: Wiley
Shen, S. F., Chen, H. C. 1976. *Arch. Mech.
Stosowanej.* In press
Shen, S. F. 1975. *Finite Elements in Fluids,*
ed. R. H. Gallagher, J. T. Oden, C. Taylor,
O. C. Zienkiewicz, Vol. 2, pp. 179–204.
London: Wiley
Shen, S. F., Habashi, W. G. 1976. *Int. J. Num.
Meth. Eng.* 10:565–77
Strang, G., Fix, G. J. 1973. *An Analysis of the
Finite Element Method.* Edgewood Cliffs,
NJ: Prentice-Hall
Taylor, C., Hood, P. 1973. *Computers and
Fluids.* 1:73–100
Taylor, G. I., Sharman, C. F. 1928. *Proc. R.
Soc. London Ser. A* 121:194–217
Tong, P. 1974. *Finite Element Methods in
Flow Problems,* ed. J. T. Oden, O. C.
Zienkiewicz, R. H. Gallagher, C. Taylor,
pp. 57–66. Huntsville, Ala. UAH Press
Vainberg, M. M. 1964. *Variational Method
of the Study of Nonlinear Operators.* San
Francisco: Holden-Day
Wang, C. T. 1948. *J. Aeronaut. Sci.* 15:
675–85
Weinbaum, S. 1966. *J. Fluid Mech.* 33:39–63
Zienkiewicz, O. C. 1971. *The Finite Element
Method in Engineering Science.* London:
McGraw-Hill

Ann. Rev. Fluid Mech. 1977. 9 : 447–68
Copyright © 1977 by Annual Reviews Inc. All rights reserved.

AEROACOUSTICS   ✕8110

J. E. Ffowcs Williams
Emmanuel College, Cambridge, England

INTRODUCTION

The 1960s witnessed a phenomenal growth in air transportation at a time of unparalleled awareness of our environment's frailty. New aircraft were propelled by powerful high-specific-thrust turbojet engines, each generating at takeoff more noise than the combined shouting power of the world's population! More economical fan engines were under development, and they made possible realistic plans for the control of aircraft noise, plans that were incorporated into international standards when this decade began. The public expectation of continued progress spurred administrators to plan more stringent environmental controls, and the air transport industry backed by governmental research funding responded by launching the most concentrated program of acoustics research ever conceived. The supersonic transports with their even more powerful high-specific-thrust engines were in advanced production, a development that mesmerized the industry's symbiotic protesters, who focused the public attention on noise control.

Concorde is now in passenger service, and public concern for dwindling resources is deflecting attention from environmental issues, particularly those, such as aircraft noise, that respond to new technology. The research programs that have so far seemed to be ever expanding are likely to level off; we are probably at the peak period for the scientific popularity of the subject. Even so, the rate of scientifically significant progress continues to grow, and being frequently associated with project development, much of the progress is not reported in the scientific press; the inevitable pressure for early exploitation leaves little time for documenting results. This review is intended to include those aspects of the industrial programs that have given rise to fundamental new insight and is intended also to convey the author's impression that the science of aeroacoustics has reached its most exciting stage. New developments are transforming our understanding of the problem and leading to a far more penetrating analysis of noise-control prospects; some of these, such as active control (cf Swinbanks 1973, Kempton 1976a, Moore 1976, Bechert & Pfizenmaier 1975), are full of both scientific challenge and commercial possibilities.

Sources of Sound Cannot Be Uniquely Determined

One of the most fascinating issues has arisen from having to grasp the nettle of positively quantifying the sources of aerodynamic sound. The question, Where are

447

the sources and how strong are they?, which has been asked since the subject began, is now seen to have no one answer; rather, there are an infinite number of equally true but different answers. Only time will tell whether this is a significant observation, but it is certain that source-location theories and experiments will, from now on, be seen more as points of view rather than fact—some points of view being obviously better than others! The source-location debate came to its climax at the 1973 AGARD Specialist Meeting on Noise Mechanisms. Many experimental-source surveys were reported there (*AGARD-CP-131*), all of them plausible, and from the debate on their relative merits arose the full realization that sources are essentially nonunique, an issue later expanded by Ffowcs Williams (1974a,b; 1975). In no way should this observation discourage the use of wave-field analyzers for the manufacture of plausible acoustic images, no more than the knowledge that any one light field may have been generated by a laser-hologram combination discourages the use of the camera. Image interpretation is only probabilistic, and confidence in any one interpretation must rest on additional extraneous facts. Telescope techniques of Grosche's (1974) reflecting-mirror-type, discrete-element arrays (Kinns 1975, Billingsley & Kinns 1976, Flynn & Kinns 1976) and those based on the wave's statistical properties (Fisher, Harper-Bourne & Glegg 1975) are all used in the various experimental programs. These techniques appear likely to prove useful, but it is far too early to draw any firm conclusion on their real significance.

Sound is Modified by Flight

That motion affects aircraft noise is obvious from the level and frequency shifts that occur during flypast. Most of these effects were once thought to arise from the well-understood classical principles described by Lorentz (1904). Convective amplification of Lighthill's (1952, 1954) quadrupoles is an example of this effect and is the sole element in that theory that accounts for any fore and aft bias of the jet's noise field. Until recently, the technical task of predicting the noise of aircraft in flight from measurements made on the ground had seemed straightforward. It is now one of the most puzzling aspects of the problem and the center of active investigation. Important elements of the noise of aircraft powered by large fan engines have turned out to be some ten decibels less in flight than had been predicted from the results of a thorough ground test program. The cause of this change (Cumpsty & Lowrie 1973, Hanson 1974a,b; 1975) is now known to be due to the differing quality of "inlet" flow, the ground-level situation being such that large atmospheric eddies ingested by the engine are spun-up into long sausage-like flow distortions and are noisily "chopped" by the engine fan. The smoother inlet flow in flight is devoid of such distortions, so that the fan operates more quietly. Jet noise, on the other hand, is sometimes amplified by flight—despite the alleviating influence of a reduced relative jet velocity, an influence that had in the past been expected to guarantee a substantial jet-noise reduction. This important (industrial) experimental result has placed greater emphasis on flight simulation techniques and on the theoretical study of moving sources. That study has cast the whole problem in a different light and has served again to reemphasize the need for a proper specification of the source process. The textbook situation in which a moving, small pulsating sphere is erroneously modeled by a convected monopole typifies the linear thinking that has so far clouded the picture. Dowling (1976) has pointed to an essential difference between these two problems, a difference that has its origin in the force induced on the fluid by any convected pulsatile body and also in the fact that even for the smallest of physical sources retarded-time differences for separate elements of the source are not negligible. Convective effects depend on source geometry and are not entirely contained in Doppler factors. The influence of convection seems to be more acute for the physically realizable sources than for the less relevant model problems described in the literature. There is also a tendency for convection to amplify the sound that is radiated perpendicular to the direction of motion, an aspect totally at variance with the classical models. However, the most significant result of Dowling's analysis is probably the conclusion that convection alters the field in a complicated way that is determined by source parameters that are not easily quantified. If this is the case, source motion will have to be studied experimentally, and confident prediction of flight effects on the noise of real aircraft can follow only from flight simulation experiments; such experiments are in their infancy, and the techniques are difficult to handle.

Forward-arc amplification in flight together with the unexpected convective change at 90° to the flight path has been shown by Crighton (1975a) to be an inevitable feature for sound scattered from turbulence by sharp-edged surfaces. This amplification, like the field itself (Crighton 1972a), is highly dependent on the manner in which the Kutta condition is applied to ensure finiteness at the diffracting edge. Some of these results were arrived at through the solution of aeroacoustic-model problems by the method of matched expansions. This method, and indeed all other techniques for embedding diffraction within the aerodynamic sound problem, have really blossomed since the subject was first treated in the *Annual Review of Fluid Mechanics*. It is exciting to follow the subject's development in this respect. There is no proof that the solution obtained through the matched-expansion scheme is, in fact, a solution of the real problem—but it usually is, and most of the serious workers believe that the method has a general validity. It is certainly a powerful generator of results which are later checked by other techniques. Sometimes the checks fail, although so far they have revealed only mistakes of algebra, but these mistakes had given rise to what seemed at the time to be important predictions. For example, there is now a possibility that the enormous acoustic emphasis placed on the Kutta condition is one such erroneous result, but only time will tell whether the error is that of algebra, of principle, or of the check! Howe (1976a) convincingly argues the unreasonableness of free modes of the inner solution, which grow with distance from the scattering edge and are matched there with an acoustic field whose consequent strength increases with the extent of the inner field. Such modes were (in effect) used by Crighton (1972a) and Davis (1975) to fulfill the more stringent Kutta conditions, and from these modes the prediction was made that edge flows are even more noisy when finiteness is insisted upon than they are in the presence of the usual square-root potential singularity (Ffowcs Williams & Hall 1970).

The Changing Emphasis on Fan Noise

The wide-bodied aircraft with its low-specific-thrust, large fan engines has focused attention on fan noise, since there is little reason to expect the low-velocity jet streams to cause much of a mixing-noise problem. Fan-tip speeds are high enough in these engines that their noise is in no way similar to the surface-dipole-dominated sound of low-Mach-number bodies supporting unsteady loads. Supersonically moving blades radiate their own "sonic bangs," and conventional steady aero-dynamics rather than acoustics provides the techniques necessary for analysis (cf Hawkings 1970). In such "peaky" waveforms even the slightest blade-to-blade variations totally destroy the regularity of the signal (cf Ffowcs Williams 1974d), which nominally repeats at the blade passing rate; the noise of supersonic rotors is therefore of lower frequency but extremely rich in harmonics, the waveform being a collection of sharp transients repeating at the rotation period of the fan, typically at some 10^2 cycles per second (cf Sofrin & Pickett 1974). Even subsonic rotors of high tip speed generate a sound that is uninfluenced by unsteady blade loading; the analytical technique most suitable for this problem is Ffowcs Williams & Hawkings' (1969a) generalization of the Lighthill-Curle theory to aerodynamic surfaces in high-speed motion. Dimensional arguments (Ffowcs Williams & Hawkings 1969b) have shown fan noise cannot be dominated by surface effects at high enough frequency, since there is an important direct radiation from the quadrupoles that are acoustically equivalent to the high-speed flow about the blade section. This effect was quantified by Morfey (1971) for a particular cascade geometry. Of the more direct surface sources, the thickness noise term appears by far the most significant at high speed, and from a study of definite model problems guidelines for minimizing the sound of high-speed machinery have been devised by Hawkings & Lowson (1974) and Farassat (1975). The thickness distribution, blade planform, and velocity are all parameters that can be modified to influence sound. Farassat & Brown (1976) and Hanson (1976) have devised schemes to evaluate numerically the linear surface terms of the theory of Ffowcs Williams & Hawkings (1969a) and they show excellent agreement with experiment.

The emphasis today is on aspects of the problem that were hardly foreseen five years ago when the subject was dominated by modal analysis in a cylindrical geometry, the source of sound being clearly rooted in the unsteady blade loading. The modern, large engine ducts are not long enough in comparison with their diameter for the cylindrical coordinate system to be any more relevant than the spherical or free-space geometry. Also, the fan dimensions are so large on the acoustic wavelength scale that geometrical acoustics, or ray theory, can give an extremely important first-order estimate of the acoustic intensity distribution. This technique is much easier to handle than the formal modal expansion, particularly in the high Helmholtz number case when the character of the modal sum bears little resemblance to that of any individual member of the series; this is probably the case for modern, "stubby," large-diameter engine inlets. However, this fact has yet to be applied to engine technology.

Acoustic Liners Absorb Sound—But Do They Destabilize the Flow?

Internal engine surfaces are commonly lined with sound-absorbent materials that comprise a porous surface layer, sometimes backed by resonant acoustic cavities. Their acoustic performance in the presence of flow can be quite different from that in still air, a difference originating in two specific effects over and above the boundary-layer action that modifies liner response in a manner that is not easily modeled analytically. First, the relatively silent small-scale boundary-layer turbulence can be scattered into sound by the inhomogeneous surface, a problem modeled by Ffowcs Williams (1972) and Leppington & Levine (1973) in studies of the diffracting properties of a porous screen. Second, the effectively compliant surface can flutter and admit boundary-layer instabilities of a type not permitted on the rigid surface (Benjamin 1963). These instabilities may or may not be noisy; again it is extremely difficult to pose tractable analytic models of the process, and although some surface liners are known to generate noise and "howl" (Tester 1973, Doak 1972a), it seems that most treated ducts do not display these effects and that surface treatments are in practice very effective (Lowson 1975). Some highly idealized problems where flow is shown to substantially modify the acoustics of a flexible surface were treated by Lovely (1974) and Ffowcs Williams & Lovely (1975); the flexible surface provides new modes for converting the energy of mean flow into sound, and this aspect must be balanced against the benefit of sound absorption in any real situation.

This surface-instability problem, and its possible acoustical significance, has much in common with the role of instabilities in the propagation of sound through a nonuniform, parallel laminar flow. Möhring (1972) demonstrated that even in the absence of exponentially growing modes acoustically important algebraically growing waves can dominate the field, making the acoustic-propagation problem entirely different in the presence of flow from that in still air. Fortunately it seems that these modes cannot be excited, a most interesting fact brought out by Swinbanks (1975). This fact, together with Dowling's (1976) demonstration that "real" sources do not display the characteristics of conveniently handled multipoles, warns that the question of which motions are permitted by physical constraints is becoming just as important as the determination of the basic solutions, only some of which have a physical existence.

Flow-Acoustic Interaction is Important

The interaction of sound and its sources with inhomogeneous turbulent flow is a much more complex and far more important aspect of the aeroacoustic problem; it is at the heart of the jet-noise problem, which is today's most préssing aeroacoustic issue. Even the noise of large turbofan engines is largely determined by jet noise. Such noise threatens the continued existence of the low-bypass-ratio engines that power the supersonic transport and the bulk of today's long-range commercial jet fleet. This noise is now known to be highly sensitive to details of the mean jet flow and to conditions at, or upstream of, the propulsion nozzle. These are aspects quite apart from those usually included in Lighthill's (1954) modeling of the problem, and it is with these aspects that the remainder of this review is concerned.

The Lighthill model allows an unambiguous analysis of the flow's acoustic field in the low-Mach-number regime where acoustic wavelengths are larger than the characteristic dimension of the source flow. Of course there must be an interaction of sound with the mean flow, but such an interaction is not nearly so simply described as would be the case in the high-frequency regime where ray theory applies. This "geometrical acoustics" situation is still devoid of any theory of aerodynamic-sound production. The prediction that refraction, by any shear of the mean flow and gradients of refractive index, forms beams and zones of relative silence is straightforward in this high-frequency regime, but the degree to which it might apply to the jet-noise problem when so much of it is known to lie in the low-frequency domain of the acoustic analogy is open to speculation. Ribner (1964) has consistently argued for at least the qualitative similarity of the two, a view that is now gaining general acceptance and which is strongly supported by experiments on the distribution of sound radiating from a source positioned inside a jet flow (Atvars et al 1966, McGregor et al 1973). Only the fact that the embedded source itself radiates an asymmetric field of a level influenced by flow prevents such tests from giving a definitive measure of refractive effects.

Lush's (1971) jet-noise experiments provided a turning point for the subject. He took care to produce a quiet, smooth nozzle flow and studied the noise field in greater detail and with closer reference to the Lighthill theory than ever before. He found some fascinating differences between theory and experiment, differences that are now the hallmark of the subject. Lush defined two regimes, one in which the Lighthill model was completely verified, and the other in which it seemed irrelevant, the parameter that determined relevance being the ratio of the mixing-layer dimension *in the direction of a sound ray* to the acoustic wavelength. If "rays" have to travel a distance greater than a wavelength through the shear layer, then they can be thought of as rays, and details of these high frequency elements of the jet-noise field can be determined only when proper account is taken of the flow-acoustic interaction. Lush's experiments dealt with cold jets, where velocity is the only "refractive" influence, an influence that is centered on those relatively un-important rays traveling parallel to the jet axis. Variations in the speed of sound have refractive properties at all angles. It was therefore obvious from the significance of the difference between the high- and low-frequency regimes in Lush's experiments and from the fact that few of the most subjectively important frequencies were free of high-frequency effects that the subject of flow-acoustic interaction had become a central issue, especially for the noise of hot-jet flows.

Excess Noise

The other and possibly the currently dominant practical matter concerns the fact that at the lower jet speeds, and even at the higher speeds in flight, jet noise fails to conform with the free-quadrupole model. This model guarantees that at low enough speed the sound of unconstrained mixing flows scales with the eighth power of relative velocity. The fact that real flows do not display this property is proof that the sound originates in aspects of the flow additional to those normally retained in the Lighthill theory and additional to the noise generated by the jet turbulence alone. The difference between the measured noise of practical jets and the lower-level U^8 noise of the "pure" mixing situation has been called "excess" noise. This difference is probably a characteristic of the particular rig or engine on which the noise is measured, and it is extremely unlikely that there is one universal cause of the excess. On the other hand, some of the difference may be fundamental to the pure mixing process, which differs from the Lighthill model because the real flow might well have more than the one characteristic scale of velocity and length assumed in this model and the quadrupole strength may not scale in the assumed strict proportion to the jet's dynamic pressure. It is certain that the search for control of the excess-noise problem has stimulated the subject enormously and given rise to a number of new techniques and extremely elegant analyses. Such studies [e.g. Crighton (1972b) and Howe (1975a)] have begun to include the determination of the unsteady motion that drives the sound as a natural part of the aeroacoustic problem, a problem that is consequently penetrated to a far greater extent than in the past. It is this aspect of the subject that at present makes it so scientifically rewarding, and from the rapid rate at which the subject is progressing one can anticipate that important practical results are not too far away.

Diffraction Radiation

Much of the search for acoustically efficient aerodynamic sources that may be dominating the "excess-noise" field of low-velocity jets has centered on the interaction of turbulence with bounding surfaces that might support powerful diffraction fields (e.g. Ffowcs Williams et al 1972). The principle that passive surfaces can greatly increase the acoustic output of a nearby source can be traced to Stokes (1868) and is at its extreme in the class of problems that Levine (1975) has termed "diffraction radiation." This class consists of fields scattered from static obstacles that are irradiated by the "silent" passage of a steady source; such a source might be supported on an eddy that is convected with the mean flow past the trailing edge of an aerofoil. The steady convection of a nonevolving eddy is absolutely silent, but the transient loads that the eddy applies to the surface as it is swept past can give rise to important sound fields. Indeed, Levine has confirmed the prediction made by Ffowcs Williams & Hall (1970) on the basis of Lighthill's acoustic analogy that sharp-edged large surfaces support much more powerful source terms than is possible in free turbulent flow, and that an eddy's acoustic output is thereby increased by a factor $M^{-3}(l/d)^3$, where M is the small flow Mach number and l and d are the eddy scale and distance from the edge, respectively.

Edge Scattering

The enhancement of aerodynamic sound by edge scattering is potentially so efficient that this type of noise may be an extremely important part of another technological problem—that caused by the flow around large aeroplanes. Some believe (Gibson 1974, Morgan & Hardin 1974) the intrinsic noise of an airframe to be already at a level comparable with that of its propulsive engines, and the author understands from Warren Burggraf of the Boeing Company that much of that noise can be correlated with trailing-edge conditions at the flaps, wings, and fin. The subject has recently been reviewed in an excellent paper by Hardin (1976). The scattering of mixing-layer turbulence by nozzle edges and by the trailing edges of screens, or aerofoils, shielding the noise of jets from below, for both noise control and source diagnostic purposes, is extremely important; and screening tests (e.g. Head & Fisher 1976) have provided some of the most direct experimental evidence of this process. Its theoretical modeling has developed rapidly.

The formal treatment of the problem whereby Lighthill's equation is assumed to specify the field, with the quadrupole characteristics being determined independently of the edge conditions, was enormously advanced by the work of Crighton (1972c) and Crighton & Leppington (1970, 1971). They developed a reciprocal argument whereby the radiation problem was reduced to that of determining the local structure of an incompressible flow in the vicinity of the edge. They showed also that details of the edge were not nearly as important as the fact that there existed an edge to a large screen, and through this result they were able to extend the theory to diffraction of aerodynamic sound by a wedge and to show that the wedge of zero thickness, the semi-infinite screen, was in fact the most efficient scatterer. Surface compliance can greatly inhibit the edge-scattered sound, and Crighton and Leppington illustrated the effect with a specific example. This class of problems was widened when Crighton (1972c) and Cannell (1975) determined the sound generated by a turbulent eddy located above an elastic half plane. The eddy drives an energetic elastic wave in the surface, which propagates (silently) toward the edge, where it is partially scattered as sound. The acoustic power of the eddy is increased over its free-space value by a factor ε^2 times the ratio of the acoustic speed to the speed of "in vacuo" surface waves, ε being the small fluid-loading parameter; this factor seems to be of the order of 18 dB at low frequencies (100 Hz).

A direct check that the Lighthill formulation provides the correct estimate of these edge-scattered fields is afforded by Crighton's (1972b) evaluation (by the method of matched asymptotic expansions) of the sound generated by a vortex filament negotiating the edge of a screen. The agreement with "acoustic analogy" estimates is complete and furthermore provides the constants of proportionality. Provided that the edge singularity is permitted, there seems to be little doubt that the screen scatters sound in a well-understood manner.

Some form of Kutta condition may be in order, however, and the analysis might have to be reworked to ensure finiteness of the velocity field in the vicinity of the edge. In such a case, vortex shedding must be admitted, and the problem becomes much more difficult. Model problems that postulate rather unphysical distributions of the shed vorticity indicate that finiteness at the edge is permitted only at the expense of an even more powerful edge-scattered acoustic field (Crighton 1972a, Jones 1972). Great doubt has been cast on the relevance of these models by Howe's (1976a) recent work that treats the vortex-shedding problem in much more detail; it (and Crighton's model) includes the possibility that the screen separates two different streams that mix downstream across a vortex sheet. Howe's conclusions are that the Kutta condition, if met, causes vortex shedding that relieves the scattering process, the sound generated by the shed vorticity being exactly opposite, and cancelling, that caused by scattering. Whether viscous effects are rapid enough to bring about unsteady vortex shedding at acoustically important frequencies is very much a matter of debate, but it seems probable now that the simple-scattering theories must be regarded as upper-bound estimates on edge-noise fields. Well-controlled experiments are badly needed; they alone can resolve the issue of which theoretical model is the most relevant to the various regimes of flow.

Nozzle Scattering

The sharp edges in the jet problem are provided by the nozzle termination of a pipe. This is a situation in which, at first sight, relatively powerful monopole sources might be expected whenever the nozzle mass flow is unsteady. Even those jet flows driven from a quiescent upstream plenum must fluctuate somewhat in response to the unsteady downstream pressure that results from turbulent mixing. This problem has been examined in some detail. Although nozzle response is inevitably the controlling feature for jet noise at low enough velocity, the nozzle-scattering process is not nearly as acoustically efficient as might be anticipated from either edge-scattering models or from the expectation of a monopole inducing unsteady mass flow. Jacques (1975) modeled this problem by examining the unsteady linear motion of a doubly infinite jet flow bounded by a vortex sheet, the displacement of which was maintained at zero at a downstream position corresponding to the nozzle lip. Jacques' basic assumption was that the real semi-infinite jet was equivalent to the downstream half of this continuous flow. Jet flows support instability waves, and Jacques determined from their study the unsteady mass flow at the nozzle exit station. He equated this to the value determined from a second problem, the motion permitted in a narrow (on an acoustic scale) pipe terminated by an unsteady nozzle flow. This matching produced the result that the acoustic monopole was essentially weak, and although it is the most important feature in determining the sound generated by the unsteady jet, the radiation intensity scales in proportion to the sixth power of Mach number and not the fourth power usually associated with monopoles. The Jacques source term obviously disappears altogether as the jet becomes choked. If the unsteadiness is provoked by sound, then the wave energy generated by this effect is smaller than the acoustic energy incident on the nozzle plane; the jet producing nozzle flow has a small scattering cross section.

These conclusions are consistent with two other investigations in which the problem was treated more formally. In the first, Leppington (1972) solved (by Lighthill's acoustic analogy) the diffraction problem of a circular pipe irradiated by a quadrupole source, the equivalent of the nozzle-less jet. Again Leppington confirmed that the unsteady volume flux excited in the pipe acts as an acoustic monopole, but one so weak that its effect is of the same order as that of the unsteady thrust; both terms generate a scattered field that scales in proportion to the sixth power of Mach number, the small parameter in the problem. This result is again consistent with the specific model problem examined by Cannell & Ffowcs Williams (1973). They used a singular-perturbation technique to determine the sound generated when line vortices exhaust from a two-dimensional duct. The field of any symmetrically excited duct is a dipole generating (in three dimensions) a "sixth power law" acoustic field. On the other hand, from this problem it can be seen that asymmetric nozzle fields remain subject to the more potent edge-scattering sound, which scales in proportion to the fifth power of velocity.

Nozzle-scattering sources are thus well modeled theoretically, and the models indicate their superior acoustic performance at the lower flow speeds. No direct experimental verification of these sources is yet available however, and only experiment can establish at precisely what jet condition nozzle interference effects dominate. It seems likely on purely theoretical grounds that these conditions correspond to important aspects of the excess-noise problem.

The Acceleration of Inhomogeneous Flow is Noisy

The search for the excess-noise source has introduced another source mechanism, one that had been studied twenty years ago in connection with the instability of rocket combustion and nozzle systems (Tsien 1952, Crocco 1953). Flow accelerating through a nozzle, if contaminated by density inhomogeneities, will tend to an unsteadiness acoustically equivalent, at first sight, to a potent monopole source of sound. The density, or entropy, fluctuations are coupled to an acoustic mode, and the degree of coupling has been examined in several interesting model situations. Candel (1972) was the first to examine the issue, and he chose to evaluate the coupling of weak inhomogeneities, traveling through a slowly varying, "parallel" flow nozzle, to one-dimensional acoustic modes; the powerful coupling was experimentally verified by Zukoski (1975), and every indication was then provided that the uniformity of a jet flow is an important parameter in determining its noisiness.

Cumpsty & Marble (1974) examined another aspect of the same mechanism, an aspect particularly relevant to the noise of turbines driven by inhomogeneous flow. They considered the coupling between a wavelike pattern of entropy variation incident onto an actuator disc representation of a turbine stage, and a sound wave scattered both upstream and downstream. The coupling is very effective, and Cumpsty (1974) showed that this mechanism might well account for the bulk of the excess noise measured on three different commercial engines. A similar analysis was conducted by Pickett (1974), who in addition to the Cumpsty model included features of the propagation through a gas turbine nozzle; again he concluded that this mechanism is likely to be an important part of real excess-noise problems.

The linear approximation of these theories and the Fourier decomposition that this permitted was bypassed by Ffowcs Williams & Howe (1975) and Howe (1975b) at the expense of confining the analysis to low-Mach-number flows. They treated the problem of determining what sound was generated when a "slug" of different density fluid was accelerated with the mean jet flow through a nozzle contraction, and also that associated with the passage of a small "pellet" of inhomogeneous fluid through an accelerating flow. In each case the sound is generated by the unsteady component of "thrust," scales with the sixth power of the characteristic (low) Mach number, and is acoustically more efficient than that of the jet-mixing process. The mechanism was examined experimentally by Whitfield (1975), who was able to confirm the dominance of this source process, at least at low Mach number, and to devise means of dramatically reducing its sound by careful nozzle design.

Combustion Noise

Turbojets are fueled with kerosene, whose flame front in air advances at a rate of some tens of meters per second. The flow through the combustion chamber and afterburner section proceeds at some 100 meters per second. The flame must therefore be artificially stabilized if it is not to be blown out. This is achieved by turbulent eddies, essentially unsteady and inevitably the cause of an unsteady combustion process, which adds heat unsteadily and causes noise. The degree to which this combustion noise is a feature of the aircraft-noise problem is still under debate.

Unsteady expansion is acoustically equivalent to a monopole source, and Smith & Kilham (1963) provided convincing experimental confirmation of this point. It was further documented in a beautiful series of experiments by Thomas & Williams (1966), who also argued that the maximum fraction of the energy of combustion that could be radiated as sound was 10^{-6}. The combustion-noise efficiency calculation cannot be regarded as conclusive, since the power radiated is a function of the flow geometry, but if this really is an upper bound, then combustion noise is hardly the dominant issue in the high-speed jet problem, for this noise is known to be more intense by a factor of 10 to 10^2. On the other hand, combustion could very well be an important feature of the excess noise, for it is certainly capable of producing more noise (according to the experiments of Abdelhamid et al (1973), some 15 dB more) than the subsonic jet mixing process, which converts into sound only some $10^{-4} M^5$ of the thermal energy supplied in the engine (Lighthill 1963), M being the jet Mach number.

The Thomas and Williams view of the combustion noise process was given further support in an impressive experimental program reported by Hurle et al (1968), but Strahle (1972) argued that other features such as flow refraction and convection must also be accounted for. Also the confinement of a flame increases its acoustic output; according to Strahle (1975) up to 10^{-5} of the energy of confined combustion can be radiated as sound. This process is then a leading contender for the excess noise title, and Mathews & Rekos (1976) have concluded that combustion accounts for virtually all of an engine's internal noise. Their studies indicate that this noise is determined by the rate at which the heat addition during combustion changes and that is precisely what the comprehensive experiments of Shivashankara et al (1974) show to be the hallmark of combustion noise, and the Chiu & Summerfield (1974) theory also supports this view.

An important contribution was made by Hassan (1974) who introduced a discussion of the rates at which the various chemical species of reaction are interchanged. He showed a good correlation of theoretical and experimental scaling laws, although his detailed argument seems to contain a possible logical flaw in ignoring the inevitable variability of one mass fraction when another is caused to change. This subject is advancing rapidly and one might now argue with Crighton's (1975b) observation that even though Strahle (1971) and Ffowcs Williams (1969a) have both arrived at the same expression for the combustion-noise source term in Lighthill's theory, this theory remains somewhat rudimentary.

Diffusive Sources of Sound

Lighthill (1954) emphasized that wherever the density change was not related to pressure in exactly the same way that it is in the distant linear sound field, then the difference would act as a source of sound, scattering either an acoustic or hydro-dynamic pressure field in the way described by Rayleigh (1896). This difference could be due to the speed of sound being locally different from the ambient, an effect Lighthill showed to account for an eighth-power-law radiation, which is basically less efficient than that of the turbulence in a hot-jet flow. But the difference could also be caused by nonisentropic elements of the flow. They account for a basically more efficient radiation than that of the isentropic mixing process (if such a con-tradiction can be considered!) and as such have again been the subject of study in connection with the excess-noise problem. We consider in another section the sound of turbulent nonisentropic flow in the absence of diffusion (!) and examine first the way in which diffusion might account for efficient sound production.

It is widely thought that diffusion is acoustically equivalent to efficient monopole radiation, accounting for a sound in which the intensity scales with the fourth power of flow Mach number. Tanna, Fisher & Dean (1973) and Lush & Fisher (1973) rationalised the apparent fourth-power variation in the noise of hot jets on this theoretical notion, and the idea receives strong support from Crighton's (1975b) emphasis on the inevitability of entropy production and from a matched-asymptotic-expansion treatment of the problem due to Obermeier (1975). Both of these, how-ever, seem to give a false support for linking Tanna et al's empirical data collapse with diffusive sources, the support in Obermeier's case being due to a simple mis-interpretation of his own final formula. Crighton's argument fails because (notwithstanding our later observations on Howe's nonlinear analogy) entropy is of itself not a particularly significant parameter as far as the sources of sound are concerned, and its inevitable increase has no obvious acoustic consequence. This fact can be illustrated conclusively by noting that the adiabatic mixing of two perfect gases of initially different temperature does not bring about any change of pressure at constant volume or any change of volume at constant pressure, pro-vided only that the ratio of specific heats is constant. Volumetric change is the very essence of an acoustic monopole; both are zero.

Diffusive aerodynamic sources have been examined in detail by Morfey (1976) and Kempton (1976b), who showed that the actual "entropy" monopole contained a density factor, the variation of which was ignored by Crighton, but this variation exactly counters that of entropy production to bring about a vanishing of the monopole term. The only monopole is that resulting from the small variability in the ratio of specific heats and is weak for this reason. Both Morfey's and Kempton's analyses point to the probable insignificance of this monopole as far as jet noise is concerned ... but it may be excess noise.

Of course, unsteady diffusion generates sound, and whenever there is an external source of heat, or matter, this sound originates in a highly efficient monopole source. Landau & Liftshitz (1959) determined the sound driven by diffusion of heat from a solid boundary whose temperature is caused to oscillate, and Kempton has worked several examples of this type. Howe (1975a) showed how this kind of sound-generation problem is conveniently handled through the entropy-source term. In fact he showed that without entropy variation there would be no source of sound in potential flow; entropy changes do have a fundamental role, but they do not always produce a monopole. We return to this point later.

Kempton (1976b) has also worked on several model problems of a different type to determine the characteristic features of diffusive acoustic sources. The most significant example concerns the inviscid thermal mixing of two slightly unsteady, different temperature streams, which are initially separated by a semi-infinite insulating half plane. The unsteady heat transfer from one stream to the other induces a dipole radiation. The dipole degenerates into a quadrupole in the axially symmetric jet case, a quadrupole whose mean square acoustic pressure, at distance r from the pipe exit, scales as

$$\overline{p^2} \sim \frac{a^2}{r^2} \rho_0^2 C_0^4 \left(\frac{\Delta T}{T}\right)^2 M^8 R^{-2} S^4,$$

where ρ_0 and C_0 are the mean density and speed of sound in the far field respectively, $\Delta T/T$ the fractional temperature difference between the two streams, $C_0 M$ the jet speed, and R the Peclet number, which is effectively the "Reynolds" number based on jet velocity, diameter, and thermal conductivity. S is the Strouhal number, and a the pipe radius. This dimensional dependence is derived from the usual practice of setting the unsteady velocity equal to a constant fraction of the mean jet speed. Even though this field scales on the sixth power of jet velocity, it does so only through the appearance of the very small factor R^{-2}; such diffusive sources are likely to be negligible in engineering practice. Kempton argued that this result for the particular model problem (in which the ratio of specific heats is constant) should be amplified by the usually large factor $M^{-2} R S^{-1}$ in order to properly represent this source in turbulent flow where the symmetry is destroyed and sharp interfaces are continually regenerated by the mixing-layer eddies; this implies a fifth-power velocity dependence, the same as Crighton's calculation would have provided if proper recognition had been given to the vanishing of the monopole term.

The most probable location of a significant diffusive source is in the internal flow of an engine where unsteady heat transfer between thermally inhomogeneous flow and turbine blading of high heat capacity induces a monopole. This sound scales on the cube of jet velocity (in intensity), although even here Kempton has argued that the coefficient is so small as to make the source negligible. A treatment accounting for real turbulence is difficult and might give a different result. Nonetheless, at low enough Mach number, such diffusive sources must be dominant, and it is possible (the accurate computation of magnitude being extremely difficult) that they already form an important part of the excess-noise problem on engines, a noise that is sometimes referred to as "core noise."

Howe's Reformulation of the Governing Equations

Lighthill's theory, whereby the problem of determining the sound radiated from an aerodynamic flow is reduced to an exactly analogous problem in classical acoustics, has long been criticised for its inability to display explicitly the inevitable subsequent interaction of sound with its parent flow. Theories designed to overcome this difficulty by postulating different linear analogies (e.g. Phillips 1960, Doak 1973, Lilley 1974) gain what advantage they have at the cost of being either approximate, noncausal, or analytically intractable. Lighthill's (1952) analogy has the enormous advantage of being an exact reformulation of the equations of motion in terms of a well-documented and understood linear operator for which uniqueness properties guarantee that the inhomogeneities of Lighthill's equation qualify as sources of sound; it is certain that no sound could exist without them. Lighthill's theory is extremely difficult to apply when extensive flow-acoustic interaction is expected, however; the obviously artificial prediction of singular forward scattering by a line vortex is one example (O'Shea 1971). Howe (1975a) has evolved a different, exact, and causal theory capable of handling many of these problems unambiguously, and as such it deserves extremely careful study; Howe's own application of this new approach to determine both the flow and the consequent acoustic radiation in the Flute might well be called the first deductive theory of aerodynamic sound. This method joins that of matched asymptotic expansions and Lighthill's master-piece in being the only procedures able so far to permit physically self-consistent aerodynamic model problems to be solved!

Howe arranges the exact equations of continuity and momentum for a perfect gas into a form where B, the specific stagnation enthalpy, appears as a field driven by variations of entropy S and vorticity ω, $Q(S, \omega)$, say.

$$H(B) = Q$$

is a symbolic representation of Howe's fundamental equation (4.14), H being a second-order nonlinear differential operator with variable coefficients. $H(B) = 0$ is in a class of equations with unique solutions, so that B equals a constant, or absolute silence is guaranteed by the vanishing of Q, a step that promotes the view that Q is the "source" of the B fluctuations that comprise the acoustic field at large distances. Unbounded irrotational isentropic flows complying with the radiation condition are silent, but since it is difficult to see how they could even be unsteady, this is hardly a significant statement! The clear localisation of the source to vortical regions of isentropic flows is extremely significant; it formalizes a step initially taken by Powell (1964), whose unproven, but attractive, conclusions are now seen to be essentially correct. The vortex-scattering problem that O'Shea had such difficulty in handling through the Lighthill analogy becomes trivially easy in this new framework, and Crighton's (1972b) solution for the sound generated by a vortex negotiating the edge of a semi-infinite screen is accomplished with far greater ease and with a more explicit description of the sound-production process.

It is perhaps surprising that Howe's nonlinear operator can be handled at all, but it conforms to a reverse-flow reciprocal theorem (Howe 1975b) at small mean flow Mach number, a fact that permits the radiation problem to be reduced to that of determining local features of potential flow near a singularity. Such problems are relatively straightforward, and Howe's (1975a) paper contains a wide variety of solved problems that leave the reader in no doubt as to the new theory's versatility and power.

The emphasis that has been placed on the entropy source terms in Howe's theory will probably change as the subject is developed, for though it is certain that entropy can act as a source-identifying label, there are the examples already referred to in which these terms integrate instantaneously to zero; the source is then of higher order. As a matter of fact the same is true of Powell's vorticity term that appears as a dipole distribution, and one would obtain the wrong impression of a flow's acoustic output if its inevitable cancelling neighboring source distribution went unrecognised. The examples worked out by Howe, other than those associated with an unsteady heat supply, all involve density inhomogeneities in accelerating flow, and their description in mechanical rather than thermodynamic terms will be more appealing to many. Such descriptions are straightforward in the more developed forms of Lighthill's analogy and are discussed now in reference to the noise of hot-jet flows.

The Noise of Hot Jets at Low Mach Number

There is now ample experimental evidence on jet-flow rigs uncontaminated by internal noise sources that the noise produced by the mixing of low Mach number jet flows with a different density environment is more efficient than that produced by the mixing of equal density streams. The first definitive experiments were reported by Hoch et al (1973), and the experimental range was widened by Tanna, Fisher & Dean (1973). The variation of this noise with jet velocity is more gradual than the eighth-power dependence of Lighthill's unsteady Reynolds-stress quadrupoles, and a clear reason for this was provided by Morfey (1973). He examined the sound generated by the isotropic elements of Lighthill's stress tensor, those caused by the failure of pressure fluctuations in the source region to be exactly balanced by C_0^2 times the density variations. Morfey's analysis is for nondiffusive flows at low Mach number where the variation in density of any one fluid particle is negligible. The source mechanism can be clearly visualized if one considers a "blob" of fluid of different density from its environment being accelerated in the turbulent mixing flow, an acceleration requiring a different force than that needed to accelerate the environmental fluid; this difference is acoustically equivalent to a dipole. There is little possibility of the dipole strength integrating to zero, since this would require an improbably uniform correlation of the density inhomogeneity and acceleration field. This view of the source was described by Ffowcs Williams [1969b, equation (12), p. 204], but it was Morfey who first recognized its significance to the hot-jet-noise problem and pointed out that hot jets would consequently be relatively noisy at low Mach number. This too is an aspect of "excess noise."

Long-Wavelength Flow-Acoustic Interaction

The acoustic analogy is certainly valid whenever the source region is small on the wavelength scale, and therefore, one should expect to elicit from the analogy whatever the consequences of mean flow/acoustic interaction might be in that limit. In fact this feature has been slow to emerge, the subject being thoroughly confused by the apparently logical step of distinguishing between the linear and quadratic terms of Lighthill's stress tensor. Lighthill (1954) was the first to do this, pointing out that in the presence of high mean velocity shear $\partial(\rho u_i u_j)/\partial t$ could be approximated by $p(\partial U_i/\partial y_j)$, a linear term that seemed to illustrate the "amplifying effect of shear." From then until now many have followed Lighthill's lead and examined separately the so-called "turbulence-turbulence" term (quadratic in the unsteady quantity) and the "mean flow-turbulence" source term that is linear. Ribner (1964) described the same type of source decomposition, calling them the "self-noise" and "shear noise" components. The distinction is confusing. Ribner's shear term has a double-lobed field beamed onto the jet axis at a 45° angle to the peak in Lighthill's four-lobed pattern. One at least must be wrong!

In an ingenious exact rearrangement of the Lighthill quadrupoles, Michalke & Fuchs (1975) derived "shear amplified" terms in which an effect of source convection at the instantaneous velocity of a fluid particle is apparent. For large enough shear, they argued that these linear terms are dominant; it is this reviewer's opinion that even though that view is extremely reasonable at first sight, it is in fact untenable in the long run. Extensive linear terms in the analogy must indicate that the basic field conforms to some differential equation other than the acoustic-wave equation, and as such they are more likely an indication of refractive, or propagation, effects than of sound-producing elements.

A distinct and tractable model problem was needed to put the issue in its proper perspective.

Berman (1974) was the first to produce new evidence, which he did by examining the intensity of sound radiated from a source embedded in a uniform jet flow bounded by a vortex sheet. Goldstein (1975) extended this type of analysis to cover jet flows with arbitrary velocity profile, requiring only that the flow be "compact" on the wavelength scale. They both deduced that the embedding greatly amplifies the effects that Lighthill attributed to source convection, the intensity of low-frequency-sound scaling in proportion to the inverse tenth power of the Doppler factor based on mean flow. This conclusion is consistent with Mani's (1972) deduction that convective amplification was underestimated by simple convected quadrupole models, a view further supported by Mani's (1974) investigation of hot-jet flows through a vortex-sheet analysis. None of these studies, in which the mean shear is *extremely* large emphasized the features previously associated with a shear-amplification term. In fact, Ffowcs Williams (1974c) argued that such terms were entirely spurious, leading to secular elements responsible for changing the character of the radiation; they arise only because the problem is viewed from an awkward angle!

The large "mean gradient" terms are avoided if sources are referred to a Lagrangian coordinate system, in which most of the features identified by Berman and Goldstein are readily apparent from the acoustic analogy. The Lagrangian description avoids embarrassingly high rates of change as high vorticity particles drift past the observation point. Ffowcs Williams & Hawkings (1969), or equivalently equation (17) of Ffowcs Williams (1974c), described the far-field sound exactly as follows:

$$\rho(\mathbf{x}, t) = \frac{1}{4\pi x^3} \frac{x_i x_j}{C_0^4} \int \left[\rho^* \frac{D}{Dt} \left\{ \frac{1}{(1-M_r)} \frac{D}{Dt} \left(\frac{T_{ij}}{\rho(1-M_r)} \right) \right\} \right] \frac{d^3\eta}{(1-M_r)}.$$

Here $\rho(\mathbf{x}, t)$ is the density fluctuation heard at time t at a distant point $\mathbf{x}$; C_0 is the speed of sound at $\mathbf{x}$; ρ^* is the density of a fluid particle at the time when the Lagrangian coordinate system, η, was coincident with the fixed axes; and the brackets indicate that the integration is to be conducted at retarded time. T_{ij} is Lighthill's stress tensor,

$$T_{ij} = \rho u_i u_j + p_{ij} - C_0^2(\rho - \rho_0)\delta_{ij}.$$

Fluid particles move instantaneously with a velocity u_i that has a component $C_0 M_r$ in the direction of $\mathbf{x}$. The operator D/Dt is the Lagrangian rate-of-change operator and acts on all elements of the integrand.

When the differential operators in this equation are applied to the unsteady Doppler factors, $(1 - M_r)$, the resulting term is the most sensitive to convective acceleration; it scales in mean square in proportion to $(1 - M_r)^{-10}$, which is the Berman-Goldstein result. From this view it is clear that mean gradients do not of themselves produce powerful source terms, but only in as much as particles traveling through such gradients suffer a rapid change in property. It is easy to imagine how such changes are dramatic enough to act as centers of potent sound waves. From this Lagrangian view one can also see clearly how elements of the stress tensor signifying a density inhomogeneity can act as powerful acoustic emitters. In fact Morfey's result is an immediate consequence of the equation (cf Ffowcs Williams & Howe 1975) that provides further the dependence of this sound on source motion, and shows it to be beamed along the jet axis by interaction with the mean flow at low frequencies.

Flow-Acoustic–Interaction Models

Not only does sound interfere with the flow that produced it, but it can be shielded and otherwise modified by the presence of nearby secondary flows. The optimization of these effects offers practical scope for jet-noise control, a potential that is already evident in the experimental results reported by Morris, Richarz & Ribner (1973) and Hoch & Hawkins (1974). The effective theoretical modeling of this flow-acoustic process has aroused much debate and is one of today's most important issues.

The Phillips (1960) theory whereby the sound is modeled through a convective-wave equation has been greatly advanced by Pao (1973), who determined from the Phillips theory both familiar properties and new predictions for high-Mach-number sound. Many (e.g. Doak 1972b) involved in the debate on this problem point to the fact that the basic wave operator isolated by Phillips is not that governing weak perturbations about a laminar flow and, as such, fails to account properly for refractive and convective effects. It is far from certain, however, that this is a weakness of the Phillips theory; it may be an inspired reordering of the equation, for through it he avoided completely the possibility that the presence of eigenfunctions of the linear operator destroy the freedom to identify the inhomogeneity of the equation as the source.

If the flow acoustic interaction problem is forced into a form where the linear equation describes accurately the behavior of small perturbations of an inviscid laminar flow, then the compressible form of the Rayleigh equation is inevitable and unforced exponentially growing, noncausal modes must be anticipated. It has become customary to refer to this reordered form of the equation after Lilley (1974), who first advocated its use in this context. It is known as the Lilley equation, and there has been considerable effort applied to determine characteristics of its solutions. Contributing most to this end, Tester & Morfey (1975) have examined the motions induced by concentrated model sources in a weakly perturbed, laminar axisymmetric jet flow; they deliberately suppress the instability waves that would actually dominate any physical realization of that model. The analysis is largely numerical, and from it are deduced several of the features recognizable in experimental jet-noise fields, particularly those that fall under the general heading of "refraction."

Many of these experimental features are also consistent with the results of another computational procedure that involves the deliberate suppression of instabilities. Mani (1972, 1974), Balsa (1975), Berman (1974), Graham & Graham (1974), and Dash (1976) have all analyzed aspects of the jet-noise problem on the basis of vortex-sheet flow models, with some of the vortex sheets having a definite thickness. It seems that the computational results are relatively insensitive to shear-layer details (Knott 1974, p. 54), indicating the variation to be as little as 1 or 2 dB. If this is indeed so, then the vortex-sheet models have an enormous advantage over the Lilley equation in admitting to analytical solutions; they are also known to be exactly analogous to the real problem as far as the distant sound field is concerned (Ffowcs Williams 1974c).

Are Large Eddies Part of the Noise Problem?

A discussion of large eddies and their role in the jet-noise problem has been deliberately left until last, for it is far too early to put the issue into its proper perspective, and scope for error is enormous. The large eddies may yet turn out to be an acoustically irrelevant local feature of jet development, attracting attention only because they can be seen clearly. At the other extreme, however, they are more likely the very essence of the source process, and the experimental evidence that they are controllable might well be the beginning of a completely new principle of noise suppression. To admit their importance is to require some considerable rethinking of the ideas already discussed where the source dynamics are decoupled from the wave field, which is then sensibly studied separately. Turbulence structure is influenced by sound, however, and Crow & Champagne's (1971) experimental evidence has now received ample confirmation from later tests. The most substantial survey is due to Moore (1976), who showed that only a small-percentage-fraction variation of the mean flow is needed to lock the large eddy structure onto the driving frequency and phase. Moore's clear visualization of these eddies that are produced in a controllable sequence cannot help but give observers the feeling that the whole subject is at a turning point. This feeling is reinforced by the results produced independently by Bechert & Pfizenmaier (1975) and Moore (1976) (which at first seem incredible) that the broad-band noise of a jet can be adjusted over a wide range by weak harmonic acoustic excitation at the nozzle.

The proper description of flow-acoustic interaction must account for acoustic/ turbulence interference in addition to and separate from mean flow effects, and sometimes the turbulence interference is the more important. Howe (1976b) found this to be true when analyzing the experimental shielding results of Norum (1973), as did Dowling (1974) when modeling the shear-layer-scattering problem. Certainly the large eddies can be visualized clearly (Merle & Fragassi 1971, Rockwell 1972), and their coupling to the sound field is reaffirmed by Poldervaart's (1974a,b) splendid films. The idea that large eddies form an important part of the jet-noise problem (e.g. Bishop et al 1971) has led to wider interest in modeling them, and such modeling is very diverse.

Damms & Küchemann (1972) developed a model of Prandtl's where eddies are formed by the "rolling up" of vortex sheets. This process is treated numerically by Davies et al (1975), Hardin (1974), and Acton (1976), whose calculations show features in close agreement with Moore's experimental results. Laufer et al (1974) treated large eddies as essentially deterministic events and pointed to eddy inter- action as a powerful source of sound. At the other extreme, the large eddies are regarded as the immediate by-products of the shear-layer instability (Landahl 1967) that are highly coupled to the sound wave (e.g. Jones 1975, Michalke 1969) and show that details of scale, growth rate, and phase velocity computable from models of weakly disturbed laminar flows (Michalke 1971). To include effects of the shear-layer growth, Liu (1974), Morris (1971), Ko et al (1970), and Chan (1974) used integral formulations involving a turbulence-energy equation, while Crighton & Gaster (1976) solved the problem in fine detail through a stability analysis of the "slowly varying" layer. Again many of the predictions of these small-amplitude theories are, perhaps surprisingly, remarkably in accord with experiment.

Progress on this front has been extremely rapid, and the prospect of its being central to the noise-producing process puts the subject now at perhaps its most exciting stage. So far these studies are confined to jets of simple geometry, and the cynic might well observe that they are therefore technologically irrelevant since the round parallel jet is patently too noisy to be a lasting feature of the aeronautical scene! Progress with simple geometries, however, is the natural prerequisite for understanding and exploiting the noise-control potential of the quieter flows of complex geometry. So far their development rests on strictly ad hoc procedures; they might soon be amenable to rational thought!

Literature Cited

Abdelhamid, A. N., Harrje, D. T., Plett, E. G., Summerfield, M. 1973. *AIAA Paper 73-189*
Acton, E. 1976. The modeling of large eddies in a two-dimensional shear layer. *J. Fluid Mech.* 76:561
Atvars, J., Schubert, L. K., Grande, E., Ribner, H. S. 1966. *NASA Contract. Rep. CR-494*
Balsa, T. F. 1975. *J. Fluid Mech.* 70:17
Bechert, D., Pfizenmaier, E. 1975. On the amplification of broad band jet noise by pure tone excitation. *J. Sound Vib.* 43:581
Benjamin, T. B. 1963. *J. Fluid Mech.* 16:436
Berman, C. H. 1974. *AIAA Paper 74-2*
Billingsley, J., Kinns, R. 1976. The acoustic telescope. *J. Sound Vib.* Vol. 48, No. 4: In press
Bishop, K. A., Ffowcs Williams, J. E., Smith, W. 1971. *J. Fluid Mech.* 50:21
Candel, S. M. 1972. PhD thesis. Calif. Inst.

Tech., Pasadena

Cannell, P. C. 1975. *Proc. R. Soc. London Ser. A* 347:213

Cannell, P., Ffowcs Williams, J. E. 1973. *J. Fluid Mech.* 58:65

Chan, Y. Y. 1974. *Phys. Fluids* 17:9, 1667

Chiu, H. H., Summerfield, M. 1974. *Acta Astronaut.* 1, 7–8:967–84

Crighton, D. G. 1972a. *Proc. R. Soc. London Ser. A* 330:185

Crighton, D. G. 1972b. *J. Fluid Mech.* 51:357

Crighton, D. G. 1972c. *J. Sound Vib.* 22:25

Crighton, D. G. 1975a. *J. Fluid Mech.* 72:209

Crighton, D. G. 1975b. *Prog. Aerosp. Sci.* 16(1):31

Crighton, D. G., Gaster, M. 1976. Stability of slowly diverging jet flow. *J. Fluid Mech.* 77:397

Crighton, D. G., Leppington, F. G. 1970. *J. Fluid Mech.* 43:721

Crighton, D. G., Leppington, F. G. 1971. *J. Fluid Mech.* 46:577

Crocco, L. 1953. *Aerotechnica* 33:46

Crow, S. C., Champagne, F. H. 1971. *J. Fluid Mech.* 48:547

Cumpsty, N. A. 1974. Excess noise from gas turbine exhausts. Cambridge Univ. Eng. Dep.

Cumpsty, N. A., Lowrie, B. W. 1973. *ASME Pap. 73-WA/GT-4*

Cumpsty, N. A., Marble, F. E. 1974. The generation of noise by the fluctuations in gas temperature into a turbine. Cambridge Univ. Eng. Dep.

Damms, S., Küchemann, D. 1972. *R. Aircr. Establ. Tech. Rep. 72139*; *Phil. Trans. R. Soc. London* 339:4451

Dash, R. 1976. Radiation from a source near a jet. *J. Sound Vib.* Vol. 49, No. 3: In press

Davies, P. O. A. L., Hardin, J. C., Edwards, A. V. J., Mason, J. P. 1975. *AIAA Paper 75-441*

Davis, S. S. 1975. *AIAA J.* 13:375

Doak, P. E. 1972a. *I.S.V.R. Tech. Rep. 55*

Doak, P. E. 1972b. *J. Sound Vib.* 25:263

Doak, P. E. 1973. *J. Sound Vib.* 26:91

Dowling, A. 1974. *R. Aircr. Establ. Tech. Rep. 74001*; *Rep. Memo. A.R.C.* No. 3770

Dowling, A. 1976. Convective amplification of real simple sources. *J. Fluid Mech.* 74:529

Farassat, F. 1975. *NASA Tech. Rep. TR-R-451*

Farassat, F., Brown, T. J. 1976. *AIAA Pap. 76-563*

Ffowcs Williams, J. E. 1969a. *AFOSR-UTIAS Symp.*, ed. H. S. Ribner, Univ. Toronto

Ffowcs Williams, J. E. 1969b. *Ann. Rev. Fluid Mech.* 1:197

Ffowcs Williams, J. E. 1972. *J. Fluid Mech.* 51:737

Ffowcs Williams, J. E. 1974a. *Proceedings, Eighth International Congress on Acoustics*, Vol. 3. Trowbridge, Engl.: Goldcrest

Ffowcs Williams, J. E. 1974b. *AGARD-CP-131. Pap. 1*

Ffowcs Williams, J. E. 1974c. *J. Fluid Mech.* 66:791

Ffowcs Williams, J. E. 1974d. *NASA Spec. Publ. SP-304*, pp. 425–35

Ffowcs Williams, J. E. 1975. *Aerodynamic Noise Notebook*. AIAA

Ffowcs Williams, J. E., Hall, L. H. 1970. *J. Fluid Mech.* 40:657

Ffowcs Williams, J. E., Hawkings, D. L. 1969a. *Philos. Trans. R. Soc. London Ser. A* 1151:264, 321

Ffowcs Williams, J. E., Hawkings, D. L. 1969b. *J. Sound Vib.* 10:11

Ffowcs Williams, J. E., Howe, M. S. 1975. *J. Fluid Mech.* 70:605

Ffowcs Williams, J. E., Leppington, F. G., Crighton, D. G., Levine, H. 1972. *Aeronaut. Res. Counc. CP 1195*

Ffowcs Williams, J. E., Lovely, D. J. 1975. *J. Fluid Mech.* 71:689

Fisher, M. J., Harper-Bourne, M., Glegg, S. A. L. 1975. Jet noise source location using polar correlation. *I.S.V.R. Contract. Rep. 75-18*

Flynn, O. E., Kinns, R. 1976. Multiplicative signal processing for sound source location on jet engines. *J. Sound Vib.* Vol. 46, No. 1:137–50

Gibson, J. S. 1974. *I.C.A.S. Paper 74-59*

Goldstein, M. E. 1975. *J. Fluid Mech.* 70:595

Graham, E. W., Graham, B. B. 1974. *NASA Contract Rep. CR-2390*

Grosche, F. R. 1974. *AGARD-CP-131. Pap. 4*

Hanson, D. B. 1974a. The spectrum of rotor noise caused by atmospheric turbulence. Hamilton Standard

Hanson, D. B. 1974b. *J. Acoust. Soc. Am.* 56:110

Hanson, D. B. 1975. *AIAA Pap. 75-468*

Hanson, D. B. 1976. *AIAA Pap. 76-565*

Hardin, J. C. 1974. *AIAA Pap. 74-550*

Hardin, J. C. 1976. *NASA TMX 73908*

Hassan, H. A. 1974. *J. Fluid Mech.* 66:445

Hawkings, D. L. 1970. *Loughborough Symp. Aerodyn. Noise*

Hawkings, D. L., Lowson, M. V. 1974. *J. Sound Vib.* 36:1

Head, R. W., Fisher, M. J. 1976. *AIAA Pap. 76-502*

Hoch, R. G., Duponchel, J. P., Cocking, B. J., Bryce, W. D. 1973. *J. Sound Vib.*

28:649
Hoch, R. G., Hawkins, R. 1974. *AGARD-CP-131. Pap. 19*
Howe, M. S. 1975a. *J. Fluid Mech.* 71:625
Howe, M. S. 1975b. *J. Fluid Mech.* 67:597
Howe, M. S. 1976a. The influence of vortex shedding on the generation of sound by convected turbulence. Cambridge Univ. Eng. Dep.
Howe, M. S. 1976b. *Proc. R. Soc. London Ser. A* 347:513
Hurle, I. R., Price, R. B., Sugden, T. M., Thomas, A. 1968. *Proc. R. Soc. London Ser. A* 303:409
Jacques, J. R. 1975. *J. Sound Vib.* 41:13
Jones, D. S. 1972. *J. Inst. Math. Its Appl.* 9:114
Jones, D. S. 1975. *J. Inst. Math. Its Appl.* 15:33
Kempton, A. J. 1976a. The ambiguity of acoustic sources—a possibility for active control. *J. Sound Vib.* Vol. 48, No. 4: In press
Kempton, A. J. 1976b. Heat diffusion as a source of aerodynamic sound. *J. Fluid Mech.* In press
Kinns, R. 1975. *J. Sound Vib.* 44:275
Knott, P. R. 1974. *Air Force Aero-Propul. Lab. TR-74-25*
Ko, D. R. S., Kubota, T., Lees, L. 1970. *J. Fluid Mech.* 40:315
Landahl, M. T. 1967. *J. Fluid Mech.* 29:441
Landau, L. D., Lifshitz, E. M. 1959. *Fluid Mechanics.* Reading, Mass: Addison-Wesley
Laufer, J., Kaplan, R. E., Chu, W. T. 1974. *AGARD CP-131. Pap. 21*
Leppington, F. G. 1972. Section 5 in *Aeronaut. Res. Counc. CP 1195*
Leppington, F. G., Levine, H. 1973. *J. Fluid Mech.* 61:109
Levine, H. 1975. *Phillips Res. Rep.* 30:240
Lighthill, M. J. 1952. *Proc. R. Soc. London Ser. A* 211:564
Lighthill, M. J. 1954. *Proc. R. Soc. London Ser. A* 222:1
Lighthill, M. J. 1963. *AIAA J.* 1:1507
Lilley, G. M. 1974. *AGARD CP-131. Pap. 13*
Liu, J. T. C. 1974. *J. Fluid Mech.* 62:437
Lorentz, H. A. 1904. *Encycl. Math. Wiss.* 5:14, 184
Lovely, D. J. 1974. Some novel aspects of flow driven aerodynamic noise. PhD thesis. Imperial College, Cambridge, Engl.
Lowson, M. V. 1975. Aircraft noise generation emission and control; duct acoustics and mufflers. *AGARD Lect. Ser. 77*
Lush, P. A. 1971. *J. Fluid Mech.* 46:477
Lush, P. M., Fisher, M. J. 1973. *AGARD CP-131. Pap. 12*
Mathews, D. C., Rekos, N. F. 1976. *AIAA Pap. 76-579*
Mani, R. 1972. *J. Sound Vib.* 25:337
Mani, R. 1974. *J. Fluid Mech.* 64:611
McGregor, G. R., Ribner, H. S., Lam, H. 1973. *J. Sound Vib.* 27:437
Merle, M., Fragassi, J. P. 1971. *J. Soc. Motion Pict. Tel. Eng.* 80:282
Michalke, A. 1969. *Dtsch. Luft. Raumfahrt* FB 69–90
Michalke, A. 1971. *Z. Flugwiss.* 19:319
Michalke, A., Fuchs, H. V. 1975. *J. Fluid Mech.* 70:179
Möhring, W. 1972. *Symp. Acoust. Flow Ducts, Southampton*
Moore, C. J. 1976. The role of shear layer instability waves in jet exhaust noise. Derby, Engl.: Rolls-Royce Adv. Res. Lab.
Morfey, C. L. 1971. *J. Acoust. Soc. Am.* 49:1690
Morfey, C. L. 1973. *J. Sound Vib.* 31:391
Morfey, C. L. 1976. Sound radiation due to unsteady dissipation in turbulent flows. *J. Sound Vib.* 48:95
Morgan, G. H., Hardin, J. C. 1974. *AIAA Pap. 74-949*
Morris, P. 1971. The structure of turbulent shear flow. PhD thesis. Univ. Southampton, Southampton, Engl.
Morris, P. J., Richarz, W., Ribner, H. S. 1973. *J. Sound Vib.* 29:443
Norum, T. D. 1973. *NASA Tech. Note TN D7230*
Obermeier, R. 1975. *J. Sound Vib.* 41:463
O'Shea, S. 1971. Scattering of sound by aerodynamic flows. PhD thesis. Univ. London, London, Engl.
Pao, S. P. 1973. *J. Fluid Mech.* 59:451
Phillips, O. M. 1960. *J. Fluid Mech.* 9:1
Pickett, G. F. 1974. *Proceedings, Eighth International Congress on Acoustics,* Vol. 2. Trowbridge, England: Goldcrest
Poldervaart, L. J. 1974a. Modes of vibration. Film available from Technische Hogeschool, Eindhoven
Poldervaart, L. J. 1974b. Sound pulse—boundary layer interaction studies. Film available from Technische Hogeschool, Eindhoven
Powell, A. 1964. *J. Acoust. Soc. Am.* 36:1, 177
Rayleigh, Lord 1896. *The Theory of Sound.* London: Macmillan
Ribner, H. S. 1964. *Adv. Appl. Mech.* New York: Academic
Rockwell, D. O. 1972. *J. Appl. Mech. Trans. ASME* December, 39E:883
Shivashankara, B. N., Strahle, W. C., Handley, J. C. 1974. *AIAA Pap. 74-47*
Smith, T. J. B., Kilham, J. K. 1963. *J. Acoust. Soc. Am.* 35:5, 715
Sofrin, T. G., Pickett, G. F. 1974. *NASA Spec.*

Publ. SP-304, 435

Stokes, G. 1868. *Philos. Trans. R. Soc. London* 158:447–63

Strahle, W. C. 1971. *J. Fluid Mech.* 49:399

Strahle, W. C. 1972. *AIAA Pap. 72-198*

Strahle, W. C. 1975. *AIAA Pap. 75-522*

Swinbanks, M. A. 1973. *J. Sound Vib.* 29:411

Swinbanks, M. A. 1975. *J. Sound Vib.* 40:51

Tanna, H. K., Fisher, M. J., Dean, P. D. 1973. *AIAA Pap. 73-991*

Tester, B. J. 1973. *J. Sound Vib.* 28:151

Tester, B. J., Morfey, C. L. 1975. *AIAA Pap. 75-477*; *J. Sound Vib.* 46:79

Thomas, A., Williams, G. 1966. *Proc. R. Soc. London Ser. A* 294:449

Tsien, H. S. 1952. *J. Am. Rocket Soc.* 22:139

Whitfield, O. J. 1975. Novel schemes for jet noise control. PhD thesis. Cambridge Univ., Cambridge, Engl.

Zukoski, E. E. 1975. *ASME Pap. 75-GT-40*

Ann. Rev. Fluid Mech. 1977. 9:469–94

STUDY OF THE UNSTEADY ×8111
AERODYNAMICS OF LIFTING
SURFACES USING THE
COMPUTER

S. M. Belotserkovskii

N. E. Joukowski Central Aero-Hydrodynamic Institute, Moscow, USSR

1 INTRODUCTION

With the development of aviation, shipbuilding, and other technical domains, unsteady aerodynamic and hydrodynamic data are used in an expanding range of problems, and the demand for them increases. In unsteady aerodynamics these problems can be classified into the following three groups:

1. Steady processes and those changing slowly with time (aeroelastic, divergence, control reversal, long-period oscillations, auto-oscillations occurring at small values of the dimensionless frequency p_j^*, and so on). Here we require in the first place aerodynamic derivatives for $p_j^* \to 0$.
2. Flows in which the dependence on time can be considered harmonic, but the dimensionless frequency varies within wide limits (high-frequency flutter, the effect of a turbulent atmosphere). In this case in order to determine the so-called transfer functions it is, generally speaking, necessary to consider the entire range of variation of the dimensionless frequency ($0 \leqq p_j^* \leqq \infty$).
3. Arbitrary dependence on time (violent maneuvers of an aircraft, rapid deflection of control surfaces, envelopment by a gust, effect of a stream past a shock wave).

The appearance and development of electronic computers and progress in the domains of theoretical aerodynamics and computational mathematics have considerably enlarged the possibilities of theoretical methods of investigation. A comprehensive review of achievements in this field has been made by Ashley & Rodden (1972), but it does not reflect investigations carried out in the Soviet Union.

The nature of the development of unsteady aerodynamics is largely determined by the evolution of those sciences and applied fields that are its main consumer. The analysis of three-dimensional aircraft motion required a whole series of new aerodynamic properties (Byushgens & Studnev 1967). Many specific problems were introduced by aeroelasticity (Bisplinghoff, Ashley & Halfman 1955; Försching 1974).

A new step was the development of automatic flight-control systems, including the suppression of elastic vibrations. It has now become necessary to consider the requirements associated not only with dynamic flight and elastic vibration of designs, but also with automatic control systems (Krasovskii 1973).

In simultaneous consideration of these questions there arises one fundamental difficulty. It results from the fact that in unsteady motion the aerodynamic characteristics depend upon the entire previous history of the flow, which is determined by all the laws of motion, deformation, and so on. The memory in which this previous history is stored is the aerodynamic wake forming on the surface of the body and behind it. At subsonic flight speeds (Mach number $M < 1$) the effect of the wake appears for all time, and a transient process has asymptotic character. If $M > 1$ and the body does not intersect the wake or approach it, only part of the wake influences the flow past the body, and a transient process is completed in finite time. On the other hand, the motion of the body and the deformation of its surface depend also on the aerodynamic loading. In the general case all these problems must be solved simultaneously, and consecutively in time, determining the aerodynamic and elastic characteristics and the work of the control system. Such an approach to the problem makes it immense in practice. Linearization of the aerodynamic part of the problem permits rigorous closure of the general equations (Belotserkovskii 1972).

The main features of contemporary understanding of the matter, arising from analysis of the requirements of practice and of the possibilities of theory and experiment, allowing for the inevitable compromises, can be summarized in the following form:

1. Fundamental for the practical determination of the unsteady aerodynamic characteristics of aircraft and their components must be numerical methods, based on the wide application of electronic digital computers. This is particularly important at the preliminary design stage of aircraft.

2. With an accurate choice of calculating scheme and a carefully adjusted computing process, linear theory permits abundant and accurate information to be obtained. In the absence of any sort of critical phenomenon (general breakaway, local separation, or strong local shock waves), the calculated data are found to agree satisfactorily with experiments in the whole range of variation of the kinematic parameters ε_j, so long as a linear dependence of the aerodynamic characteristics on ε_j is maintained. This results in particular from the following. Aircraft shapes tend to become such that in the basic flight regime there is a linear dependence upon the ε_j of the lift and side forces, moments, and aerodynamic loads on supporting elements. For applications, these characteristics should be found first from unsteady aerodynamics, especially since their experimental determination is difficult.

3. The role of experiment remains very important in purely aerodynamic investigations as well as in the development and refinement of aircraft. Its role includes such important functions as (a) the substantiation of theoretical schemes, (b) the verification and sometimes the refinement of the results of calculations, (c) the

choice of a whole series of modifications in the actual arrangement of a device in order to eliminate sudden changes and ensure linear dependence on the ε_j in the basic regimes, (d) checking whether the results of final tests turn out to show desirable effects on a flying device, and so on.

Further, three important aspects of the current approach to the investigation of linear aerodynamic characteristics are subjected to analysis: first, the method of schematizing a flying device that is used in the calculations; second, the choice of the basic dependence upon time of the kinematic parameters ε_j that characterize the general unsteady problem; third, particularities of the method of solution of the aerodynamic boundary-value problem, including the basic numerical scheme.

Today, in connection with the development of the applied problems mentioned above that arise from machine methods of design, an increasing role is played by mathematical models of a flying device as an aerodynamic object. The creation and introduction of such models, which should be effective and adapted to the needs and opportunities of the consumer, is already a new stage in the development of contemporary aerodynamics. Here the question should be of a specific general scientific concept and of a computational system, for whose development possible paths are also analyzed.

However, linear aerodynamic characteristics alone will not completely satisfy the requirements of practice. The stages of take-off and landing, flight at heavy loading, and many operating regimes of turbines and compressors take place at significant angles of attack, with large angles of deflection of control surfaces. Greater application is being made of wings of triangular and related planforms with sharp leading edges. Separation at these edges begins at small angles of attack, and the resulting steady vortex sheet leads to a significant increase in the lifting properties of a wing (Legendre 1965, Chang 1970). The region of application of nonlinear and separated aerodynamics is very extensive, and at the present time is only beginning to be developed. Here too the development of digital computers opens up completely new possibilities for investigations of fundamental as well as applied character.

A very important direction in the investigation of separated flows is based on examination of a viscous medium. It is possible to become acquainted with the state of affairs in this field from the monograph of Chang (1970), for example, and with the help of the survey of Gogish, Neiland & Stepanov (1975). We merely remark that numerical realization of an approach using the Navier-Stokes equations runs, at high Reynolds numbers, against a series of difficulties that are evidently not only of computational but also of a fundamental character.

Of great theoretical and practical interest is the question of the possibility of studying nonlinear characteristics and separated flows on the basis of a scheme for an ideal medium. This problem is now once again in the center of interest of many scientists, as it was at the beginning of the present century. Then the basic ideas of Zhukovskii (Joukowski), Prandtl, and Chaplygin determined, in a few decades, the course of development of aerodynamics. (See, for example, Joukowski 1949, Prandtl & Tietjens 1931, and Chaplygin 1948.) Here contemporary investigators (Nikol'skii 1957a,b; Sedov 1973; Lavrent'ev & Shabat 1973; O. M. Belotserkovskii et al 1974) are looking for new possibilities by various paths, and much progress has been achieved in this field.

It is still early to expect extensive theoretical results for lifting devices as a whole. But for thin lifting surfaces, typical of many contemporary airplanes, including wings of complicated planform, real practical possibilities have opened up. We consider the flow past a thin lifting surface of an ideal incompressible fluid, a problem encountered in the most advanced aerodynamics. The restriction to thin wings is important also from the point of view that it eliminates another problem—determination of the separation point of the vortex sheet from a lifting surface. This question evidently cannot be solved without considering a viscous medium.

A study of nonlinear characteristics and separated flow still to a considerable extent requires a combination of theoretical and experimental approaches. What is more, in this field theoretical investigations themselves all begin to a great extent to have the nature of numerical experiments.

2   SCHEMATIZATION OF THE AIRCRAFT

We consider the basic computing schemes that are currently in use.

Wing-Cylinder

The most commonly used version is a slender wing and a circular cylinder, or more rarely an elliptic one. This is one of the oldest schemes, which made it possible to obtain valuable data at subsonic and supersonic speeds. It has many attractive features: it is simple (described by a small number of parameters; boundary conditions are easier to satisfy on the surface than in the general case), and as a rule it is sufficiently accurate to model the flow conditions on a wing and body. It is not infrequently employed also in modern approaches, in whole or in part (Giesing, Kalman & Rodden 1972; Marino, Chen & Sucio 1975).

It must be kept in mind that in three-dimensional problems the method of inversion, usually applied for $M < 1$, does not provide exact satisfaction of the boundary conditions on the cylinder. But the main deficiency of the scheme is that it it gives only a partial description of the actual configuration.

Nonplanar Bodies

A highly developed example is that of two slender lifting surfaces, modeling a wing and horizontal tail. Such a scheme for supersonic speeds was thoroughly examined by Appa Kari & Jones (1975), and also by Cunningham (1974). To study the mutual influence of wing and suspensions at subsonic speeds Roos, Bennekers & Zwaan (1975) used a slender wing-body scheme.

These approaches give a partial description of the apparatus, and are directed toward the study of only certain interference questions.

Body Scheme

Some of the first authors who began to systematically apply this scheme were Hess & Smith (1964) and then Maslov (1966). They considered the circulation-free problem in an incompressible fluid, when the scheme corresponds completely to the formulation of the problem: no linearization with respect to any boundary condition is carried out. Woodward (1973) obtained a body scheme based on solution of the steady linear problem with circulation for both subsonic and supersonic airplane speeds (Figure 1), and Johnson & Rubbert (1975) studied the flow with circulation past an airplane and wing in an incompressible medium. This schematization is in part encountered in a series of other works: Giesing, Kalman & Rodden (1972) for $M < 1$; Roos, Bennekers & Zwaan (1975) for $M < 1$; and Marino, Chen & Sucio (1975) for $M < 1$ and $M > 1$.

The scheme has the advantage of providing full modeling of the complete airplane, in principle permitting an exact description of the most complicated points of intersection of different surfaces (wing and fuselage, tail and fuselage, etc.). However, this makes the price too expensive. The awkwardness of the computations increases, which for limited machine capacity may give a further negative effect. Usually one cannot avoid inaccuracies in modeling the vortex wake. Much supplementary information on the form of the surface is required, which is unnecessary in the majority of problems. It is also disadvantageous that the problem is not solved within the strict framework of linear theory, but on the basis of a mixed approach: the linearized continuity equation is used and the method of superposition, but the boundary conditions on the apparatus are taken in nonlinear form. The resulting solution will then not strictly satisfy the general relations and theorems proved for a fully linearized problem. All this prevents the investigator from creating a rigorous and economical system for calculating and analyzing the linear unsteady aerodynamic characteristics and controlling the resulting computations.

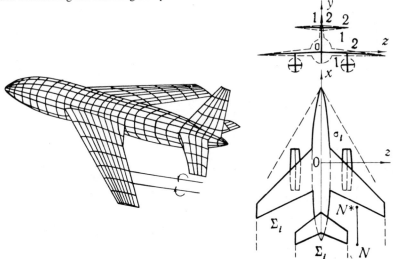

Figure 1 Different schemes for an airplane (Woodward's on left). The Σ_i are vortex surfaces, 1 and 2 are opposite side of the base elements s_i, the σ_i are diaphragms for supersonic flow.

Scheme with Plane Base Elements

It is convenient to construct the flow past an apparatus by distributing gas-dynamic singularities (vortices, dipoles, sources) that satisfy the continuity equation over so-called plane base elements. These are parallel to the longitudinal axis Ox of the airplane, and are chosen so that the corresponding part of the surface of the apparatus lies nearby. The boundary conditions on the surface are transferred to them; and the vortex wake lies on the base elements or their extensions. In Figure 1, which shows one fundamental variant of the schematization of an airplane (Scheme 2), the base elements are shown by solid lines. The simplest variant of this (Scheme 1) is obtained when a single base plane is introduced, which in the study of symmetric flow is taken as the projection of the apparatus onto the Oxz-plane. This scheme proved to be sufficiently accurate and economical, and permitted a comparatively simple generalization of the numerical method first developed for wings (Belotserkovskii 1972, 1975b; Belotserkovskii & Skripach 1975).

A fundamental defect in the scheme consists in inexact modeling of such regions as the wing-fuselage. Here errors are introduced in the transfer of the boundary conditions that do not arise, strictly speaking, from the linearization. To eliminate these defects one must somewhat complicate the scheme by introducing supplementary base elements in regions of intersecting surfaces.

Let 1 and 2 be the opposite sides of any base element s_i, where the positive normal n_o^i goes from 1 to 2. We represent the shapes of the parts of the apparatus, its dimensionless deformation, and the deflections of the controls in the form

$$\Delta n_{o1,2}^i = \sum_v \delta_v^i(\tau) f_{\delta,1,2}^i(\bar{x}, \bar{y}, \bar{z}), \quad \bar{x} = \frac{x}{b}, \quad \bar{y} = \frac{y}{b}, \quad \bar{z} = \frac{z}{b}, \quad \tau = \frac{tU_0}{b}, \tag{1}$$

where b is a characteristic linear dimension, U_0 the average speed of the center of mass, and t the time. The thickness of parts of the apparatus is characterized by the sum, and their curvature by the half-difference, of the functions in (1):

$$f_{\delta_v}^{*i} = f_{\delta,2}^i + f_{\delta,1}^i, \quad 2f_{\delta_v}^i = f_{\delta,2}^i - f_{\delta,1}^i. \tag{2}$$

We represent the dimensionless velocities of atmospheric nonhomogeneity or of the stream behind a weak shock wave in a form analogous to (1):

$$\frac{W_{x,y,z\Delta}}{U_0} = \sum_k \Delta_{x,y,z}^k(\tau) w_{x,y,z\Delta}(\bar{x}, \bar{y}, \bar{z}). \tag{3}$$

Then the whole set of kinematic parameters $\varepsilon_j(\tau)$ describing the general linear unsteady problem will be determined by the boundary condition of no flow through the surface of the body, and reduce to the following:

$$\varepsilon_j(\tau) \in \alpha, \beta, \omega_{x,y,z}; \delta_v^i, \dot{\delta}_v^i; \Delta_{y,z}^k,$$

$$\omega_{x,yz} = \frac{\Omega_{x,y,z}b}{U_0}, \quad \dot{\delta}_v^i = \frac{d\delta_v^i}{d\tau}. \tag{4}$$

Here α and β are the angles of attack and sideslip, and $\omega_{x,y,z}$ are the dimensionless angular velocities.

The basic problem in linear unsteady aerodynamics is the determination of the dimensionless loadings $\Delta p^i = p_1^i - p_2^i$ on the base elements, the section characteristics, and the total coefficients of force and moment c_y, c_z, m_x, m_y, m_z. The Schemes 1 and 2 already mentioned are also adapted to this purpose.

The following fact helps to give a proper understanding of the possibilities. Exact solutions for the aerodynamic derivatives of rectangular and triangular wings of very small aspect ratio (neglecting thickness) and the corresponding expressions obtained from slender-body theory (including thickness) give coincident results, independent of the Mach number M.

3 BASIC TIME DEPENDENCES

Let the basic kinematic parameters vary according to a harmonic law

$$\varepsilon_j(\tau) = \varepsilon_j^* \cos\left(p_j^*\tau + \theta_j\right). \tag{5}$$

Then any linear aerodynamic characteristic can be expressed in terms of aerodynamic derivatives that do not depend on time, but are functions of p_j^*:

$$c_\mu(\tau) = \sum_j \left[c_\mu^{\varepsilon_j}(p_j^*)\varepsilon_j(\tau) + c_\mu^{\dot\varepsilon_j}(p_j^*)\dot\varepsilon_j(\tau) \right]. \tag{6}$$

If the functions $c_\mu^{\varepsilon_j}(p_j^*)$ and $c_\mu^{\dot\varepsilon_j}(p_j^*)$ are known, then passing to any periodic dependence $\varepsilon_j(\tau)$ is accomplished with series, and to an aperiodic dependence with the Fourier integral.

It is simple enough to determine the aerodynamic derivatives $c_\mu^{\dot\varepsilon_j}$ as $p_j^* \to 0$ (Belotserkovskii 1955a,b; Fursov 1961). The problem is actually solved as a steady one, but with new boundary conditions. However for large p_j^* numerical methods become ineffective, which is to a considerable degree connected with the fact that the functions $\sin p_j^*\bar x$ and $\cos p_j^*\bar x$ appear in the expressions for the downwash and in the conditions on the vortex sheet. Furthermore, in certain problems the formulation of equivalent conditions in frequency form is difficult (for example, in the study of the effect of the stream behind a weak shock wave).

In another approach, a time-dependent one, the fundamental variation is taken to be a step function:

$$\frac{\varepsilon_j}{\varepsilon_j^*} = 1(\tau) = \begin{cases} 0, & \tau < 0 \\ 1, & \tau \geq 0. \end{cases} \tag{7}$$

Here transient functions are determined, which for each ε_j are conveniently represented in the form

$$[c_{\mu\varepsilon_j}(\tau)] = I_{c_\mu}^{\varepsilon_j}(\tau) + c_\mu^{\varepsilon_j}(0). \tag{8}$$

It is not difficult to see that as $\tau \to \infty$ the unsteady part $I_{c_\mu}^{\varepsilon_j}$ of the transient function vanishes.

The transition to an arbitrary continuous dependence $\varepsilon_j(\tau)$ is effectively achieved with integral equalities obtained with the use of the Duhamel integral (Belotserkovskii, Skripach & Tabachnikov 1971; Belotserkovskii 1975b)

$$c_{\mu\varepsilon_j}(\tau) = c_\mu^{\varepsilon_j}(0)\varepsilon_j(\tau) + \int_0^\tau \dot\varepsilon_j(\tau - \tau_1) I_{c_\mu}^{\varepsilon_j}(\tau_1) d\tau_1 + v\Delta^* c_\mu^{\dot\varepsilon_j} \dot\varepsilon_j(\tau)$$

(9)

$$c_\mu = \sum_j c_{\mu\varepsilon_j}, \qquad v = \begin{cases} 0, & M \ne 0 \\ 1, & M = 0. \end{cases}$$

The aerodynamic derivatives for arbitrary p_j^* are found from the relations

$$c_\mu^{\dot\varepsilon}(p_j^*) = c_\mu^{\dot\varepsilon}(0) + p_j^* \int_0^\infty I_{c_\mu}^{\varepsilon_j}(\tau) \sin p_j^* \tau\, d\tau$$

and

(10)

$$c_\mu^{\dot\varepsilon}(p_j^*) = v\Delta^* c_\mu^{\dot\varepsilon_j} + \int_0^\infty I_{c_\mu}^{\varepsilon_j}(\tau) \cos p_j^* \tau\, d\tau.$$

Here the $\Delta^* c_\mu^{\dot\varepsilon_j}$ are the solutions of the linear problem without circulation in an incompressible fluid.

The relations (9) represent an explicit functional expression describing the dependence of the c_μ on the entire previous history of the flow.

At present the second approach has provided solutions for all the basic problems of linear wing theory (Lomax, Heaslet, Fuller & Sluder 1952; Miles 1959; Vorob'ev 1959; Golubinskii 1961, 1963, 1968; Belotserkovskii 1966a; Belotserkovskii & Kolesnikov 1969; Belotserkovskii & Popytalov 1970; Shamshurin 1974). It has proved effective also for airplanes as a whole (Belotserkovskii 1975b, Popytalov 1976).

4   METHODS OF SOLUTION

The general linear problem of unsteady aerodynamics can be reduced to the determination of transfer functions for the velocity potential $[\varphi_{\varepsilon_j}]$ from the following conditions on the canonical problems for the ε_j.

Outside the body and vortex wake the function $[\varphi_{\varepsilon_j}]$ satisfies the linearized continuity equation

$$(1 - M^2)\frac{\partial^2[\varphi_{\varepsilon_j}]}{\partial\bar x^2} + \frac{\partial^2[\varphi_{\varepsilon_j}]}{\partial\bar y^2} + \frac{\partial^2[\varphi_{\varepsilon_j}]}{\partial\bar z^2} + 2M\frac{\partial^2[\varphi_{\varepsilon_j}]}{\partial\bar x\partial\tau} - M^2\frac{\partial^2[\varphi_{\varepsilon_j}]}{\partial\tau^2} = 0.$$

(11)

On the base surface s_i the condition of zero flux is satisfied:

$$\frac{\partial[\varphi_{\varepsilon_j}]}{\partial n_0^i} = 1(\tau) F_{\varepsilon_j}^i(\bar x, \bar y, \bar z),$$

(12)

where the $F_{\varepsilon_j}^i$ are known functions.

On any vortex sheet Σ_i lying in the plane s_i the condition of conservation of circulation yields an equation for the difference in potential (Figure 1):

$$\Delta[\varphi_{\varepsilon_j}]_{N,\tau} = \Delta[\varphi_{\varepsilon_j}]_{N^*,\tau^*}, \qquad \tau^* = \tau - (\bar x^* - \bar x).$$

(13)

Here N^* denotes the point lying immediately behind the trailing edge of s_i. The condition (13) also ensures the absence of excess pressure on Σ_i.

For $M < 1$ the Chaplygin-Joukowski hypothesis of finite velocity must be satisfied on the trailing edges of the s_i. At subsonic trailing edges for $M > 1$ it is sufficient to require that the potential itself be continuous (but not derivatives). Since the velocity potential is continuous also at weak shock waves, the same conditions will be satisfied also at supersonic trailing edges. On diaphragms there is no vortex wake, and the velocity potential is continuous:

$$\Delta[\varphi_{\varepsilon_j}]_{\sigma_i} = 0. \tag{14}$$

The values of the gas parameters at infinity are assumed known; and the dimensionless pressure is found from the Cauchy-Lagrange integral:

$$[p_{\varepsilon_j}] = 2\left(\frac{\partial[\varphi_{\varepsilon_j}]}{\partial \bar{x}} - \frac{\partial[\varphi_{\varepsilon_j}]}{\partial \tau}\right). \tag{15}$$

In problems without circulation at $M = 0$ (incompressible medium) (13) is replaced by the requirement that there be no vortex wake.

For harmonic time dependence (5) the canonical problems for φ^{ε_j} and $\varphi^{\dot{\varepsilon}_j}$ will differ only in the form in which the conditions (11)–(13) and (15) are written.

In all current numerical methods of solving the boundary-value problem formulated above, Equation (11) is satisfied exactly. This is achieved in one of two ways. In the first, one uses some integral representation of the velocity potential at any point in space in terms of the values of the potential and its derivatives on the boundary. In the second, the body is replaced by a continuous or discrete distribution of gas-dynamic singularities. The first path usually leads to an integral equation, and the second either to the same or directly to a system of algebraic equations.

The basic approaches to numerical solution of the problems described above are appropriately divided into four groups.

Method of Integral Equations

These are obtained on the basis of the approaches mentioned and relations (12)–(14); the solutions are constructed with consideration of the remaining conditions of the problem.

One of the oldest methods of solution is the representation of the desired functions by series or by approximation with linear combinations of known functions, whose coefficients are determined from a system of algebraic equations (Golubev 1949). Similar approaches find application still at the present time (Rowe, Winther & Redman 1973), especially the popular collocation method, in which the boundary conditions are satisfied at discrete points.

An intensively developed method is that of so-called finite elements, in which the surface of the apparatus or the base surfaces are in one way or another partitioned into elements. Inside each of them the unknown functions and their derivatives are assumed constant (or are represented by other simple laws). The boundary conditions are also satisfied at a discrete number of points, and in place of integral equations one obtains a system of algebraic equations. The method was employed with various refinements by Hess & Smith (1964) for the solution of problems without circulation, by Appa Kari & Jones (1975) at supersonic speeds; by Roos, Bennekers & Zwaan

(1975) at subsonic speeds; by Marino, Chen & Sucio (1975) for $M < 1$ and $M > 1$, etc. The accuracy of the idea and the apparent generality of the approach make it very attractive.

In the indicated methods, aside from a number of technical difficulties, there arises a fundamental one: it is necessary to ensure stability of the numerical calculations with increase in the number of terms in the series or of elements, for which the system of algebraic equations should be well determined. But it is

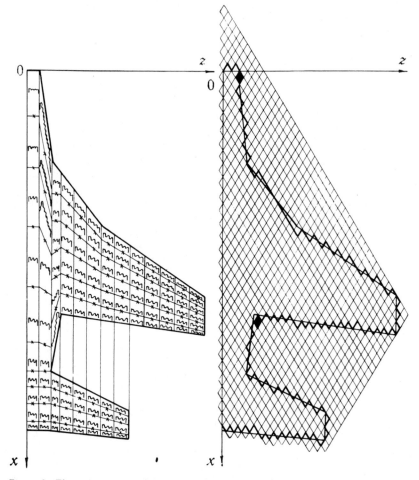

Figure 2 Elementary computing schemes for an airplane (*left M < 1, right M > 1*). In problems without circulation ($M = 0$) the sum of the cirulations of the vortices for each section $z = $ const is equal to zero. In problems with circulation ($M < 1$) the last vortices of the sections have zero circulation. The crosses are control points at which the boundary conditions are satisfied.

impossible to restrict oneself to a small number of them, especially in many problems of deformation, with large p_j^*, and others.

Another path is solution of the integral equations by the method of iteration. For problems without circulation an effective approach was developed by Maslov (1966). Polyahov & Shesternina (1975) used an iteration method in the problem of flow past a wing with circulation. Popytalov (1976) has proposed a promising method of integral equations for the study of supersonic flow past three-dimensional configurations. In the steady case the system he obtained is solved by a method of successive approximations. The rapid convergence of the process is explained by the fact that the basic contribution to the expression for the potential is given by integrals over the base surfaces s_i of the normal derivatives of the potential, which are known from (12). The unsteady problem is solved by alternating solution of two integral relations at each computing step in time.

Panel Distributions of Singularities

The surface of the apparatus or the base surfaces are approximated with the aid of standard elements. On each of them are considered elementary distributions of gas-dynamic singularities (sources, dipoles, vortices), and the vortex wake is constructed in a corresponding way. In contrast to the finite-element method, here there are no integral equations, and the investigator can construct his computing scheme quite freely, adapting it only to the conditions of the problem and computational convenience (Woodward 1973). This is especially important when not only linear but also nonlinear problems are investigated in an incompressible medium (Johnson & Rubbert 1975). The basic difficulties here are the same as in the finite-element method.

Discrete-Vortex Methods

These have great clarity and accuracy of the physical representations that are essential for the construction of complicated vortex systems (Figure 2), and especially of an unsteady vortex wake (Figure 3). These are based on the ideas of Joukowski about the mechanism of formation of lift and the associated vortices (Joukowski 1949) and Prandtl's model explaining the origin of circulation on a wing in an ideal medium (Prandtl & Tietjens 1931).

In steady problems the numerical method of discrete vortices was first systematically employed by Falkner (1953). Next appeared other variants of the method in connection with steady and unsteady linear problems, at first for wings of arbitrary planform (Belotserkovskii 1955a,b; 1966a,b; 1967; Kolesnikov 1967; Belotserkovskii & Kolesnikov 1969), and then also for airplanes as a whole. [For this see the book of Belotserkovskii & Skripach (1975).]

In its ideas the discrete-vortex method is near to the panel method. In contrast to the latter, the Chaplygin-Joukowski condition and the nature of the singularities at leading edges, discontinuities, etc., are not imposed explicitly. This greatly simplifies the calculations, but requires further verification of the arrangement of vortices and control points, which at first was done by heuristic reasoning and numerical experimentation. But recently rigorous mathematical proofs of the basic facts and bases for the approaches developed have begun to appear (Lifanov & Polonskii 1975).

The method has a number of good numerical properties (simplicity of the basic element, flexibility in the construction of the vortex system, and stability of the system of algebraic equations because of the dominant values of the diagonal terms in the basic determinant of the system). In its direct form it has proved to be very effective for linear and nonlinear problems of an incompressible fluid. For small τ and certain combinations of the geometric parameters of the lifting surface and the Mach number M oscillations are observed in the calculated values of the transfer functions. They are due to the fact that the boundary of the perturbed region from discrete vortices at an instant of calculation passes close to a point where the boundary conditions are satisfied. To eliminate this, certain modifications were allowed, based on the selection of additional terms in the conversion of integrals over the surface into sums. As τ increases the additional terms vanish, and the basic variant of the method is again obtained.

Discrete-vortex methods have permitted solution of all the basic linear subsonic problems of steady and unsteady wing theory (Belotserkovskii, Skripach & Tabachnikov 1971).

Direct Methods

In these methods the conditions on the boundary-value problem are directly used and satisfied. They are based on the fundamental properties of linear supersonic flows

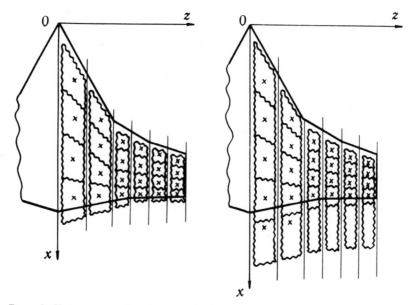

Figure 3 Vortex system of a wing in unsteady motion (first and second time steps of the computation).

(Ward 1949, Evvard 1950, Krasil'schikova 1952). They have been developed in wing theory and in an elementary scheme for airplanes (Scheme 1). The base surface and the region of influence are divided by Mach lines into elementary cells (Figure 2). In each of these the velocity potential and slope are taken independent of the coordinates. On the base surface the slope is known from (12); and on apertures σ_i the condition (14) has the form $[\varphi_{\varepsilon_j}] = 0$. Determination of the slope is carried out sequentially in cells, beginning with the σ_i, and then on the vortex sheet Σ_i according to condition (13), beginning with the cells shown in black. There is a whole series of modifications (Pines, Dugundji & Neuringer 1955; Belotserkovskii, Kudryavtseva & Fedotov 1969). It proved to be convenient also for arbitrary time dependence (Belotserkovskii & Popytalov 1970, Shamshurin 1974).

The method is sufficiently simple and effective with a large number of cells; its shortcoming is the complexity associated with extending it to three-dimensional configurations (Scheme 2). Furthermore, in order to calculate the pressure numerically from Equation (15) it is necessary to resort to numerical differentiation, which is characteristic of all methods in which the velocity potential is determined.

5   EXACT RELATIONS AND THE SYSTEM OF CALCULATION

Numerical methods are the basic method of obtaining unsteady aerodynamic characteristics. It is therefore particularly important to create accurate and economical computing systems and to guarantee control of their results. In this connection the value of exact solutions, relations, and theorems is at present growing rather than decreasing, because they now serve the following functions.

They are used for analysis of the scheme of phenomena and for study of the properties of solutions, based on a general mathematical formulation of the problem (Golubev 1949, Miles 1959, Andrew & Stanton 1967, Sedov 1973). They help establish similarity parameters and self-similar properties for aerodynamic characteristics (Sedov 1954). They give asymptotic solutions when one or another parameter tends to a limiting value (Van Dyke 1964) or help obtain them on the computer (Van Dyke 1975). They provide standards for numerical calculations. The most important of these are exact solutions for wings of infinite span, for the circular wing in an incompressible fluid (Kochin 1949a), for wings with supersonic edges (Fursov 1961), for a triangular wing with subsonic edges (Ribner & Malvestuto 1948), and others. Mathematically they justify numerical methods and demonstrate their convergence and the fulfilment of the boundary conditions of the problem, as was done for the vortex method by Lifanov & Polonskii (1975). Important means of control in the hand of the computing aerodynamicist are such general relations as the reciprocity theorem and its corollaries (Flax 1953, Ul'yanov 1961). At present such approaches have become an essential part of computing systems (Belotserkovskii 1967, 1975b; Belotserkovskii, Skripach & Tabachnikov 1971; Belotserkovskii & Skripach 1975).

We dwell on some equations, exact in the framework of linear theory, that have proved particularly useful for that purpose. We consider a step law of variation of the kinematic parameters (7). At the initial moment of the transient process $(\tau \to 0, \tau > 0)$ the perturbations in a compressible medium are not able to spread, and each element of the surface of the body creates them independently of the others. Thus all the assumptions of piston theory are satisfied, and it gives exact values of the transient function. Taking the boundary conditions (12) into account, it is not difficult to obtain for the aerodynamic loads:

$$[\Delta p_{\varepsilon_j}^i]_{\tau \to 0} = -\frac{4}{M} F_{\varepsilon_j}^i(\bar{x}, \bar{y}, \bar{z}). \tag{16}$$

The second limiting value, as $\tau \to \infty$, of the transfer function for any linear aerodynamic characteristic coincides with the steady value corresponding to the aerodynamic derivative:

$$[c_{\mu\varepsilon_j}]_{\tau \to \infty} = c_{\mu}^{\varepsilon_j}(0). \tag{17}$$

In addition, the integral formula (9) also permits the total impulse to be found from the unsteady part of this function:

$$\int_0^\infty I_{c_{\mu}^{\varepsilon}}^{\varepsilon}(\tau)d\tau = c_{\mu}^{\dot{\varepsilon}_j}(0) - v\Delta^* c_{\mu}^{\dot{\varepsilon}_j}. \tag{18}$$

In using frequency properties it is important to know their limiting values for very large values of dimensionless frequency, which satisfy

$$c_{\mu}^{\varepsilon}|_{p_j \to \infty} = [c_{\mu\varepsilon_j}]_{\tau \to 0}, \qquad c_{\mu}^{\dot{\varepsilon}}|_{p_j \to \infty} = v\Delta^* c_{\mu}^{\dot{\varepsilon}_j}. \tag{19}$$

The reciprocity theorem establishes an integral relation between the boundary conditions and the corresponding aerodynamic loads on an airplane for direct $(+)$ and reverse $(-)$ motions with respect to the main velocity U_0. We denote by φ^* that part of the potential whose normal velocity changes discontinuously across the base surfaces s_i. Then for arbitrary functions $\varepsilon_j(\tau)$ we have

$$\sum_i \int \int_{\bar{s}_i} \Delta p_+^i \frac{\partial \varphi_-^*}{\partial n_0^i} d\bar{s}_i = \sum_i \int \int_{\bar{s}_i} \Delta p_-^i \frac{\partial \varphi_+^*}{\partial n_0^i} d\bar{s}_i, \tag{20}$$

where the $\bar{s}_i = s_i/s$ are the dimensionless areas of the base elements.

Let the kinematic parameters in the direct and reverse motions coincide identically: $\varepsilon_{j^+}(\tau) = \varepsilon_{j^-}(\tau)$. Then the coefficients of the linear aerodynamic forces and moments for the parameters $\alpha, \beta, \omega_x, \omega_y, \omega_z$ in direct and reverse motion of an airplane can be expressed in terms of one another. For the right-handed system of coordinates shown in Figure 1 one can obtain from (20):

$$c_{y\alpha+}(\tau) = c_{y\alpha-}(\tau), \qquad c_{z\beta+}(\tau) = c_{z\beta-}(\tau),$$

$$m_{x\omega_x+}(\tau) = m_{x\omega_x-}(\tau), \qquad m_{y\omega_y+}(\tau) = m_{y\omega_y-}(\tau), \qquad m_{z\omega_z+}(\tau) = m_{z\omega_z-}(\tau), \tag{21}$$

$$m_{x\beta+}(\tau) = c_{z\omega_x-}(\tau), \qquad m_{y\beta+}(\tau) = -c_{z\omega_y-}(\tau),$$

$$m_{y\omega_x+}(\tau) = -m_{x\omega_y-}(\tau), \qquad m_{z\alpha+}(\tau) = c_{y\omega_z-}(\tau).$$

Other consequences of the reciprocity theorem permit one to greatly simplify the calculation of total effects caused by changes in curvature of each part of the airplane:

$$[c_{y\delta_v}^i]_+ = \int\int_{\bar{s}_i} [\Delta p_z^i]_- \left(\frac{\partial f_{\delta_v}^i}{\partial \bar{x}}\right)_+ d\bar{s}_i$$

$$\cdots\cdots\cdots\cdots\cdots\cdots\cdots\cdots\cdots\cdots\cdots\cdots\cdots\cdots\cdots\cdots \qquad (22)$$

$$[m_{z\delta_v}^i]_+ = -\int\int_{\bar{s}_i} [\Delta p_{\omega_z}^i]_- f_{\delta_v+}^i \, d\bar{s}_i.$$

In order to find any transient function (22) it is sufficient to solve the problems for α, β, ω_x, ω_y, ω_z in the reverse motion, after which the calculation for arbitrary deformations is reduced to the computation of integrals.

From Equations (9) and (10) and what has been said above it follows that the following problems are fundamental:

Steady and nearly steady problems, in which $c_{\mu}^{\varepsilon_j}$ and $c_{\mu}^{i_j}$ are determined for $p_j^* \to 0$.

Unsteady problems for a step time dependence (7), where the transient functions $[c_{\mu\varepsilon_j}]$ are found.

Auxiliary linear problems of flow without circulation past an airplane of an incompressible fluid (the functions $\Delta^* c_{\mu}^{\varepsilon_j}$ are determined).

For each ε_j the problem is solved independently, where the problems of thickness and curvature are also separated. If the boundary conditions are imposed on the base surfaces, the solution of the first of these is obtained at once by means of a distribution of sources. In Scheme 1 the sources do not induce additional loads on the s_i. They do in Scheme 2, and here there arises the additional problem of compensating the normal velocities induced on one base element by sources distributed over the others.

Adjustment of the system of numerical computation and some methods of control are illustrated in Figures 4 and 5. We consider first the progressive entry of a triangular wing with supersonic edges into a step gust at $M = 2$ (Figure 4). Comparison is made with the exact solution for the transient function of the lift coefficient $[c_{y\Delta y}]$ (solid line) and the steady value of the derivative c_y^z (broken line). The practical convergence of the numerical method (points) is also studied for increase in the number of cells indicated in Figure 2. Here N denotes the number of sections into which the half-span of the wing is divided.

The next diagram in Figure 4 is concerned with an instantaneous change in

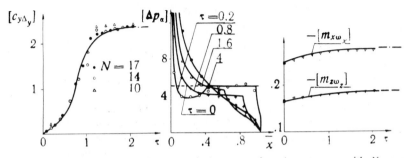

Figure 4 Methods of controlling numerical computations (convergence with N, comparison with exact and known solutions, reciprocity theorem).

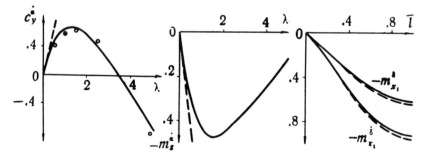

Figure 5 Methods of controlling numerical computations (impulse theorem, limiting solutions for $\lambda \to 0$, reciprocity theorem).

angle of attack α at $\tau = 0$ on a plate of infinite span ($M = 0.8$). Theoretical values are shown for the loading, which is plotted with solid lines (Lomax, Heaslet, Fuller & Sluder 1952). The results of numerical calculations by the discrete-vortex method are shown by points (the number of vortices on the chord being $n = 10$).

The third diagram illustrates the use of the reciprocity theorem as a control tool. For triangular wings with subsonic edges the moment characteristics in direct and reverse motion are compared for changes in the angular velocities ω_x and ω_z according to the law (7). The solid line shows the characteristics in direct motion (vertex ahead), and points those in reverse motion ($M = 2.0$). Figure 5 shows examples of flow of an incompressible fluid ($M = 0$) past rectangular wings of various aspect ratios λ. The aerodynamic derivatives obtained by direct numerical calculation for $p_j^* \to 0$ are shown by solid lines. The straight dashed lines extending from the origin represent the solutions of the problem for $\lambda \to 0$. The points are obtained by impulses of the unsteady parts of the transient function (18) and terms without circulation. Equations (22) allow determination of the rolling moments caused by aileron deflection by the second method (*dashed curves*). These diagrams in Figure 5 apply to a rectangular wing of aspect ratio $\lambda = 2.5$; and $\bar{l}$ is the relative span of the aileron, whose relative chord is equal to 0.5.

6   NONLINEAR CHARACTERISTICS

Study of the nonlinear characteristics of slender wings of small aspect ratio is one of the most difficult and important problems in aerodynamics. The first report in which specific numerical results were obtained was published by Bollay (1939). On the basis of an elementary vortex scheme for a rectangular wing of very small aspect ratio he obtained steady properties in satisfactory agreement with experiment.

All subsequent investigations have, from the point of view of the manner of describing the phenomena, developed along the following three directions. The first considers the steady problem without the formation of a bow sheet on the wing (Gersten 1961, Algazin & Gorelov 1974, Pavlenko 1975). In the second, steady flow is also postulated, but the formation of a bow sheet is assumed. At first the problem mainly considered was that of a triangular wing of very small aspect ratio (usually in the approximation of slender-body theory): Legendre (1965), Smith (1968), Nikol'skii, Betyaev & Malyshev (1971), and Jones (1975). Next appeared investigations based on extensive use of computers, where calculations are carried out for flow past wings of various planforms (Belotserkovskii 1968; Minoru & Tsutomu 1972; Perrier & Vitte 1970; Rehbach 1971, 1973; Kandil, Mook & Nayfeh 1974; Johnson & Rubbert 1975).

The third direction studies the problem as unsteady and considers the entire process of formation of the vortex wake (Djojodihardjo & Widnall 1969, Belotserkovskii & Nisht 1974a). The problem is solved in a variant that is interpreted as unseparated flow (vortices do not spring from the leading edge). The article of Belotserkovskii, Nisht & Sokolova (1975) is devoted to separated flow, when free vortices spring from all the edges of a slender wing including the leading edges.

Among the numerical approaches, discrete-vortex methods have today proved most effective. One of the few examples of specific realizations obtained in the three-dimensional case by other means (the method of panel singularities) was published by Johnson & Rubbert (1975).

We dwell in detail on those approaches that are organically connected with numerical experimentation on digital computers and have so far given the greatest volume of information. Their particularity consists in the fact that the entire process of formation of the flow is studied, and not only the limiting regime. The flow model is constructed on the basis of a scheme of an ideal incompressible medium, where boundary conditions are advanced that have clear physical significance (no flow through the surface of the body, finiteness of the velocity and pressure in the entire space). To ensure this last requirement at all sharp edges and breaks on a lifting surface (the Chaplygin-Joukowski hypothesis), it is necessary to assume free vortices springing from them. The problem is formulated and solved on the basis of a discrete representation in the coordinates and in time.

This numerical approach reflects the physical process of formation of the flow structure, which is very important in numerical experiments in general (O. M. Belotserkovskii et al 1974). It links up well with the possibilities and peculiarities of digital computers. And, finally, the discrete mode admits very flexible and broad possibilities for describing flows in which the vortex surface loses stability and changes to the formation of a type of Kármán vortex street.

The development of these approaches was at first carried out for plane-parallel flows (Belotserkovskii & Nisht 1972a, 1973, 1974b). Next basic problems for slender wings of arbitrary planform were considered (Belotserkovskii & Nisht 1972b, 1974a; Belotserkovskii, Nisht & Sokolova 1975).

We illustrate the above assertions with specific examples, based on slender lifting plate surfaces of various planforms (Belotserkovskii 1975a). The basic unsteady regime considered below is obtained by suddenly subjecting the body to a translational velocity:

$$\frac{v}{v_0} = \begin{cases} 0, & \tau < 0 \\ 1, & \tau \geqq 0. \end{cases} \tag{23}$$

Here the angle of attack α remains constant. The root chord is chosen as the characteristic linear dimension.

Figure 6 shows the process of development of the vortex wake behind a plate of infinite span. On the left are the results of calculations for unseparated flow ($\alpha = 20°$, no leading-edge sheet, the Chaplygin-Joukowski condition not satisfied at the leading edge, free vortices shown by points). This is a numerical reproduction of the process

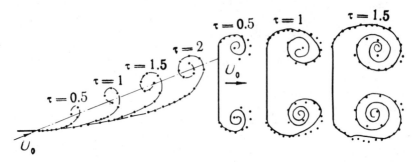

Figure 6   Development of vortex sheet on plate of infinite span. *Left*: starting vortex of Prandtl ($\alpha = 20°$, computed); *right*: $\alpha = 90°$, points are computations, curves are experiments (Wedemeyer).

Figure 7   Velocity field for symmetrical regime of flow past a plate ($\alpha = 90°$, $\tau = 20$, computation).

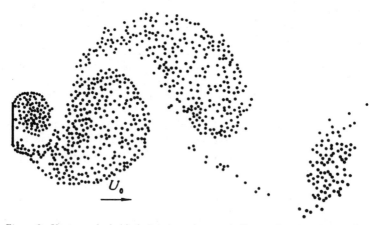

Figure 8   Vortex wake behind plate in unsymmetric flow regime ($\alpha = 90°$, $\tau = 25$, computation, points are free vortices).

of development of Prandtl's starting vortex, which maintains its stability sufficiently long. Corresponding experimental results are given in the book of Prandtl & Tietjens (1931). On the right ($\alpha = 90°$) the calculated values are plotted as points, and the experimental data (Wedemeyer 1961) as solid lines.

Figures 7 and 8 show that it is not always permissible to postulate one or another limiting flow in problems with separation. It may depend, for example, on the initial conditions of the problem (Belotserkovskii & Nisht 1973). If the condition of symmetry is included in the initial conditions, the vortex sheet loses stability, but symmetry of the flow is preserved in the calculations. The velocity field on the upper half of the plate is shown to scale in Figure 7 ($\tau = 20$). Experiments in a hydraulic channel show that for such a flow to be stable it is necessary

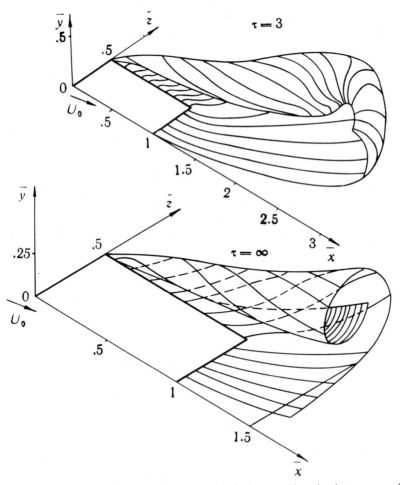

Figure 9 Development of vortex wake on right half of rectangular wing in unseparated flow ($\alpha = 30°$, $\lambda = 1$, $\tau = 3$ and $\tau = \infty$, computation).

to set a small splitter plate at a definite place in the wake, parallel to U_0. Otherwise the symmetry of the flow is broken, and the near wake is converted into a checker-board Kármán street. Creation of unsymmetric initial conditions (for example, starting at angle of attack $\alpha = 90°$ by rotating the plate from a horizontal position) produces also in the calculations a different limiting regime. The form of such a vortex wake, obtained theoretically, is shown in Figure 8, where the free vortices are shown by points. This vortex formation is stable, which also resolves the well-known paradox of the instability of the Kármán vortex street noted by Kochin (1949b). Theoretical values of the parameters of the street and of the drag coefficient correspond well to known experimental values.

In an analogous way it has been possible to construct a model that qualitatively and quantitatively describes the phenomenon of so-called traveling separation in lattices of profiles (Belotserkovskii, Gulyaev & Nisht 1975).

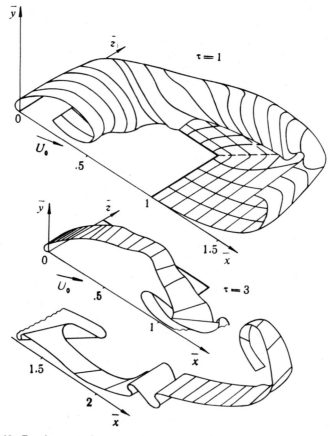

Figure 10 Development of vortex wake on rectangular wing in separated flow ($\alpha = 30°$, $\lambda = 1$, $\tau = 1$ and $\tau = 3$, computation).

Of main interest for aerodynamics are problems of flow past wings of finite span. The steady nonlinear theory already showed the possibility of obtaining calculations for such effects as the displacement of the side sheets into rolled-up vortices (Belotserkovskii 1968). Figure 9 shows the formation of the vortex wake behind a rectangular wing of small aspect ratio ($\lambda = 1$) at angle of attack $\alpha = 30°$ and unseparated flow (no bow sheet, Chaplygin-Joukowski condition not enforced at the leading edge, shown for the right half of the wing). Here Prandtl's starting vortex is again stable, and the limiting process $\tau \to \infty$ leads to the same regime as is obtained by solving the steady problem.

A different picture is observed for separated flow past the same wing (Figure 10, $\alpha = 30°$, a bow sheet exists). At first ($\tau = 1$) the free vortices springing from all edges of the wing represent a rather smooth surface. But then, mainly because of inter-action of the bow and stern vortex wakes, the stability of the sheet begins to break down, as is evident already at $\tau = 3$.

A proper qualitative description of these complex phenomena, and agreement of calculated and experimental data for aerodynamic loads and integral properties were reduced to the attempt to describe in analogous fashion the significant mechanism of loss of stability of the bow sheet on wings of triangular and nearby

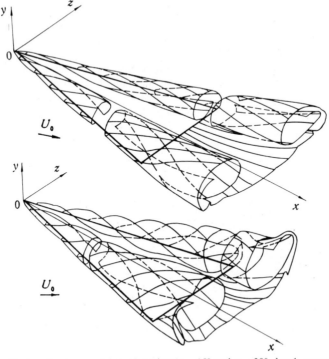

Figure 11   Vortex structure on triangular wing ($\alpha = 15°$ and $\alpha = 30°$, $\lambda = 1$, computation according to steady theory).

planforms (Aparinov et al 1976). Here it was necessary to apply a combination of approaches, based on steady and unsteady theory. At moderate angles of attack, when the bow sheet is stable, one can succeed in obtaining the solution of the problem in a steady formulation, as is indicated by convergence of the iterations in the numerical process. Calculations according to unsteady theory gave practically the same limiting regime. But at a certain angle of attack, when the bow sheet loses stability near the trailing edge of the wing or especially on the whole wing, a steady solution no longer exists (the iterations do not converge). Here solution of the unsteady problem at large τ leads to a periodic dependence on time.

As examples, Figures 11 and 12 show the results of calculations for a triangular wing with $\lambda = 1$. For $\alpha = 15°$ the vortex sheet in the region of the wing is stable, a tendency to destroy itself being observed relatively far behind the wing (Figure 11). Calculations according to unsteady theory confirmed this result. Figure 12 gives an analysis of the normal-force coefficient c_n and pitching moment m_z of the wing and the normal-force coefficients c'_n for sections of this wing. The solid lines give the results of calculations according to steady theory, and the asterisks show mean values of the coefficients found in the transient process at $\tau = 6$. The dot-dash curves show data from nonlinear steady theory without a bow sheet. The points indicate experimental values obtained by balance tests and from the pressure diagram on a slender triangular wing (of thickness ratio about 1.5%).

Evidently the main features of the mechanism of collapse of the vortex sheet on slender lifting surfaces is described fairly well by this approach. Attempts have been made in an analogous way to describe to some extent the large-scale turbulent properties of the near wake (Belotserkovskii & Nisht 1974b). But today it is still too soon to make definite conclusions about the limits of application of the models considered here. Only trial of their systematic application, and simultaneous analysis of the results of numerical and physical experiments can give the final answer to this important question.

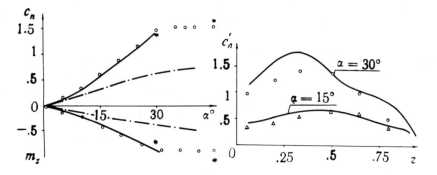

Figure 12 Comparison of computed data with experiments for slender triangular wing ($\lambda = 1$). Steady theory with bow sheet ———; steady theory without bow sheet — · —; mean values according to unsteady theory *; experiment $\triangle$ $\bigcirc$.

7 CONCLUSIONS

The contemporary period of development of theoretical aerodynamics is characterized first of all by the appearance and formation of new research ideologies, related to thorough and extensive application of digital computers. On the one hand, this permits the possibilities of aerodynamics to be significantly expanded, and on the other hand to fully satisfy the demands of practice. This is a problem of great magnitude; and the present review does not in any degree pretend to a complete examination of it. [On this point see also Krause (1974).] Using one of the branches of aerodynamics as an example, an attempt has been made here to delineate the specific features of developing approaches, and to a greater or lesser extent the systems of methods and viewpoints that are taking shape.

Numerical aerodynamics is a new scientific and applied research methodology, and not simply a translation of existing or even new equations and methods into the language of computers. It leads to reconsideration, extension, and refinement of basic schemes in aerodynamics. A new method of research has appeared: numerical experiment, which is closely connected with the physical content of the phenomenon, the possibilities and features of digital computers, and numerical methods of solving problems. Important trends have been the extension of unsteady techniques and the development of discrete ways of describing the conditions of the problem (in the coordinates and in time).

Success in the formation and development of the ideology that has been discussed demands professional long-term development in each specific branch of aerodynamics. Specialists capable of doing this must combine erudition in the field of aerodynamics with a subtle knowledge of numerical methods and the capabilities of digital computers. Stability of the numerical methods used, their careful selection and adaptation, the creation of a system of research and computation—all these are essential conditions for success.

Literature Cited

Algazin, V. A., Gorelov, D. N. 1974. On arbitrary motion of finite wing in incompressible fluid. *Izv. Sib. Otd. Akad. Nauk SSSR, Ser. Tekh. Nauk* 3(1):90–98

Andrew, L. V., Stanton, T. E. 1967. Unsteady aerodynamic forces on a thin wing oscillating in transonic flow. *AIAA Paper No. 67-16*

Aparinov, V. A., Belotserkovskii, S. M., Nisht, M. I., Sokolova, O. N. 1976. On mathematic simulation in ideal fluid of stalled flow past a wing and vortex sheet destruction. *Dokl. Akad. Nauk SSSR.* 227(4):820–23

Appa Kari, Jones, W. P. 1975. Integrated potential formulation of unsteady aerodynamics for interacting wings. *AIAA Paper No. 762*, pp. 1–13

Ashley, H., Rodden, W. P. 1972. Wing-body aerodynamic interaction. *Ann. Rev. Fluid Mech.* 4:431–72

Belotserkovskii, O. M., et al. 1974. *Numerical Investigation in Current Problems of Gasdynamics.* Moscow: Nauka. 398 pp.

Belotserkovskii, S. M. 1955a. Horseshoe vortices in nonsteady motion. *Prikl. Mat. Mekh.* 19(2):159–64

Belotserkovskii, S. M. 1955b. Spatial nonsteady motion of lifting surface. *Prikl. Mat. Mekh.* 19(4):410–20

Belotserkovskii, S. M. 1966a. Calculating the effect of gusts on an arbitrary thin wing. *Izv. Akad. Nauk SSSR, Mekh. Zhidk. Gaza* 1:51–60; *Fluid Dyn.* 1(1):34–40

Belotserkovskii, S. M. 1966b. Some aspects

of calculating subsonic flow over wings of complex planform. *Izv. Akad. Nauk SSSR, Mekh. Zhidk. Gaza* 6:74–80; *Fluid Dyn.* 1(6):46–49

Belotserkovskii, S. M. 1967. *The Theory of Thin Wings in Subsonic Flow.* New York: Plenum. 230 pp.

Belotserkovskii, S. M. 1968. Calculation of the flow around wings of arbitrary planform over a wide range of angle of attack. *Izv. Akad. Nauk SSSR, Mekh. Zhidk. Gaza* 4:32–44; *Fluid Dyn.* 3(4):20–27

Belotserkovskii, S. M. 1972. Mathematical model for nonstationary linear aeroautoelasticity. *Dokl. Akad. Nauk SSSR* 207(3):557–60; *Sov. Phys. Dokl.* 17(11):1048–50

Belotserkovskii, S. M. 1975a. Stalled flows and nonlinear characteristics of thin lifting surfaces of incompressible fluid. *Dokl. Konf.sov.-fran.Podgr.Aerodyn., Aviatsyon. Akust. Prochnosti.*, pp. 1–44. Moscow: Tsent. Aero. Gidr. Inst. im. Prof. Zhukovskogo

Belotserkovskii, S. M. 1975b. A mathematical model of aircraft for the investigation of nonstationary aerodynamic characteristics. *Prikl. Mat. Mekh.* 39(5):934–41; *Appl. Math. Mech.* 39(5):899–906

Belotserkovskii, S. M., Gulyaev, V. V., Nisht, M. I. 1975. Investigation of separated flow regimes in arrays of profiles. *Dokl. Akad. Nauk SSSR* 221(3):555–58; *Sov. Phys. Dokl.* 20(3):177–78

Belotserkovskii, S. M., Kolesnikov, G. A. 1969. Gust effects on wings of complex planform at subsonic speeds. *Izv. Akad. Nauk SSSR, Mekh. Zhidk. Gaza* 5:129–38; *Fluid Dyn.* 4(5):86–91

Belotserkovskii, S. M., Kudryavtseva, N. A., Fedotov, B. N. 1969. Method for calculating aerodynamic characteristics of wings of complex planform with subsonic leading and trailing edges. *Izv. Akad. Nauk SSSR, Mekh. Zhidk. Gaza* 2:128–32; *Fluid Dyn.* 4(2):88–91

Belotserkovskii, S. M., Nisht, M. I. 1972a. Analysis of nonstationary separation flow around a thin profile. *Izv. Akad. Nauk SSSR, Mekh. Zhidk. Gaza* 3:177–82; *Fluid Dyn.* 7(3):525–29

Belotserkovskii, S. M., Nisht, M. I. 1972b. Analysis of nonstationary stalled flow around a slender wing in a nonviscous incompressible fluid stream. *Int. Congr. Theor. Appl. Mech., 13th, Moscow. Book of Abstracts.* 33 pp.

Belotserkovskii, S. M., Nisht, M. I. 1973. Two regimes of stalled flow around a plate. *Dokl. Akad. Nauk SSSR* 213(4):804–8; *Sov. Phys. Dokl.* 18(12):782–84

Belotserkovskii, S. M., Nisht, M. I. 1974a.

Nonstationary nonlinear theory of a thin wing of arbitrary planform. *Izv. Akad. Nauk SSSR, Mekh. Zhidk. Gaza* 4:100–8; *Fluid Dyn.* 9(4):583–89

Belotserkovskii, S. M., Nisht, M. I. 1974b. Investigation of a turbulent wake behind a plate. *Dokl. Akad. Nauk SSSR* 216(6):1240–44; *Sov. Phys. Dokl.* 19(6):333–35

Belotserkovskii, S. M., Nisht, M. I., Sokolova, O. N. 1975. Design of stalled flow past thin wing of finite span. *Izv. Akad. Nauk SSSR, Mekh. Zhidk. Gaza* 2:107–12

Belotserkovskii, S. M., Popytalov, S. A. 1970. Action of a gust on a wing with subsonic leading and trailing edges. *Izv. Akad. Nauk SSSR, Mekh. Zhidk. Gaza* 2:92–100; *Fluid Dyn.* 5(2):252–58

Belotserkovskii, S. M., Skripach, B. K. 1975. *Aerodynamic Derivatives of Vehicle and Wing at Subsonic Speeds.* Moscow: Nauka. 424 pp.

Belotserkovskii, S. M., Skripach, B. K., Tabachnikov, V. G. 1971. *Wing in Unsteady Gas Flow.* Moscow: Nauka. 768 pp.

Bisplinghoff, R. L., Ashley, H., Halfman, R. L. 1955. *Aeroelasticity.* Cambridge Mass.: Wesley. 860 pp.

Bollay, W. 1939. A non-linear theory and its application to rectangular wings of small aspect ratio. *ZAMM* 19(1):21–35

Byushgens, P. S., Studnev, R. V. 1967. *Dynamics of Spatial Motion of Aircraft.* Moscow: Mashinostroenie. 226 pp.

Chang, P. K. 1970. *Separation of Flow.* Oxford: Pergamon. 777 pp.

Chaplygin, S. A. 1948. *Results of Theoretical Investigations of Airplane Motion.* Collected Works 2:230–45. Moscow, Leningrad: Gostekhizdat. 644 pp.

Cunningham, A. M. 1974. Oscillatory supersonic kernel function method for interfering surfaces. *J. Aircr.* 11(11):664–70

Djojodihardjo, R. H., Widnall, S. E. 1969. A numerical method for the calculation of nonlinear, unsteady lifting potential flow problems. *AIAA J.* 7:2001–9

Evvard, J. C. 1950. Use of source distribution for evaluating theoretical aerodynamics of the finite wing at supersonic speeds. *NACA Rep. No. 951.* 32 pp.

Falkner, V. M. 1953. The solution of lifting-plane problems by vortex-lattice theory. *Aeronaut. Res. Counc. Rep. Mem. 2591.* 30 pp.

Flax, A. N. 1953. Reverse flow and variational theorems for lifting surfaces in nonstationary compressible flow. *J. Aeronaut. Sci.* 20(2):120–26

Försching, H. W. 1974. *Grundlagen der Aeroelastik.* Berlin, Heidelberg, New York: Springer. 374 pp.

Fursov, M. K. 1961. On design of coefficients of wing rotary derivatives at supersonic speeds. *Izv. Akad. Nauk SSSR, Mekh. Mashinostr.* 2:89–96

Gersten, K. 1961. Nichtlineare Tragflächentheorie insbesondere für Tragflügel mit kleinem Seitenverhältnis. *Ing.-Arch.* 30(1): 431–52

Giesing, J. P., Kalman, T. P., Rodden, W. P. 1972. Subsonic unsteady aerodynamics for general configurations. *AIAA Paper No. 72-26*, 13 pp.

Gogish, L. V., Neiland, V. Ya., Stepanov, G. Yu. 1975. Theory of two-dimensional stalled flows. *Itogi Nauki: Tekhniki. Gidrodinamika* 8: 5–75. Moscow: VINITI. 132 pp.

Golubev, V. V. 1949. *Lectures on Wing Theory.* Moscow, Leningrad: Gostekhizdat. 480 pp.

Golubinskii, A. I. 1961. On a flow past a moving plate by a moving shock wave. *Inzh. Zh.* 1(2): 26–30

Golubinskii, A. I. 1963. Thin profile lift and moment at arbitrary unsteady flow. *Inzh. Zh.* 3(3): 442–46

Golubinskii, A. I. 1968. Impulse of forces and moments on a slender wing in unsteady flow. *Izv. Akad. Nauk SSSR, Mekh. Zhidk. Gaza* 6: 114–17; *Fluid Dyn.* 6(3): 78–79

Hess, J. L., Smith, A. M. O. 1964. Calculation of potential flow about arbitrary body shapes. *International Symposium on Analogue and Digital Techniques Applied to Aeronautics, Liége, 1963*, pp. 109–21

Johnson, F. T., Rubbert, P. E. 1975. Advanced panel-type influence coefficient methods applied to subsonic flows. *AIAA Paper No. 75-50*, 10 pp.

Jones, J. P. 1975. Flow separation from yawed delta wings. *Comput. Fluids* 3: 155–77

Joukowski, N. E. 1949. *On Bound Vortices.* Collected Works 4: 69–91. Moscow, Leningrad: Gostekhizdat. 652 pp.

Kandil, O. A., Mook, D. T., Nayfeh, A. H. 1974. Nonlinear prediction of the aerodynamic loads on lifting surfaces. *AIAA Paper No. 503.* 14 pp.

Kochin, N. E. 1949a. *On Steady-State Oscillations of Wing of Circular Plan Form.* Collected Works 2: 386–421. Moscow, Leningrad: Akad. Nauk SSSR. 588 pp.

Kochin, N. E. 1949b. On instability of Kármán vortical chains. See Kochin 1949a, pp. 479–85

Kolesnikov, G. A. 1967. Method of calculating unsteady aerodynamic characteristics of a lifting surface at high subsonic speed. *Izv. Akad. Nauk SSSR, Mekh. Zhidk. Gaza*

6: 118–25; *Fluid Dyn.* 2(6): 81–84

Krasil'schikova, E. A. 1952. *Finite Wing in Compressible Gas Flow.* Moscow, Leningrad: Gostekhizdat. 160 pp.

Krasovskii, A. A. 1973. *Automatic Flight Control Systems and Its Analytical Design.* Moscow: Nauka. 560 pp.

Krause, E. 1974. Application of numerical techniques in fluid mechanics. *J. Aeronaut.* 78(764): 333–54

Lavrent'ev, M. A., Shabat, B. V. 1973. *Problems of Hydrodynamics and Its Mathematical Models.* Moscow: Nauka. 416 pp.

Legendre, R. 1965. Vortex sheets rolling-up along leading edges of delta wings. *Prog. Aeronaut. Sci.*, pp. 7–33

Lifanov, I. K., Polonskii, Ya. E. 1975. Proof of the numerical method of "discrete vortices" for solving singular integral equations. *Prikl. Mat. Mekh.* 39(4): 742–46; *Appl. Math. Mech.* 39(4): 713–18

Lomax, H., Heaslet, M. A., Fuller, F. B., Sluder, L. 1952. Two- and three-dimensional unsteady lift problems in high-speed flight. *NACA Rep. No. 1077.* 55 pp.

Marino, L., Chen, L. T., Sucio, E. O. 1975. Steady and oscillatory subsonic and supersonic aerodynamics around complex configurations. *AIAA J.* 13(3): 368–74

Maslov, L. A. 1966. Arbitrary motion of an elongated body in an ideal fluid. *Izv. Akad. Nauk SSSR, Mekh. Zhidk. Gaza* 6: 81–86; *Fluid Dyn.* 1(6): 50–53

Miles, J. W. 1959. *The Potential Theory of Unsteady Supersonic Flow.* Cambridge, Engl: The University Press. 220 pp.

Minoru, O., Tsutomu, T. 1972. Predictions of vortex-life characteristics by an extended vortex-lattice method. *J. Jpn. Soc. Aeronaut. Space Sci.* 20(226): 635–41

Nikol'skii, A. A. 1957a. On "second" form of motion in ideal fluid near streamlined body (investigation of stalled vortical flows). *Dokl. Akad. Nauk SSSR* 116(2): 193–96

Nikol'skii, A. A. 1957b. On force action of "second" form of hydrodynamical motion on two-dimensional bodies (dynamics of two-dimensional stalled flows). *Dokl. Akad. Nauk SSSR* 116(3): 365–68

Nikol'skii, A. A., Betyaev, S. K., Malyshev, I. P. 1971. On marginal form of stalled automodel stream of ideal fluid. *Prob. Prikl. Mat. Mekh.*, pp. 262–68. Moscow: Nauka

Pavlenko, V. S. 1975. Nonlinear characteristics of thin wing at unshocked flow past nose. *Izv. Akad. Nauk SSSR, Mekh. Zhidk. Gaza* 5: 183–85

Perrier, P., Vitte, W. 1970. Éléments de calcul d'aérodynamique tridimensionelle en fluide parfait. *AFIIAE 7 colloque d'aérodynamique appliquée.* 37 pp.

Pines, S., Dugundji, J., Neuringer, J. 1955. Aerodynamic flutter derivatives for a flexible wing with supersonic and subsonic edges. *J. Aeronaut. Sci.* 22(10): 693–700

Polyahov, N. N., Shesternina, Z. N. 1975. On existence of solution of an equation of lifting surface. *Vestn. Leningr. Univ., Mat., Mekh., Astron.* 19(4): 102–7

Popytalov, S. A. 1976. Design of unsteady aerodynamic characteristics of spacial lifting systems. *Izv. Akad. Nauk SSSR, Mekh. Zhidk. Gaza* 1: 173–76

Prandtl, L., Tietjens, O. 1931. *Hydro-und Aeromechanik,* Vol. 2. Berlin: Springer. 294 pp.

Rehbach, C. 1971. Étude numérique de l'influence de la forme de l'extremité d'une aile sur l'enroulement de la nappe tourbillonnaire. *Rech. Aérosp.* 6: 567–68

Rehbach, C. 1973. Calcul d'écoulements autour d'ailes sans épaisseur avec nappes tourbillonnaires évolutives. *Rech. Aérosp.* III–IV(2): 53–61

Ribner, H. S., Malvestuto, F. S., Jr. 1948. Stability derivatives of triangular wings at supersonic speeds. *NACA Rep. No. 908.* 9 pp.

Roos, R., Bennekers, B., Zwaan, R. I. 1975. A calculation method for unsteady subsonic flow about harmonically oscillating wing-body configurations. *AIAA Paper No. 864,* 10 pp.

Rowe, W. S., Winther, B. A., Redman, M. C. 1973. Numerical method for predicting unsteady aerodynamic loadings caused by control surface motions in subsonic flow. *AIAA Paper No. 73-315,* 11 pp.

Sedov, L. I. 1954. *Similarity and Dimensional Methods in Mechanics.* Moscow: Nauka. 328 pp.; English transl., 1959. New York: Academic

Sedov, L. I. 1973. *A Course in Continuum Mechanics,* Vols. 1, 2. Moscow: Nauka. 536 pp., 584 pp.; English transl., 1971–1972. Vols. 1, 2, 3, 4. Groningen, Wolters-Noordhoff.

Shamshurin, A. D. 1974. Effect of a weak shock on a wing of complex planform at subsonic speeds. *Izv. Akad. Nauk SSSR, Mekh. Zhidk. Gaza* 2: 173–76; *Fluid Dyn.* 9(2): 311–14

Smith, J. H. B. 1968. Improved calculations of leading-edge separation from slender, thin, delta wings. *Proc. R. Soc. London Ser. A* 306: 67–90

Ul'yanov, B. I. 1961. Some generalizations and consequences of reverse theorem for unsteady motions. *Izv. Akad. Nauk SSSR, Mekh. Mashinostr.* 1: 70–74

Van Dyke, M. 1964. *Perturbation Methods in Fluid Mechanics.* New York, London: Academic. 229 pp.

Van Dyke, M. 1975. Computer extension of perturbation series in fluid mechanics. *SIAM J. Appl. Math.* 28(3): 720–34

Vorob'ev, N. F. 1959. Unsteady motion of finite-span wing in supersonic flow in the case of discontinuous speed change. *Izv. Akad. Nauk SSSR, Mekh. Mashinostr.* 1: 167–70

Ward, G. N. 1949. Supersonic flow past thin wings. *Q. J. Mech. Appl. Math.* 11(2): 136–52. 11(3): 374–84

Wedemeyer, E. 1961. Ausbildung eines Wirbelpaares an den Kanten einer Platte. *Ing. Arch.* 30(3): 187–200

Woodward, F. A. 1973. An improved method for the aerodynamic analysis of wing-body-tail configurations in subsonic and supersonic flow. Part I. Theory and application. *NASA CR-2228.* 126 pp. Part II. Computer program description. *NASA CR-2228.* 313 pp.

AUTHOR INDEX

495

CUMULATIVE INDEXES

CONTRIBUTING AUTHORS VOLUMES 5-9

CHAPTER TITLES VOLUMES 5-9